IHDP/Future Earth—Integrated Risk Governance Project Series

About this Series

This book series, entitled "IHDP/Future Earth—Integrated Risk Governance Project Series" for the International Human Dimensions Programme on Global Environmental Change—Integrated Risk Governance Project (IHDP/Future Earth—IRG Project), is intended to present in monograph form the most recent scientific achievements in the identification, evaluation and management of emerging global large-scale risks. Future Earth is a flagship initiative of the Science and Technology Alliance for Global Sustainability. It aims to provide critical knowledge required for societies to understand and address challenges posed by global environmental change (GEC) and to seize opportunities for transitions to global sustainability. Future Earth identifies three research themes, i.e., Dynamic Planet, Global Development and Transition toward Sustainability in its plan and adopts a new approach of "Co-designing and co-producing" to incorporate GEC researchers with stakeholders in governments, industry and business, international or intergovernmental organizations, and civil society.

Books published in this series are mainly collected research works on theories, methods, models and modeling, and case analyses conducted by scientists from various disciplines and practitioners from various sectors under the IHDP/Future Earth—IRG Project. It includes the IRG Project Science Plan, research on social-ecological system responses, "Entry and Exit Transition" mechanisms, models and modeling, early warning systems, understanding regional dynamics of vulnerability, as well as case comparison studies of large-scale disasters and paradigms for integrated risk governance around the world. This book series, therefore, will be of interest not only to researchers, educators and students working in this field but also to policy-makers and decision-makers in government, industry and civil society around the world.

The series will be contributed by the international research teams working on the six scientific themes identified by the IHDP/Future Earth—IRG Project science plan, i.e., Social-Ecological Systems, Entry and Exit Transitions, Early Warning Systems, Models and Modeling, Comparative Case Studies, and Governance and Paradigms, and by six regional offices of the IRG Project around the world.

More information about this series at http://www.springer.com/series/13536

Qiuhong Tang · Quansheng Ge
Editors

Atlas of Environmental Risks Facing China Under Climate Change

Springer

Editors
Qiuhong Tang
Key Laboratory of Water Cycle and Related Land Surface Processes, Institute of Geographic Sciences and Natural Resources Research
Chinese Academy of Sciences
Beijing
China

Quansheng Ge
Institute of Geographic Sciences and Natural Resources Research
Chinese Academy of Sciences
Beijing
China

ISSN 2363-4979 ISSN 2363-4987 (electronic)
IHDP/Future Earth—Integrated Risk Governance Project Series
ISBN 978-981-10-4198-3 ISBN 978-981-10-4199-0 (eBook)
DOI 10.1007/978-981-10-4199-0

Library of Congress Control Number: 2017934449

No. of licensed maps: JS(2016)01-143 (Star Map Press)

Printed on acid-free paper

This Springer imprint is published by Springer Nature
The registered company is Springer Nature Singapore Pte Ltd.
The registered company address is: 152 Beach Road, #21-01/04 Gateway East, Singapore 189721, Singapore

Preface

Climate change is one of the most serious threats and the greatest challenges facing human society today. According to the Fifth Assessment Report (AR5) of the Intergovernmental Panel on Climate Change (IPCC), each of the last three decades has been successively warmer at the Earth's surface than any preceding decade since 1850. The unequivocal warming in the climate system has caused impacts on natural and human systems over the world and is bringing environmental risks having important implications for sustainable development. The environmental risk induced by climate change could be highly regional or even local. Although the IPCC reports have assessed the future impacts of climate change and risks at the global scale and/or from a regional perspective, it is not sufficient for decision-making due to its deficiency in risk assessment at regional and local scales. Assessment of regional environmental risks under climate change is urgently needed to support climate change adaptation at the regional scale. China is a developing country with relatively high level of agricultural population and relatively low level of urbanization but rapid rural-to-urban migration. There are several modernized cities and relatively less developed rural areas in China. Having a tremendously large territory, China has a complex and diverse terrain and a variety of ecosystems ranging from alpine tundra to evergreen tropics and from desert to forest. The climate of China is also diverse, wide-ranging from tropical climate in the south to subarctic climate in the north. With respect to these differences, regional vulnerabilities to climate change are different. Thus, different regions would face environmental risks differently under climate change. Understanding the different regional risks is essential for climate change adaptation in China.

In order to assess the risk of future climate change impacts, the most common means is to use climate-impact models driven by various climate change projections from the general circulation models (GCMs). The model-run scenarios for the IPCC AR5, collected by the Coupled Model Intercomparison Project Phase 5 (CMIP5) archive, are widely used as input forcing data for the climate-impact models. The Inter-Sectoral Impact Model Intercomparison Project (ISI-MIP), which is a community-driven climate-impact modeling initiative aimed to provide a quantitative and cross-sectoral synthesis of the differential impacts of climate change, has offered a framework for modeling of the impacts of climate change by sharing a daily-resolution, downscaled, and bias-corrected climate data from the CMIP5 archive. Many ISI-MIP outcomes have been used by or cited in the IPCC AR5. This atlas leverages the ISI-MIP framework to assess the environmental risks faced by China under climate change. While the common climate change projections provided by ISI-MIP are adopted, the vulnerabilities are assessed using the region-specific information in China. Uncertainty of the assessment is rooted in many aspects of the modeling framework. The spreads of both climate and impact models are large, in particular at the regional and local scales. Despite the irreducible model spread, the uncertainties should not prevent decision-making from using the assessment to support climate change adaptation actions. Indeed, many decisions, for example in insurance industry, have been made with huge uncertainty in future prediction. The multi-model ensembles, including multiple-impact model-runs driven by climate change projections from multiple climate models, are used to estimate future risks and to provide the basis for uncertainty estimation. The atlas illustrates many multi-model ensembles, which allow decision-makers to know the implications of a wide range of future projections so they can manage the uncertainty and make adaptation policy arguments bearing the uncertainties.

The objective of the atlas is to provide the most comprehensive and accurate illustrations of environmental risks relating to climate change vulnerability and adaptation in China. It addresses the agricultural, ecosystem, and heat wave-related human health risk posed by climate change and presents the projected environmental risks in the twenty-first century under various climate change and socioeconomic scenarios. The detailed and concise risk assessments are mapped in grid units, allowing easy identification of the environmental risk for specific locations. The atlas contains six chapters. Chapter 1 presents the geographic, social, demographic and economic features in China. Chapter 2 illustrates the geographical distribution of temperature and precipitation and their long-term changes in the twenty-first century. Chapters 3–5 demonstrate the risks posed by climate change in the human health, agricultural, and ecosystem sectors with a focus on mortality due to heat wave, food production, and ecosystem shift over naturally vegetated land, respectively. Chapter 6 addresses the integrated environmental risk posed by climate change, which is a combined risk evaluation of climate change impacts on the basis of the risk assessments in the human health, agricultural, and ecosystem sectors. The maps in the atlas are with No. of licensed maps: JS(2016)01-143 (Star Map Press).

I would like to thank Dr. Ge who agreed to be listed as a co-editor of this atlas. Thanks also go to the National Basic Research Program of China (Grant No. 2012CB955403), the National Natural Science Foundation of China (Grant No. 41425002), and the National Youth Top-notch Talent Support Program that supported funding to make this atlas a reality. I would like to extend my thanks to the staff of the Star Map Press (Beijing) and Springer-Verlag Berlin Heidelberg, for smoothly and professionally handling all aspects of the production processes.

I hope this atlas would provide the reader clear pictures of the environmental risks across different sectors posed by climate change and their evolution in the twenty-first century over China. I further hope that the atlas will contribute to the knowledge base for climate change adaptation in China and provide a valuable resource for students and professionals in the fields of geographic sciences and climate change.

Beijing, China Qiuhong Tang

Editorial Committee

Editors
Qiuhong Tang, Quansheng Ge

Associate Editors
Yuanyuan Yin, Wenxiang Wu, Xingcai Liu, Xinchuang Xu, Xuexia Zhang, Xueqin Zhang, Huijuan Cui

Lead Authors
Qiuhong Tang, Yuanyuan Yin, Xingcai Liu, Quansheng Ge, Wenxiang Wu, Xinchuang Xu, Xueqin Zhang, Xuexia Zhang

Text Editors
Huijuan Cui, Pedram Attarod, Qiuhong Tang

Contents

List of Maps

Chapter 1 Geography of China

Chapter 2 Climate Change Projections in the Twenty-First Century

Chapter 3 Heat Health Risks

Chapter 4 Agricultural Risks

Chapter 5 Ecosystem Risks

Chapter 6 Integrated Environmental Risks

Geography of China

1

Qiuhong Tang, Yuanyuan Yin, Xuexia Zhang, Tong Zhu, Jingyao Zhao and Wei Yang

1.1 Topography

China contains five kinds of landform types, including mountain, plateau, hill, basin, and plain. Among the five landform types, mountains and plateaus are most extensive, accounting for 59% of the total China land area followed by basins (19%), plains (12%), and hills (10%) (Chinese Academy of Sciences 1985).

China topography varies greatly from highly mountainous regions to inhospitable desert zones and flat, fertile plains visualizing in form of steps of a stairway descending from west to east. The first topographic step is the Qinghai–Tibetan Plateau with an average elevation of more than 4500 m. There are a series of east–west or northwest–southeast trending mountain ranges with elevation over 5000 m. The Qaidam Basin is embedded inside these mountains. In the south, the Himalayan mountain range has a number of peaks over 7000 m, with the tallest in the world, Mount Everest, at the Sino–Nepalese border.

The second topographic step contains a series of high mountains with elevation ranging from 1500 to 2500 m, as well as plateaus with elevation ranging from 1000 to 2000 m. The Inner Mongolia and Yunnan–Guizhou Plateau as well as the Tarim, Sichuan, and Junggar basins are located in this topographic step.

The third topographic step consists of plains and peneplains. From north to south, the Northeast Plain, the North China Plain, and the Middle-lower Yangtze River Plain are located with elevation mostly below 200 m. The vast area extending to the south of the Yangtze River is mainly occupied by hills with elevation less than 500 m. Finally, the coastal zones, stretching from Vietnam to North Korea and bordering the South China, East China and Yellow Seas, join the Pacific Ocean. Many Islands are distributed here such as the Hainan Island and Taiwan Island.

1.2 Climate

China is located at the east coast of the largest continent (the Eurasia Continent) bordered by the world's largest ocean (the Pacific Ocean). The mean annual precipitation ranges spatially from less than 50 mm to more than 1800 mm with low precipitation in the northwest and high precipitation in the southeast. The mean temperature ranges spatially from −12 to 26 °C. In general, the mean temperature in the southeast is higher than that in the northwest. The mean daily minimum temperature is greater than 0 °C in the south and is less than 0 °C in the north. The mean daily maximum temperature is greater than 10 °C in most parts of China.

China's climate is mainly dominated by dry seasons and wet monsoons, which lead to pronounced precipitation and temperature differences between winter and summer. Due to the vast land size and various landforms, the types of climates are complex and diverse in China. China can be divided into four wet/dry climate regions and six climate zones. The four wet/dry climate regions are humid region, sub-humid region, sub-arid region, and arid region. The six climate zones include the cool temperature, the middle temperature, the warm temperature, the subtropical, the tropical, and the highland climates.

Climate change regionalization could represent the regional difference of climate change, and can reflect the disasters and environmental risks incurred by climate changes. The tendency and fluctuation values of temperature and precipitation from 1961 to 2010 are the key indicators for regionalization of climate change in China (1961–2010) (Shi et al. 2014). Based on the temperature tendency value and the precipitation tendency value (1961–2010), China is divided into five tendency zones during level-I regionalization, i.e.,

Q. Tang (✉) · Y. Yin
Key Laboratory of Water Cycle and Related Land Surface Processes, Institute of Geographic Sciences and Natural Resources Research, Chinese Academy of Sciences, Beijing, 100101, China
e-mail: tangqh@igsnrr.ac.cn

X. Zhang · T. Zhu · J. Zhao · W. Yang
Beijing Forestry University, Beijing, 100083, China

Q. Tang and Q. Ge (eds.), *Atlas of Environmental Risks Facing China Under Climate Change*, IHDP/Future Earth-Integrated Risk Governance Project Series, DOI 10.1007/978-981-10-4199-0_1

Northeast China–North China warm-dry trend zone, East China–Central China wet-warm trend zone, Southwest China–South China dry-warm trend zone, Southeast Tibet–Southwest China wet-warm trend zone, and Northwest China-Qinghai-Tibet Plateau warm-wet trend zone. Level-II regionalization refers to fourteen fluctuation regions based on level-I regionalization according to the temperature and precipitation fluctuations.

1.3 Rivers

There are more than 50,000 rivers across China of which 1500 rivers have a drainage area over 1500 km^2. Based on climatic and geomorphological conditions, China's rivers can be divided into two large systems: oceanic and inland (Hong and Blackmore 2015). Almost all large rivers in China are located in the oceanic system (humid southeast) of the country, which directly or indirectly emptying into the seas. The oceanic system, accounting for 64% of the country's land area, can be subdivided into Pacific, Indian, and Arctic drainage basins. Most rivers flow into the Pacific Ocean, but some drain into the Indian Ocean and Arctic Ocean as well. Because China's terrain is high in the west and low in the east, most of the rivers flow eastward and empty into the Pacific Ocean (e.g., Yangtze, Yellow, Heilong, Pearl, Liaohe, and Haihe rivers). The inland system accounts for the remaining 36% of the total land area of China and includes the Tarim He, Yili He, and Hei He rivers, all of which flow into saline inland lakes or peter out in sandy deserts or salt marshes. The watershed boundary between the out-flowing river basins and the inland river basins generally coincides with the 400 mm isohyet (Hartmann and Wang 2014).

The volume of annual runoff in China is about 2700 billion m^3, equivalent to a depth of 280 mm over the whole land area. The total volume of river runoff in the six major river basins in north China (namely, Songhua Jiang, Liao He, Hai He, Yellow River, Huai He, and Northwest Inland rivers) account for less than 21% of the national total runoff, but the four major river basins in south China (namely, Yangtze River, Southwest Rivers, Pearl River, and Southeast Rivers) account for more than 79% of the national total (Ministry of Water Resources People's Republic of China 2013). The mean annual runoff depth ranges from almost 0 mm in the desert to over 200 mm in the humid area (Zhang et al. 2014). From a spatial perspective, the mean annual runoff depth is usually more than 200 mm in the south, and less than 50 mm in the northwest. The runoff coefficient is close to 0 in the arid area, while it can be more than 0.80 in the humid area. The runoff coefficient is larger than 0.45 in Yangtze River, Southeast Rivers, Pearl River, and Southeast Rivers. The coefficient is between 0.25 and 0.45 in Songhua Jiang and Liao He rivers.

1.4 Vegetation and Soil

China has a variety of vegetation types ranging from alpine tundra to evergreen tropics and from desert to forest. According to the classification system in the Vegetation Map of the People's Republic of China (1: 1,000,000) published by the Geology Press in 2007, there are eleven vegetation type groups covering 55 vegetation forms, 960 vegetation formations and sub-formations, and more than 2000 communities of dominant species. The eleven vegetation type groups are: (1) coniferous forest, (2) coniferous and broad-leaf mixed forest, (3) broad-leaf forest, (4) bush, (5) meadow, (6) steppe, (7) grass, (8) alpine vegetation, (9) desert, (10) crop land, and (11) wetland (Edition Board of Vegetation Map of China 2007). The latitudinal zonal distribution is obvious in the East Monsoon Zone from south to north, including the tropical seasonal rainforest or rainforest, subtropical evergreen broadleaf forest, warm temperature deciduous broadleaf forest, temperate broadleaf and coniferous mixed forest, and the cold temperature deciduous forest (Edition Board of Vegetation Map of China 2007).

With obvious longitude zonal distribution, the vegetation types in northern China ordered from east to west are as follows: forest, steppe, and desert. There are obvious horizontal and vertical zonal features on the Qinghai–Tibet Plateau. With increasing altitude and decreasing rainfall from southeast to northwest, vegetation types ordered from southeast to northwest are: mountain forest, alpine shrub, alpine meadow, alpine steppe, and alpine desert. According to the vegetation regionalization map of China, the country is divided into eight vegetation regions and 26 vegetation zones.

China, with vast and diverse climatic conditions, topography, vegetation, and human development, has a wide variety of soils. According to the classification system in the Soil Map of the People's Republic of China (1:1,000,000) published by the National Soil Survey Office in 1995, there are 12 orders, 61 great groups, 235 subgroups, and 909 families in the database (Shi et al. 2004). In the East Monsoon Zone, the latitudinal zones from south to north include latosol, lateritic red soil, red soil and yellow soil, yellow-brown soil, burozem and drab soil, dark brown forest soil, and podzolic soil. In the Northwest Non-monsoon Zone, the zonal soil types from northeast to northwest are chernozem, chestnut soil, brown soil, sierozem, gray brown desert soil, brown desert soil. On the Qinghai–Tibet Plateau, the soil types from east to west are alpine meadow soil, alpine steppe soil, alpine desert soil, and alpine frozen soil. According to the soil regionalization map of China, the country is classified into three soil regions and 17 soil zones.

1.5 Land Use/Land Cover and Eco-Geographical Regionalization

According to the classification system in the Land Use and Land Cover Map of the People's Republic of China published by the Chinese Academy of Sciences (Liu et al. 2002) (Table 1.1), there are six land cover types, namely, crop land, woodland, grassland, water body, built-up land, and unused land. The land use and land cover distribution in China are summed up as follows: Forests in northeastern China, southeastern China and parts of Tibet region; Grassland in Inner Mongolia region, and Southwest China; Farmland in the eastern part of the country; and large areas of sparsely vegetated land and bare land in Northwest China (Liu et al. 2002).

Eco-geographical regionalization considered the basis for spatial differentiation of biodiversity research is a regional system formed by dividing or merging geographical space according to natural geographic differentiation principles (Zheng 2008). Eco-regions is aimed to reflect the spatial heterogeneity of natural features, including temperature, moisture, soil, and organisms based on eco-regions, as well as the characteristics of land cover types (Jin et al. 2016). The homogeneity within an eco-region and the heterogeneity between eco-regions are important means of identifying the features and differences of the regional environment. According to the Eco-geographical Regionalization of China promulgated by the Ministry of Environmental Protection and the Research Center for Eco-Environmental Sciences, Chinese Academy of Sciences, China has three ecological regions and 50 ecological sub-regions. Based upon the topography, warm-humid combination, and the soil and vegetation, the three ecological regions are East Monsoon Ecological Region, West Arid Ecological Region, and Qinghai–Tibet Plateau Alpine Region.

1.6 Population and Economy

Population density varies strikingly in China, with the greatest contrast occurring between the southeast and the northwest. The Heihe–Tengchong Line, which is also known internationally as the "Hu Line," marks a striking difference in the distribution of China's population. The southeast to the "Hu Line" accounting only for 42.9% of the land captures 95.4% of the country's population, while the northwest to the line has only 5.6% of the population. China's population distribution has the following characteristics: (1) higher population densities in the plains and lower population densities in the mountains and plateaus; (2) higher population densities in warm and humid regions and lower in cold and dry areas; (3) population distribution is closely related to cultivated land and urban areas (Hartmann and Wang 2014). Exceptionally high population densities occur in the Huang-Huai-Hai Plain, the Chengdu Plain, and the Yangtze Delta and the Pearl River Delta. On contrary, sparsely low population densities occur in the isolated, extensive western and frontier regions, such as the extremely high northern part of Tibet, the sandy wastes of the central Tarim and eastern Junggar basins in Xinjiang.

The distribution of the gross domestic product (GDP) shows huge geographic imbalances in China. The imbalances stem from the extreme diversity of the climate and relief, the large variations in population density, in natural resources and communication facilities, and the political power. The GDP varies noteworthy. The GDP in the coast of China is generally higher than interior. The coast area, which includes seven provinces, namely, Fujian, Guangdong, Hainan, Hebei, Jiangsu, Shandong, Zhejiang, and three municipalities at provincial level (Beijing, Shanghai, Tianjin), is home to 38% of the Chinese population, and creates about half China's GDP (Lemoine et al. 2014). High GDP occurs on the Huang-Huai-Hai Plain, the Guanzhong Plain and Chengdu Plain, and the Yangtze Delta and Pearl River Delta.

According to the Regulation on Classification of Three Industries promulgated by the National Bureau of Statistics of China (NBSC) in 2012, agriculture includes farming, forestry, animal husbandry, and fishery. As a fundamental sector of national economy, the added value by agriculture is continually increasing during the past 60 years. It is up to about 6000 billion Chinese Yuan in 2015 (Fig. 1.1). However, the share of agriculture in the national economy

Table 1.1 The land cover classification system

Land cover types	Description
Crop land	Paddy land, Dry land
Woodland	Forest, Shrub, Woods, Others
Grassland	Dense grass, Moderate grass, Sparse grass
Water body	Stream and rivers, Lakes, Reservoir, and ponds, Permanent ice and snow, Beach and shore, Bottom land
Built-up land	Urban built-up, Rural settlements, Others
Unused land	Sandy land, Gobi, Salina, Swampland, Bare soil, Bare rock, Others

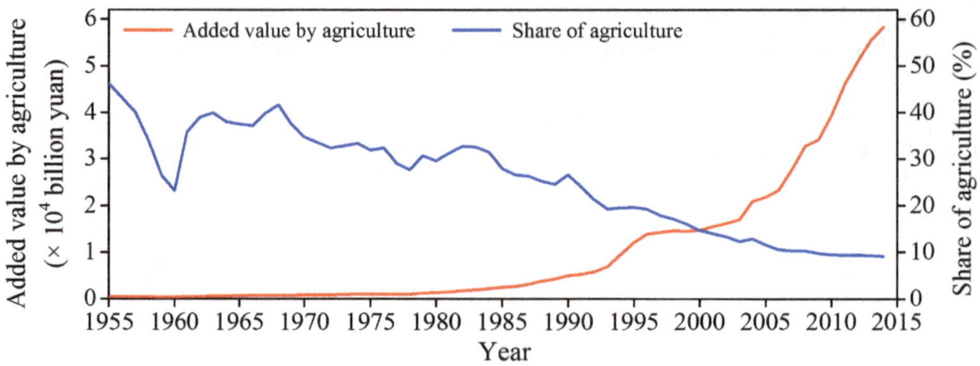

Fig. 1.1 Added value by agriculture and share of agriculture in China (*Source* National Bureau of Statistics of China, China Statistical Yearbook)

has been decreasing for many years along with rapid development of industrialization and urbanization. At the middle of the twentieth century, the share of agriculture in Chinese GDP is greater than 40%. Up to 2015, the share of agriculture in total GDP declined to less than 10%. The high productions of agriculture occur in Heilongjiang, Hebei, Shandong, Henan, Jiangsu, and Sichuan provinces.

1.7 List of Maps

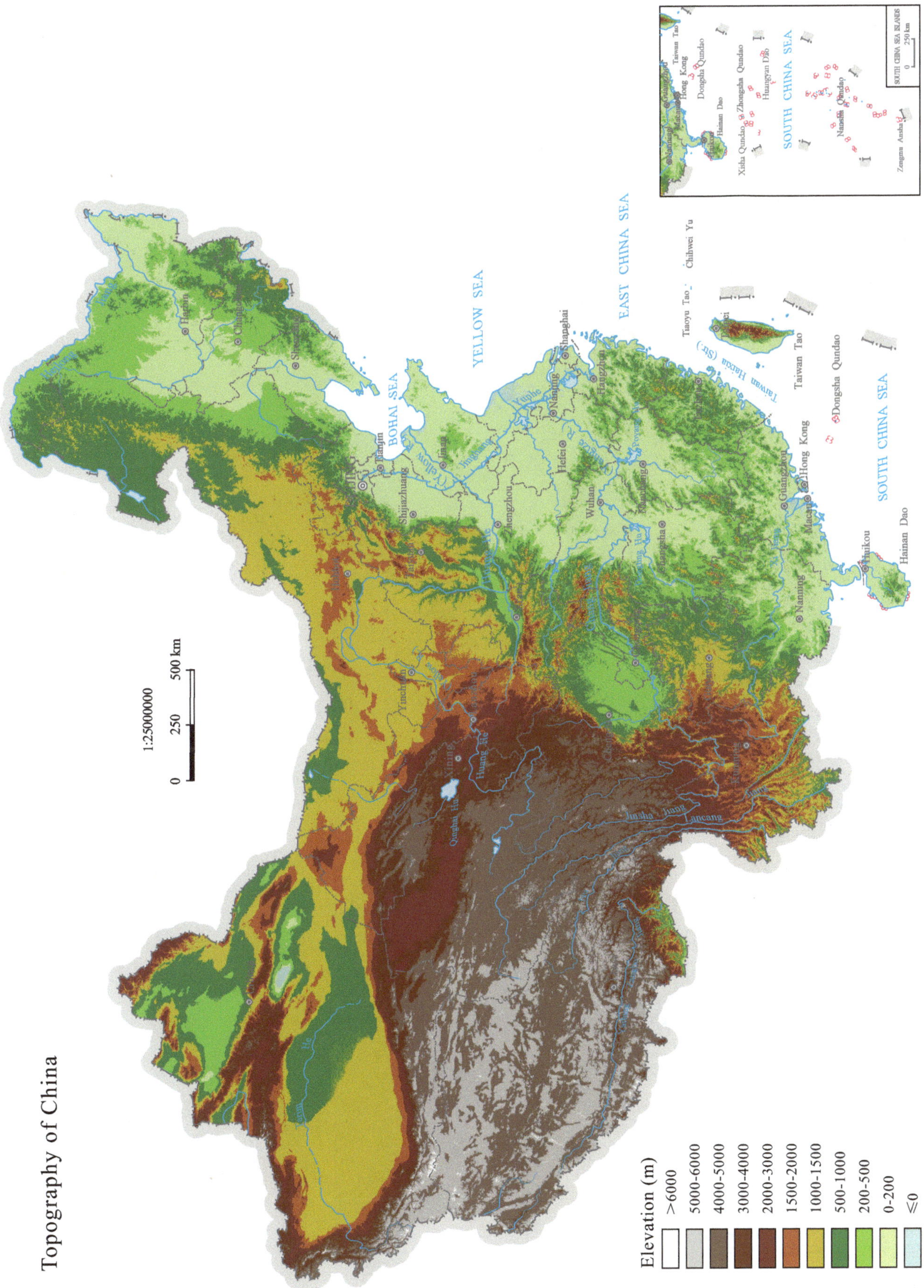

Topography of China

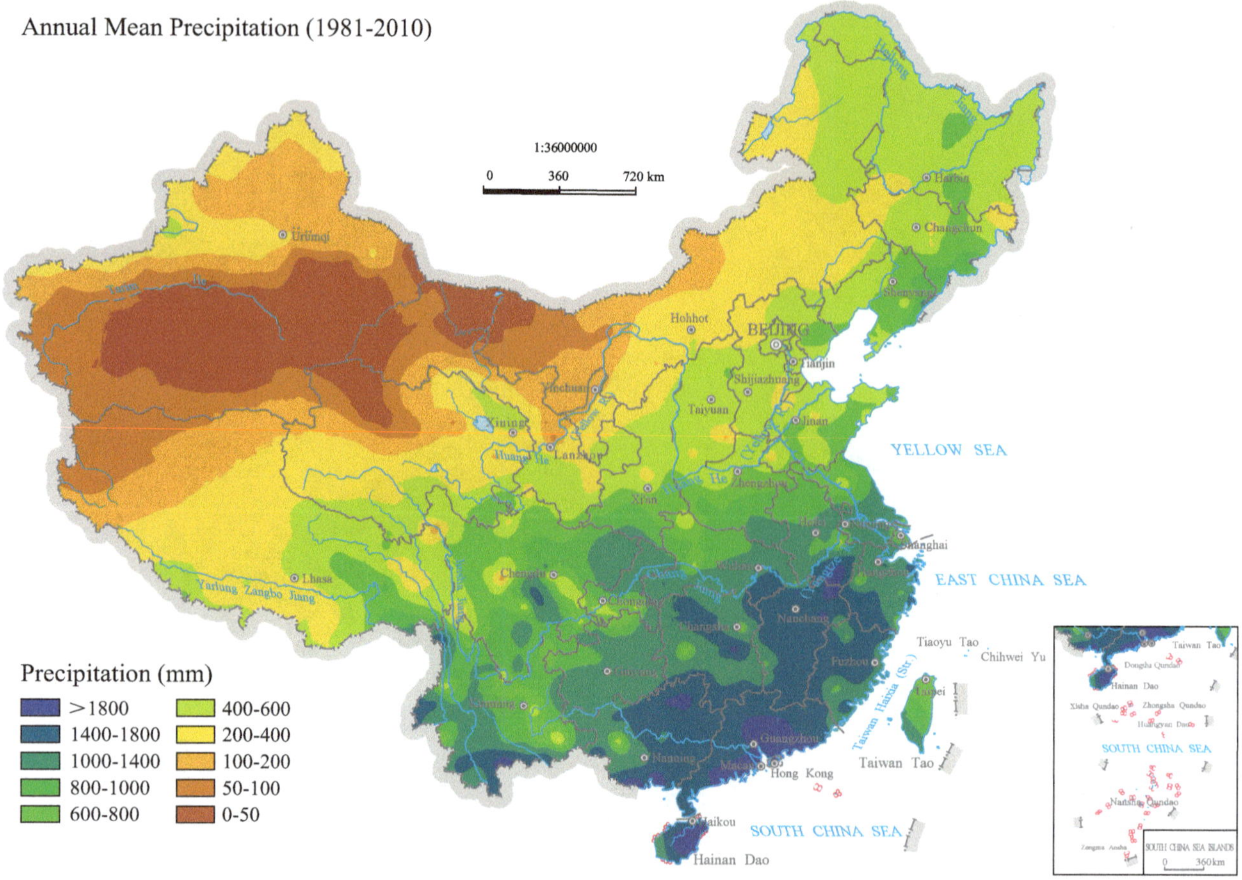
Annual Mean Precipitation (1981-2010)
1:36000000
0 360 720 km
Precipitation (mm)
>1800
1400-1800
1000-1400
800-1000
600-800
400-600
200-400
100-200
50-100
0-50
Ürümqi
Hohhot
BEIJING
Tianjin
Shijiazhuang
Taiyuan
Yinchuan
Xining
Lanzhou
Jinan
Zhengzhou
Xi'an
Harbin
Changchun
Shenyang
Shanghai
Lhasa
Yarlung Zangbo Jiang
Nanchang
Fuzhou
Guangzhou
Hong Kong
Nanning
Haikou
Hainan Dao
Taipei
Taiwan Tao
Taiwan Haixia (Str.)
Tiaoyu Tao
Chihwei Yu
YELLOW SEA
EAST CHINA SEA
SOUTH CHINA SEA
SOUTH CHINA SEA ISLANDS
0 360km

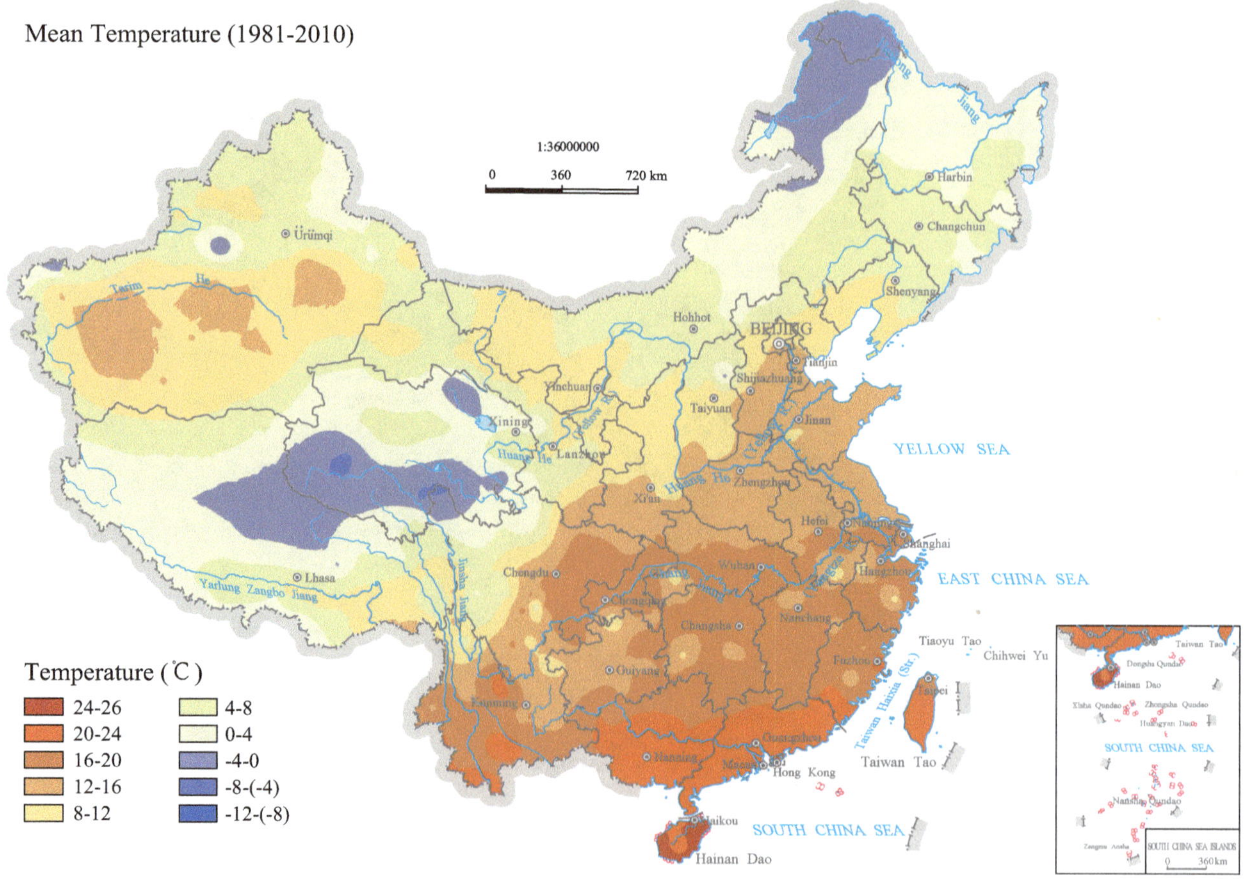
Mean Temperature (1981-2010)
1:36000000
0 360 720 km
Temperature (℃)
24-26
20-24
16-20
12-16
8-12
4-8
0-4
-4-0
-8-(-4)
-12-(-8)
Ürümqi
Tarim He
Hohhot
BEIJING
Tianjin
Shijiazhuang
Taiyuan
Yinchuan
Xining
Lanzhou
Jinan
Zhengzhou
Xi'an
Hefei
Harbin
Changchun
Shenyang
Shanghai
Hangzhou
Wuhan
Chengdu
Chongqing
Changsha
Nanchang
Fuzhou
Guiyang
Kunming
Lhasa
Yarlung Zangbo Jiang
Jinsha Jiang
Guangzhou
Nanning
Hong Kong
Haikou
Hainan Dao
Taipei
Taiwan Tao
Taiwan Haixia (Str.)
Tiaoyu Tao
Chihwei Yu
YELLOW SEA
EAST CHINA SEA
SOUTH CHINA SEA
SOUTH CHINA SEA ISLANDS
0 360km

Mean Daily Minimum Temperature (1981-2010)

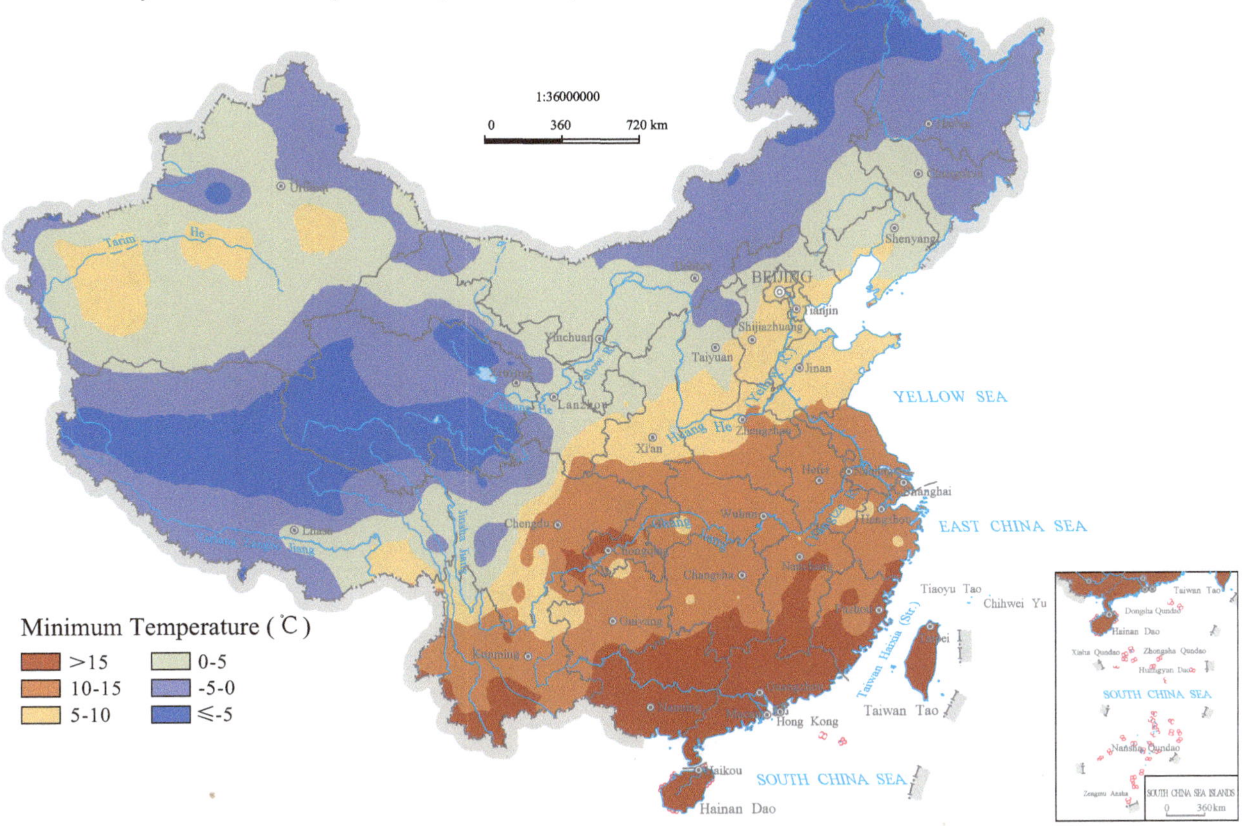

Mean Daily Maximum Temperature (1981-2010)

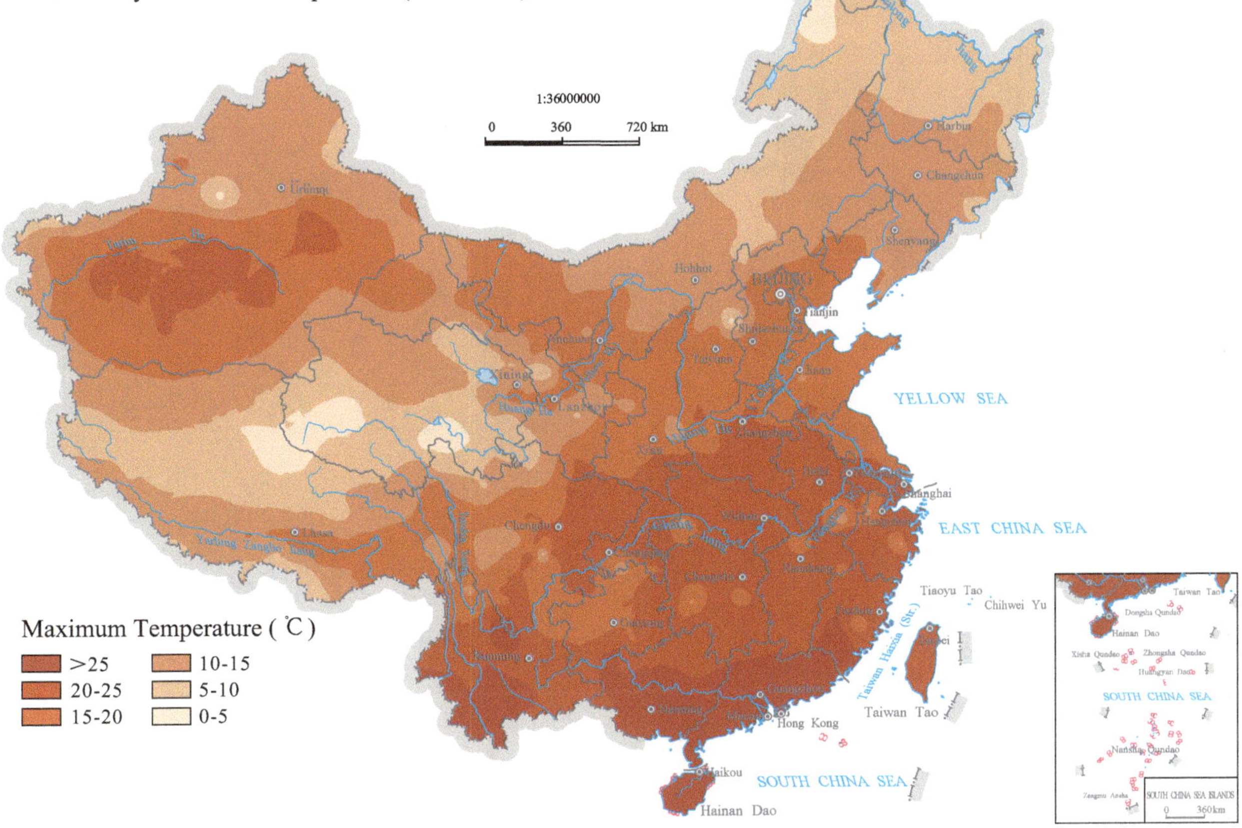

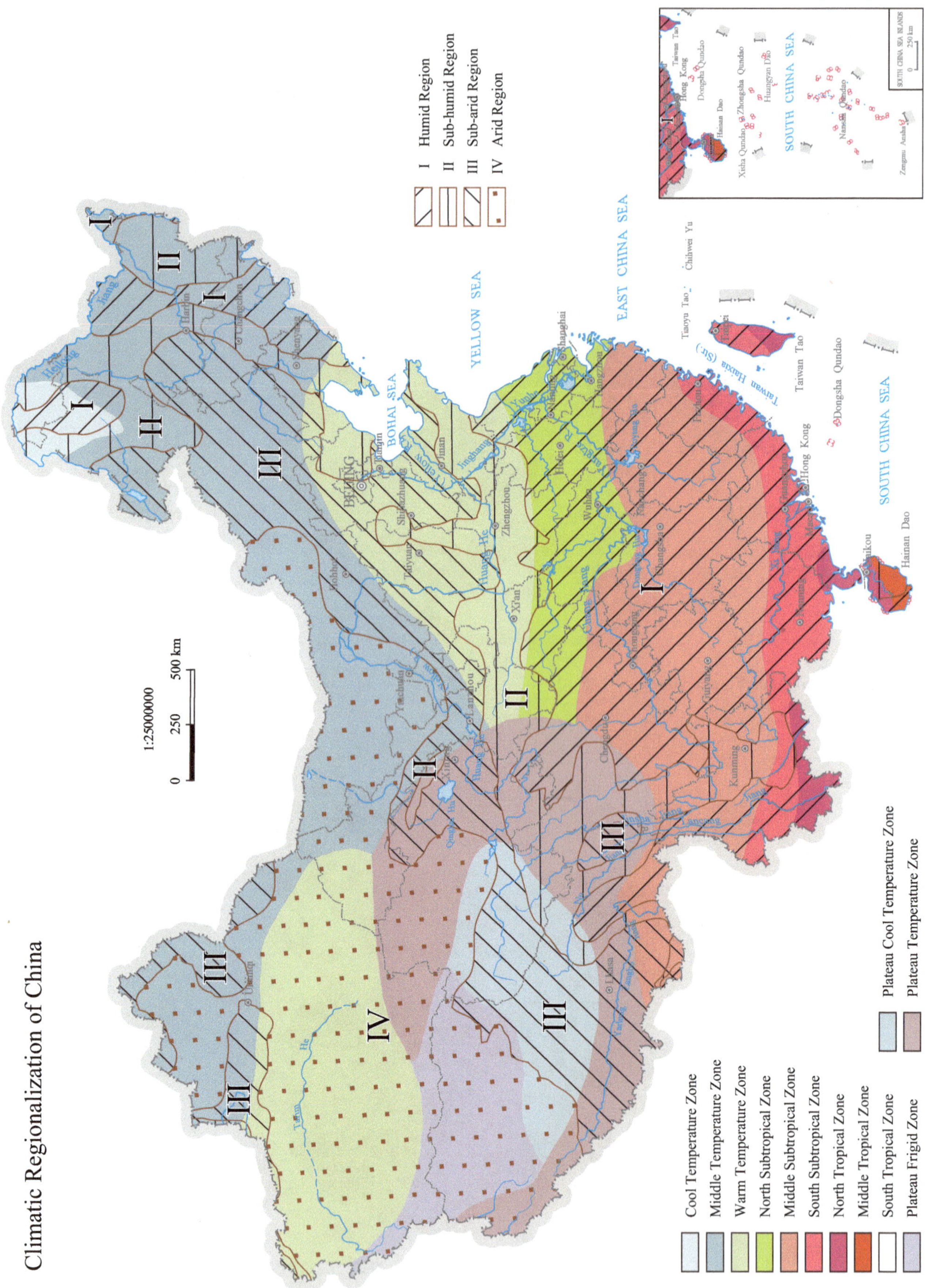
Climatic Regionalization of China
1:25000000
0 250 500 km
I Humid Region
II Sub-humid Region
III Sub-arid Region
IV Arid Region
Cool Temperature Zone
Middle Temperature Zone
Warm Temperature Zone
North Subtropical Zone
Middle Subtropical Zone
South Subtropical Zone
North Tropical Zone
Middle Tropical Zone
South Tropical Zone
Plateau Frigid Zone
Plateau Cool Temperature Zone
Plateau Temperature Zone
BOHAI SEA
YELLOW SEA
EAST CHINA SEA
SOUTH CHINA SEA
BEIJING
SOUTH CHINA SEA ISLANDS

Climatic Change Regionalization of China

Regionalization Level I

I Northeast China-North China Warm-Dry Trend Zone

II East China-Central China Wet-Warm Trend Zone

III Southwest China-South China Dry-Warm Trend Zone

IV Tibet-Southwest China Wet-Warm Trend Zone

V Northwest China-Qinghai-Tibet Plateau Warm-Wet Trend Zone

Regionalization Level II

I1: Lesser Khingan-Changbai Mountains-Sanjiang Plain Temperature Fluctuation Increasing and Precipitation Fluctuation Decreasing Region

I2: Greater Khingan-Mountains in West Liaoning-Horqin Sandy Land Temperature Fluctuation Decreasing and Precipitation Fluctuation Increasing Region

I3: Mountains in North China-Plain-Shandong Peninsula Temperature and Precipitation Fluctuation Decreasing Region

I4: Loess Plateau-Fenhe River Valley Temperature Fluctuation Increasing and Precipitation Fluctuation Decreasing Region

II1: Huai River Basin-Yangtze Delta Plain Temperature Fluctuation Decreasing and Precipitation Fluctuation Increasing Region

II2: Plains along the Lower Reaches of Yangtze River-Zhejiang-North Jiangxi-East Hunan Precipitation Fluctuation Increasing and Temperature Fluctuation Decreasing Region

II3: Hills and Mountains in Eastern Nanling Mountains Temperature and Precipitation Fluctuation Decreasing Region

III1: Mountains in Western Qinling Mountains-Sichuan Basin Temperature Fluctuation Increasing and Precipitation Fluctuation Decreasing Region

III2: Eastern Qinling Mountains-Mountains in Western Hubei Temperature Fluctuation Increasing and Precipitation Fluctuation Decreasing Region

III3: Yunnan-Guizhou Plateau-Mountains and Hills in Western Nanling Mountains Precipitation and Temperature Fluctuation Increasing Region

IV1: Mountains in Southeast Tibet-Plateau Precipitation and Temperature Fluctuation Increasing Region

IV2: Mountains, Ridges and Valleys in Hengduan Mountainous Region Precipitation and Temperature Fluctuation Increasing Region

V1: Mountains in Xinjiang-Qilian Mountain-Inner Mongolian Plateau Temperature Fluctuation Decreasing and Precipitation Increasing Region

V2: Qinghai-Tibet Plateau Temperature and Precipitation Fluctuation Increasing Region

Note: Data unavailable in Taiwan

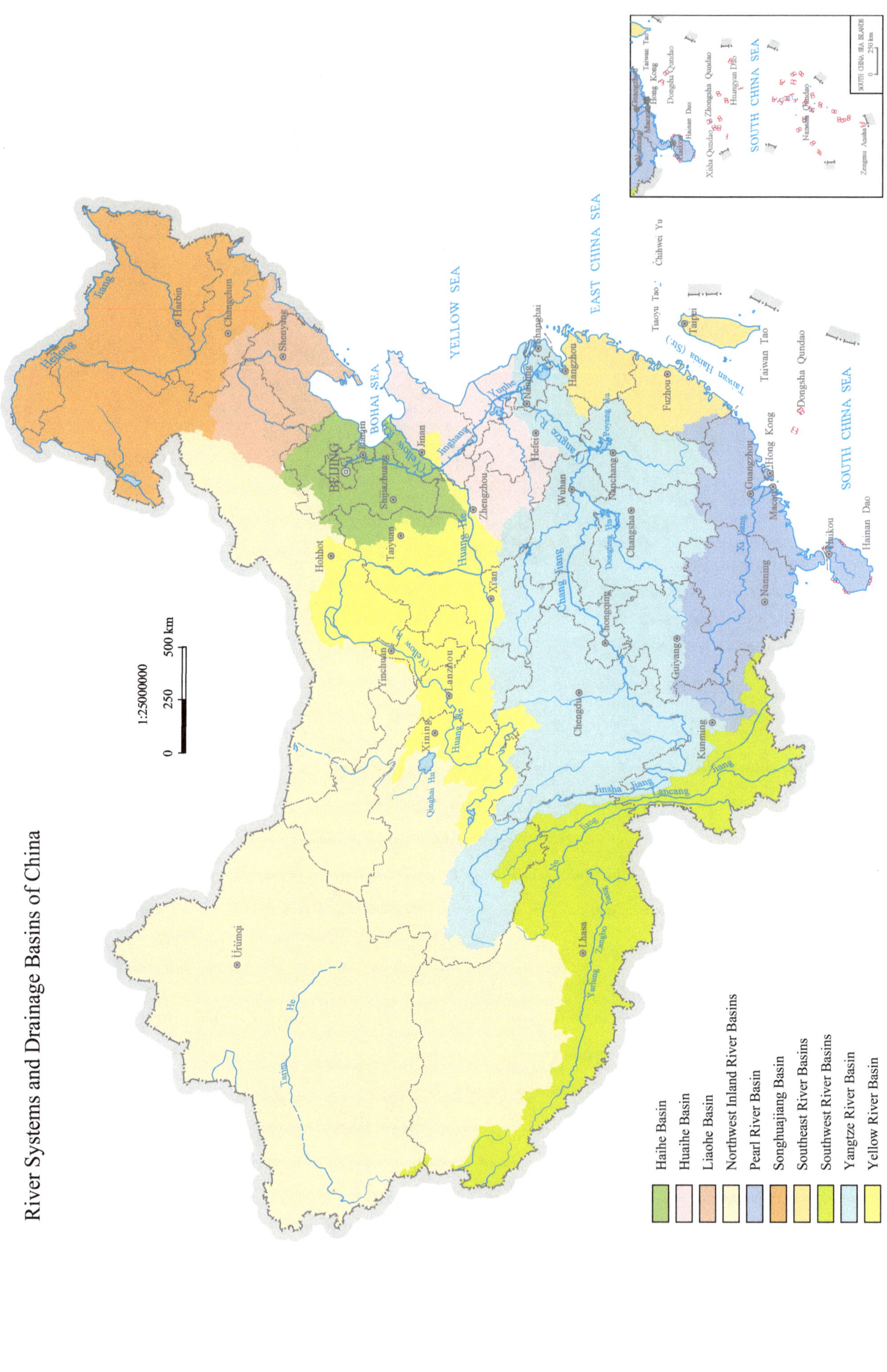
River Systems and Drainage Basins of China
Haihe Basin
Huaihe Basin
Liaohe Basin
Northwest Inland River Basins
Pearl River Basin
Songhuajiang Basin
Southeast River Basins
Southwest River Basins
Yangtze River Basin
Yellow River Basin
1:25000000
0 250 500 km
BOHAI SEA
YELLOW SEA
EAST CHINA SEA
SOUTH CHINA SEA
BEIJING
Shanghai
Wuhan
Chongqing
Chengdu
Lhasa
Ürümqi
Lanzhou
Xining
Xian
Zhengzhou
Jinan
Nanjing
Hangzhou
Fuzhou
Guangzhou
Hong Kong
Macao
Nanning
Kunming
Guiyang
Changsha
Nanchang
Hefei
Taiyuan
Shijiazhuang
Hohhot
Yinchuan
Shenyang
Changchun
Harbin
Haikou
Taipei
Hainan Dao
Taiwan Dao
Dongsha Qundao
Qinghai Hu
Tarim He
Yarlung Zangbo Jiang
Nu Jiang
Lancang Jiang
Jinsha Jiang
Chang Jiang
Huang He
Xi Jiang
Heilong Jiang
SOUTH CHINA SEA ISLANDS

Annual Runoff Depth

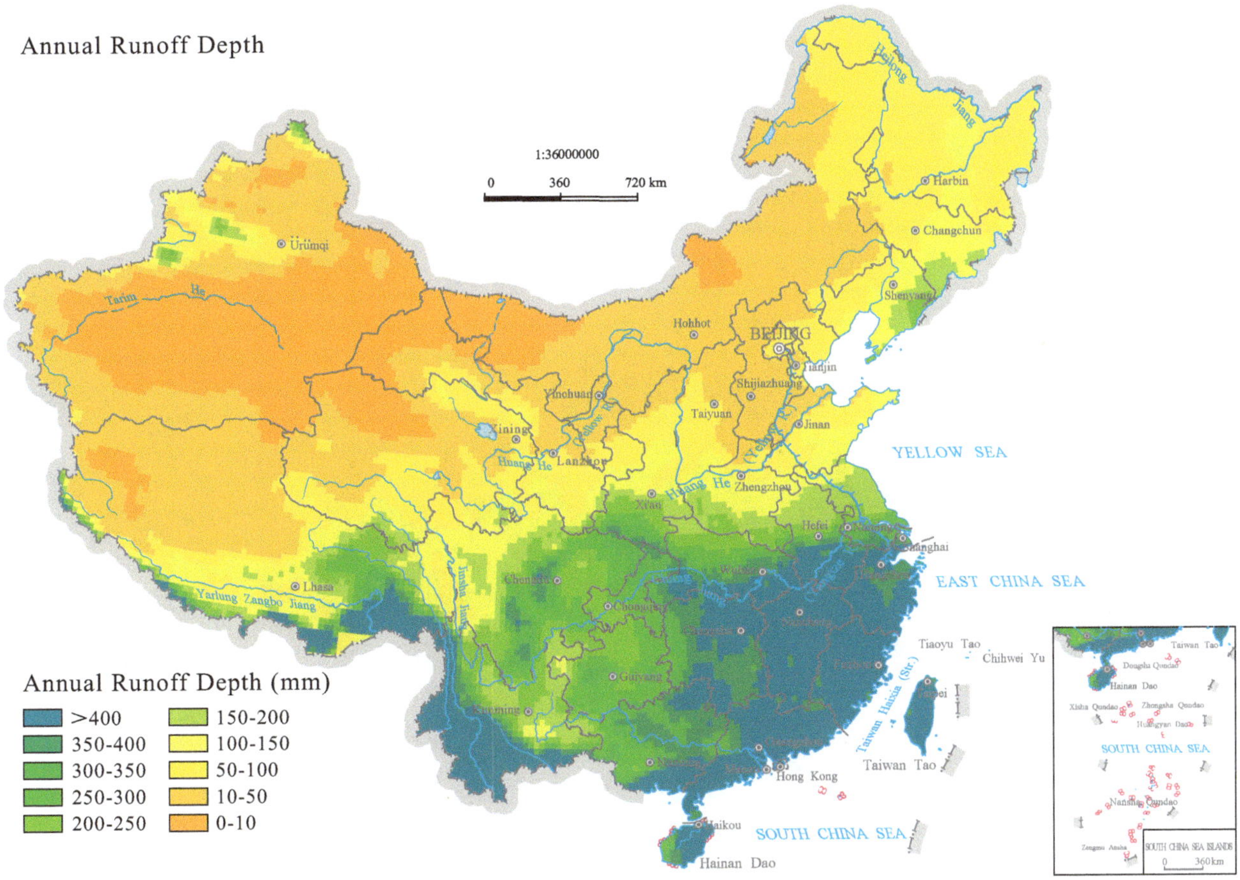

Annual Runoff Coefficient

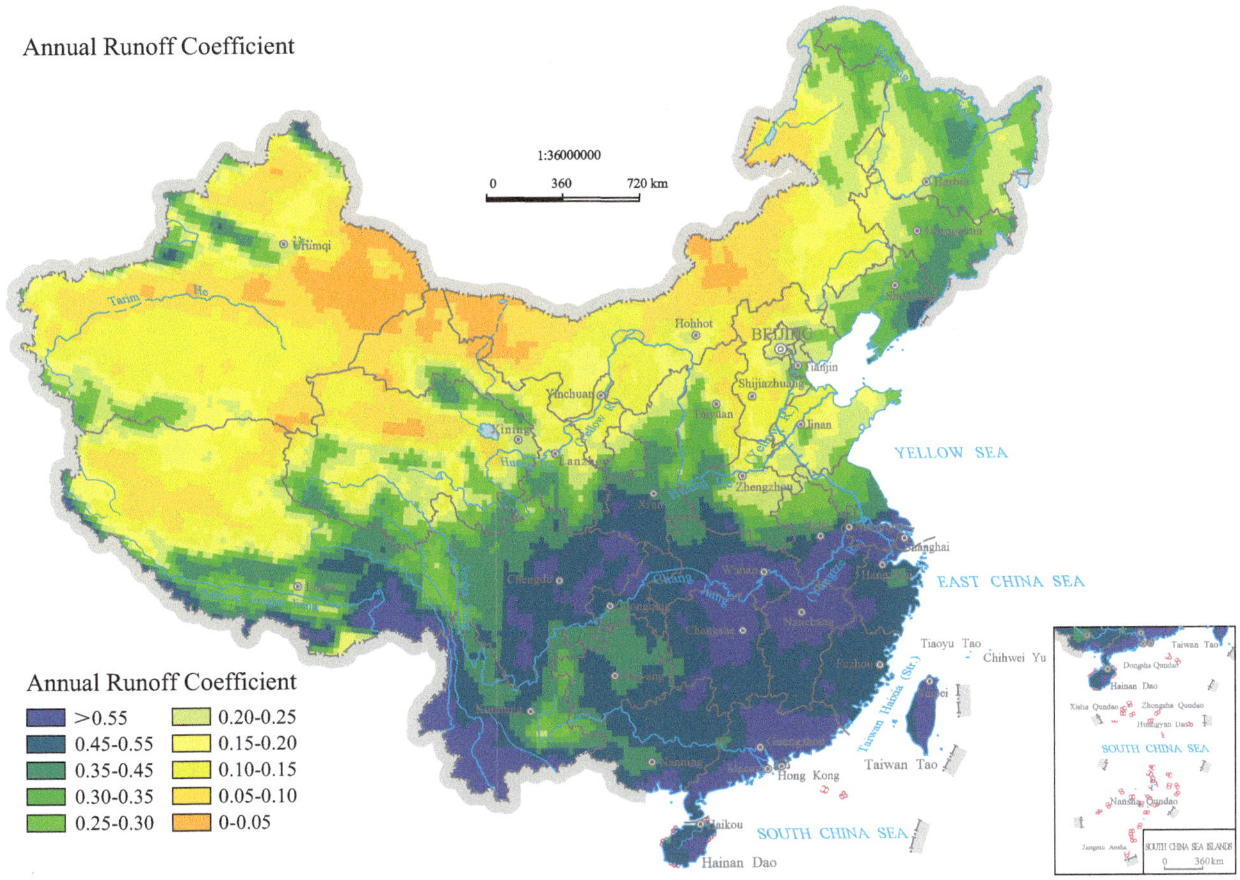

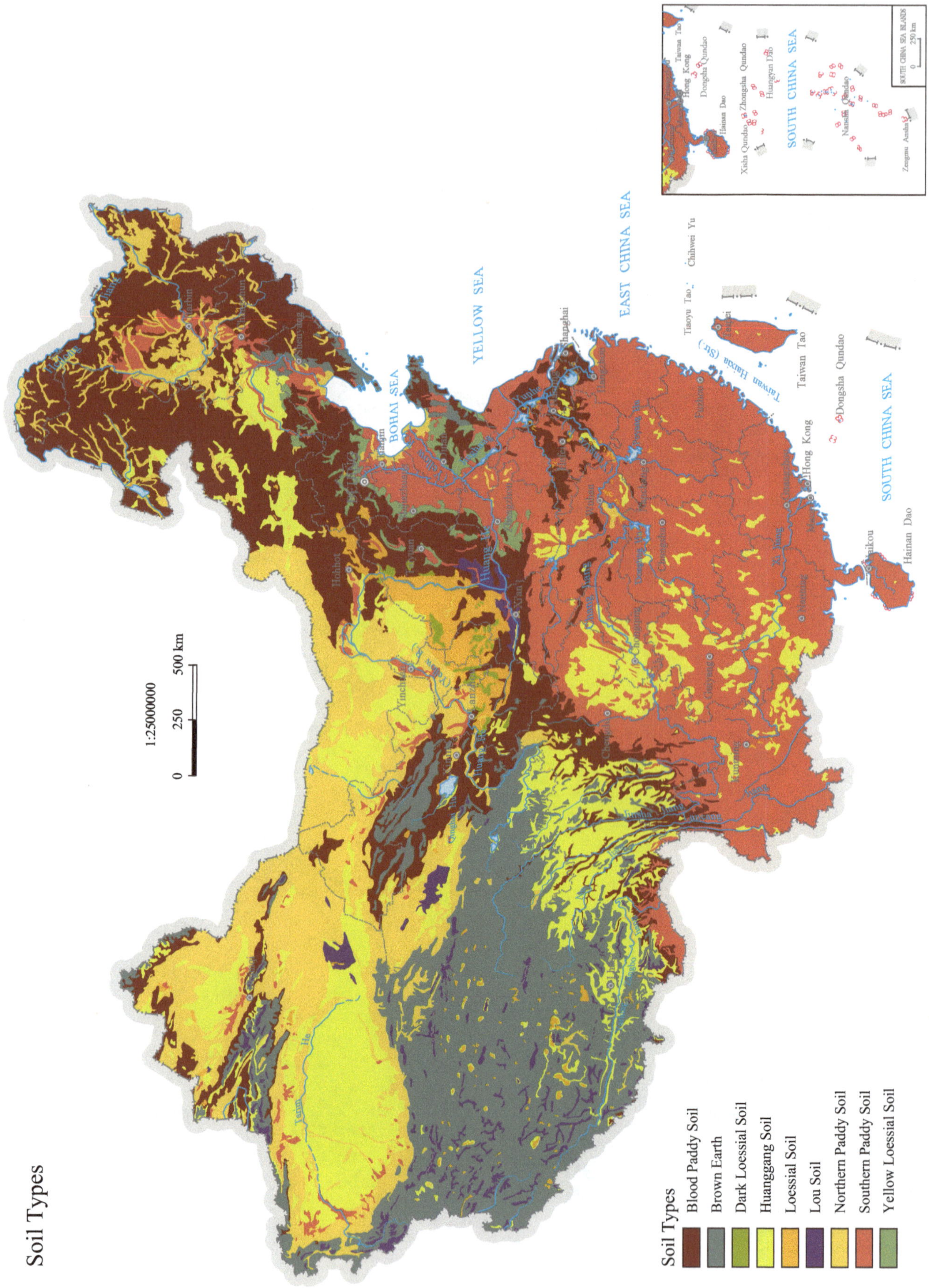
Soil Types
1:25000000
0 250 500 km
Soil Types
Blood Paddy Soil
Brown Earth
Dark Loessial Soil
Huanggang Soil
Loessial Soil
Lou Soil
Northern Paddy Soil
Southern Paddy Soil
Yellow Loessial Soil
YELLOW SEA
EAST CHINA SEA
BOHAI SEA
SOUTH CHINA SEA
Hainan Dao
Taiwan Tao
Dongsha Qundao
Hong Kong
Shanghai
Hohhot
Xi'an

Soil Regionalization of China

Regionalization Level II

- I1: Cold Temperate Zone Orthic Spodosols, Gelic Cambosols, Permagelic Gleyosols Soil Zone
- I2: Middle Temperature Zone Udic Cambosols, Udic Isohumosols, Ustic Isohumosols Soil Zone
- I3: Warm Temperate Zone Udic Argosols, Ustic Argosols, Aquic Cambosols Soil Zone
- I4: North Subtropical Zone Udic Argosols, Stagnic Anthrosols Soil Zone
- I5: Middle Subtropical Zone Udic Ferrosols, Perudic Argosols Soil Zone
- I6: South Subtropical Zone Udic Ferrosols, Udic Ferralosols Soil Zone
- I7: Torrid Zone Udic Ferralosols, Udic Ferrosols, Phosphi-Lithomorphic Isohumosols Soil Zone
- II1: Middle Temperature Zone Ustic Isohumosols, Usti-Sadic Primosols Soil Zone
- II2: Warm Temperature Zone Loessi-Orthic Primosols, Cumuli-Ustic Isohumosols, Hapli-Ustic Argosols Soil Zone
- II3: Plateau Temperate Zone Ustic Cambosols, Udic Argosols, Gelic Cambosols Soil Zone
- II4: Plateau sub-frigid Zone Gelic Cambosols, Permagelic Histosols, Orthic Gleyosols Soil Zone
- III1: Middle Temperature Zone Orthic Aridosols, Sadic Primosols Soil Zone
- III2: Warm Temperature Zone Orthic Aridosols, Orthic Halosols Soil Zone
- III3: Plateau Temperature Zone Orthic Aridosols, Orthic Halosols, Ustic Isohumosols Soil Zone
- III4: Plateau Temperature Zone Ustic Cambosols, Gelic Cambosols, Siltigi-Ustic Cambosols Soil Zone
- III5: Plateau Sub-frigid Zone Gelic Cambosols, Cryic Aridosols, Soil Zone
- III6: Plateau Frigid Zone Gelic Cambosols, Cryic Aridosols, Permi-Gelic Cambosols Soil Zone

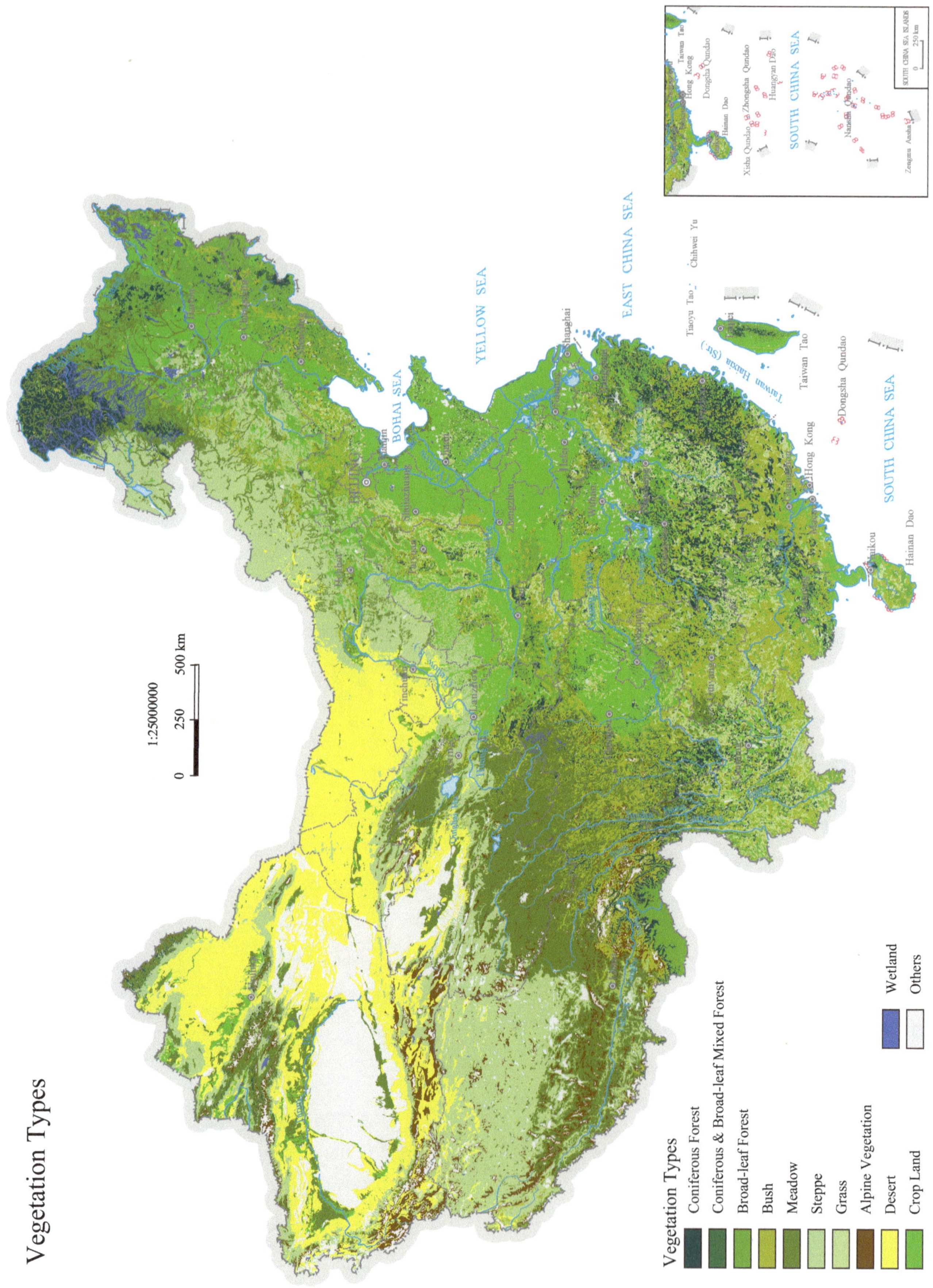
Vegetation Types
1:25000000
0 250 500 km
YELLOW SEA
BOHAI SEA
EAST CHINA SEA
SOUTH CHINA SEA
Taiwan Haixia (Str.)
Taiwan Tao
Dongsha Qundao
Hainan Dao
Hong Kong
Shanghai
Tiaoyu Tao
Chihwei Yu
Vegetation Types
Coniferous Forest
Coniferous & Broad-leaf Mixed Forest
Broad-leaf Forest
Bush
Meadow
Steppe
Grass
Alpine Vegetation
Desert
Crop Land
Wetland
Others

Vegetation Regionalization of China

Regionalization Level I

- I Cold-Temperate Needleleaf Deciduous Forest Region
- II Temperate Mixed Needleleaf and Deciduous Broadleaf Forest Region
- III Warm Temperate Deciduous Broadleaf Forest Region
- IV Subtropical Broadleaf Evergreen Forest Region
- V Tropical Monsoon Rain Forest and Rain Forest Region
- VI Temperate Steppe Region
- VII Temperate Desert Region
- VIII Qinghai-Xizang Plateau Alpine Vegetation Region

Regionalization Level II

- IVA Eastern Humid Broadleaf Evergreen Forest Sub-region
- IVB Western Semi-humid Broadleaf Evergreen Forest Sub-region
- VA Eastern Tropical Monsoon Rain Forest and Rain Forest Sub-region with Humid Bias
- VB Western Tropical Monsoon Rain Forest and Rain Forest Sub-region with Dry Bias
- VIA Eastern Steppe Sub-region
- VIB Western Steppe Sub-region
- VIIA Western Temperate Desert Sub-region
- VIIB Eastern Temperate Desert Sub-region
- VIIIA Eastern Qinghai-Xizang Plateau Sub-alpine Scrub and Alpine Meadow Sub-region
- VIIIB Middle Qinghai-Xizang Plateau Alpine Steppe Sub-region
- VIIIC North-Western Qinghai-Xizang Plateau Alpine Desert Sub-region

Regionalization Level III

- Ii: Southern Cold-Temperate Needleleaf Deciduous Forest Zone
- IIi: Northern Temperate Mixed Needleleaf and Deciduous Broadleaf Forest Zone
- IIii: Southern Temperate Mixed Needleleaf and Deciduous Broadleaf Forest Zone
- IIIi: Northern Warm Temperate Deciduous Quercus Forest Zone
- IIIii: Southern Warm Temperate Deciduous Quercus Forest Zone
- IVAi: Northern Sub-tropical Broadleaf Evergreen and Deciduous Forest Zone
- IVAii: Middle Sub-tropical Broadleaf Evergreen Forest Zone
- IVAiii: Southern Sub-tropical Monsoon Broadleaf Evergreen Forest Zone
- IVBi: Middle Sub-tropical Broadleaf Evergreen Forest Zone
- IVBii: Southern Sub-tropical Monsoon Broadleaf Evergreen Forest Zone
- IVBiii: Subtropical Mountains Cold-Temperate Needleleaf Forest Zone
- VAi: Northern Tropical Semi-evergreen Monsoon Rain Forest and Humid Rain Forest Zone
- VAii: Southern Tropical Monsoon Rain Forest, Humid Rain Forest Zone
- VBi: Northern Tropical Seasonal Rain Forest, Semi-evergreen Monsoon Rain Forest Zone
- VIAi: Northern Temperate Steppe Zone
- VIAii: Southern Temperate Steppe Zone
- VIBi: Northern Temperate Desert Steppe Zone
- VIIAi: Temperate Semi-shrubby, Dwarf Arboreous Desert Zone
- VIIBi: Temperate Semi-shrubby and Dwarf Desert Zone
- VIIBii: Warm Temperate Shrubby and Semi-shrubby Desert Zone
- VIIIAi: Sub-alpine Scrub and Alpine Meadow Zone
- VIIIAii: Alpine Meadow Zone
- VIIIBi: Alpine Steppe Zone
- VIIIBii: Temperate Steppe Zone
- VIIICi: Alpine Desert Zone
- VIIICii: Temperate Desert Zone

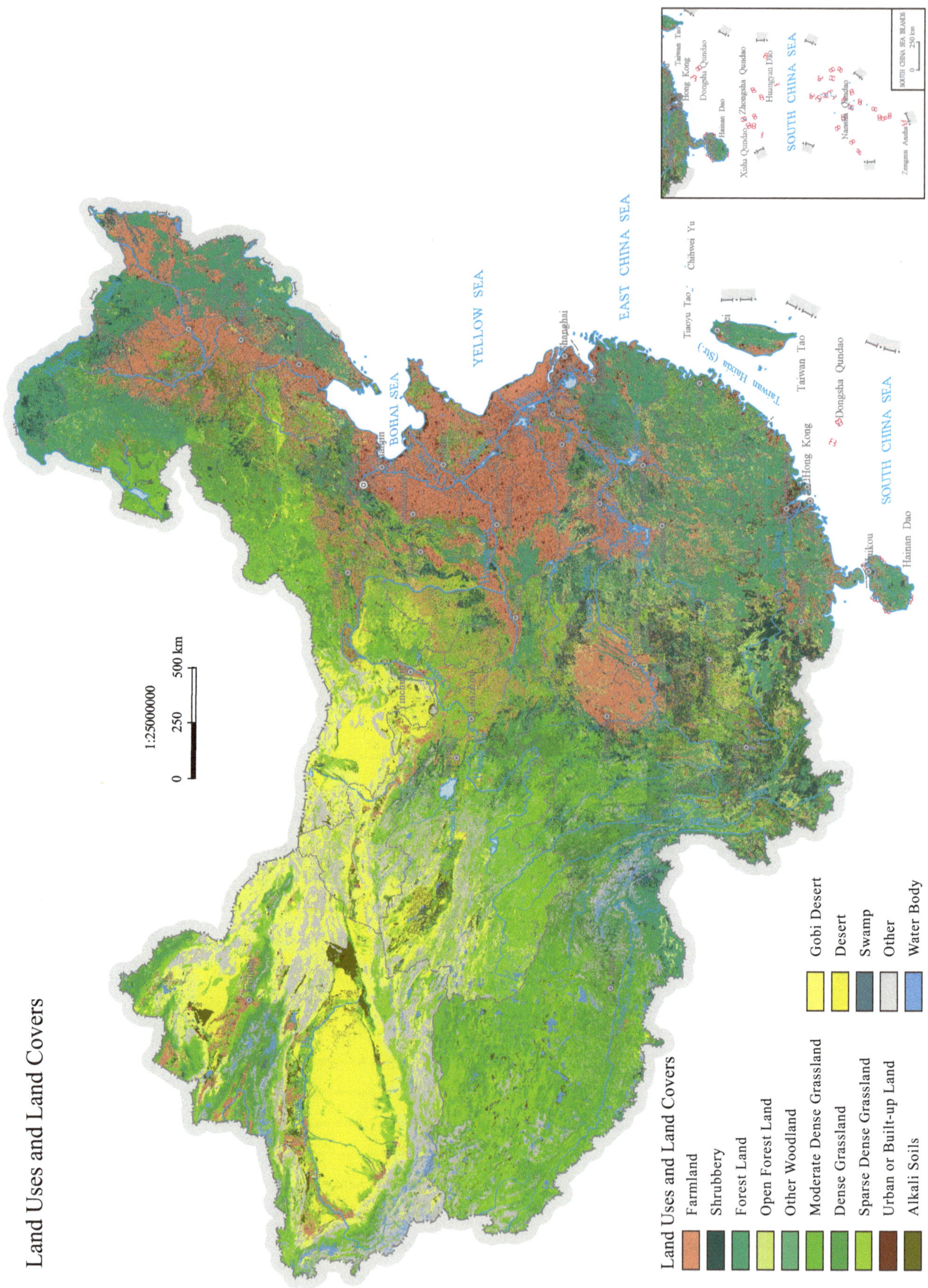
Land Uses and Land Covers
Land Uses and Land Covers
Farmland
Shrubbery
Forest Land
Open Forest Land
Other Woodland
Moderate Dense Grassland
Dense Grassland
Sparse Dense Grassland
Urban or Built-up Land
Alkali Soils
Gobi Desert
Desert
Swamp
Other
Water Body
1:25000000
0 250 500 km
BOHAI SEA
YELLOW SEA
EAST CHINA SEA
SOUTH CHINA SEA
Shanghai
Taiwan Haixia (Str.)
Taiwan Tao
Taoyu Tao
Chihwei Yu
Dongsha Qundao
Hong Kong
Hainan Dao
Haikou
Hainan Dao
Hong Kong
Taiwan Tao
Dongsha Qundao
Xisha Qundao
Zhongsha Qundao
Huangyan Dao
Nansha Qundao
Zengmu Ansha
SOUTH CHINA SEA
SOUTH CHINA SEA ISLANDS
0 250 km

Eco-geographical Regionalization of China

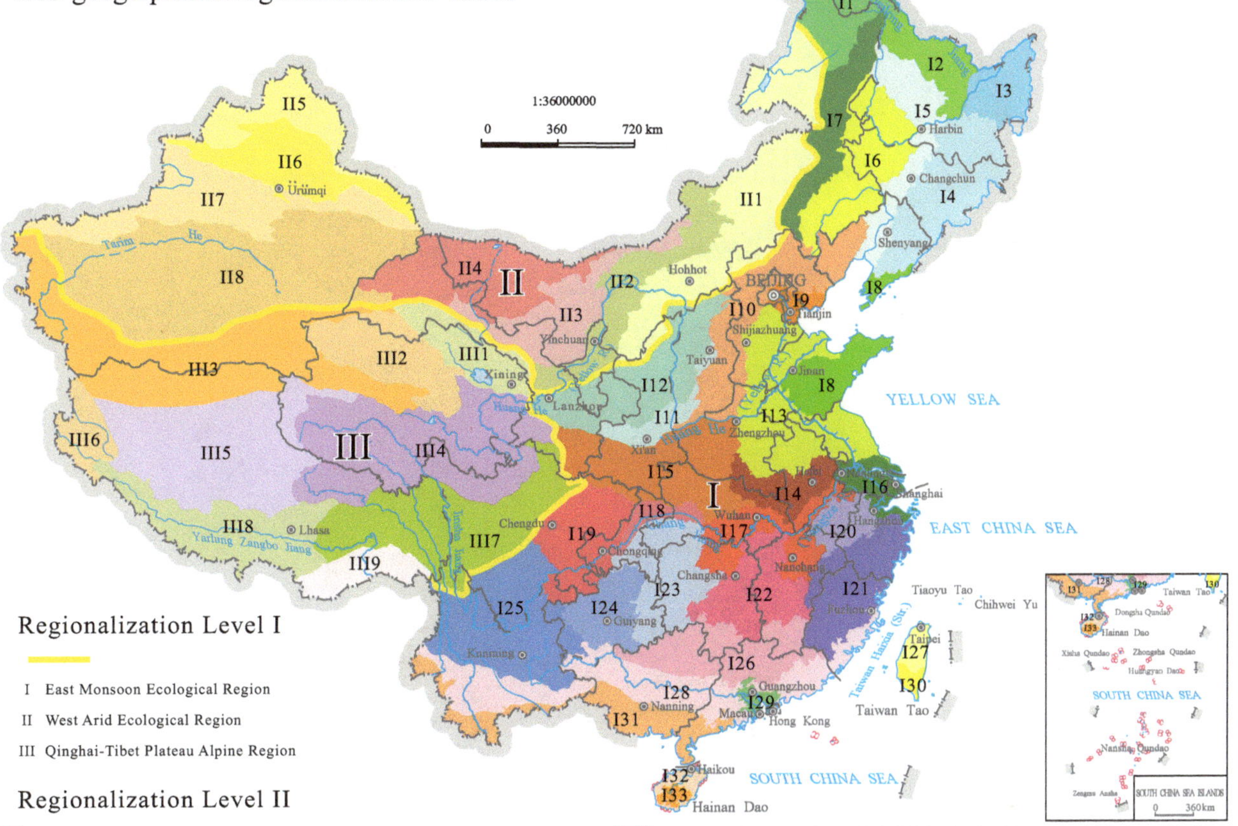

Regionalization Level I

I East Monsoon Ecological Region

II West Arid Ecological Region

III Qinghai-Tibet Plateau Alpine Region

Regionalization Level II

- I1: Deciduous Coniferous Forest Ecological Function Area in Northern Greater Hinggan Mountains
- I2: Coniferous and Broad Leaved Mixed Forest Ecological Function Area in the Lesser Xing'An Mountain
- I3: Temperate Wetland Agriculture and Wetland Ecological Function Area in Sanjiang Plain
- I4: Coniferous and Broad Leaved Mixed Forest Ecological Function Area in Changbai Mountain and Qian Mountain
- I5: Agricultural Ecological Function Area in the East of the Northeast China Plain
- I6: Meadow Steppe Ecological Function Area in the West of the Northeast China Plain
- I7: Deciduous Broad-Leaved Forest and Forest Grassland Ecological Function Area in Southern-Central Greater Hinggan Mountains
- I8: Broad Leaved Deciduous Forest Ecological Function Area in Hilly Area of Eastern Liaoning and Shandong
- I9: Urban and Suburban Agro-Ecological Function Area in Beijing-Tianjin-Tangshan
- I10: Deciduous Broad-Leaved Forest Ecological Function Area in Yan Mountain and Taihang Mountain
- I11: Agricultural Ecological Function Area in Fen-Wei Basin
- I12: Agriculture and Grassland Ecological Function Area in Loess Plateau
- I13: Agricultural Ecological Function Area in North China Plain
- I14: Evergreen Broad-Leaved Forest Ecological Function Area in Huaiyang Hills
- I15: Deciduous and Evergreen Broad-Leaved Forest Ecological Function Area in Qinling and Daba Mountain
- I16: Urban and Suburban Agro-Ecological Function Area in the Yangtze River Delta
- I17: Agro-Ecological Function Area in the Middle and Lower Reaches of Yangtze River Plain
- I18: Reservoir Ecological Function Area in Three Gorges Reservoir
- I19: Agroforestry Ecological Function Area in Sichuan Basin
- I20: Evergreen Broad-Leaved Forest Ecological Function Area in Tianmu Mountain and Huaiyu Mountain
- I21: Evergreen Broad-Leaved Forest Ecological Function Area in Mountain and Hilly Area of Zhejiang and Fujian
- I22: Evergreen Broad-Leaved Forest Ecological Function Area in Hilly Area of Hunan and Jiangxi
- I23: Evergreen Broad-Leaved Forest Ecological Function Area in Wuling Mountain and Xuefeng Mountain
- I24: Evergreen Broad-Leaved Forest Ecological Function Area in Central Karst Area of Guizhou
- I25: Evergreen Broad-Leaved Forest Ecological Function Area in Mountain Area of South-Western Sichuan and Northern-Central Yunnan
- I26: Evergreen Broad-Leaved Forest Ecological Function Area in Mountain and Hilly Area of Nanling Mountain
- I27: Evergreen Broad-Leaved Forest Ecological Function Area in Nothern Taiwan
- I28: Evergreen Broad-Leaved Forest Ecological Function Area in Hilly Area of Southern Fujian and Cental Yunnan-Guangxi-Guangdong
- I29: Urban and Suburban Agro-Ecological Function Area in Pearl River Delta
- I30: Tropical Monsoon Rainforest and Rainforest Ecological Function Area in Southern Taiwan
- I31: Tropical Monsoon Rainforest and Rainforest Ecological Function Area in Southern Yunnan-Guangxi-Guangdong
- I32: Tropical Agriculture Ecological Function Area in Hainan Island
- I33: Rainforest and Tropical Monsoon Rainforest Ecological Function Area in Central Hainan
- II1: Typical Grassland Ecological Function Area in Central and Eastern Inner Mongolia Plateau
- II2: Desert Steppe Ecological Function Area in Central Inner Mongolia Plateau and Central Gansu
- II3: Steppe Desert Ecological Function Area in Central Inner Mongolia Plateau
- II4: Desert Ecological Function Area in Western Inner Mongolia Plateau and Bei Mountain
- II5: Forest and Grassland Ecological Function Area in Altai Mountains and Western Junggar
- II6: Desert Ecological Function Area in Junggar Basin
- II7: Forest and Grassland Ecological Function Area in Tianshan Mountains
- II8: Desert Ecological Function Area in Tarim Basin and Eastern Xinjiang
- III1: Forest and Alpine Steppe Ecological Function Area in Qilian Mountain
- III2: Desert Ecological Function Area in Qaidam Basin
- III3: Alpine Desert Steppe Ecological Function Area in Pamir-Kunlun Mountain-Altun Mountain
- III4: Alpine Meadow Grassland Ecological Function Area in the Headwaters of the Yangtze and Yellow River and Southren Gansu
- III5: Alpine Desert Steppe Ecological Function Area in Northern Tibetan Plateau
- III6: Temperate Arid Desert Ecological Function Area in Ali Mountain
- III7: Cold Temperate Coniferous Forest Ecological Function Area in Eastern Tibet and Western Sichuan
- III8: Alpine Meadow Steppe Ecological Function Area in Southern Tibet
- III9: Rainforest and Tropical Monsoon Rainforest Ecological Function Area in Southestern Tibet

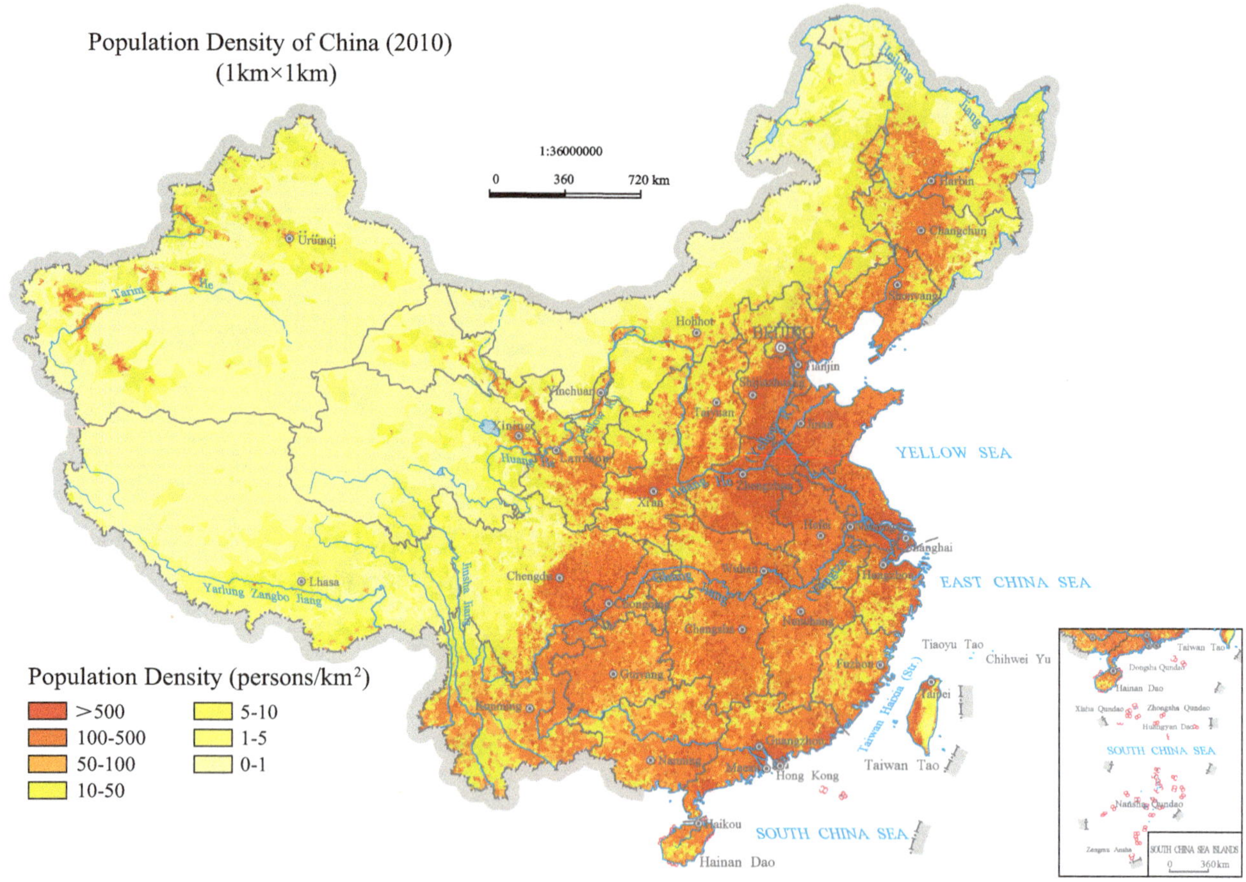
Population Density of China (2010)
(1km×1km)
1:36000000
0 360 720 km
Population Density (persons/km²)
>500
100-500
50-100
10-50
5-10
1-5
0-1
Ürümqi
Tarim He
Lhasa
Yarlung Zangbo Jiang
Jinsha Jiang
Harbin
Changchun
Shenyang
Hohhot
Beijing
Tianjin
Yinchuan
Xining
Lanzhou
Xi'an
Chengdu
Guiyang
Kunming
Nanning
Haikou
Hainan Dao
Hong Kong
Macao
Fuzhou
Taipei
Taiwan Tao
Tiaoyu Tao
Chihwei Yu
Shanghai
YELLOW SEA
EAST CHINA SEA
SOUTH CHINA SEA
SOUTH CHINA SEA ISLANDS

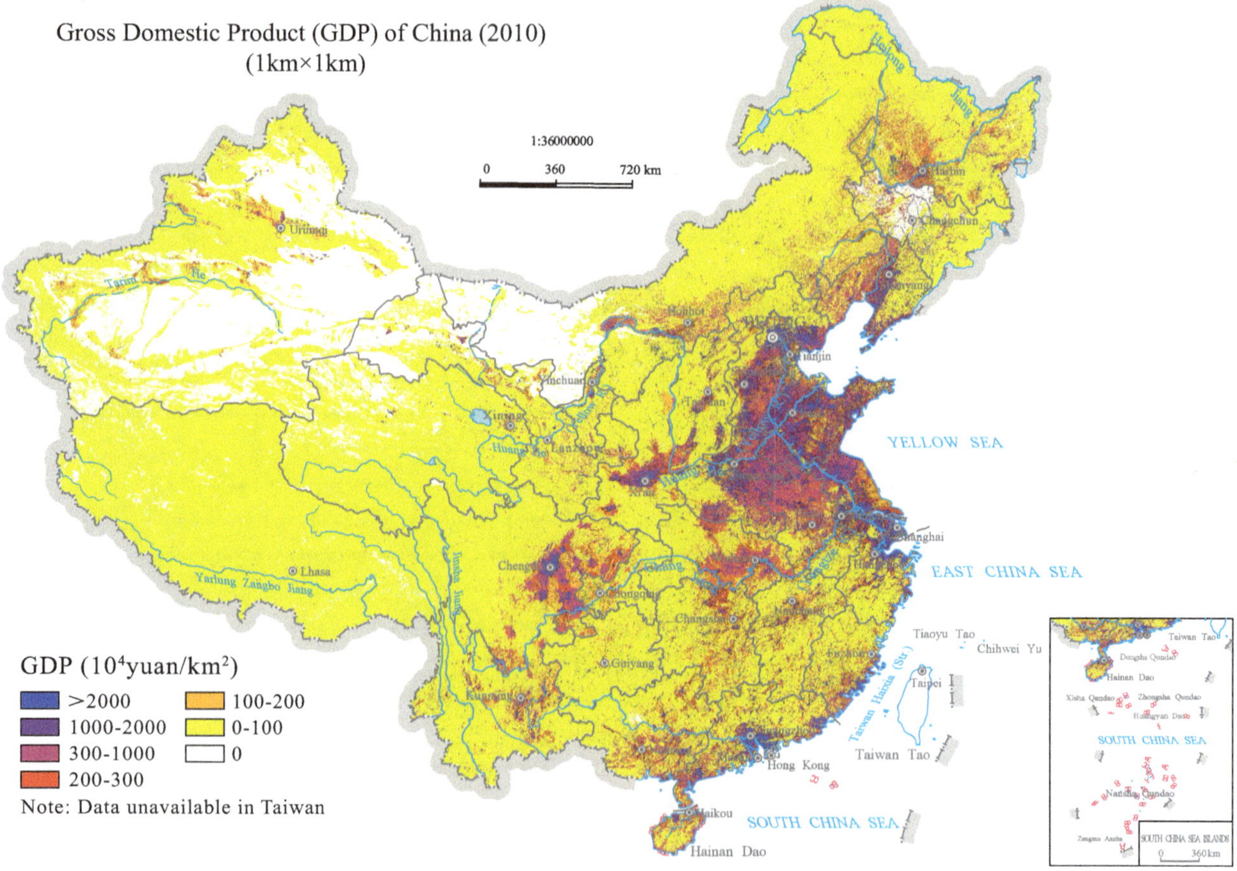
Gross Domestic Product (GDP) of China (2010)
(1km×1km)
1:36000000
0 360 720 km
GDP (10⁴yuan/km²)
>2000
1000-2000
300-1000
200-300
100-200
0-100
0
Note: Data unavailable in Taiwan
Ürümqi
Tarim He
Lhasa
Yarlung Zangbo Jiang
Jinsha Jiang
Harbin
Changchun
Hohhot
Tianjin
Yinchuan
Xining
Lanzhou
Guiyang
Haikou
Hainan Dao
Hong Kong
Taipei
Taiwan Tao
Tiaoyu Tao
Chihwei Yu
Shanghai
YELLOW SEA
EAST CHINA SEA
SOUTH CHINA SEA
SOUTH CHINA SEA ISLANDS

Products of Farming, Forestry, Animal Husbandry and Fishery (2010)

Grain Production (2010)

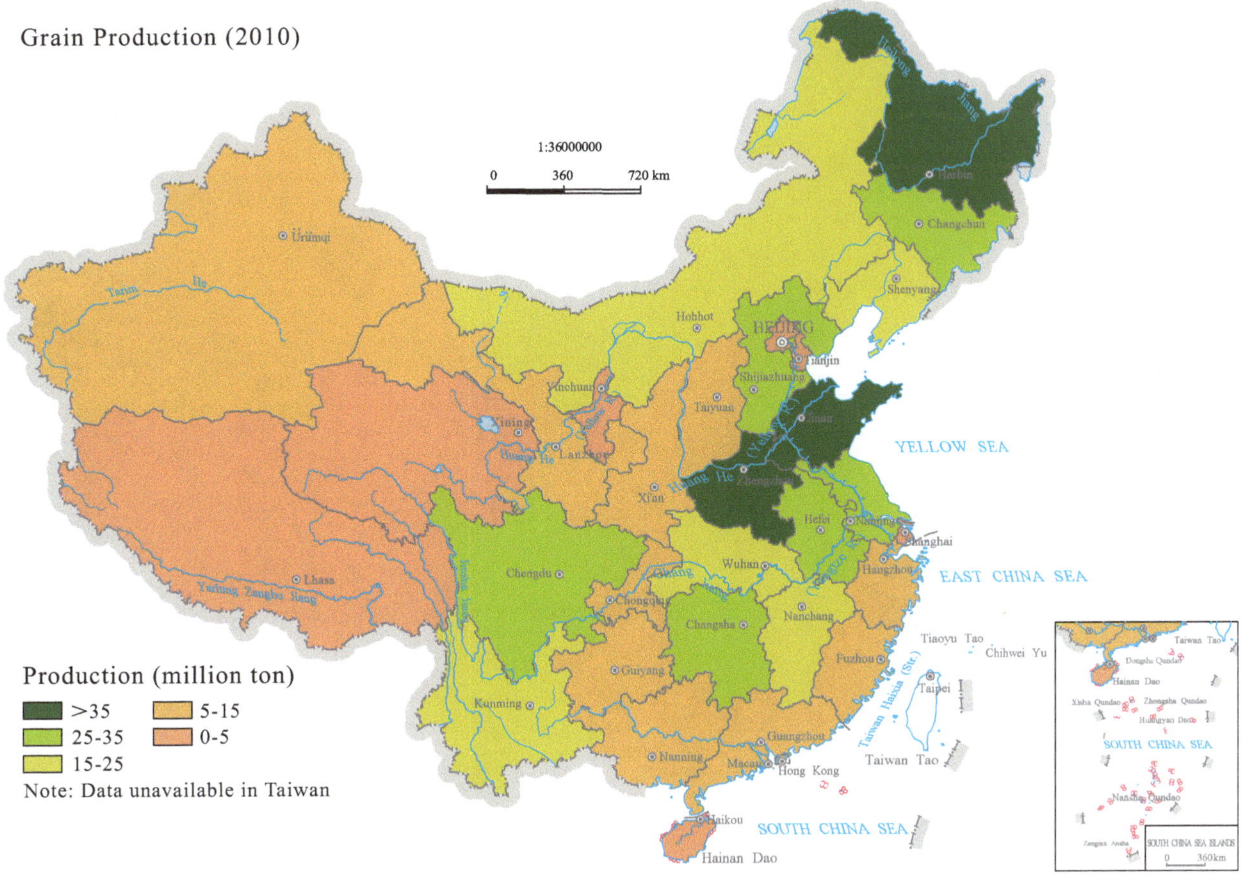

References

Chinese Academy of Sciences. (1985). *Physical geography in China Editorial Board, China Natural Geography (Zonglun)*. Beijing: Science Press.

Edition Board of Vegetation Map of China. (2007). *Vegetation Map of the People's Republic of China (1:1,000,000)*. Beijing, China: Chinese Academic of Science.

Hartmann, R., & Wang, J. A. (2014). *Comparative Geography of China and the U.S.*. Now York, USA: Springer Press.

Hong and Blackmore. (2015). *Plants of China: A companion to the Flora of China*. New York, United States: Cambridge University Press.

Jin, J., Wang, Y., Jiang, H., Kong, Y., Lu, X., & Zhang, X. (2016). Improvement of ecological geographic regionalization based on remote sensing and canonical correspondence analysis: A case study in China. *Science China: Earth Sciences, 59*(9), 1745–1753.

Lemoine, F., Mayo, G., Poncet, S., & Ünal, D. (2014). *The Geographic Pattern of China's Growth and Convergence within Industry*. Paris, French.

Liu, J., Liu, M., Deng, X., Zhuang, D., Zhang, Z., & Luo, D. (2002). The land use and land cover change database and its relative studies in China. *Journal of Geographical Sciences, 12*(3), 275–282.

Ministry of Water Resources People's Republic of China. (2013). *China Water Resources Bulletin 2013*. Beijing, China: China Water & Power Press.

Shi, P., Sun, S., Wang, M., Li, N., Wang, J., Jin, Y., et al. (2014). Climate change regionalization in China (1961–2010). *Science China: Earth Sciences, 57,* 2676–2689.

Shi, X., Yu, D., Warner, D., Pan, X., Petersen, G., Gong, Z., et al. (2004). Soil database of 1:1,000,000 digital soil survey and reference system of the Chinese genetic soil classification system. *Soil Survey Horizons, 45,* 129–136.

Zhang, X., Tang, Q., Pan, M., & Tang, Y. (2014). A long-term land surface hydrologic fluxes and states dataset for China. *Journal of Hydrometeorology, 15*(5), 2067–2084.

Zheng, D. (2008). *Study on Eco-geographic Regional System of China (in Chinese)*. Beijing, China: Commercial Press.

2 Climate Change Projections in the Twenty-First Century

Wenxiang Wu, Chenchen Xu and Xingcai Liu

2.1 Background

Climate change will inevitably continue in the next few decades. The Fifth Assessment Reports (AR5) of the Intergovernmental Panel on Climate Change (IPCC) proposed that global mean temperature, relative to preindustrial, is likely to increase by 1.5–4.0 °C by the end of the twenty-first century, accompanied with changes in rainfall patterns and an increase in climate variability (IPCC 2013). Accordingly, extreme climate events, especially extreme high temperature and precipitation, have been showing an increasing trend (He et al. 2015; IPCC 2013; Dong et al. 2015). Natural disasters such as drought and flood caused by climate change are now occurring more frequently and widely, leading to the instability of eco-environment system and restricting the socioeconomic development (Min et al. 2011).

China is one of the countries severely affected by climate change and meteorological disasters. Losses of meteorological disasters is about 70% of those caused by natural disasters, and have led to about 3–6% of Gross National Product (GNP) of direct economic loss in China (Yin et al. 2016). It is thus necessary to analyze the general features of future climate change.

The primary goal of this chapter was to present the up-to-date projected changes of climate variables, including the daily maximum, minimum, and average temperature and annual precipitation across China. In addition, we provided maps of the temperature and precipitation geographical distributions and their changing features in the twenty-first century.

2.2 Data and Method

Future climate data across China are originated from five Global Climate Models (GFDL-ESM2M, HadGEM2-ES, IPSL-CM5A-LR, MIROC-ESM-CHEM, and NorESM1-M). The model outputs were bias-corrected and downscaled to a grid with 0.5° × 0.5° resolution by the Inter-Sectoral Impact Model Intercomparison Project (ISI-MIP) (Stefan et al. 2011; Hempel et al. 2013; Piani et al. 2010; Warszawski et al. 2014). All representative concentration pathways (RCPs), namely RCP2.6, RCP4.5, RCP6.0, and RCP8.5, representing the low, middle, and high pathways for energy and industry CO_2 emissions were adopted.

To investigate the multi-decadal variability and trends of climate changes, consecutive 30-year periods throughout the twenty-first century were adopted with 1981–2010 period as baseline. Values of climate variable for each grid in 2011–2040, 2041–2070, and 2071–2010, which were the mean of five climate model outputs, were compared with those of baseline. Temperature change was evaluated by its absolute variations in daily rate and precipitation was compared with relative precipitation change in yearly rate. The standard deviation of simulated changes in climate variables from the five general circulation models (GCMs) was calculated to quantify inter-model variability (Lobell et al. 2007).

W. Wu (✉) · C. Xu
Institute of Geographic Sciences and Natural Resources Research, Chinese Academy of Sciences, Beijing, 100101, China
e-mail: wuwx@igsnrr.ac.cn

C. Xu
University of Chinese Academy of Sciences, Beijing, 100049, China

X. Liu
Key Laboratory of Water Cycle and Related Land Surface Processes, Institute of Geographic Sciences and Natural Resources Research, Chinese Academy of Sciences, Beijing, 100101, China

Q. Tang and Q. Ge (eds.), *Atlas of Environmental Risks Facing China Under Climate Change*, IHDP/Future Earth-Integrated Risk Governance Project Series, DOI 10.1007/978-981-10-4199-0_2

2.3 Results

2.3.1 Climate Change

(1) Daily maximum temperature

The daily maximum temperature is increasing across China in future periods under all RCPs scenarios. Temperature increase is greater in the areas with higher latitude or altitude, and changes in the daily maximum temperature are higher in inland regions than that of coastal areas. The area with highest increase of maximum temperature distributes near the Tibetan Plateau within ranges of 1.8–2.5, 1.8–4.0, 1.4–5.0, and 1.8–6.9 °C under RCP2.6, RCP4.5, RCP6.0, and RCP8.5 scenarios, respectively.

The slightest increase in daily maximum temperature is for 2011–2040, while the largest increase is for 2071–2100 under RCP4.5, RCP6.0, and RCP8.5 scenarios. The largest increase in temperature under RCP2.6 scenario occurs in 2041–2070.

(2) Daily minimum and mean temperatures

The daily minimum and mean temperatures change similarly in future periods under all RCPs with a greater increase in areas located in higher latitude or altitude, as well as inland areas far away from coastline. The northern Xinjiang and Heilongjiang provinces are the areas with the highest change in daily minimum and mean temperatures. Increase of the daily minimum temperature is 1.8–2.5, 1.8–4.0, 1.4–5.0, and 1.4–6.8 °C and increase of the daily mean temperature is 2.0–3.0, 2.0–4.0, 1.2–5.0, and 1.7–6.6 °C under RCP2.6, RCP4.5, RCP6.0, and RCP8.5 scenarios, respectively.

The daily minimum and mean temperatures rise by the smallest amount in 2011–2040, and the largest amount in 2071–2100 under RCP4.5, RCP6.0, and RCP8.5 scenarios. The highest change in daily minimum and mean temperatures occurs in 2041–2070 under RCP2.6 scenario.

(3) Annual precipitation

The annual precipitation exhibits a general increasing trend across China during future periods under all RCPs scenarios. Precipitation rising tends to be greater in areas having higher latitude or altitude or areas close to the coastline. The area with highest change in annual precipitation distributes near the Tibetan Plateau, the Qaidam Basin, and northern Gansu province. Under most of the RCP scenarios, minimal increase of annual precipitation is for 2011–2040 and maximal increase is for 2071–2010.

2.3.2 Intermodel Spread in Climate Changes

The standard deviations of climate variables in 2071–2010 under RCP8.5 scenarios were calculated. The spread arising from climate models for daily temperatures (maximum, minimum, and mean temperature) was generally less than 20%, indicating that their values generated from multiple climate models were in good consistency. However, the spread for annual precipitation across China showed greater regionally differences, with relative changes being less than 20% for most regions but more than 20% for eastern China.

2.4 Maps

Change in Daily Minimum Temperature for RCP2.6 (1981-2010 to 2011-2040)

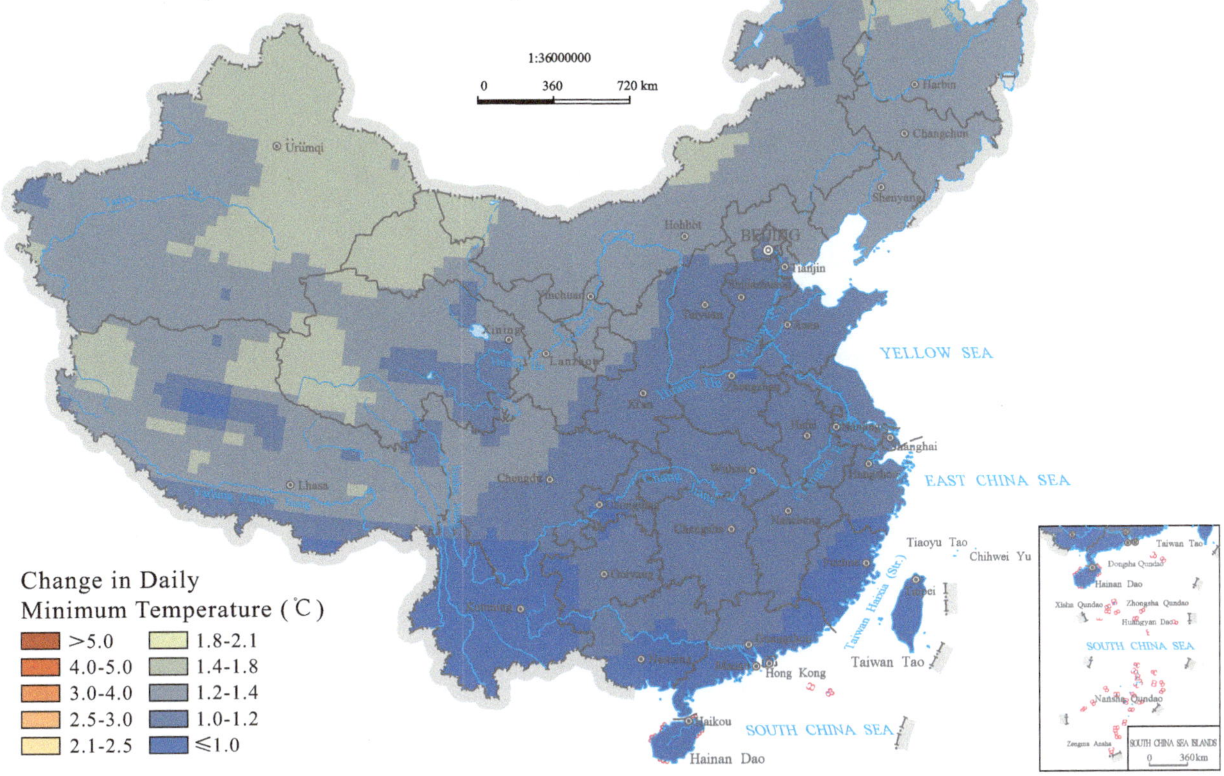

Change in Daily Maximum Temperature for RCP2.6 (1981-2010 to 2011-2040)

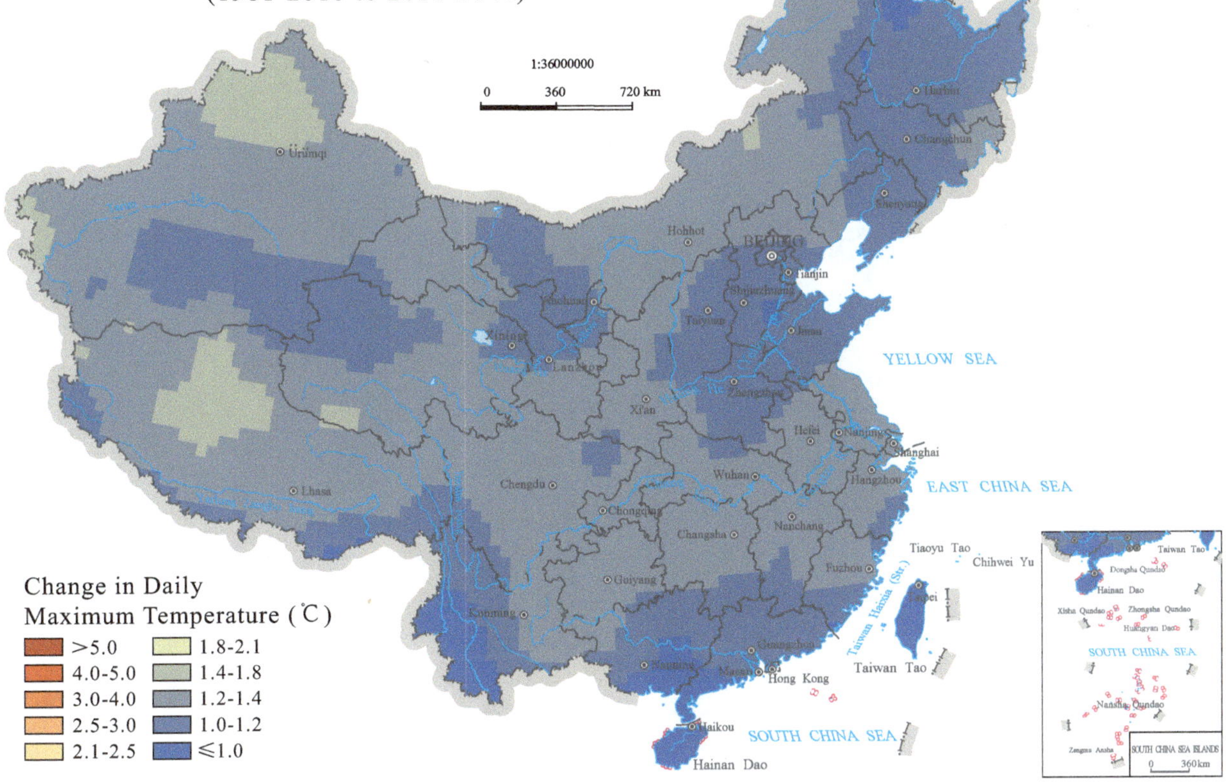

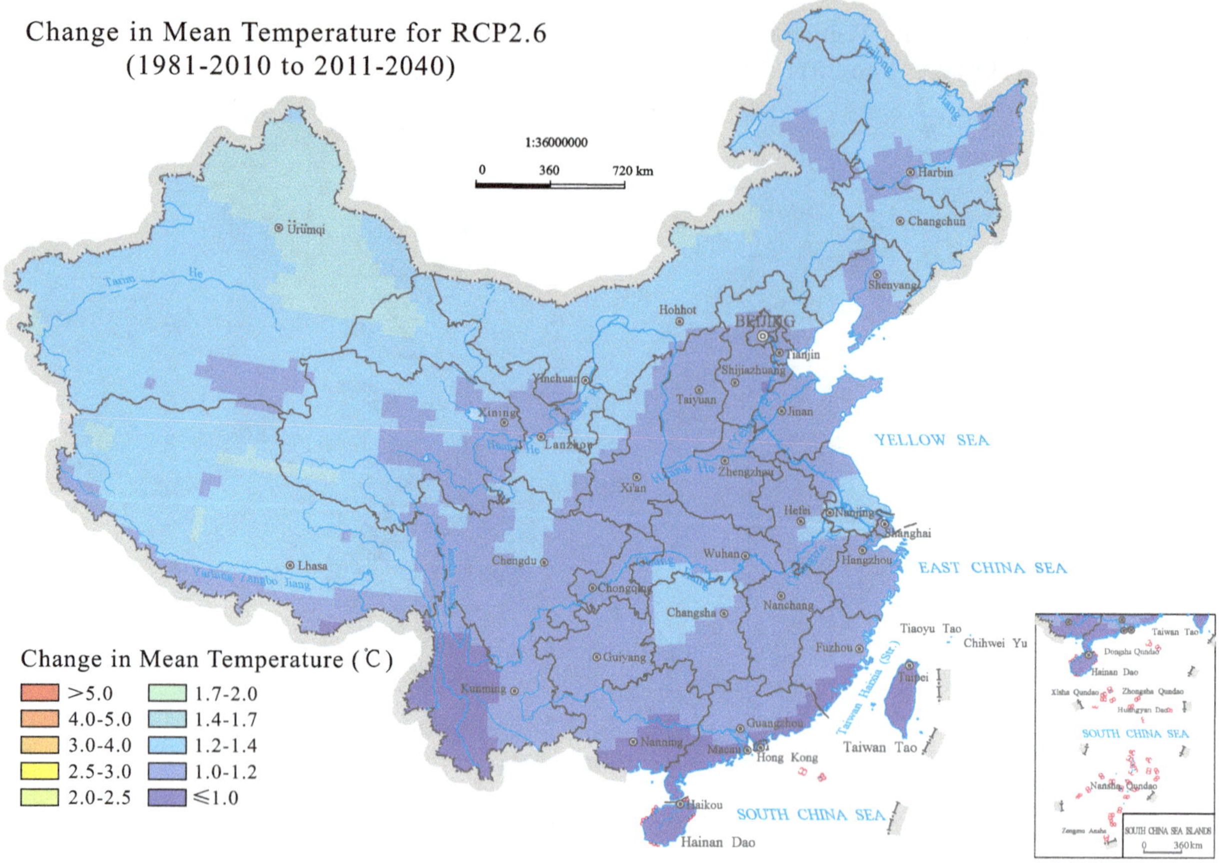
Change in Mean Temperature for RCP2.6
(1981-2010 to 2011-2040)
1:36000000
0 360 720 km
Ürümqi
Harbin
Changchun
Shenyang
Hohhot
BEIJING
Tianjin
Shijiazhuang
Taiyuan
Yinchuan
Jinan
Xining
Lanzhou
Zhengzhou
Xi'an
Hefei
Nanjing
Shanghai
Lhasa
Chengdu
Wuhan
Hangzhou
Chongqing
Changsha
Nanchang
Guiyang
Fuzhou
Kunming
Taipei
Guangzhou
Nanning
Macau
Hong Kong
Haikou
Hainan Dao
YELLOW SEA
EAST CHINA SEA
SOUTH CHINA SEA
Tiaoyu Tao
Chihwei Yu
Taiwan Tao
Change in Mean Temperature (℃)
>5.0
4.0-5.0
3.0-4.0
2.5-3.0
2.0-2.5
1.7-2.0
1.4-1.7
1.2-1.4
1.0-1.2
≤1.0
SOUTH CHINA SEA ISLANDS

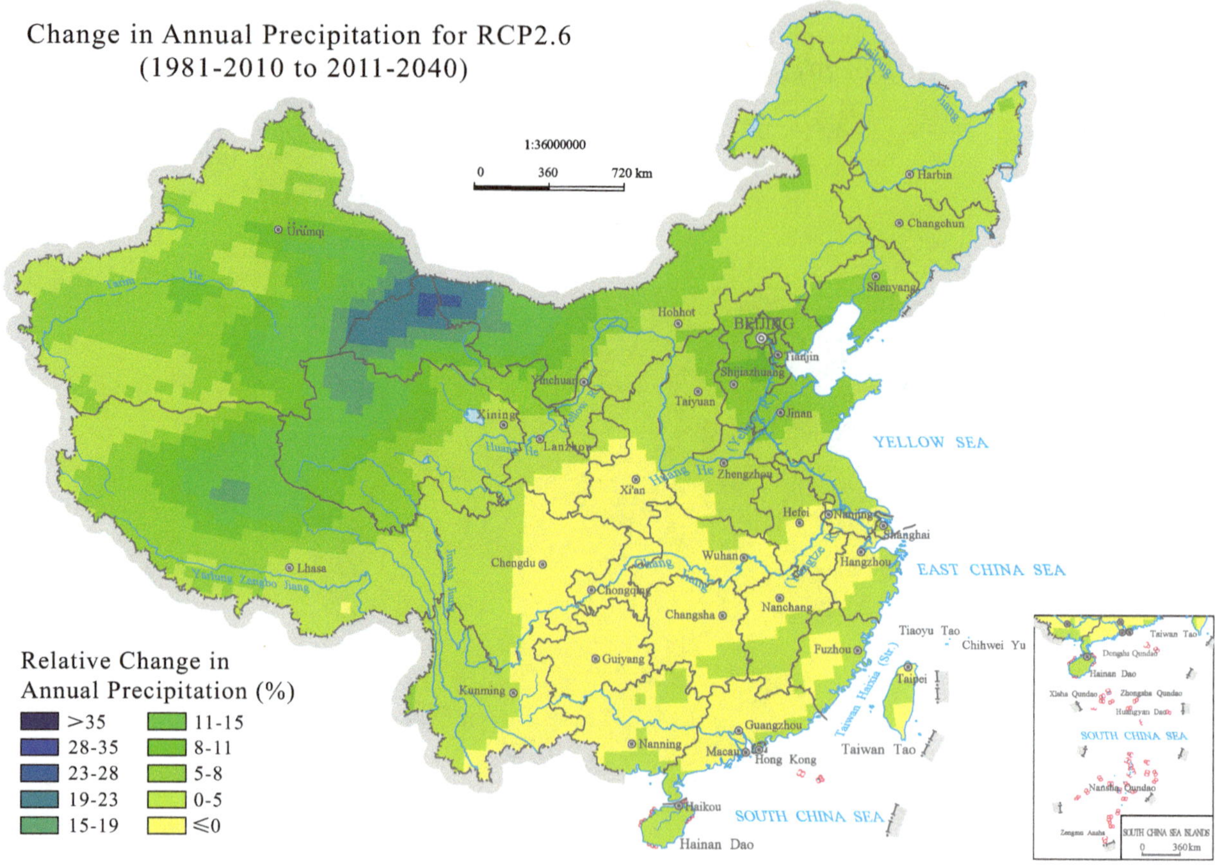
Change in Annual Precipitation for RCP2.6
(1981-2010 to 2011-2040)
1:36000000
0 360 720 km
Ürümqi
Harbin
Changchun
Shenyang
Hohhot
BEIJING
Tianjin
Shijiazhuang
Taiyuan
Yinchuan
Jinan
Xining
Lanzhou
Zhengzhou
Xi'an
Hefei
Nanjing
Shanghai
Lhasa
Chengdu
Wuhan
Hangzhou
Chongqing
Changsha
Nanchang
Guiyang
Fuzhou
Kunming
Taipei
Guangzhou
Nanning
Macau
Hong Kong
Haikou
Hainan Dao
YELLOW SEA
EAST CHINA SEA
SOUTH CHINA SEA
Tiaoyu Tao
Chihwei Yu
Taiwan Tao
Relative Change in
Annual Precipitation (%)
>35
28-35
23-28
19-23
15-19
11-15
8-11
5-8
0-5
≤0
SOUTH CHINA SEA ISLANDS

Change in Daily Minimum Temperature for RCP2.6
(1981-2010 to 2041-2070)

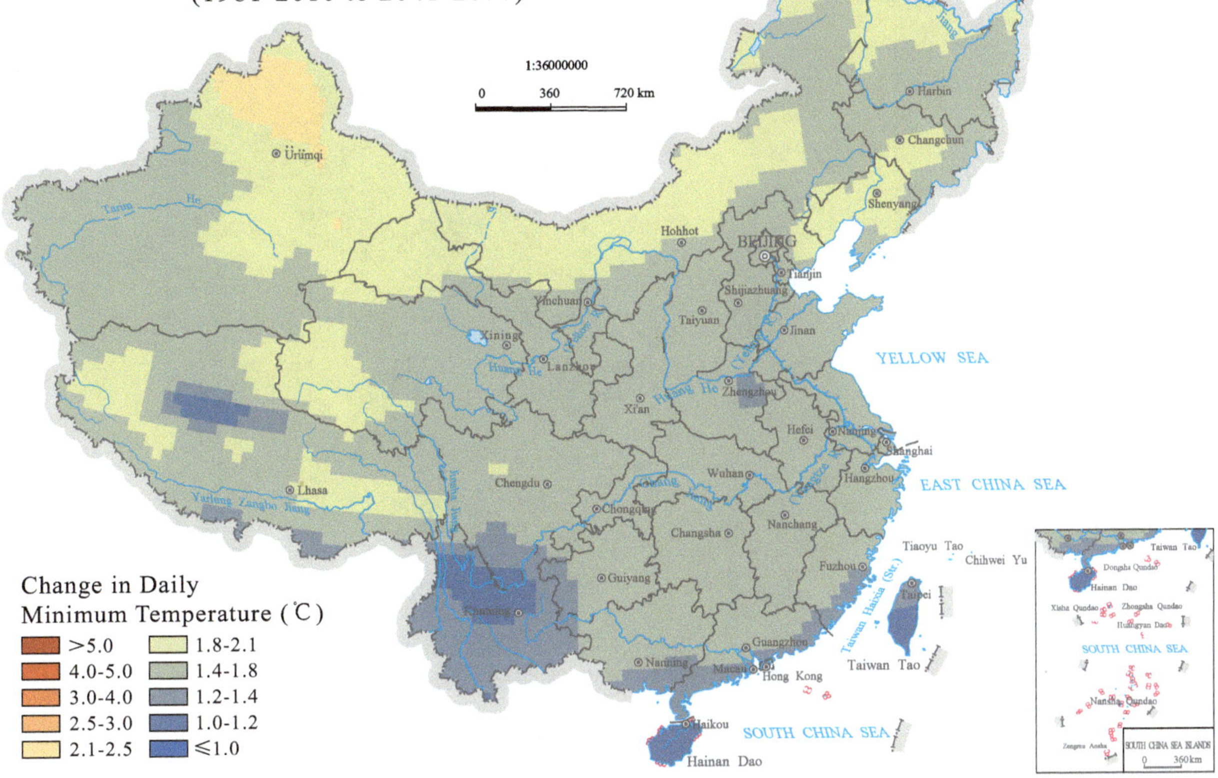

Change in Daily Maximum Temperature for RCP2.6
(1981-2010 to 2041-2070)

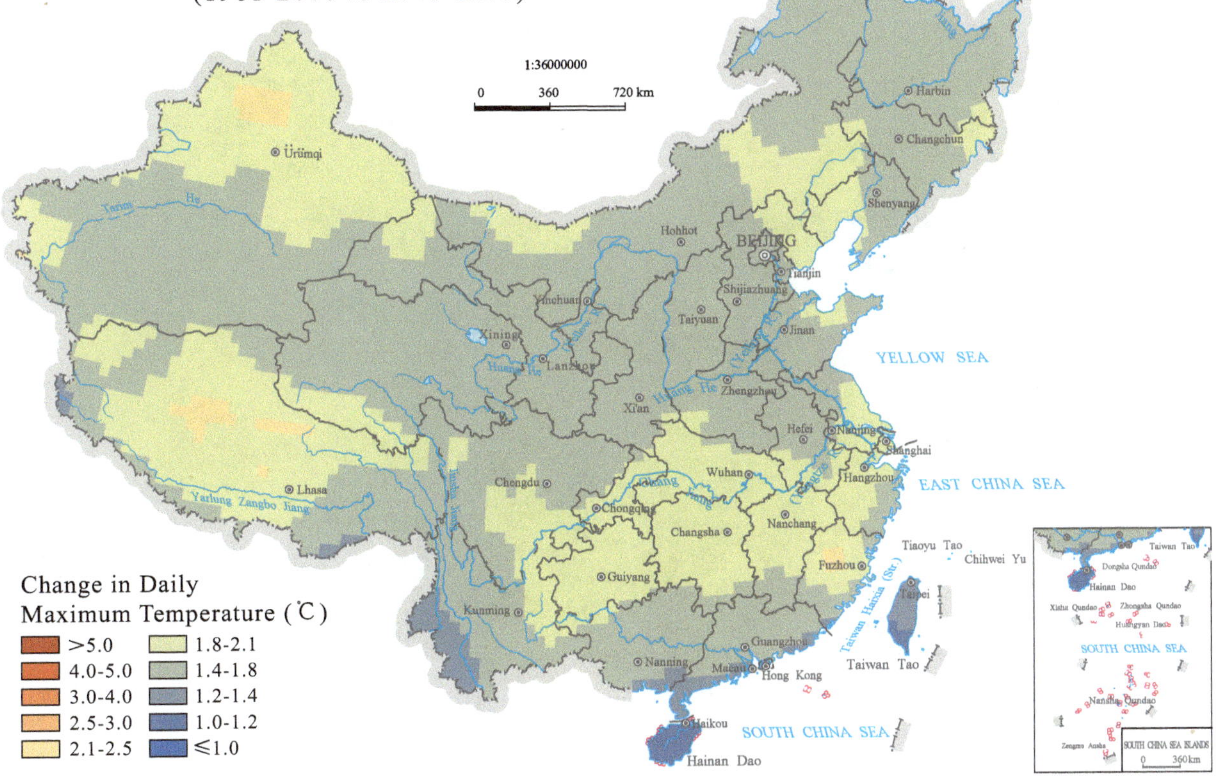

Change in Mean Temperature for RCP2.6
(1981-2010 to 2041-2070)

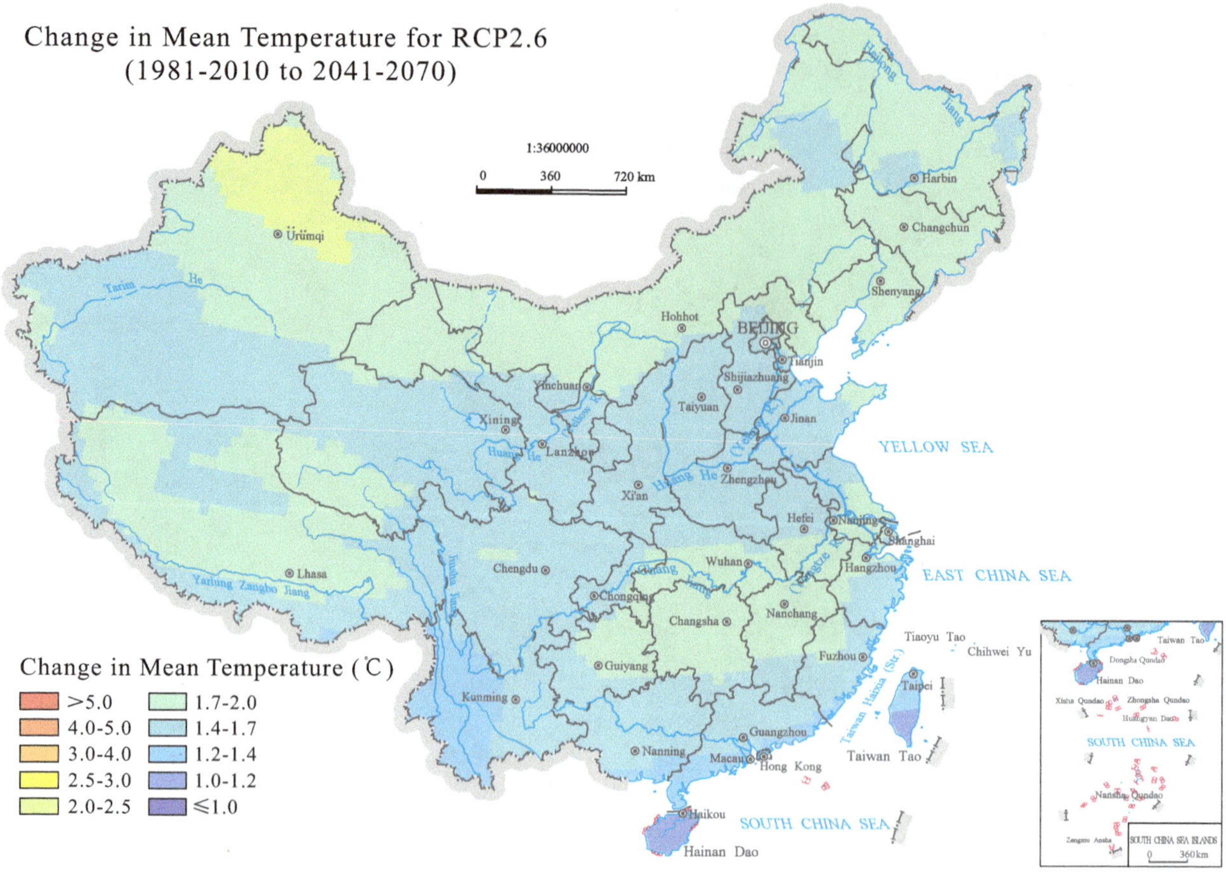

Change in Annual Precipitation for RCP2.6
(1981-2010 to 2041-2070)

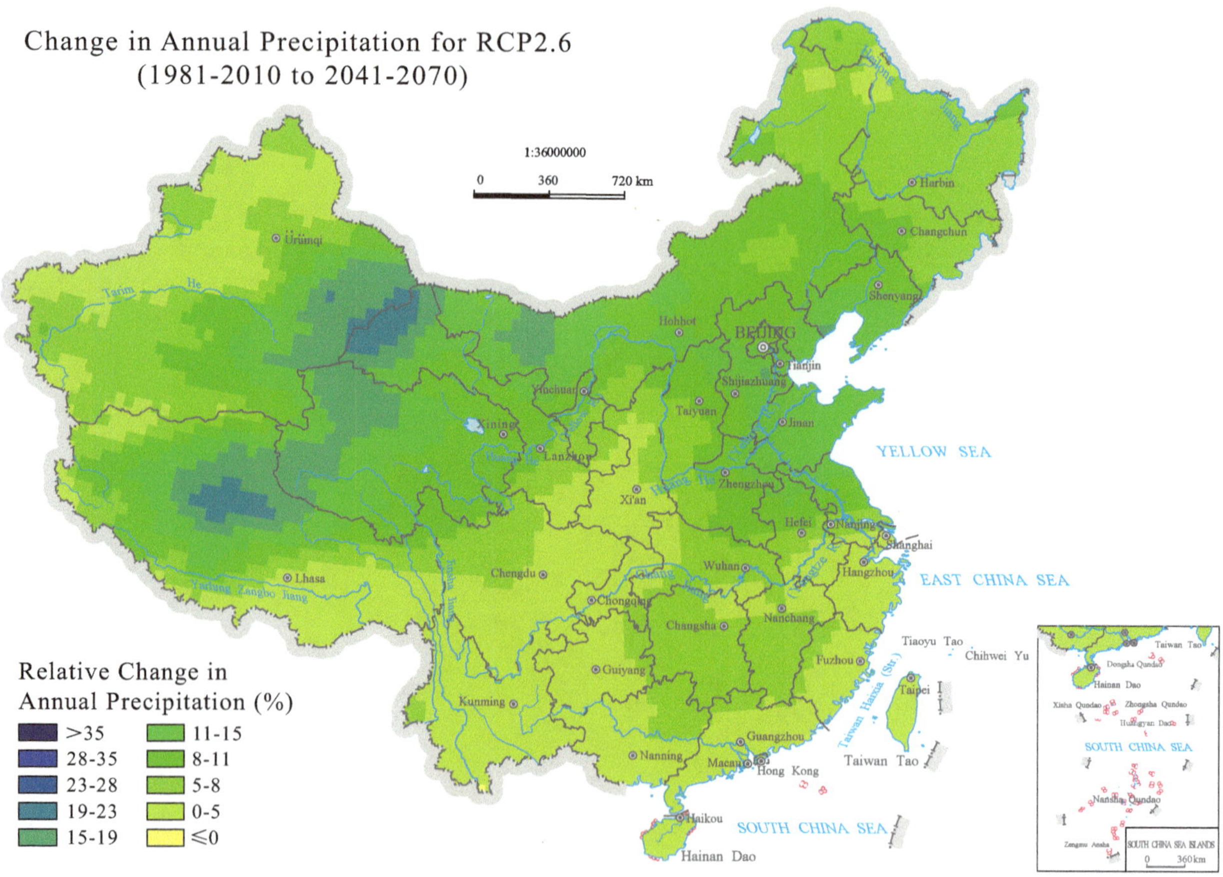

Change in Daily Minimum Temperature for RCP2.6
(1981-2010 to 2071-2100)

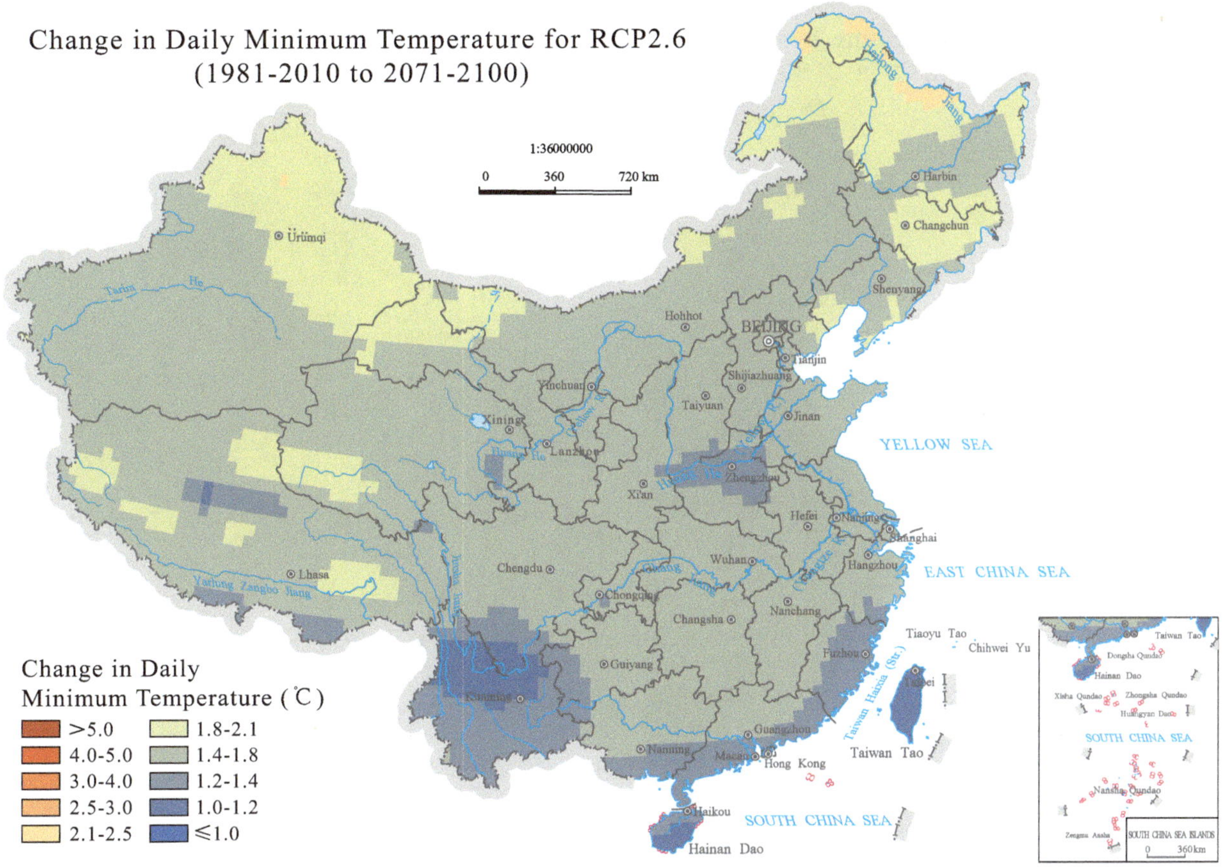

Change in Daily Maximum Temperature for RCP2.6
(1981-2010 to 2071-2100)

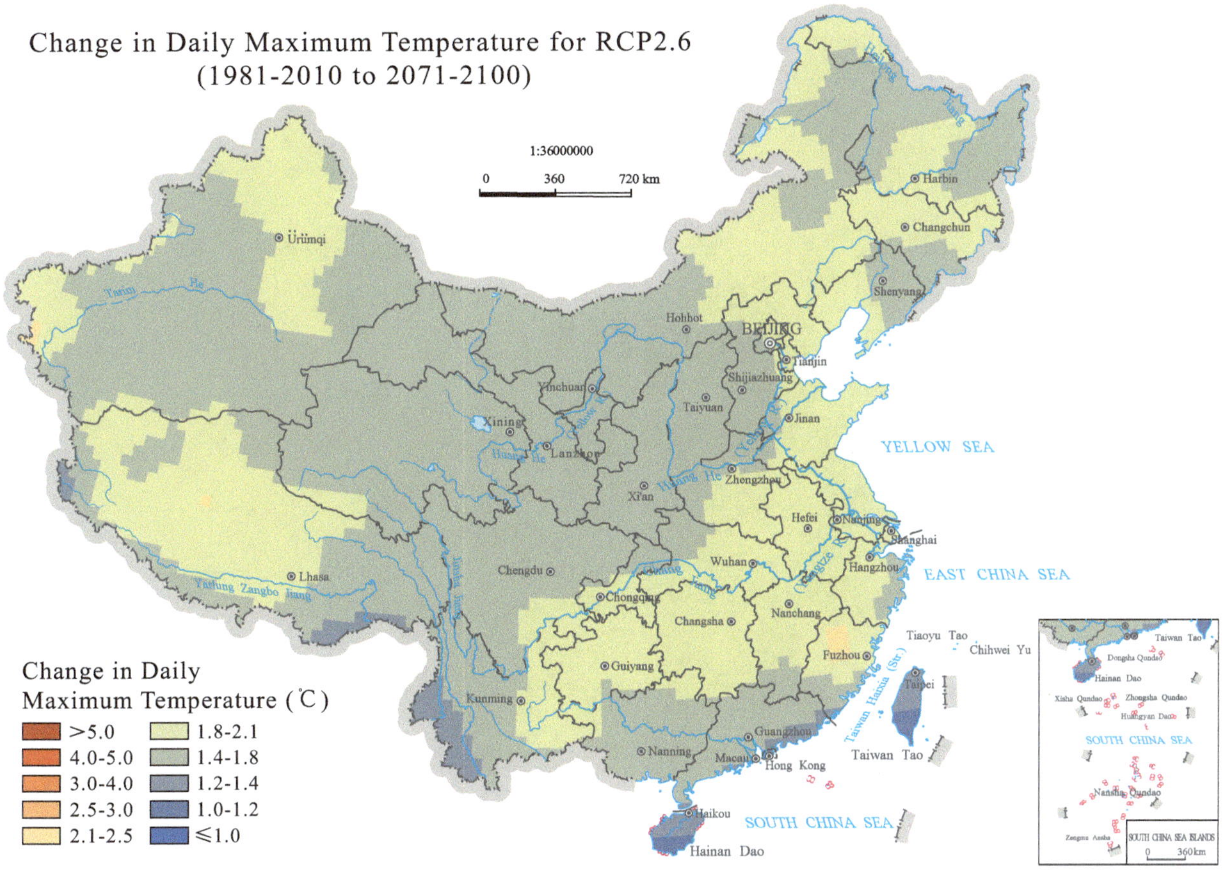

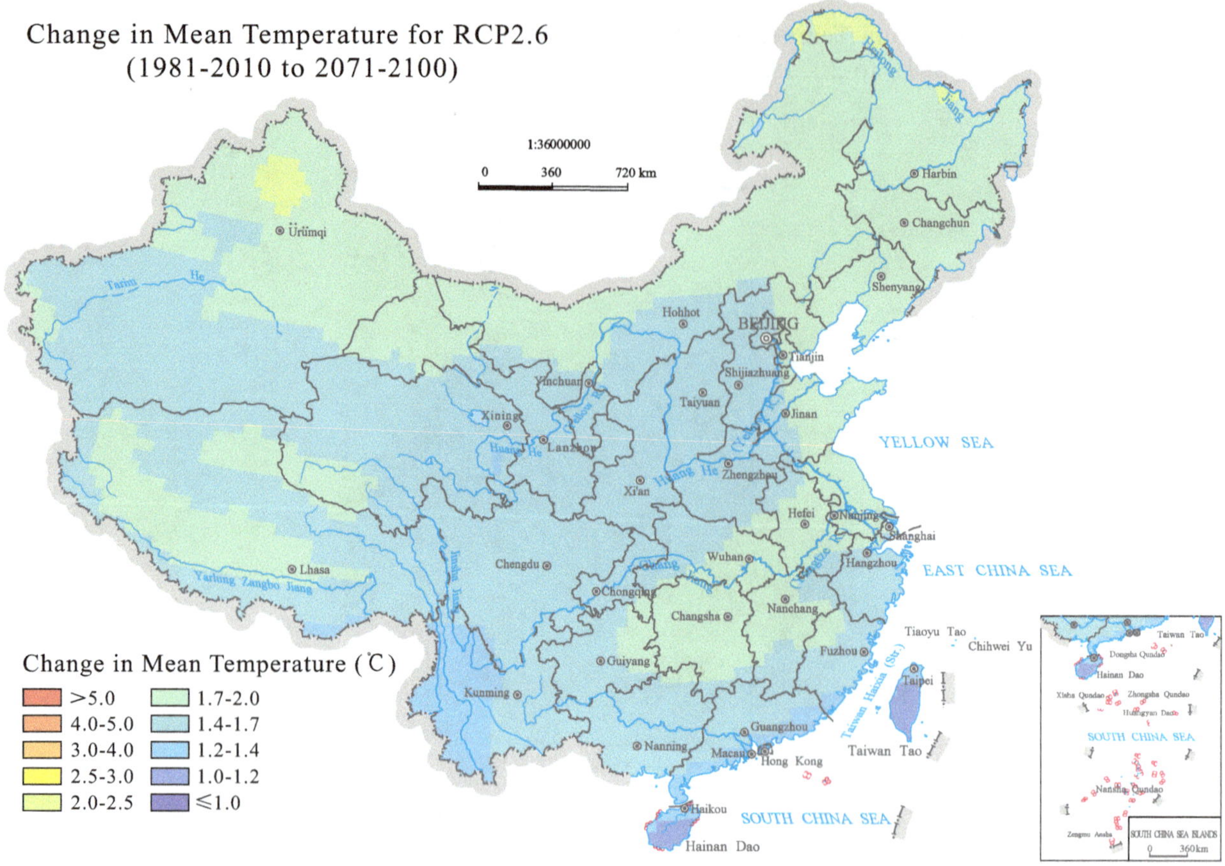
Change in Mean Temperature for RCP2.6
(1981-2010 to 2071-2100)
1:36000000
0 360 720 km
Change in Mean Temperature (℃)
>5.0
4.0-5.0
3.0-4.0
2.5-3.0
2.0-2.5
1.7-2.0
1.4-1.7
1.2-1.4
1.0-1.2
≤1.0
Harbin
Changchun
Shenyang
Ürümqi
Hohhot
BEIJING
Tianjin
Shijiazhuang
Yinchuan
Taiyuan
Jinan
Xining
Lanzhou
YELLOW SEA
Zhengzhou
Xi'an
Hefei
Nanjing
Shanghai
Lhasa
Chengdu
Wuhan
Hangzhou
EAST CHINA SEA
Chongqing
Changsha
Nanchang
Tiaoyu Tao
Chihwei Yu
Fuzhou
Guiyang
Taipei
Kunming
Guangzhou
Taiwan Tao
Nanning
Macau
Hong Kong
Haikou
SOUTH CHINA SEA
Hainan Dao
SOUTH CHINA SEA ISLANDS

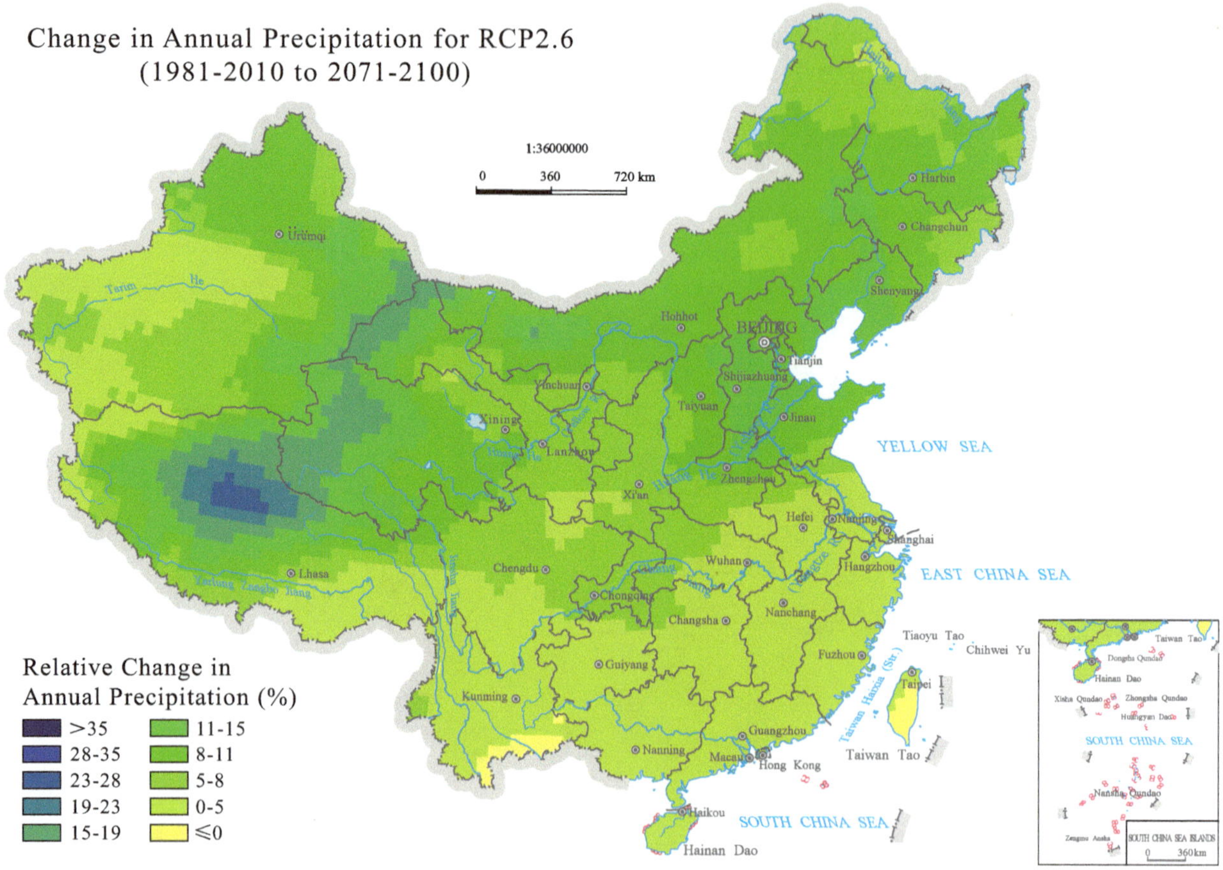
Change in Annual Precipitation for RCP2.6
(1981-2010 to 2071-2100)
1:36000000
0 360 720 km
Relative Change in
Annual Precipitation (%)
>35
28-35
23-28
19-23
15-19
11-15
8-11
5-8
0-5
≤0
Harbin
Changchun
Shenyang
Ürümqi
Hohhot
BEIJING
Tianjin
Shijiazhuang
Yinchuan
Taiyuan
Jinan
Xining
Lanzhou
YELLOW SEA
Zhengzhou
Xi'an
Hefei
Nanjing
Shanghai
Lhasa
Chengdu
Wuhan
Hangzhou
EAST CHINA SEA
Chongqing
Changsha
Nanchang
Tiaoyu Tao
Chihwei Yu
Fuzhou
Guiyang
Taipei
Kunming
Guangzhou
Taiwan Tao
Nanning
Macau
Hong Kong
Haikou
SOUTH CHINA SEA
Hainan Dao
SOUTH CHINA SEA ISLANDS

Change in Daily Minimum Temperature for RCP4.5 (1981-2010 to 2011-2040)

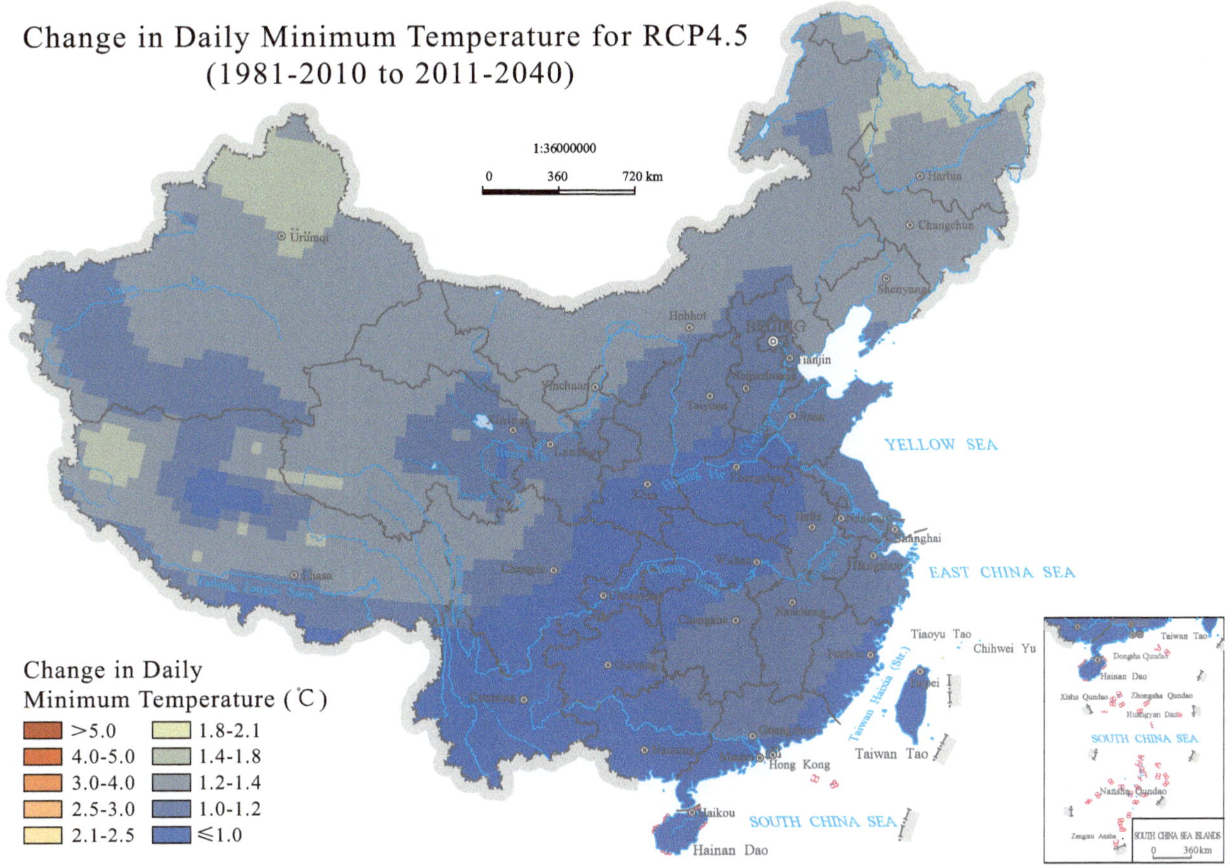

Change in Daily Maximum Temperature for RCP4.5 (1981-2010 to 2011-2040)

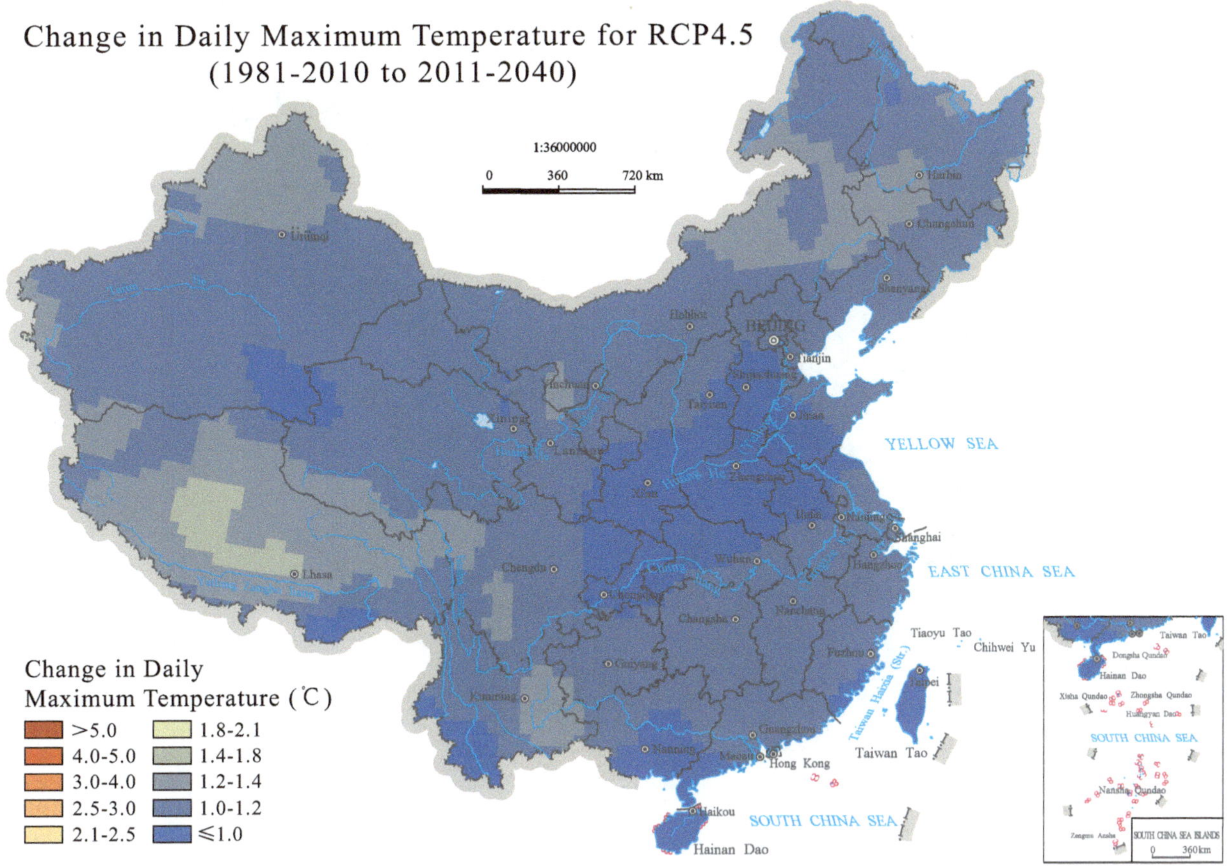

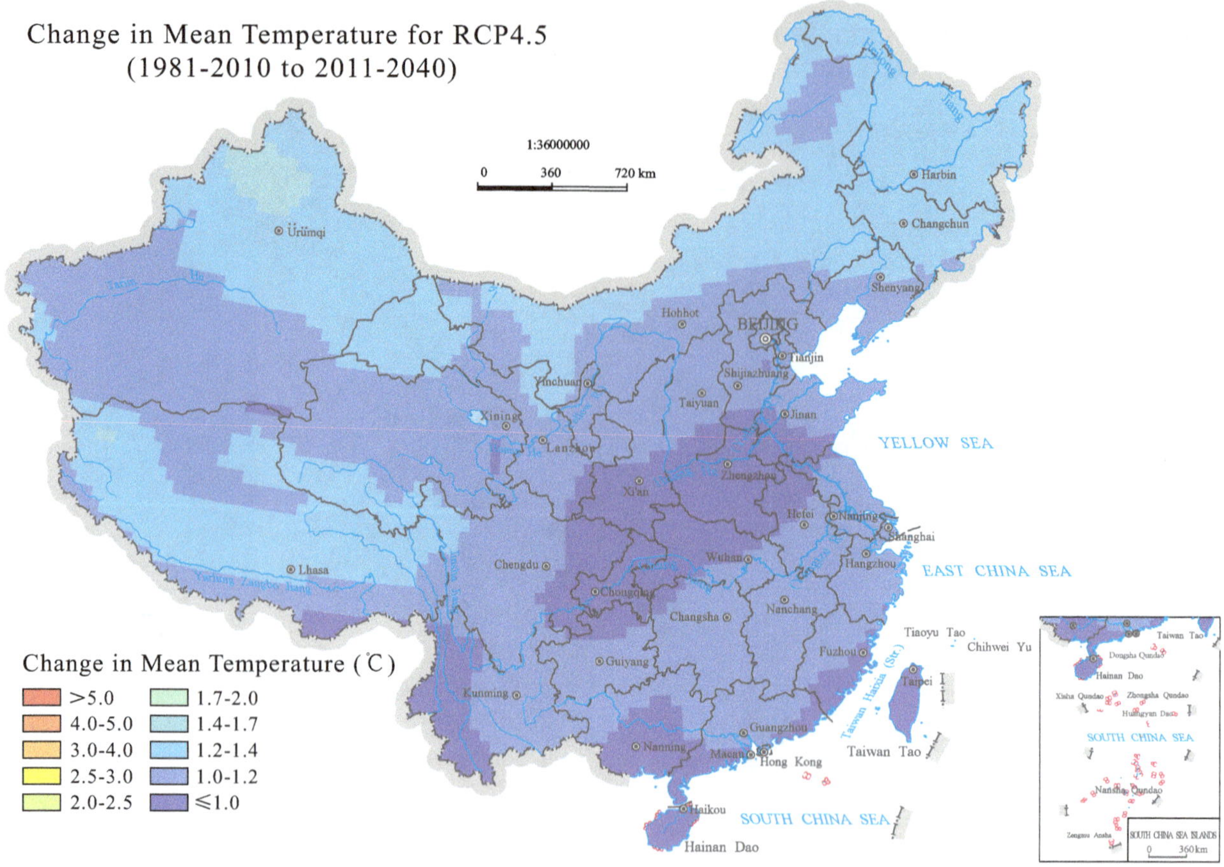
Change in Mean Temperature for RCP4.5
(1981-2010 to 2011-2040)
1:36000000
0 360 720 km
Change in Mean Temperature (℃)
>5.0
4.0-5.0
3.0-4.0
2.5-3.0
2.0-2.5
1.7-2.0
1.4-1.7
1.2-1.4
1.0-1.2
≤1.0
Ürümqi
Harbin
Changchun
Shenyang
Hohhot
BEIJING
Tianjin
Shijiazhuang
Taiyuan
Yinchuan
Xining
Lanzhou
Jinan
Zhengzhou
Xi'an
Hefei
Nanjing
Shanghai
Hangzhou
Wuhan
Chengdu
Chongqing
Lhasa
Changsha
Nanchang
Fuzhou
Guiyang
Kunming
Nanning
Guangzhou
Macau
Hong Kong
Taipei
Haikou
Hainan Dao
Taiwan Tao
Tiaoyu Tao
Chihwei Yu
YELLOW SEA
EAST CHINA SEA
SOUTH CHINA SEA
SOUTH CHINA SEA ISLANDS
360km

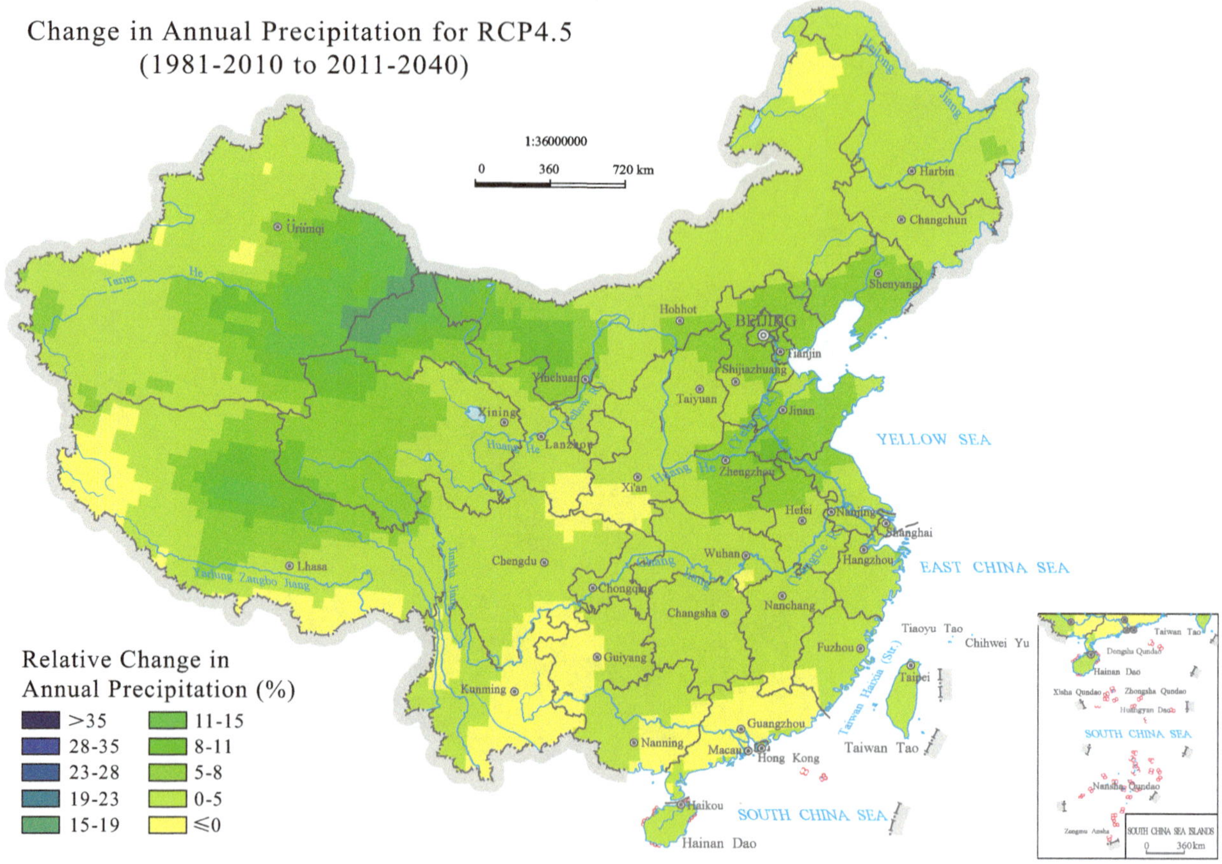
Change in Annual Precipitation for RCP4.5
(1981-2010 to 2011-2040)
1:36000000
0 360 720 km
Relative Change in
Annual Precipitation (%)
>35
28-35
23-28
19-23
15-19
11-15
8-11
5-8
0-5
≤0
Ürümqi
Harbin
Changchun
Shenyang
Hohhot
BEIJING
Tianjin
Shijiazhuang
Taiyuan
Yinchuan
Xining
Lanzhou
Jinan
Zhengzhou
Xi'an
Hefei
Nanjing
Shanghai
Hangzhou
Wuhan
Chengdu
Chongqing
Lhasa
Changsha
Nanchang
Fuzhou
Guiyang
Kunming
Nanning
Guangzhou
Macau
Hong Kong
Taipei
Haikou
Hainan Dao
Taiwan Tao
Tiaoyu Tao
Chihwei Yu
YELLOW SEA
EAST CHINA SEA
SOUTH CHINA SEA
SOUTH CHINA SEA ISLANDS
360km

Change in Daily Minimum Temperature for RCP4.5
(1981-2010 to 2041-2070)

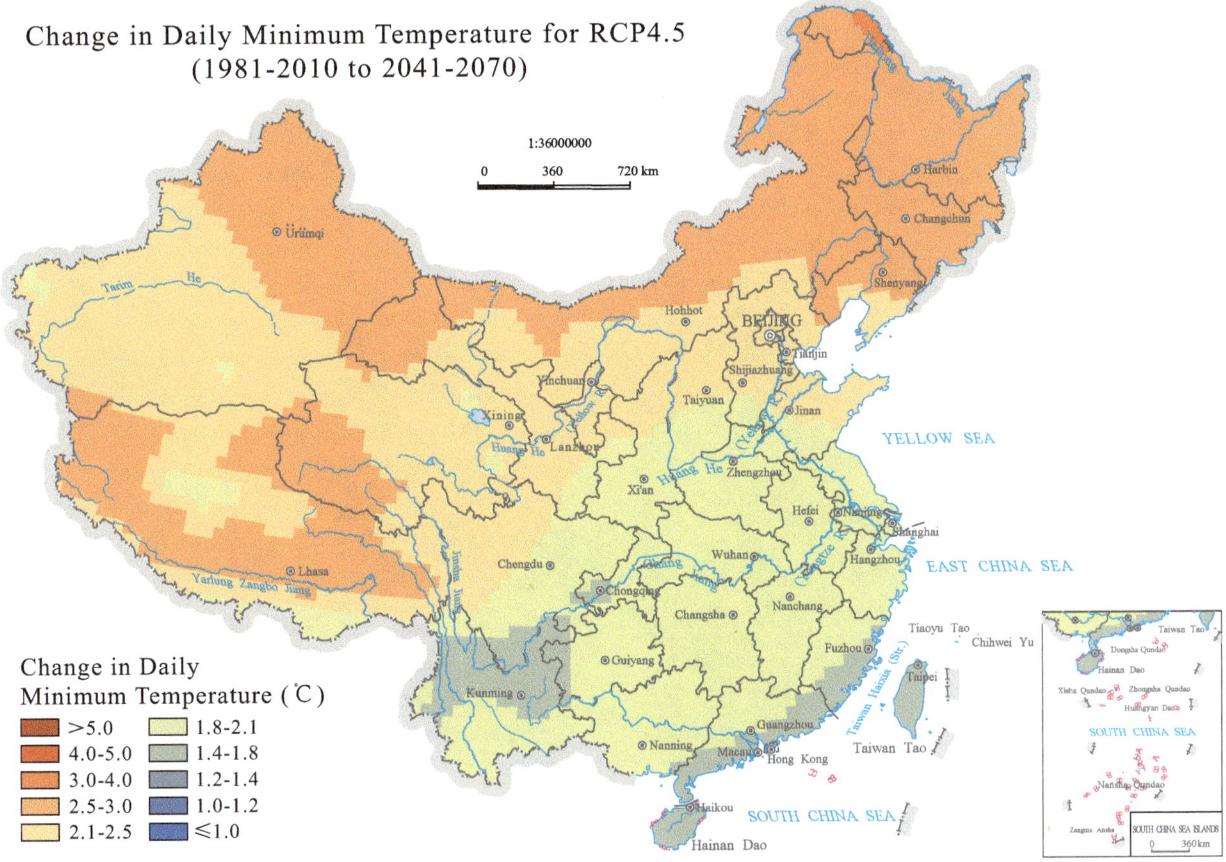

Change in Daily Maximum Temperature for RCP4.5
(1981-2010 to 2041-2070)

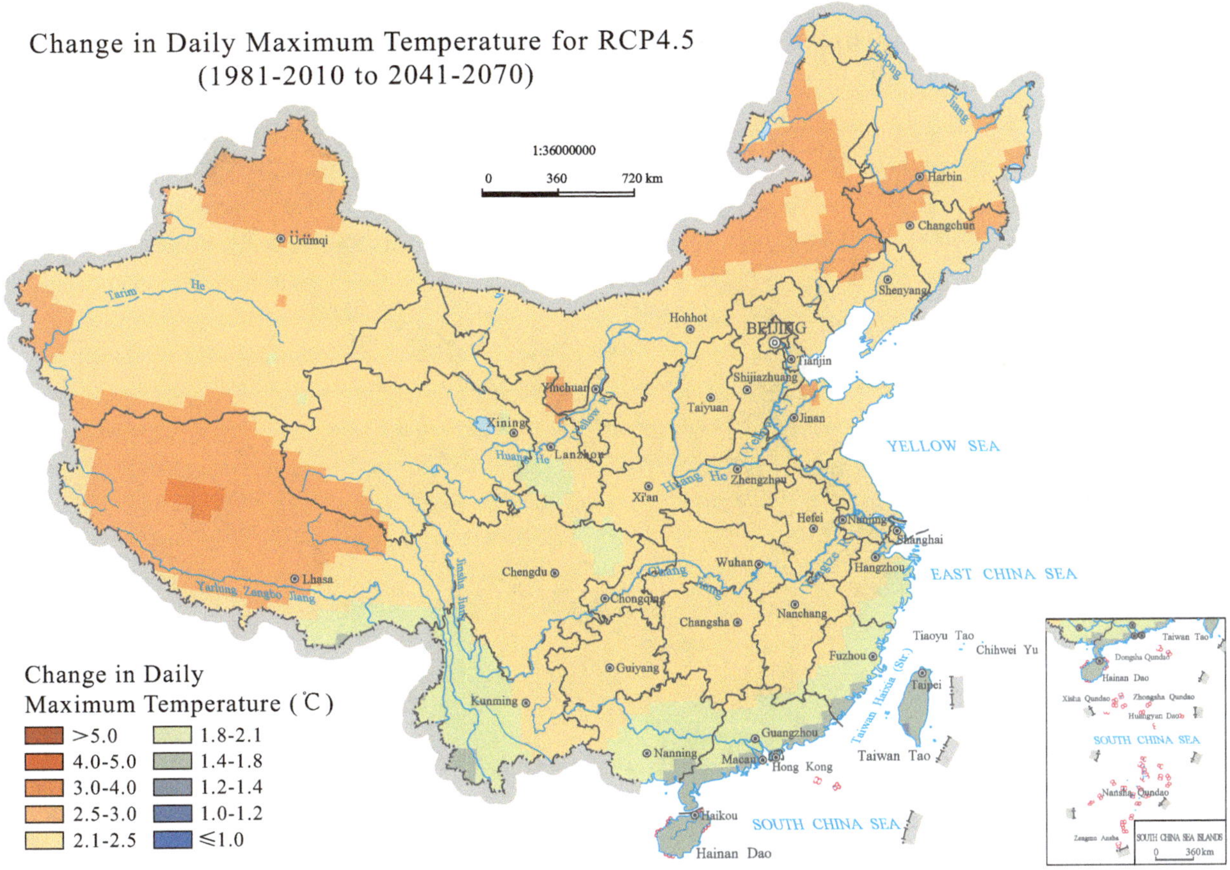

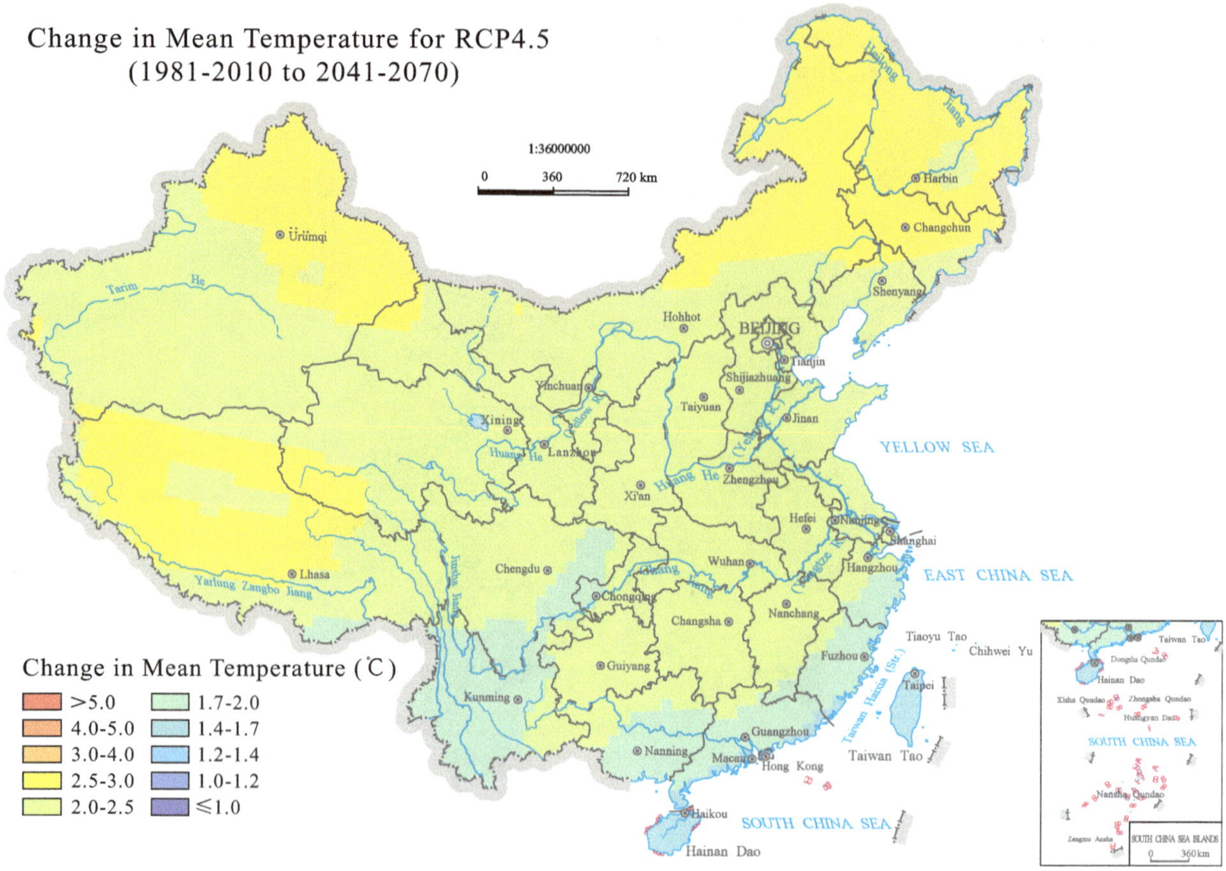
Change in Mean Temperature for RCP4.5
(1981-2010 to 2041-2070)
1:36000000
0 360 720 km
Change in Mean Temperature (℃)
>5.0
4.0-5.0
3.0-4.0
2.5-3.0
2.0-2.5
1.7-2.0
1.4-1.7
1.2-1.4
1.0-1.2
≤1.0
Ürümqi
Tarim He
Lhasa
Yarlung Zangbo Jiang
Xining
Lanzhou
Yinchuan
Hohhot
BEIJING
Tianjin
Shijiazhuang
Taiyuan
Jinan
Zhengzhou
Xi'an
Huang He
Chengdu
Chongqing
Wuhan
Hefei
Nanjing
Shanghai
Hangzhou
Nanchang
Changsha
Guiyang
Kunming
Nanning
Guangzhou
Fuzhou
Taipei
Macau
Hong Kong
Haikou
Hainan Dao
Harbin
Changchun
Shenyang
YELLOW SEA
EAST CHINA SEA
SOUTH CHINA SEA
Taiwan Tao
Tiaoyu Tao
Chihwei Yu

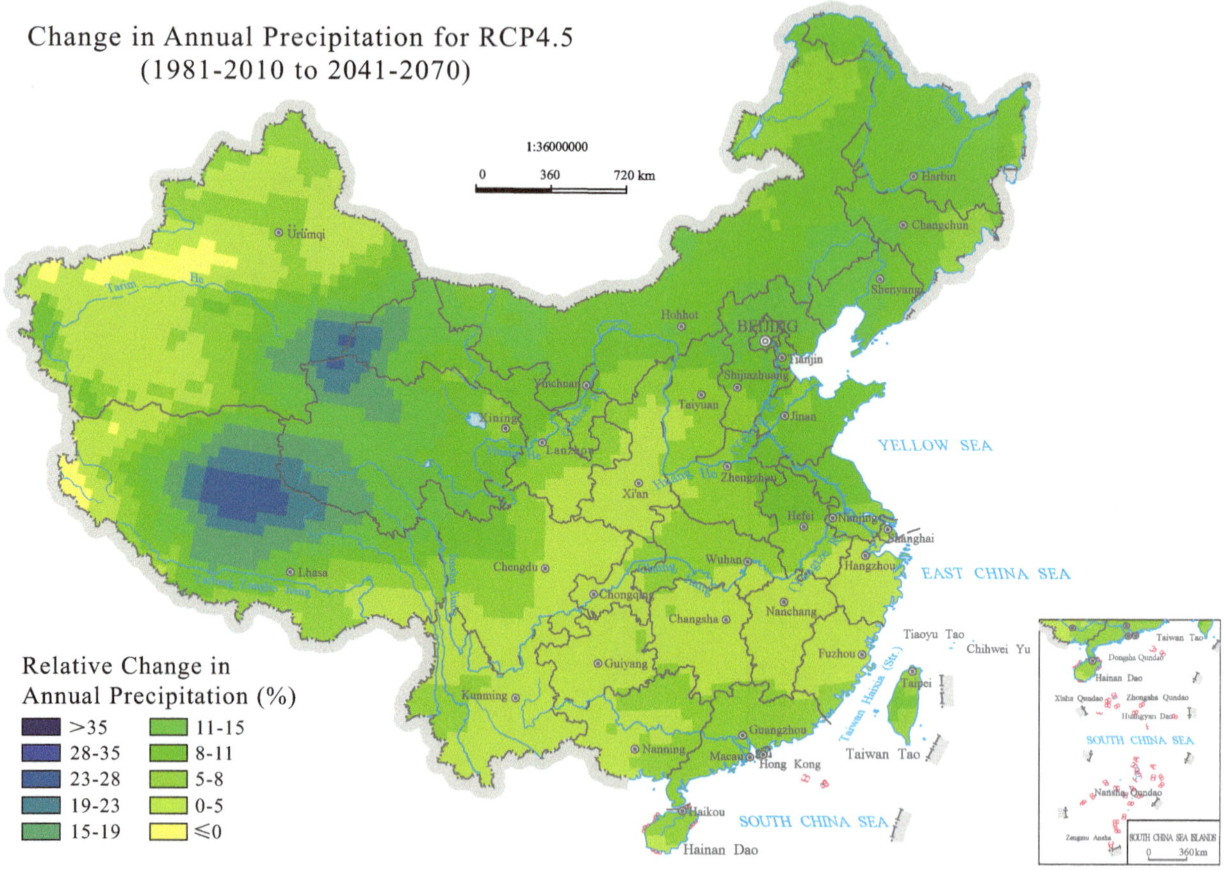
Change in Annual Precipitation for RCP4.5
(1981-2010 to 2041-2070)
1:36000000
0 360 720 km
Relative Change in
Annual Precipitation (%)
>35
28-35
23-28
19-23
15-19
11-15
8-11
5-8
0-5
≤0
Ürümqi
Tarim He
Lhasa
Xining
Lanzhou
Hohhot
BEIJING
Tianjin
Shijiazhuang
Taiyuan
Jinan
Zhengzhou
Xi'an
Chengdu
Chongqing
Wuhan
Hefei
Nanjing
Shanghai
Hangzhou
Nanchang
Changsha
Guiyang
Kunming
Nanning
Guangzhou
Fuzhou
Taipei
Macau
Hong Kong
Haikou
Hainan Dao
Harbin
Changchun
Shenyang
YELLOW SEA
EAST CHINA SEA
SOUTH CHINA SEA
Taiwan Tao
Tiaoyu Tao
Chihwei Yu

Change in Daily Minimum Temperature for RCP4.5 (1981-2010 to 2071-2100)

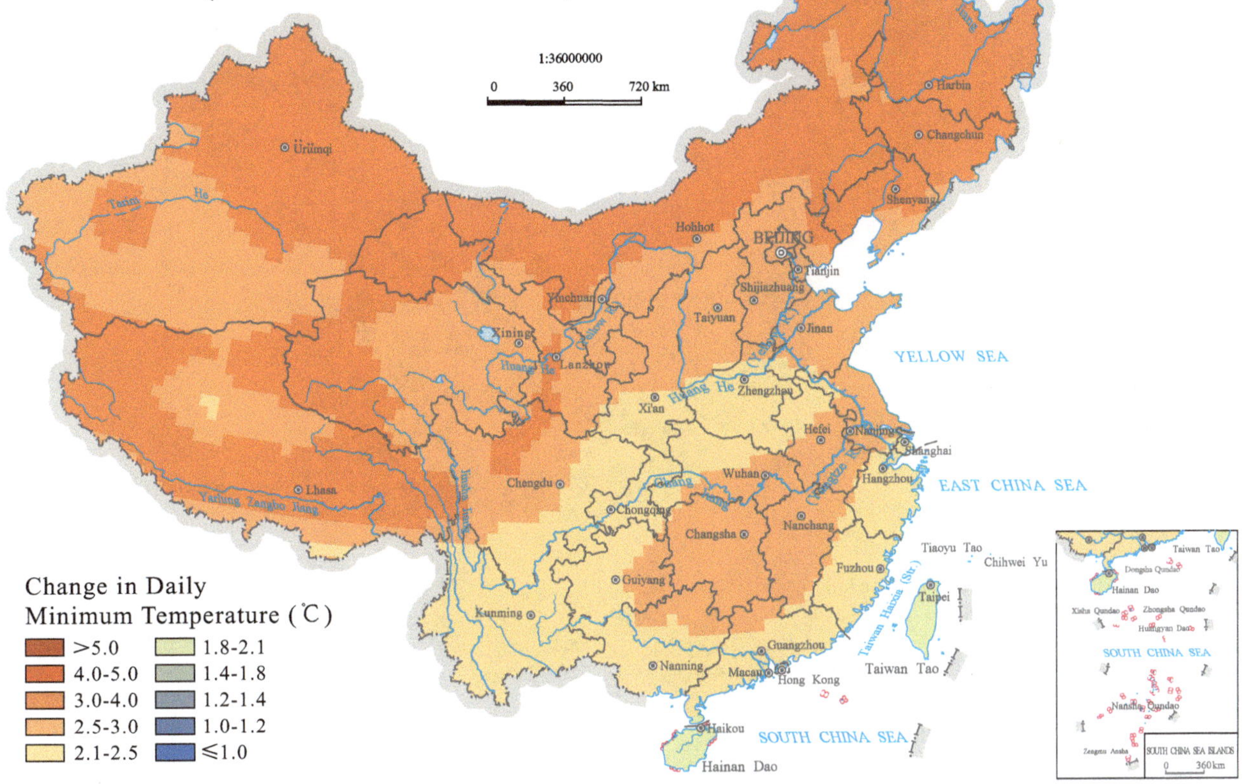

Change in Daily Maximum Temperature for RCP4.5 (1981-2010 to 2071-2100)

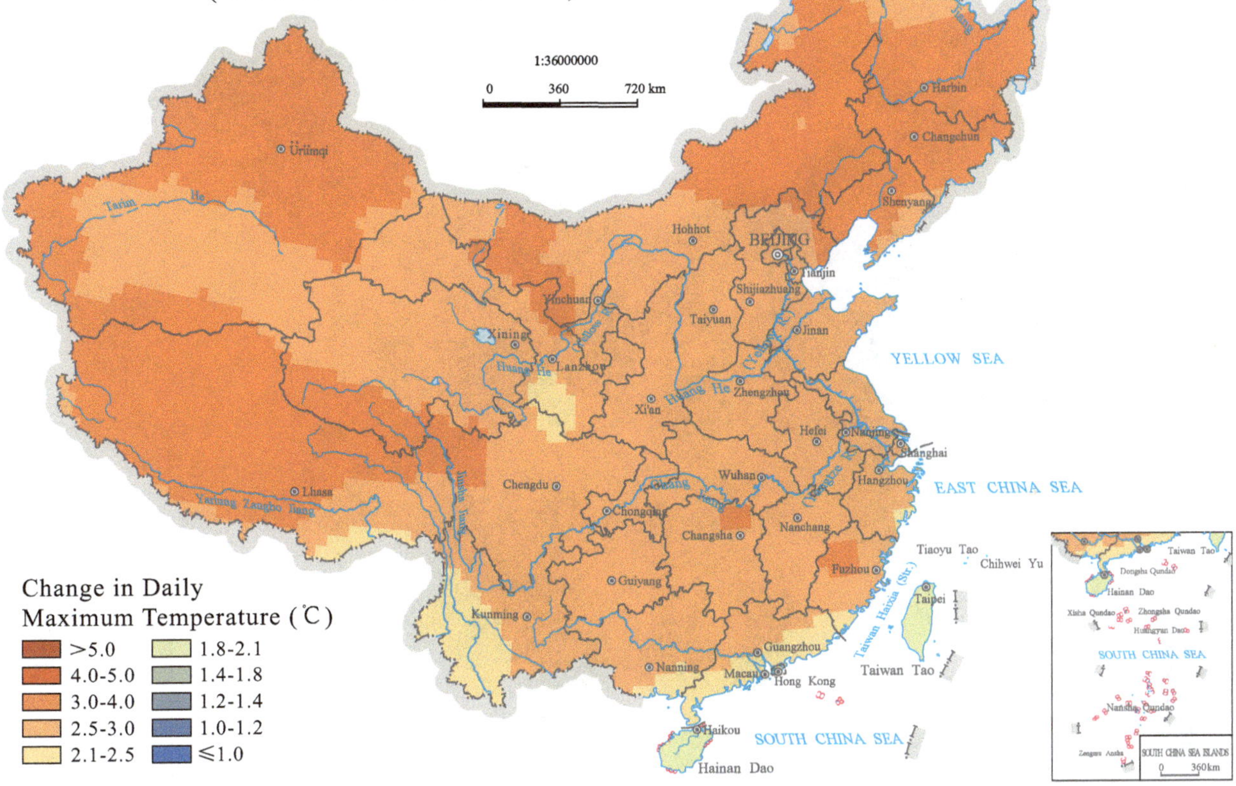

Change in Mean Temperature for RCP4.5
(1981-2010 to 2071-2100)

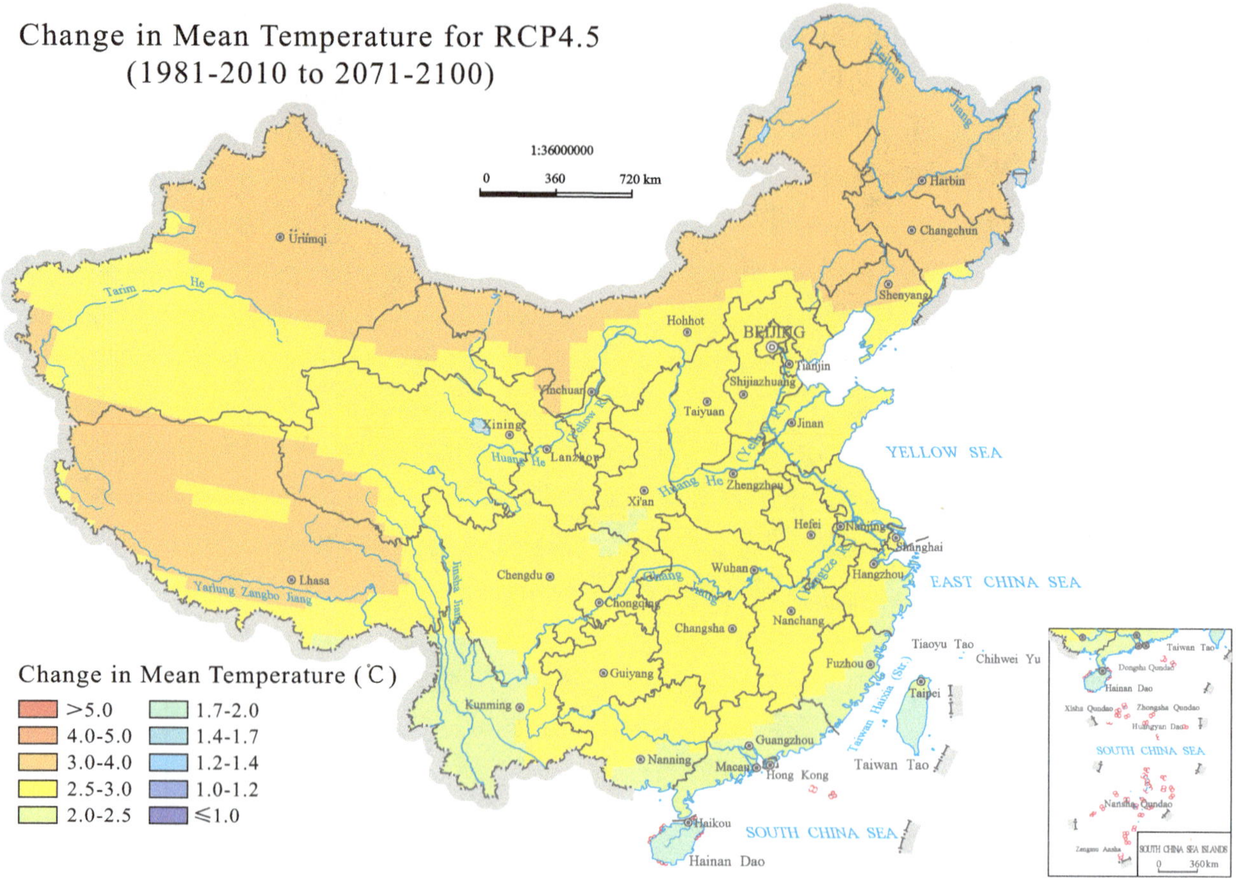

Change in Annual Precipitation for RCP4.5
(1981-2010 to 2071-2100)

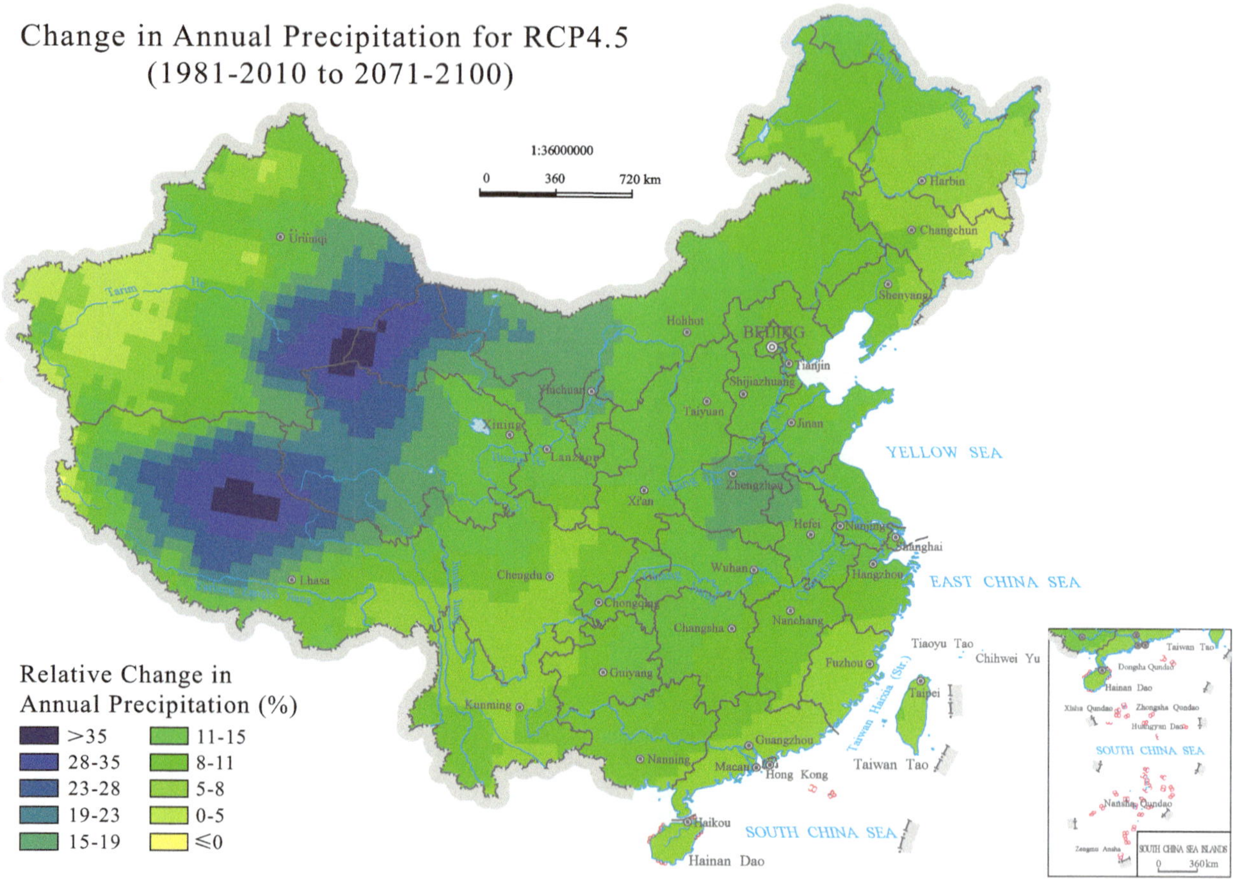

Change in Daily Minimum Temperature for RCP6.0 (1981-2010 to 2011-2040)

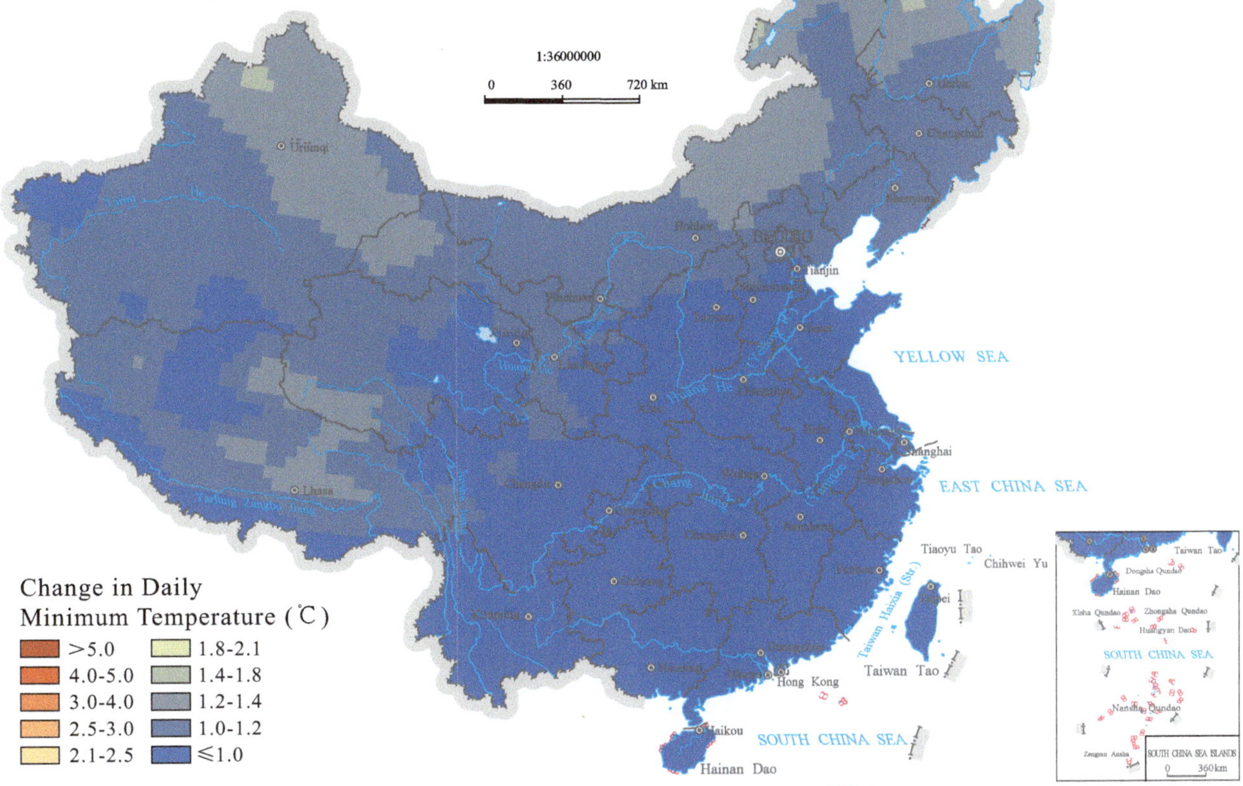

Change in Daily Maximum Temperature for RCP6.0 (1981-2010 to 2011-2040)

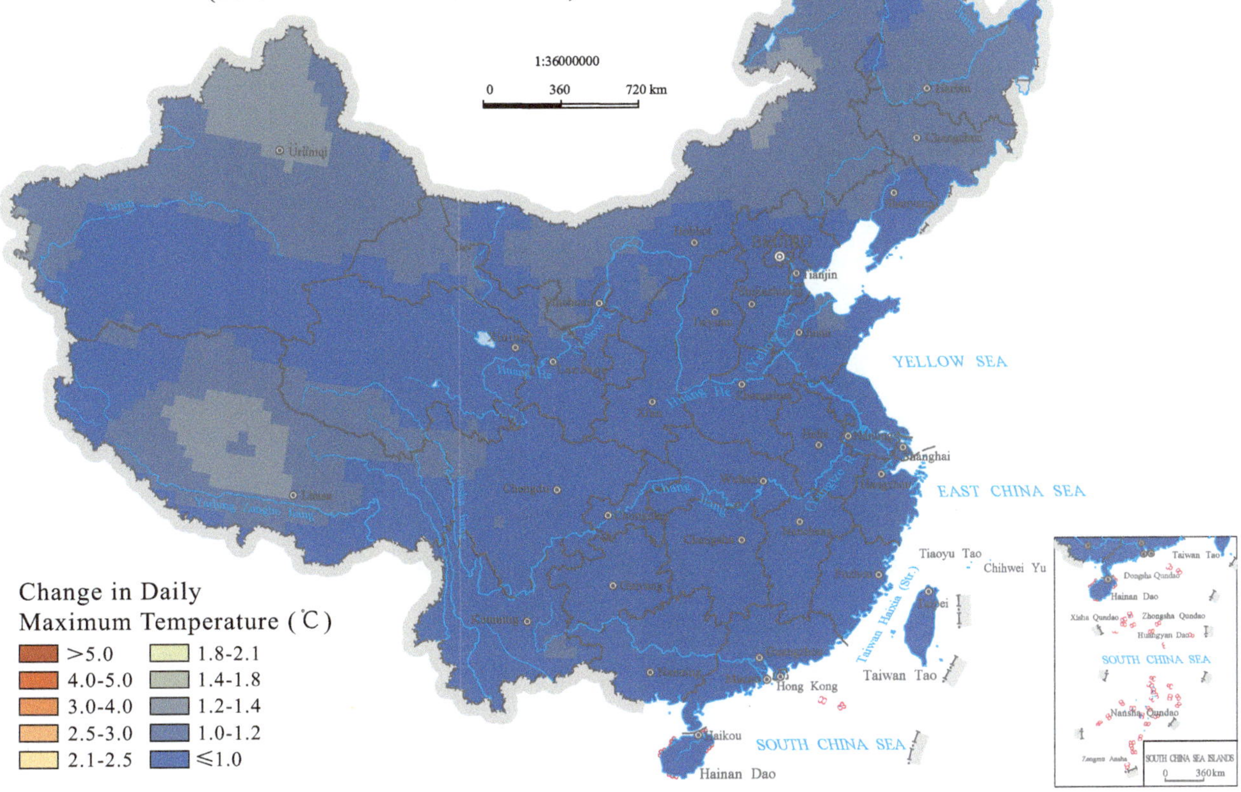

Change in Mean Temperature for RCP6.0
(1981-2010 to 2011-2040)

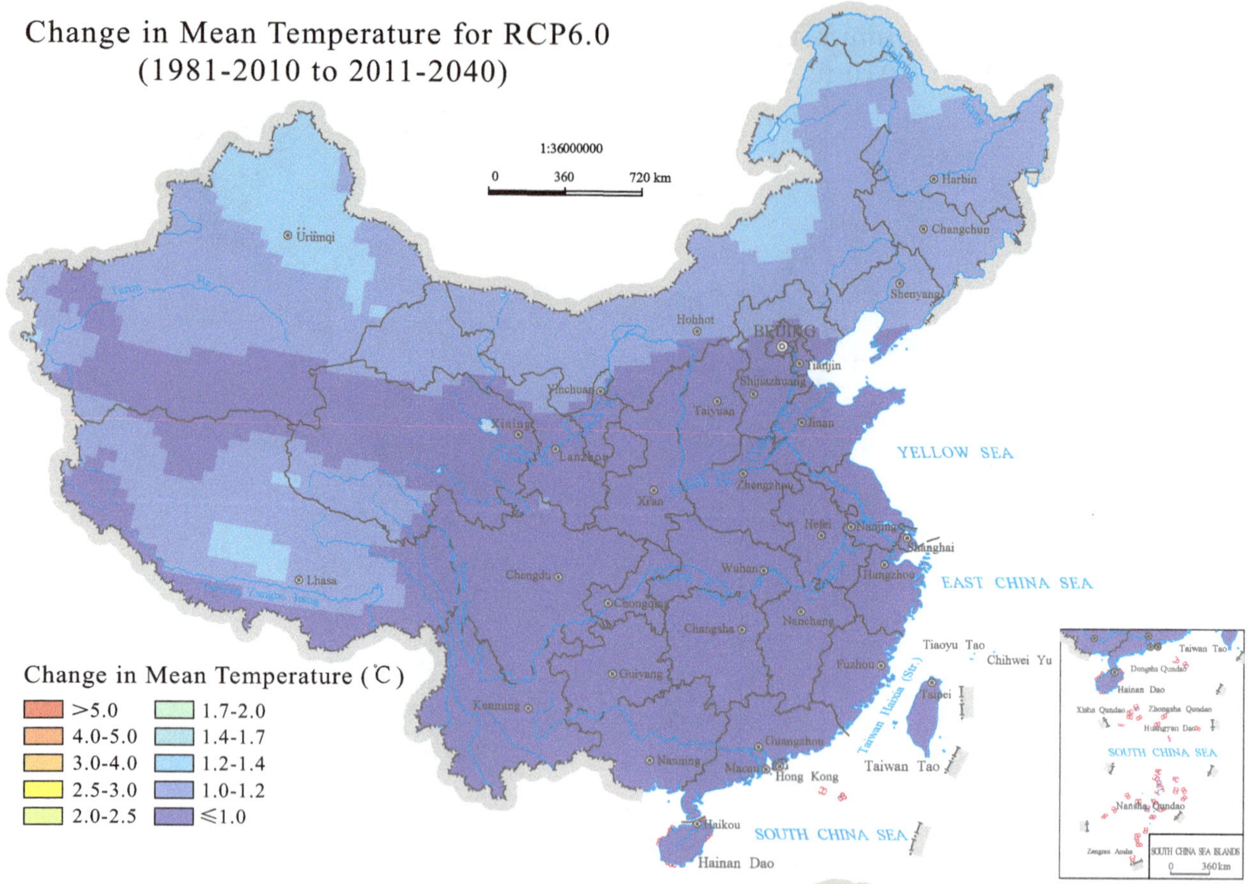

Change in Annual Precipitation for RCP6.0
(1981-2010 to 2011-2040)

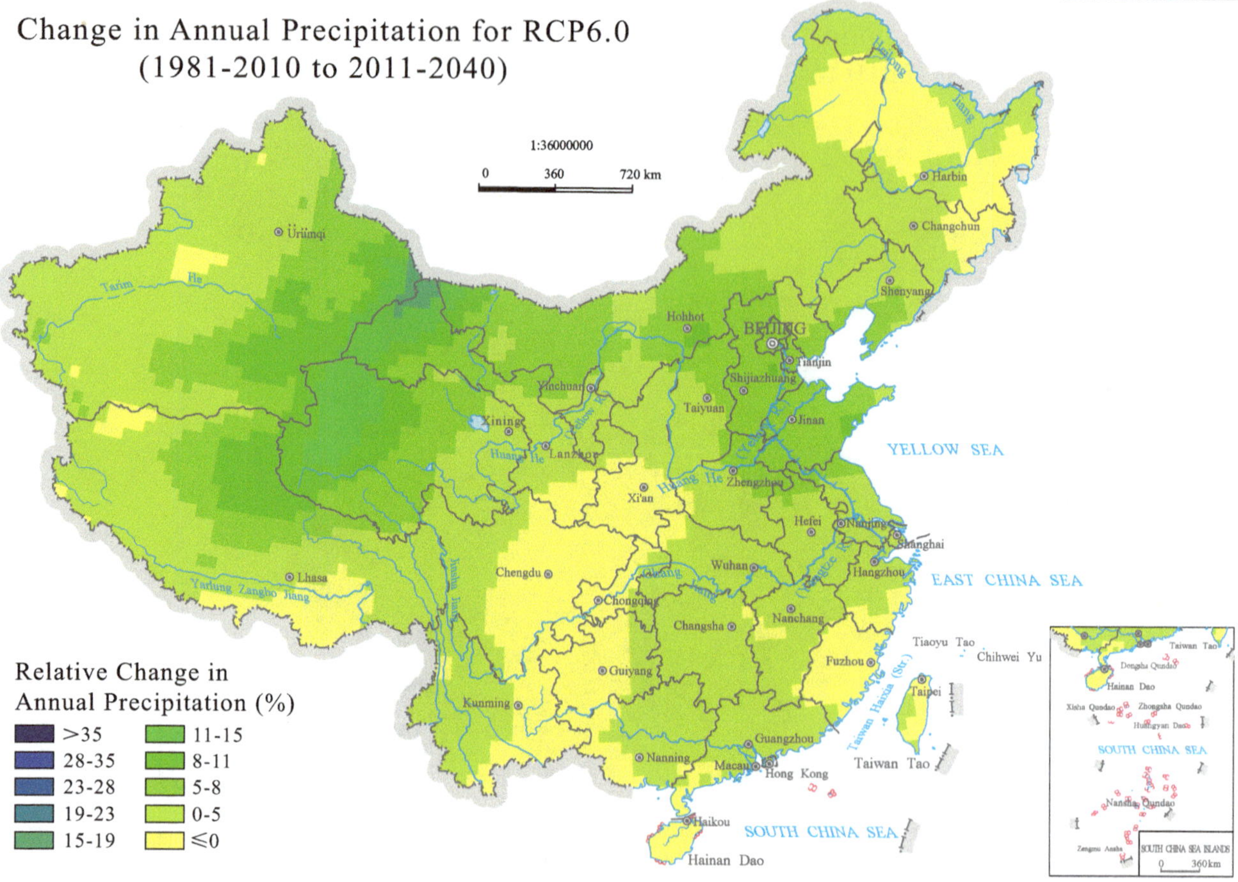

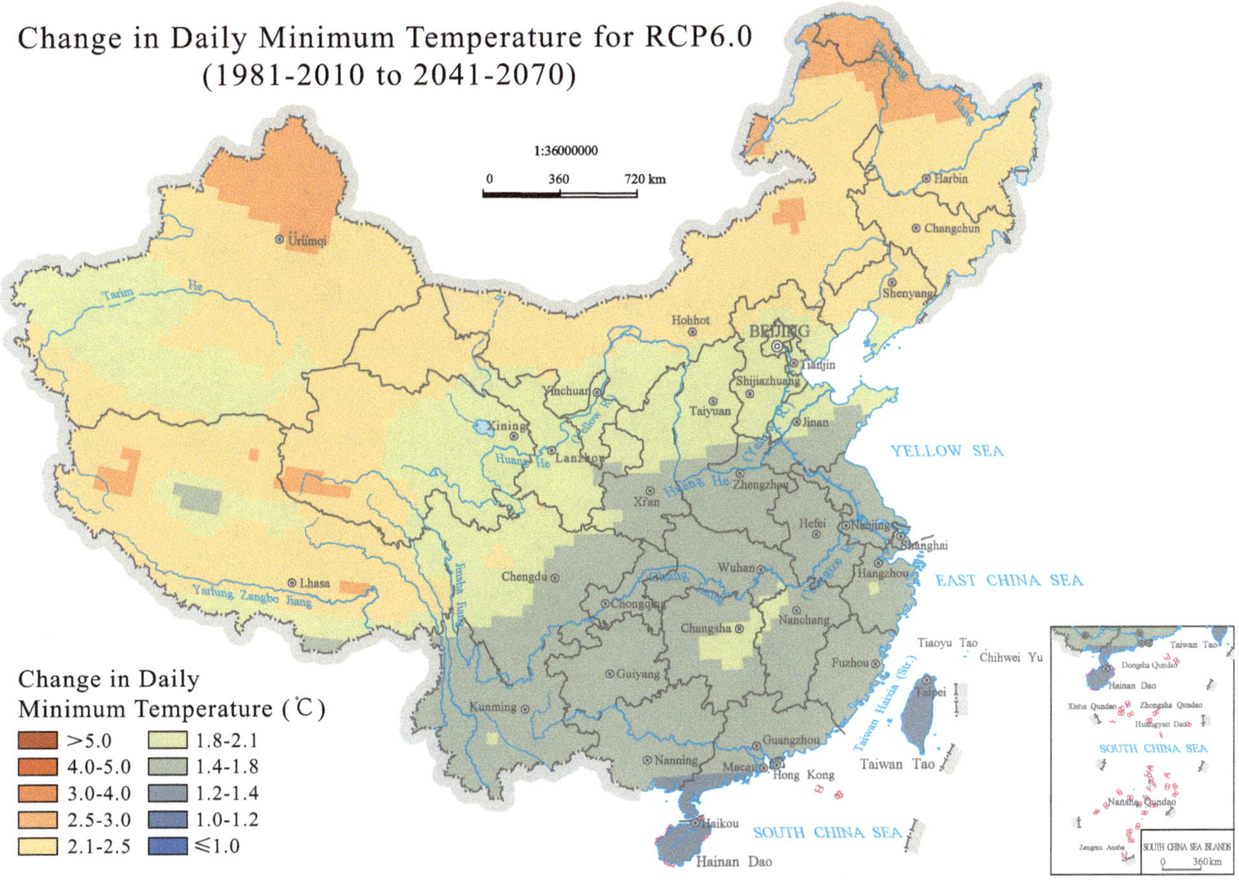
Change in Daily Minimum Temperature for RCP6.0
(1981-2010 to 2041-2070)
1:36000000
0 360 720 km
Change in Daily
Minimum Temperature (℃)
>5.0
4.0-5.0
3.0-4.0
2.5-3.0
2.1-2.5
1.8-2.1
1.4-1.8
1.2-1.4
1.0-1.2
≤1.0
Ürümqi
Tarim He
Harbin
Changchun
Shenyang
Hohhot
BEIJING
Tianjin
Shijiazhuang
Taiyuan
Jinan
Yinchuan
Xining
Lanzhou
Huang He
Xi'an
Zhengzhou
Hefei
Nanjing
Shanghai
Hangzhou
Wuhan
Chengdu
Chongqing
Changsha
Nanchang
Lhasa
Yarlung Zangbo Jiang
Jinsha Jiang
Guiyang
Kunming
Fuzhou
Taipei
Guangzhou
Nanning
Macau
Hong Kong
Haikou
Hainan Dao
Taiwan Tao
Tiaoyu Tao
Chihwei Yu
YELLOW SEA
EAST CHINA SEA
SOUTH CHINA SEA
SOUTH CHINA SEA ISLANDS
360km

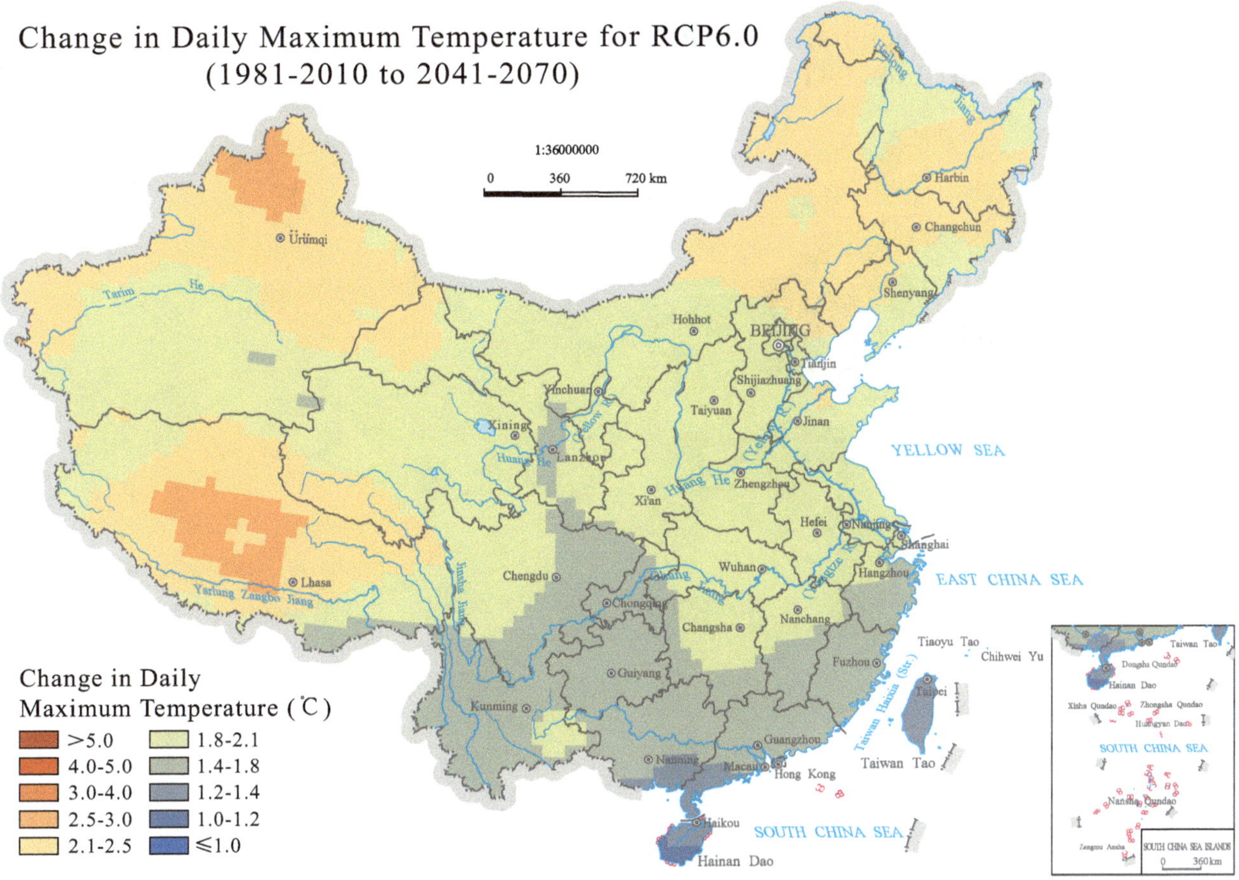
Change in Daily Maximum Temperature for RCP6.0
(1981-2010 to 2041-2070)
1:36000000
0 360 720 km
Change in Daily
Maximum Temperature (℃)
>5.0
4.0-5.0
3.0-4.0
2.5-3.0
2.1-2.5
1.8-2.1
1.4-1.8
1.2-1.4
1.0-1.2
≤1.0
Ürümqi
Tarim He
Harbin
Changchun
Shenyang
Hohhot
BEIJING
Tianjin
Shijiazhuang
Taiyuan
Jinan
Yinchuan
Xining
Lanzhou
Huang He
Xi'an
Zhengzhou
Hefei
Nanjing
Shanghai
Hangzhou
Wuhan
Chengdu
Chongqing
Changsha
Nanchang
Lhasa
Yarlung Zangbo Jiang
Jinsha Jiang
Guiyang
Kunming
Fuzhou
Taipei
Guangzhou
Nanning
Macau
Hong Kong
Haikou
Hainan Dao
Taiwan Tao
Tiaoyu Tao
Chihwei Yu
YELLOW SEA
EAST CHINA SEA
SOUTH CHINA SEA
SOUTH CHINA SEA ISLANDS
360km

Change in Mean Temperature for RCP6.0 (1981-2010 to 2041-2070)

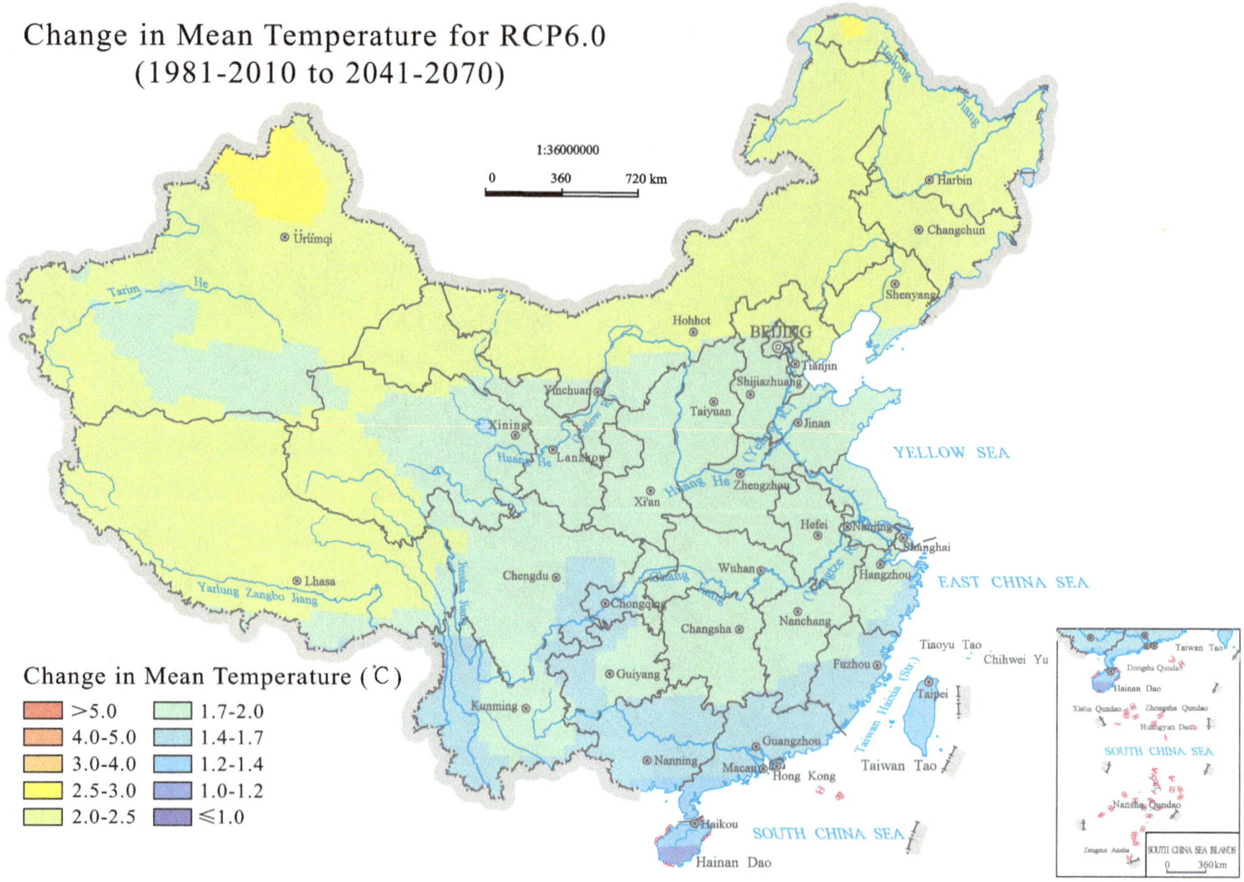

Change in Annual Precipitation for RCP6.0 (1981-2010 to 2041-2070)

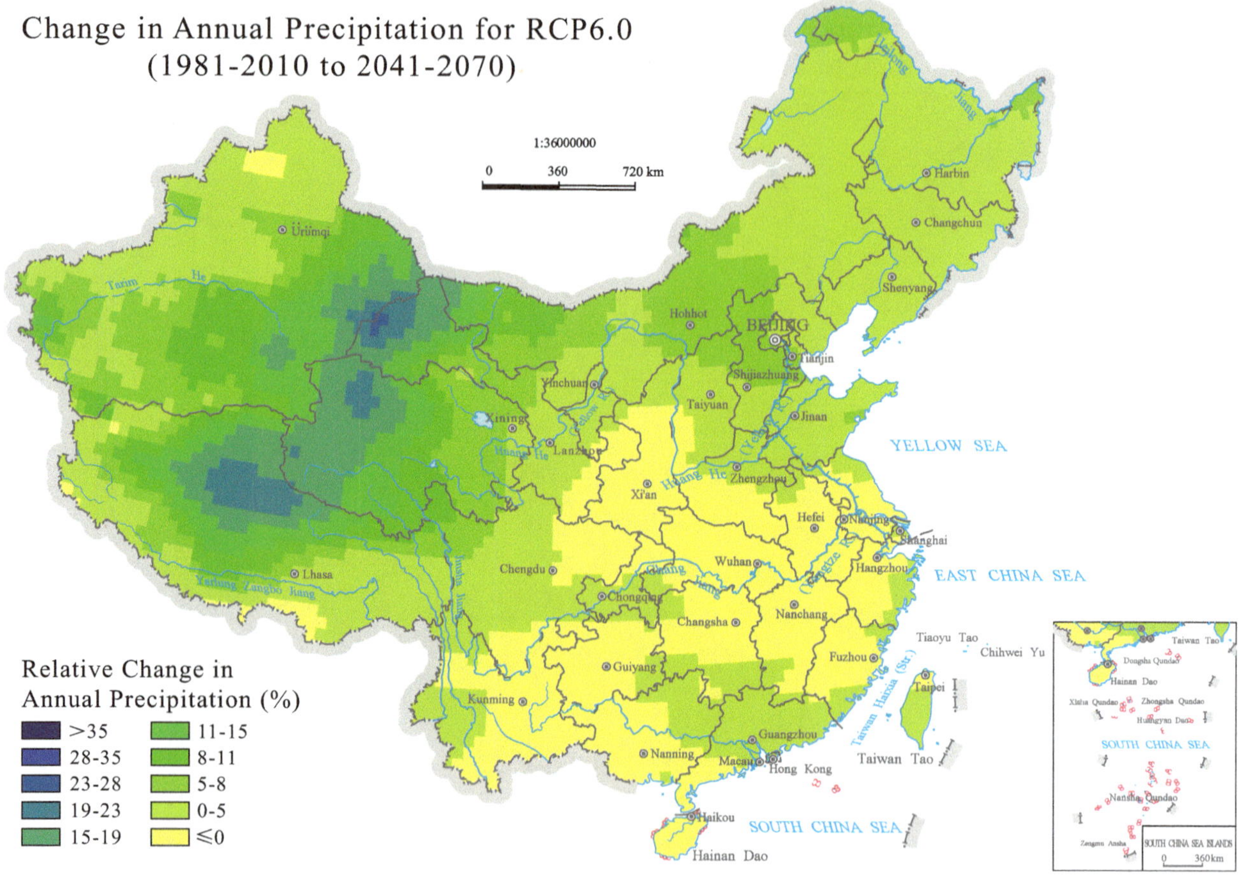

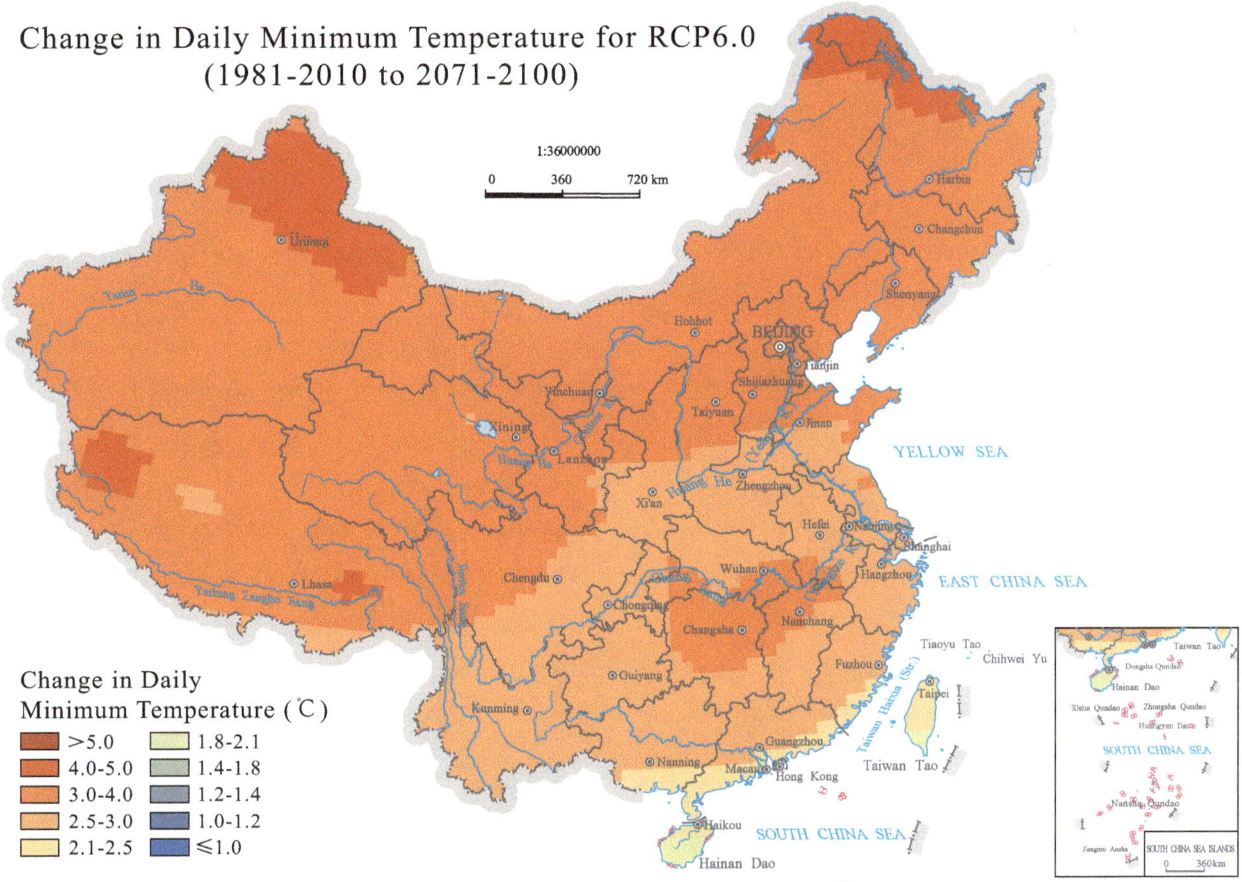
Change in Daily Minimum Temperature for RCP6.0
(1981-2010 to 2071-2100)
1:36000000
0 360 720 km
Change in Daily
Minimum Temperature (℃)
>5.0
4.0-5.0
3.0-4.0
2.5-3.0
2.1-2.5
1.8-2.1
1.4-1.8
1.2-1.4
1.0-1.2
≤1.0
Harbin
Changchun
Shenyang
Ürümqi
Hohhot
BEIJING
Tianjin
Shijiazhuang
Yinchuan
Taiyuan
Jinan
Xining
Lanzhou
Zhengzhou
Xi'an
Hefei
Nanjing
Shanghai
Wuhan
Hangzhou
Chengdu
Lhasa
Chongqing
Changsha
Nanchang
Guiyang
Fuzhou
Kunming
Taipei
Nanning
Guangzhou
Macau
Hong Kong
Haikou
Hainan Dao
Taiwan Tao
Tiaoyu Tao
Chihwei Yu
YELLOW SEA
EAST CHINA SEA
SOUTH CHINA SEA
SOUTH CHINA SEA ISLANDS
0 360km

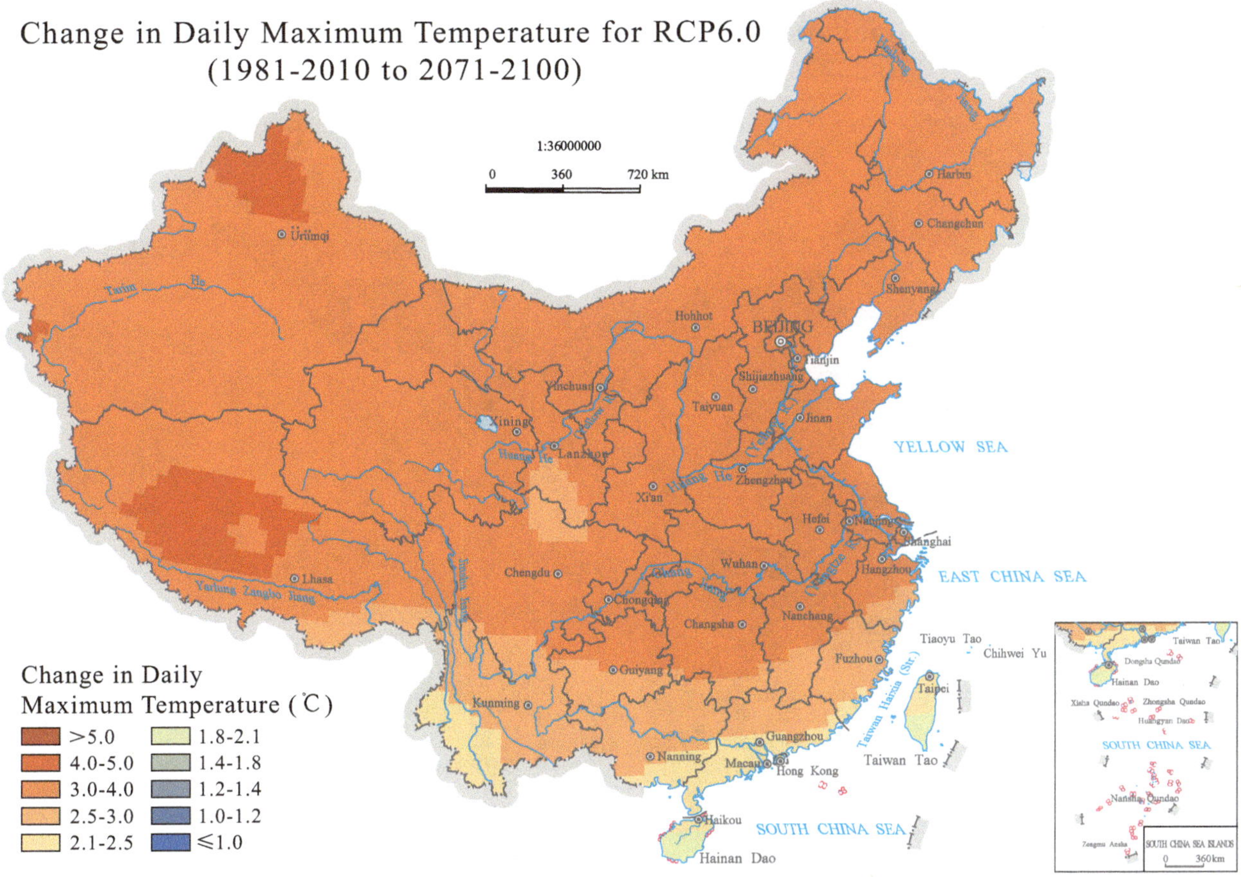
Change in Daily Maximum Temperature for RCP6.0
(1981-2010 to 2071-2100)
1:36000000
0 360 720 km
Change in Daily
Maximum Temperature (℃)
>5.0
4.0-5.0
3.0-4.0
2.5-3.0
2.1-2.5
1.8-2.1
1.4-1.8
1.2-1.4
1.0-1.2
≤1.0
Harbin
Changchun
Shenyang
Ürümqi
Hohhot
BEIJING
Tianjin
Shijiazhuang
Yinchuan
Taiyuan
Jinan
Xining
Lanzhou
Zhengzhou
Xi'an
Hefei
Nanjing
Shanghai
Wuhan
Hangzhou
Chengdu
Lhasa
Chongqing
Changsha
Nanchang
Guiyang
Fuzhou
Kunming
Taipei
Nanning
Guangzhou
Macau
Hong Kong
Haikou
Hainan Dao
Taiwan Tao
Tiaoyu Tao
Chihwei Yu
YELLOW SEA
EAST CHINA SEA
SOUTH CHINA SEA
SOUTH CHINA SEA ISLANDS
0 360km

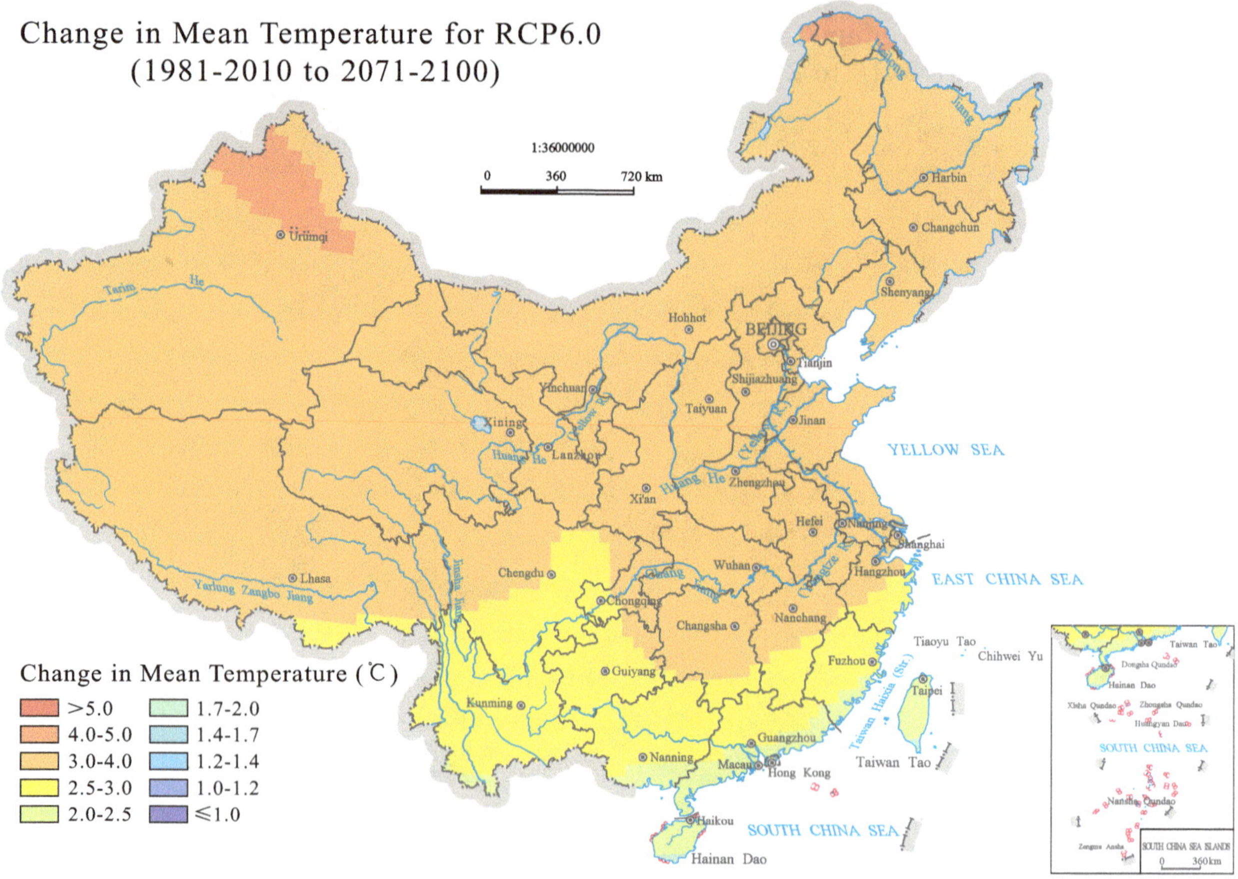
Change in Mean Temperature for RCP6.0
(1981-2010 to 2071-2100)
1:36000000
0 360 720 km
Change in Mean Temperature (℃)
>5.0
4.0-5.0
3.0-4.0
2.5-3.0
2.0-2.5
1.7-2.0
1.4-1.7
1.2-1.4
1.0-1.2
≤1.0
Ürümqi
Harbin
Changchun
Shenyang
Hohhot
BEIJING
Tianjin
Shijiazhuang
Taiyuan
Jinan
Yinchuan
Xining
Lanzhou
Zhengzhou
Xi'an
Hefei
Nanjing
Shanghai
Wuhan
Hangzhou
Chengdu
Lhasa
Chongqing
Changsha
Nanchang
Guiyang
Fuzhou
Kunming
Taipei
Guangzhou
Nanning
Macau
Hong Kong
Haikou
Hainan Dao
Taiwan Tao
Tiaoyu Tao
Chihwei Yu
YELLOW SEA
EAST CHINA SEA
SOUTH CHINA SEA
Tarim He
Yarlung Zangbo Jiang
SOUTH CHINA SEA

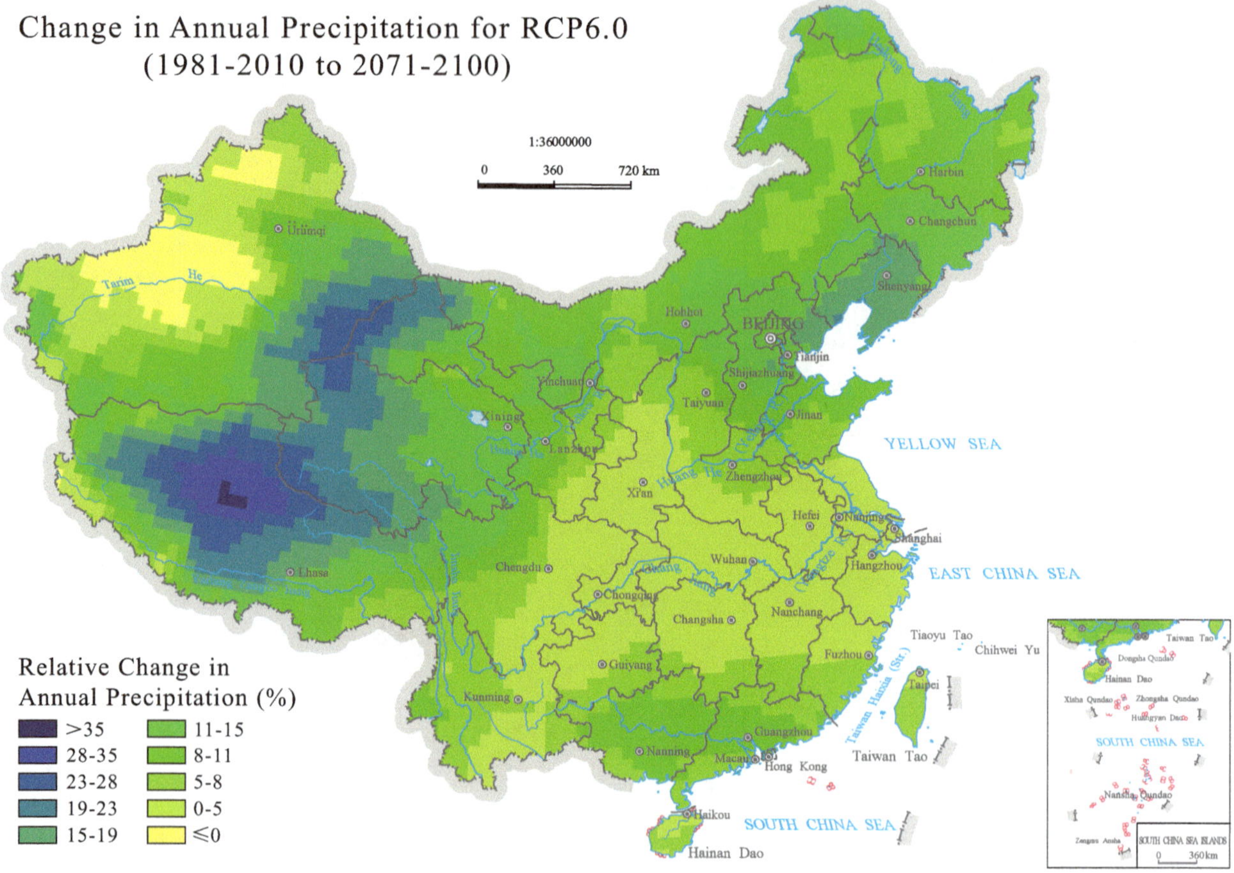
Change in Annual Precipitation for RCP6.0
(1981-2010 to 2071-2100)
1:36000000
0 360 720 km
Relative Change in
Annual Precipitation (%)
>35
28-35
23-28
19-23
15-19
11-15
8-11
5-8
0-5
≤0
Ürümqi
Harbin
Changchun
Shenyang
Hohhot
BEIJING
Tianjin
Shijiazhuang
Taiyuan
Jinan
Yinchuan
Xining
Lanzhou
Zhengzhou
Xi'an
Hefei
Nanjing
Shanghai
Wuhan
Hangzhou
Chengdu
Lhasa
Chongqing
Changsha
Nanchang
Guiyang
Fuzhou
Kunming
Taipei
Guangzhou
Nanning
Macau
Hong Kong
Haikou
Hainan Dao
Taiwan Tao
Tiaoyu Tao
Chihwei Yu
YELLOW SEA
EAST CHINA SEA
SOUTH CHINA SEA
Tarim He
SOUTH CHINA SEA

Change in Daily Minimum Temperature for RCP8.5
(1981-2010 to 2011-2040)

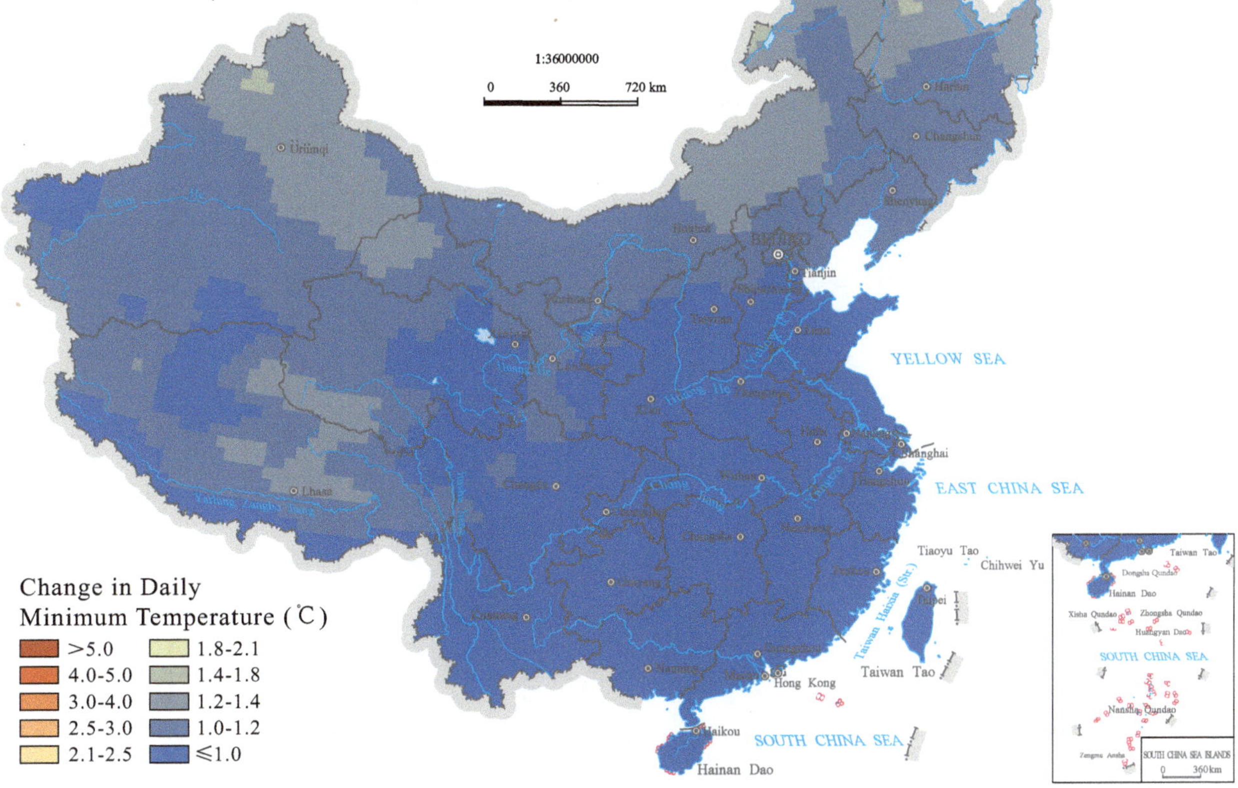

Change in Daily Maximum Temperature for RCP8.5
(1981-2010 to 2011-2040)

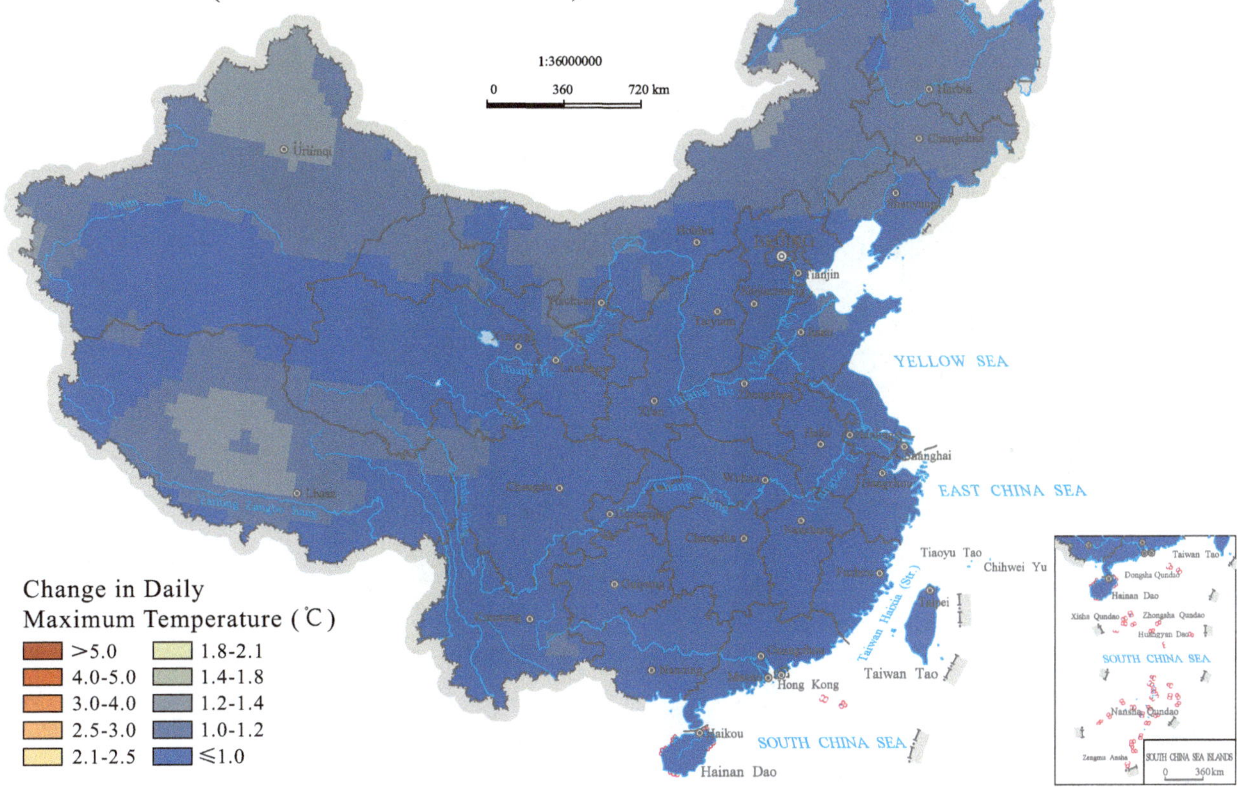

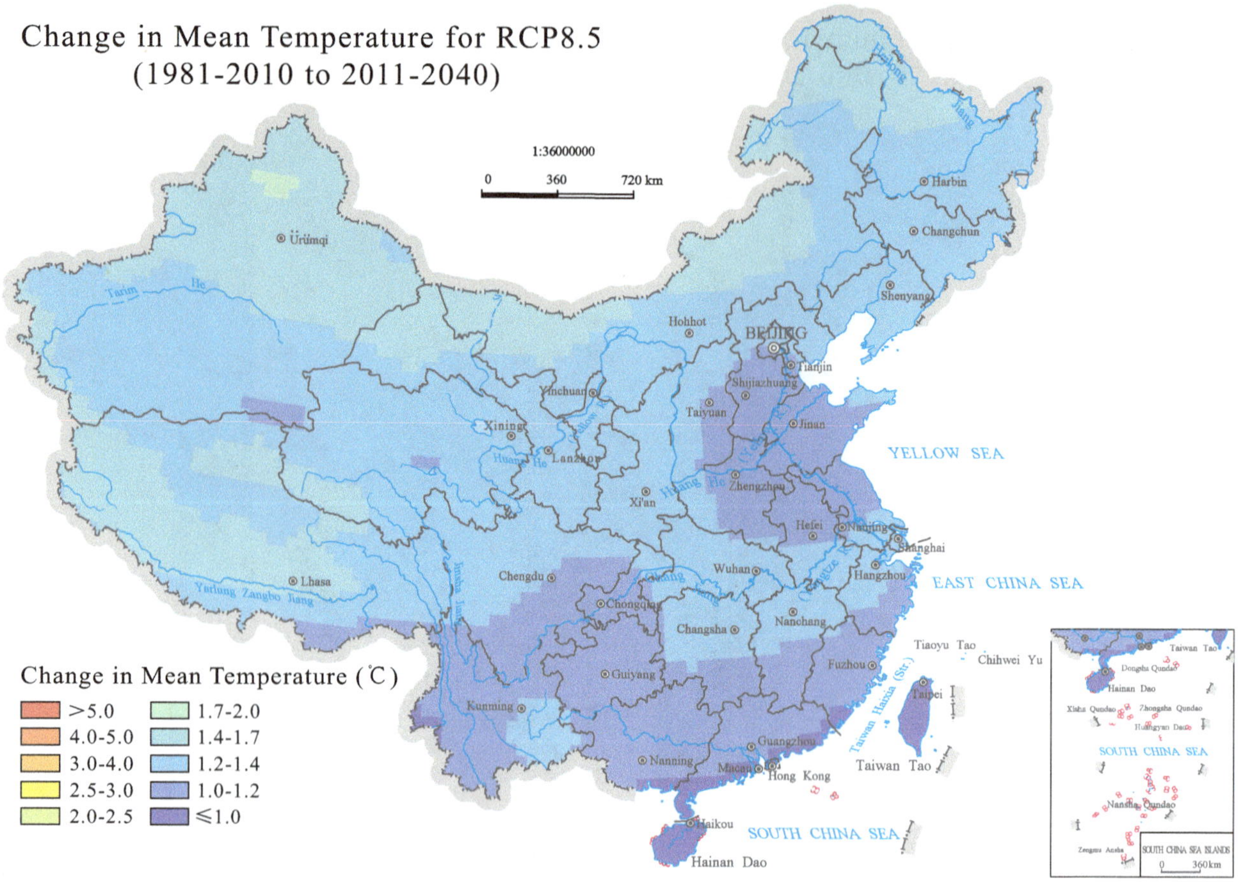
Change in Mean Temperature for RCP8.5
(1981-2010 to 2011-2040)
1:36000000
0 360 720 km
Change in Mean Temperature (℃)
>5.0
4.0-5.0
3.0-4.0
2.5-3.0
2.0-2.5
1.7-2.0
1.4-1.7
1.2-1.4
1.0-1.2
≤1.0
Ürümqi
Harbin
Changchun
Shenyang
Hohhot
BEIJING
Tianjin
Shijiazhuang
Taiyuan
Yinchuan
Xining
Lanzhou
Jinan
Zhengzhou
Xi'an
Hefei
Nanjing
Shanghai
Lhasa
Chengdu
Wuhan
Hangzhou
Chongqing
Changsha
Nanchang
Guiyang
Fuzhou
Kunming
Taipei
Nanning
Guangzhou
Macau
Hong Kong
Haikou
Hainan Dao
Taiwan Tao
Tiaoyu Tao
Chihwei Yu
YELLOW SEA
EAST CHINA SEA
SOUTH CHINA SEA
SOUTH CHINA SEA
Nansha Qundao
SOUTH CHINA SEA ISLANDS
0 360km

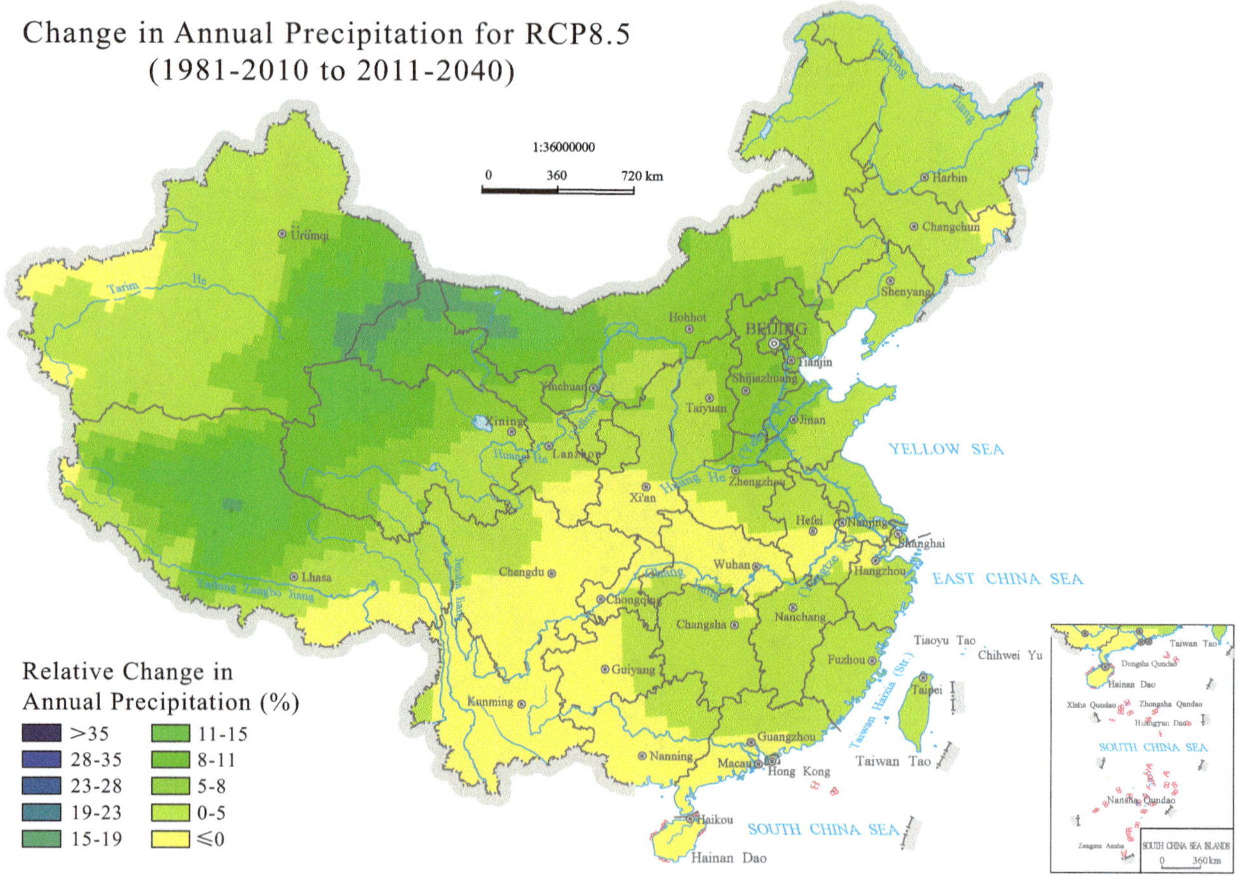
Change in Annual Precipitation for RCP8.5
(1981-2010 to 2011-2040)
1:36000000
0 360 720 km
Relative Change in
Annual Precipitation (%)
>35
28-35
23-28
19-23
15-19
11-15
8-11
5-8
0-5
≤0
Ürümqi
Harbin
Changchun
Shenyang
Hohhot
BEIJING
Tianjin
Shijiazhuang
Taiyuan
Yinchuan
Xining
Lanzhou
Jinan
Zhengzhou
Xi'an
Hefei
Nanjing
Shanghai
Lhasa
Chengdu
Wuhan
Hangzhou
Chongqing
Changsha
Nanchang
Guiyang
Fuzhou
Kunming
Taipei
Nanning
Guangzhou
Macau
Hong Kong
Haikou
Hainan Dao
Taiwan Tao
Tiaoyu Tao
Chihwei Yu
YELLOW SEA
EAST CHINA SEA
SOUTH CHINA SEA
SOUTH CHINA SEA
Nansha Qundao
SOUTH CHINA SEA ISLANDS
0 360km

Change in Daily Minimum Temperature for RCP8.5
(1981-2010 to 2041-2070)

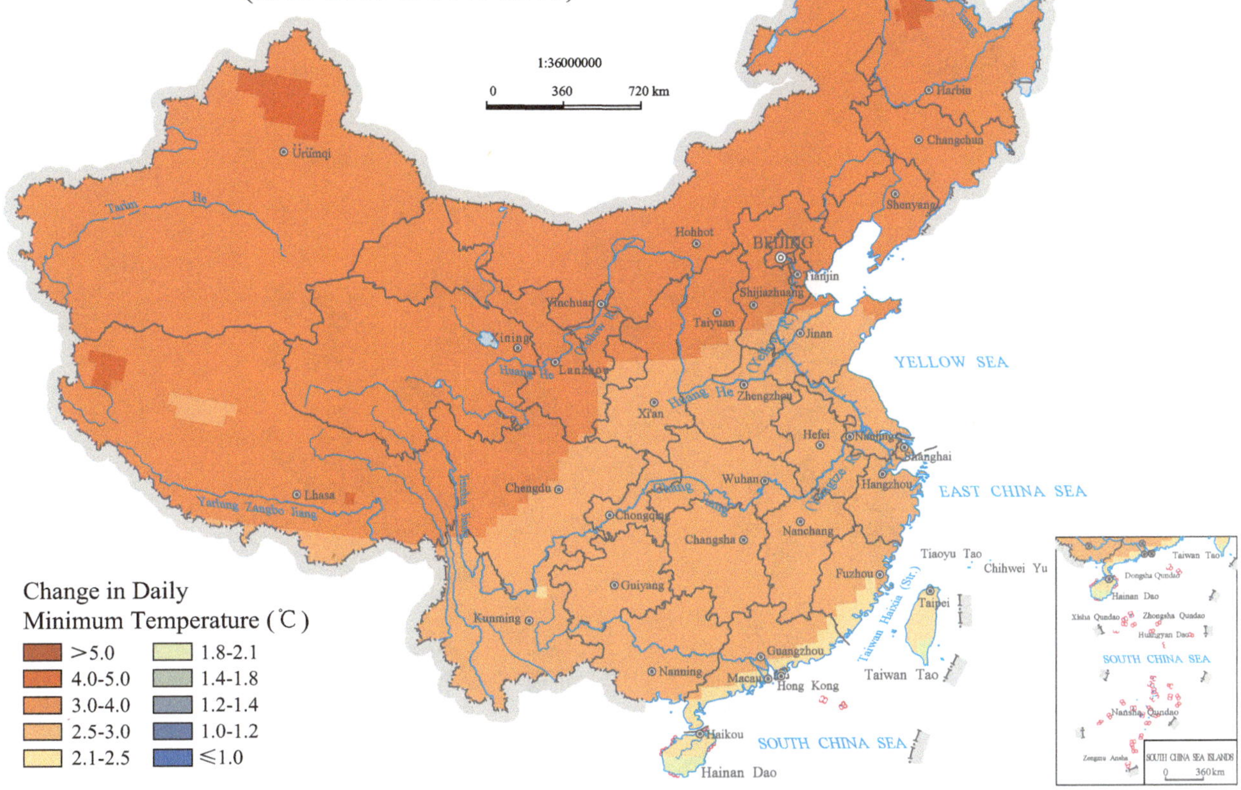

Change in Daily Maximum Temperature for RCP8.5
(1981-2010 to 2041-2070)

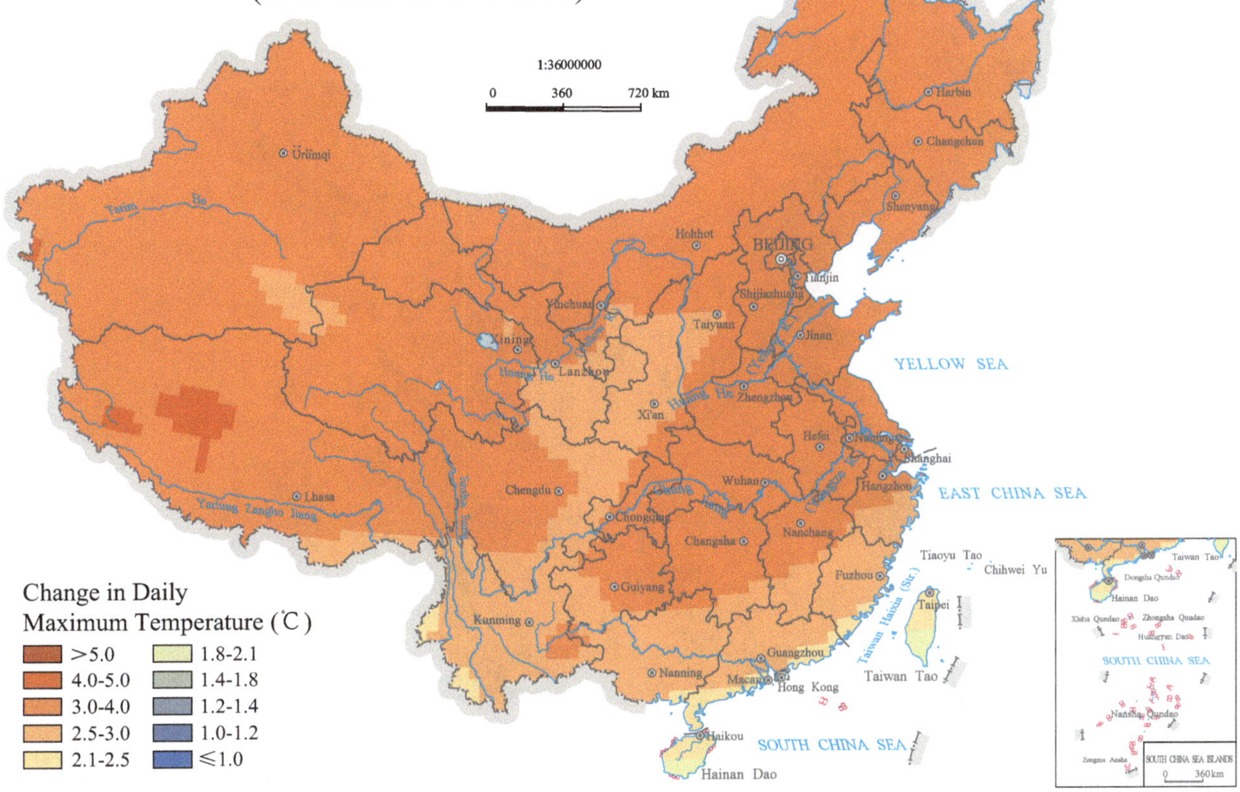

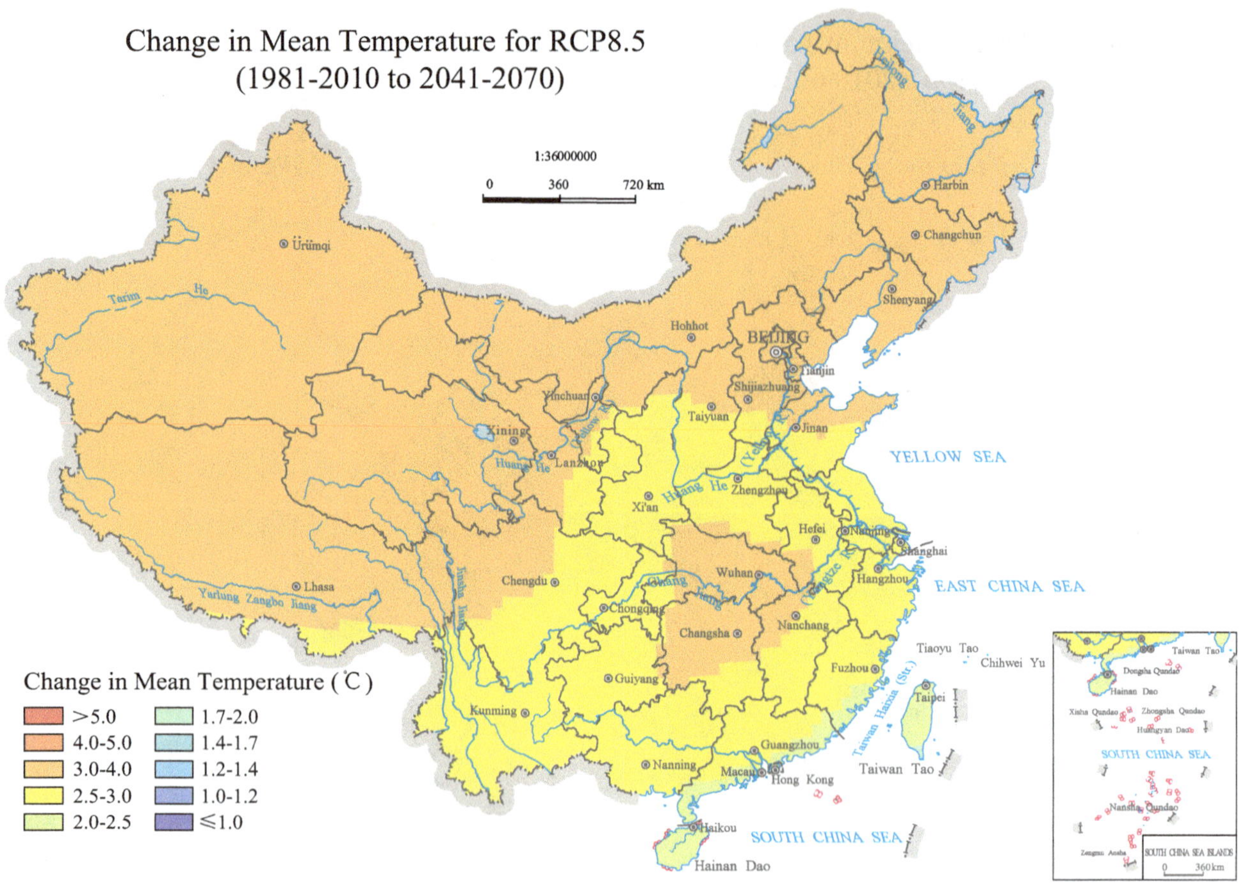
Change in Mean Temperature for RCP8.5
(1981-2010 to 2041-2070)
1:36000000
0 360 720 km
YELLOW SEA
EAST CHINA SEA
SOUTH CHINA SEA
Hainan Dao
Taiwan Tao
Tiaoyu Tao
Chihwei Yu
Hong Kong
Macau
Change in Mean Temperature (℃)
>5.0
4.0-5.0
3.0-4.0
2.5-3.0
2.0-2.5
1.7-2.0
1.4-1.7
1.2-1.4
1.0-1.2
≤1.0

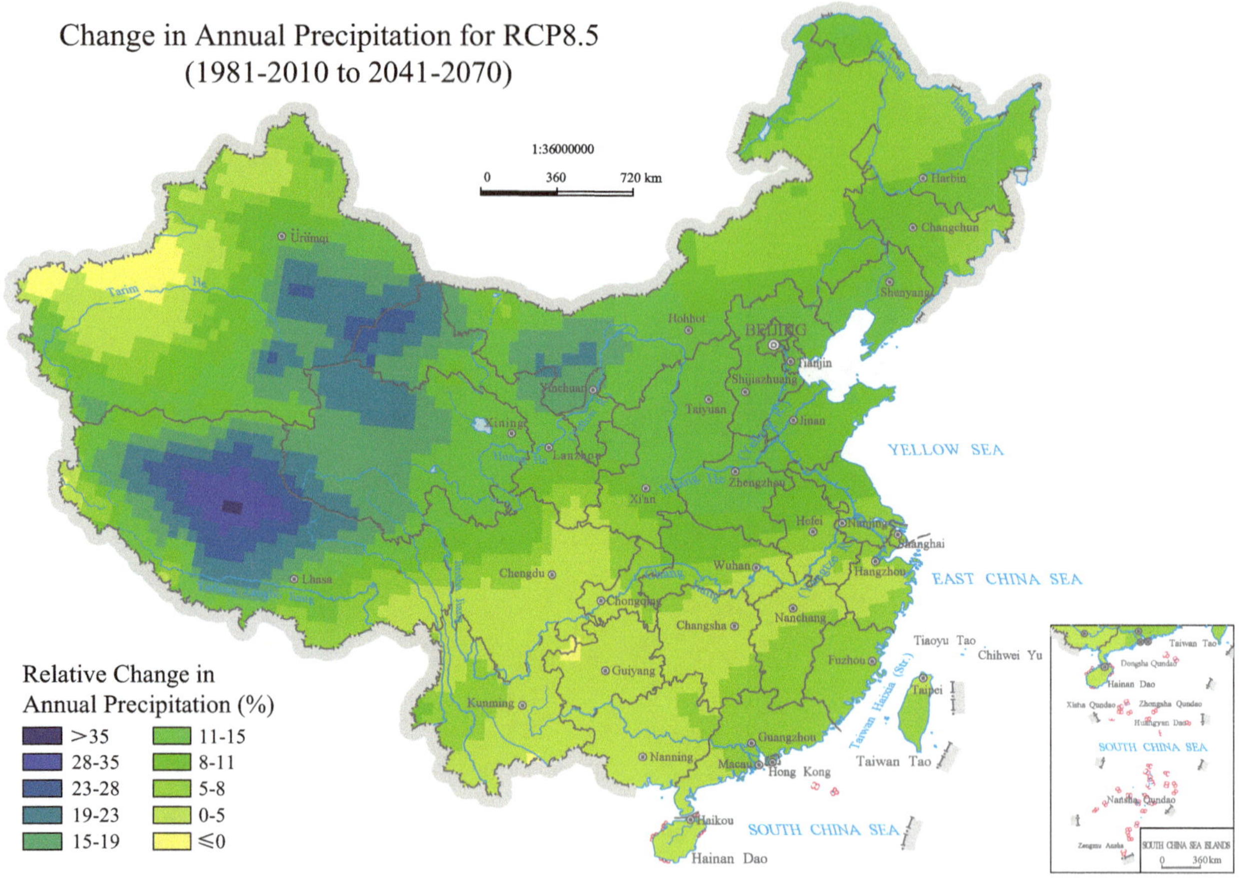
Change in Annual Precipitation for RCP8.5
(1981-2010 to 2041-2070)
1:36000000
0 360 720 km
YELLOW SEA
EAST CHINA SEA
SOUTH CHINA SEA
Hainan Dao
Taiwan Tao
Tiaoyu Tao
Chihwei Yu
Hong Kong
Macau
Relative Change in
Annual Precipitation (%)
>35
28-35
23-28
19-23
15-19
11-15
8-11
5-8
0-5
≤0

Change in Daily Minimum Temperature for RCP8.5 (1981-2010 to 2071-2100)

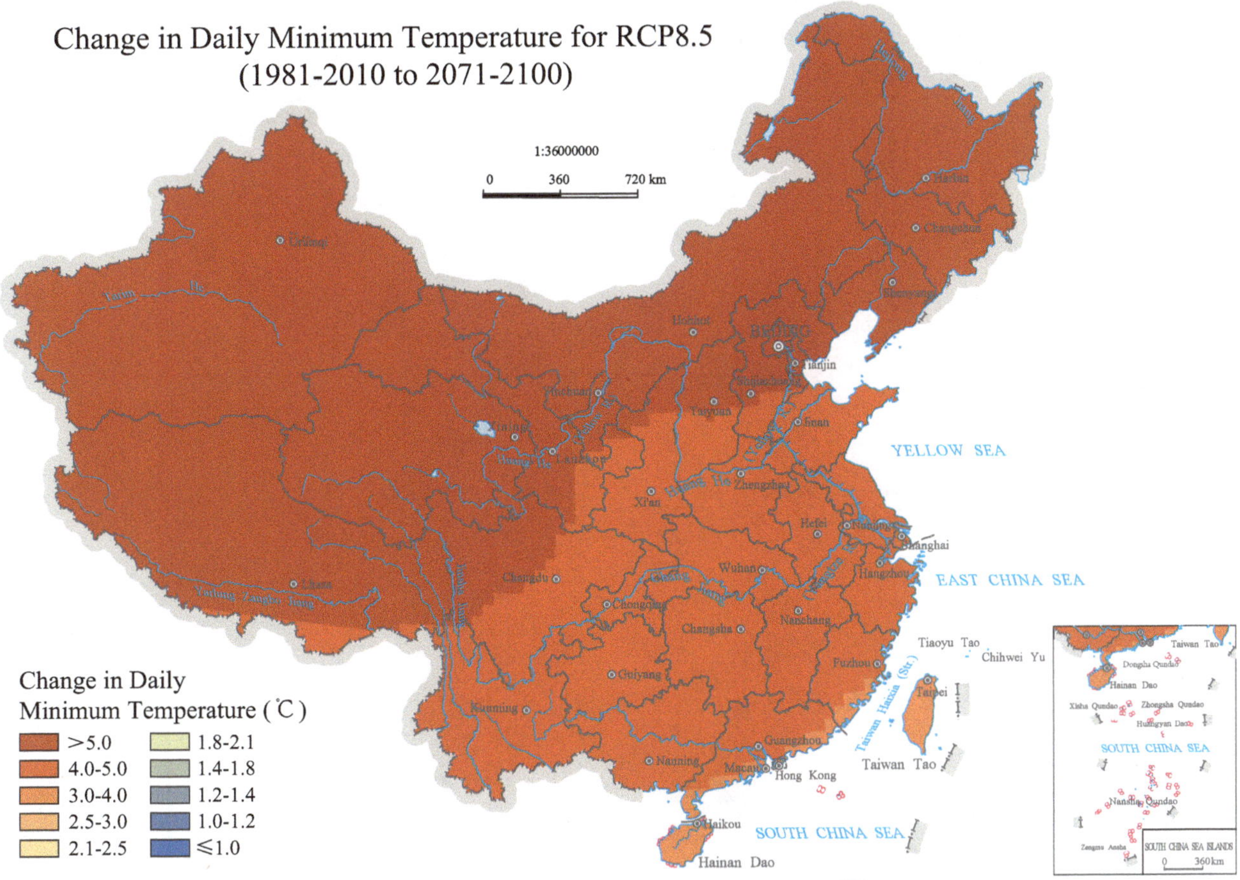

Change in Daily Maximum Temperature for RCP8.5 (1981-2010 to 2071-2100)

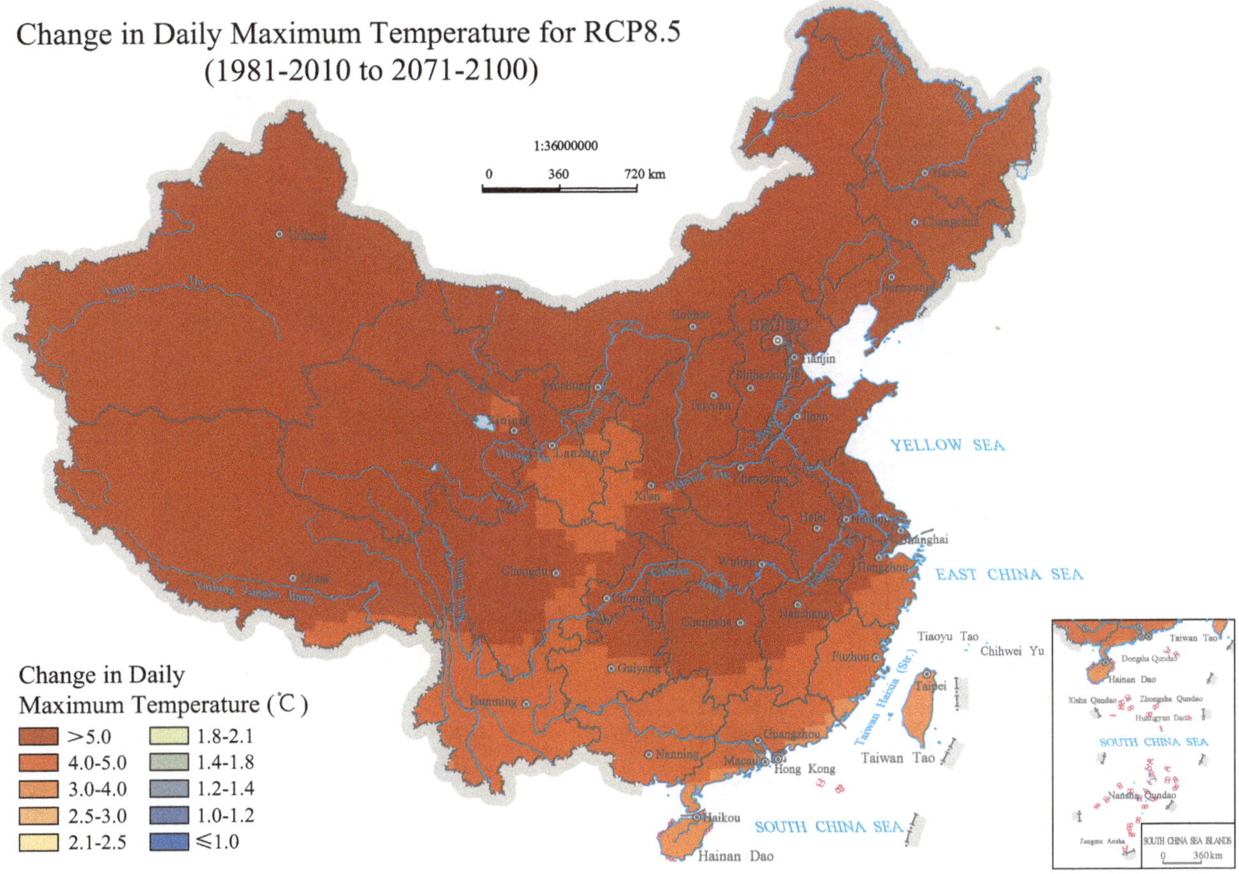

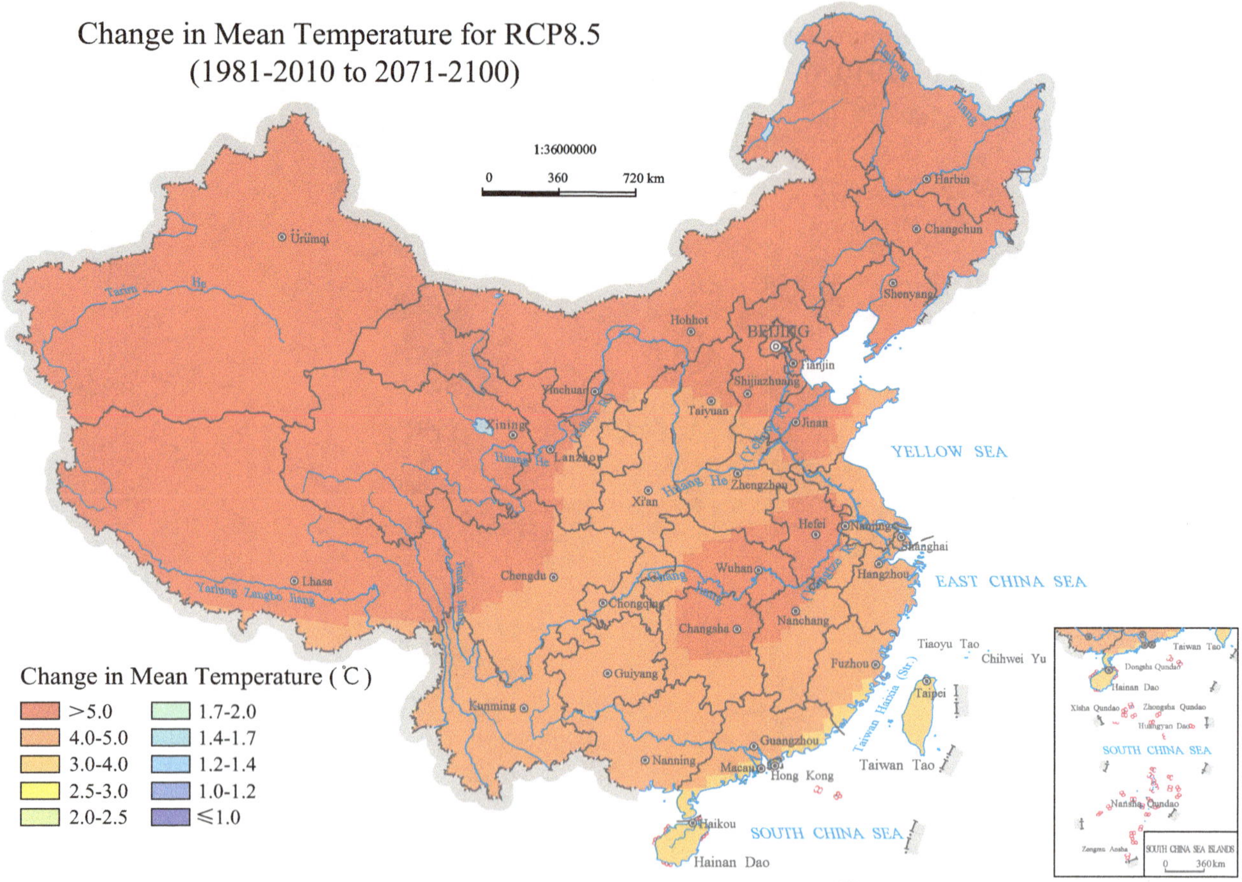

Change in Mean Temperature for RCP8.5
(1981-2010 to 2071-2100)
1:36000000
0 360 720 km
Change in Mean Temperature (℃)
>5.0
4.0-5.0
3.0-4.0
2.5-3.0
2.0-2.5
1.7-2.0
1.4-1.7
1.2-1.4
1.0-1.2
≤1.0
Ürümqi
Tarim He
Lhasa
Yarlung Zangbo Jiang
Xining
Lanzhou
Yinchuan
Hohhot
BEIJING
Tianjin
Shijiazhuang
Taiyuan
Jinan
Zhengzhou
Xi'an
Chengdu
Chongqing
Wuhan
Hefei
Nanjing
Shanghai
Hangzhou
Changsha
Nanchang
Fuzhou
Guiyang
Kunming
Nanning
Guangzhou
Macau
Hong Kong
Haikou
Hainan Dao
Taipei
Taiwan Tao
Harbin
Changchun
Shenyang
YELLOW SEA
EAST CHINA SEA
SOUTH CHINA SEA
Tiaoyu Tao
Chihwei Yu
SOUTH CHINA SEA
SOUTH CHINA SEA ISLANDS

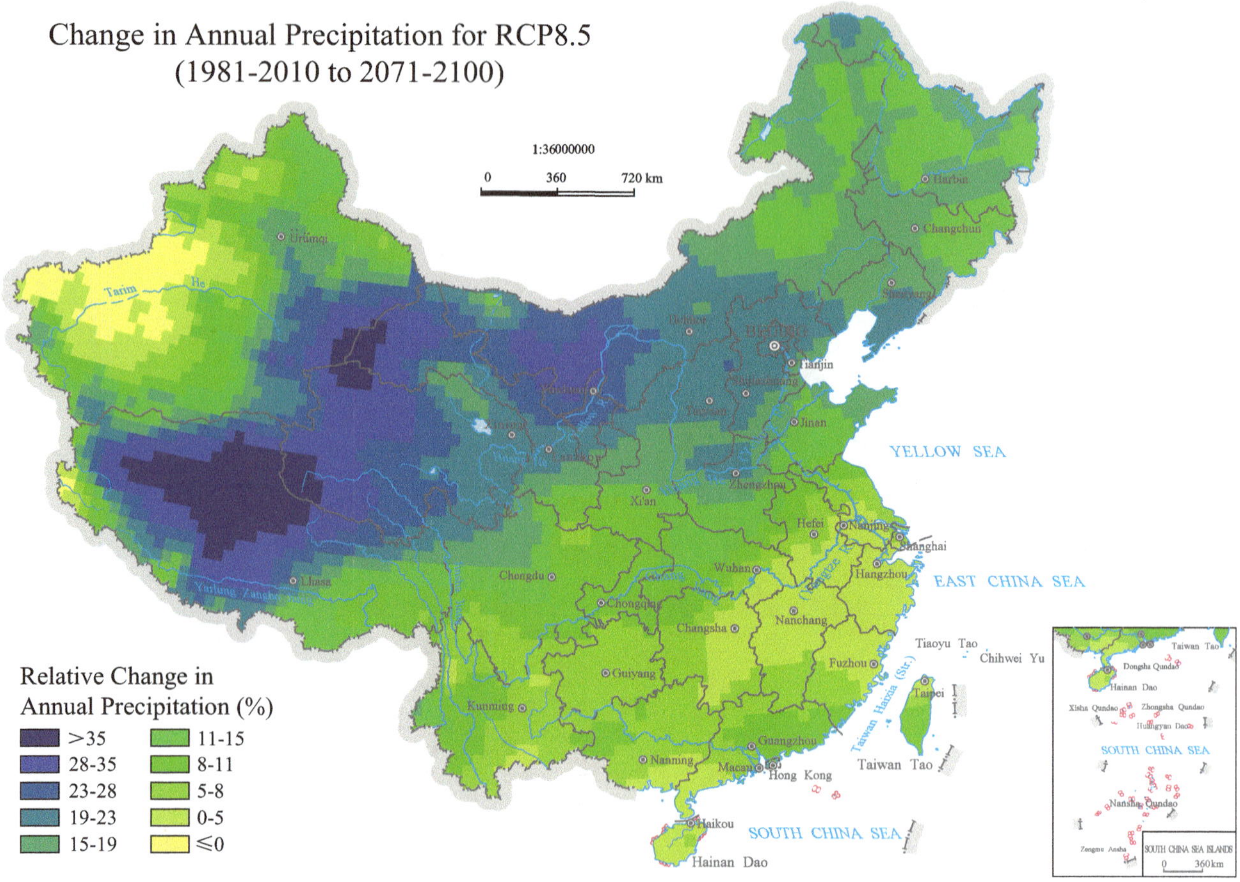

Change in Annual Precipitation for RCP8.5
(1981-2010 to 2071-2100)
1:36000000
0 360 720 km
Relative Change in
Annual Precipitation (%)
>35
28-35
23-28
19-23
15-19
11-15
8-11
5-8
0-5
≤0
Ürümqi
Tarim He
Lhasa
Xi'an
Chengdu
Chongqing
Wuhan
Hefei
Nanjing
Shanghai
Hangzhou
Changsha
Nanchang
Fuzhou
Guiyang
Kunming
Nanning
Guangzhou
Macau
Hong Kong
Haikou
Hainan Dao
Taipei
Taiwan Tao
Harbin
Changchun
Tianjin
Jinan
Zhengzhou
YELLOW SEA
EAST CHINA SEA
SOUTH CHINA SEA
Tiaoyu Tao
Chihwei Yu
SOUTH CHINA SEA
SOUTH CHINA SEA ISLANDS

Intermodel Spread in Changes in Daily Minimum Temperature for RCP8.5 (1981-2010 to 2071-2100)

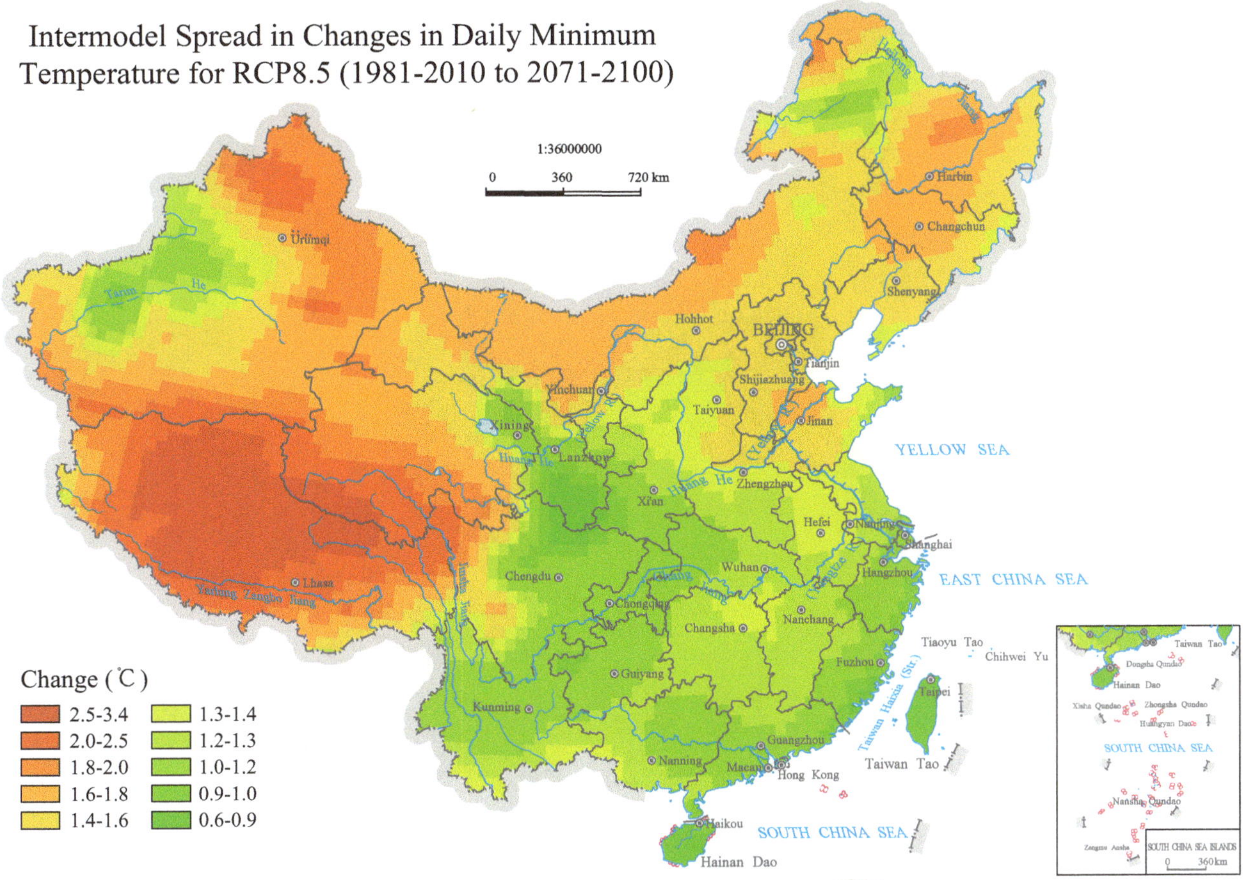

Intermodel Spread in Changes in Daily Maximum Temperature for RCP8.5 (1981-2010 to 2071-2100)

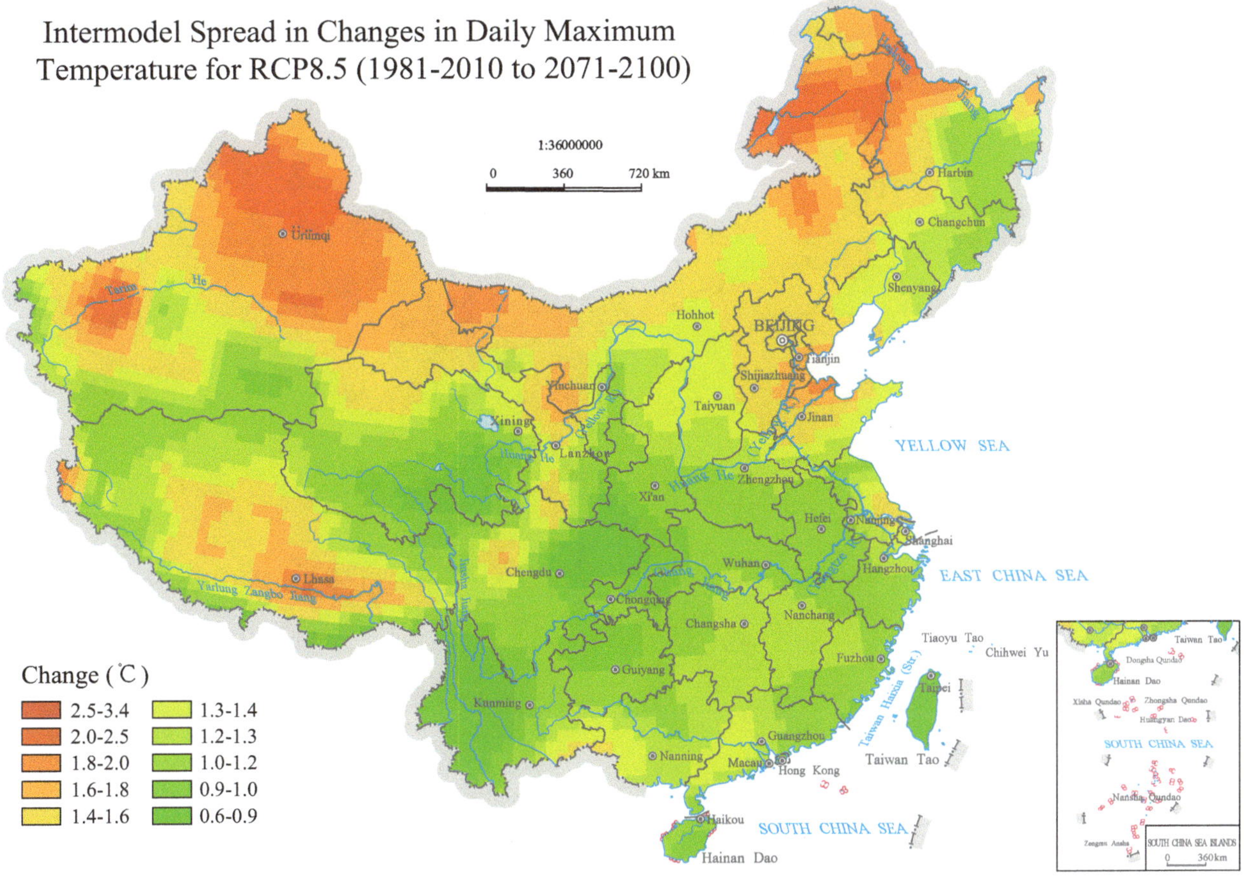

Intermodel Spread in Changes in Mean Temperature for RCP8.5 (1981-2010 to 2071-2100)

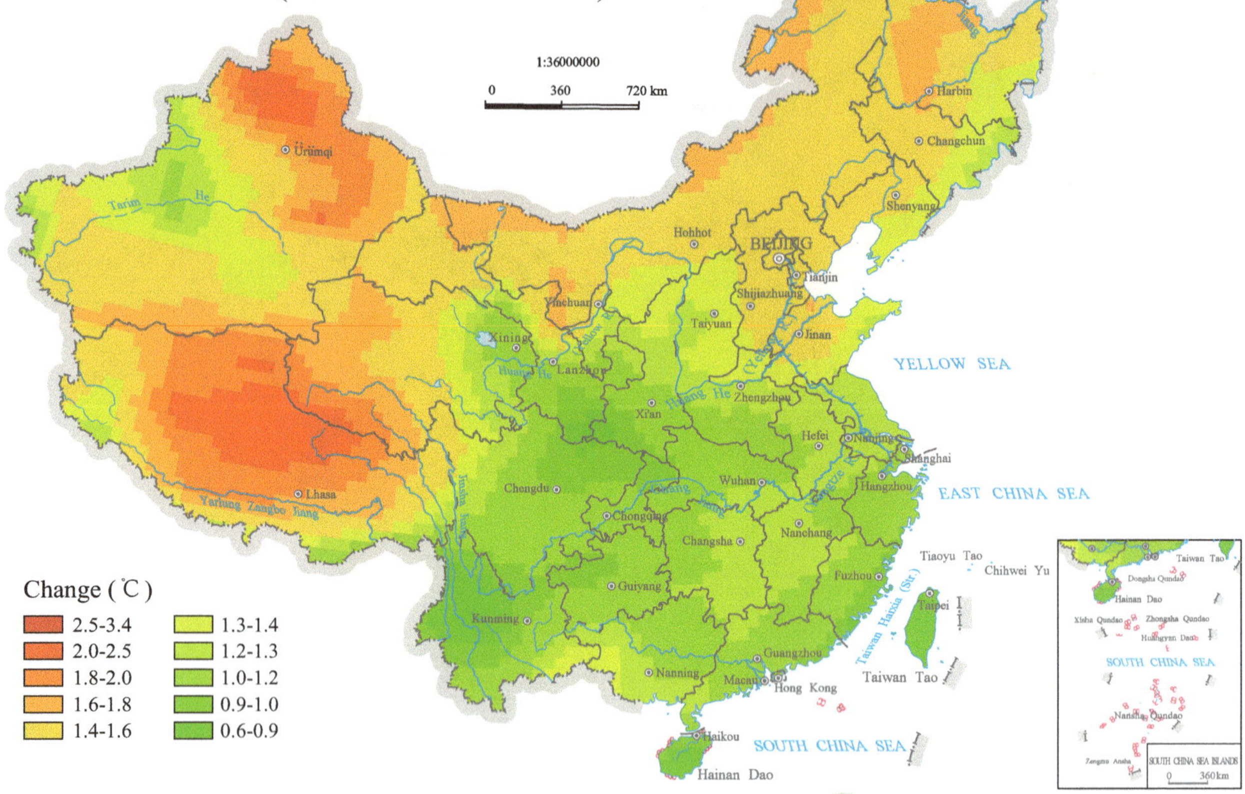

Intermodel Spread in Changes in Annual Precipitation for RCP8.5 (1981-2010 to 2071-2100)

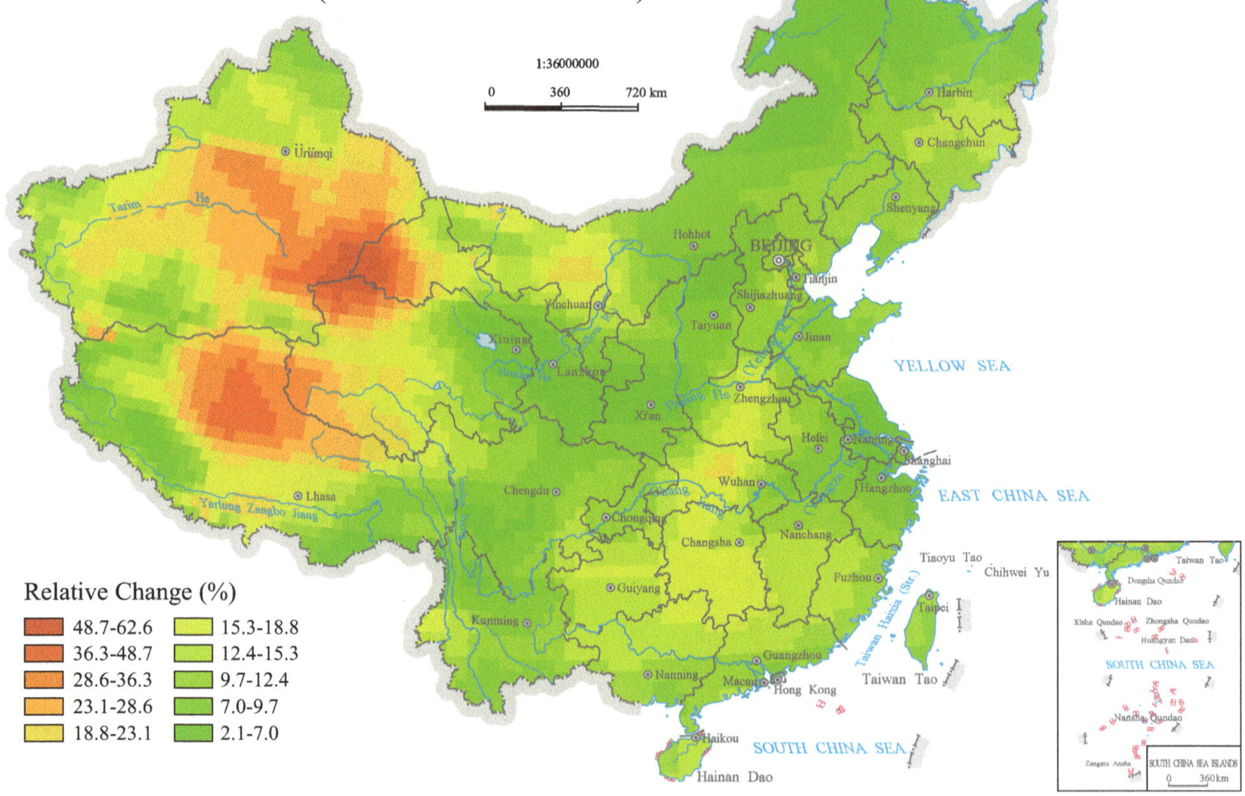

References

Dong, W., Liu, Z., Liao, H., Tang, Q., & Li, X. (2015). New climate and socio-economic scenarios for assessing global human health challenges due to heat risk. *Climatic Change, 130*(4), 505–518.

He, Y., Ye, J., & Yang, X. (2015). Analysis of the spatio-temporal patterns of dry and wet conditions in the Huai River Basin using the standardized precipitation index. *Atmospheric Research, 166*, 120–128.

Hempel, S., Frieler, K., Warszawski, L., Schewe, J., & Piontek, F. (2013). A trend-preserving bias correction: The ISI-MIP approach. *Earth System Dynamics, 4*(2), 219–236.

IPCC (Intergovernmental Panel on Climate Change). (2013). *Climate change 2013: The physical science basis*. Cambridge, England.

Lobell, D., Bonfils, C., & Duffy, P. (2007). Climate change uncertainty for daily minimum and maximum temperatures: A model inter-comparison. *Geophysical Research Letters, 34*(5), 1–5.

Min, S., Zhang, X., Zwiers, F., & Hegerl, G. (2011). Human contribution to more-intense precipitation extremes. *Nature, 470* (7334), 378–381.

Piani, C., Weedon, G., Best, M., Gomes, S., Viterbo, P., Hagemann, S., et al. (2010). Statistical bias correction of global simulated daily precipitation and temperature for the application of hydrological models. *Journal of Hydrology, 395*(3), 199–215.

Stefan, H., Cui, C., Jan, O. H., Jens, H., Dieter, G., & Claudio, P. (2011). Impact of a statistical bias correction on the projected hydrological changes obtained from three GCMs and two hydrology models. *Journal of Hydrometeorology, 12*(4), 556–578.

Warszawski, L., Frieler, K., Huber, V., Piontek, F., Serdeczny, O., & Schewe, J. (2014). The inter-sectoral impact model intercomparison projection (ISI-MIP): Project framework. *Proceedings of the National Academy of Sciences of the United States of America, 111*, 3228–3232.

Yin, J., Yan, D., Yang, Z., Yuan, Z., Yuan, Y., & Zhang, C. (2016). Projection of extreme precipitation in the context of climate change in Huang-Huai-Hai region, China. *Journal of Earth System Science, 125*(2), 417–429.

3 Heat Health Risks

Xinchuang Xu, Quansheng Ge and Xingcai Liu

3.1 Background

High temperature is a natural hazard linking with excess mortality and morbidity (Kovats and Hajat 2008). High temperature can induce heat stroke, communicable disease, cardiovascular disease, and respiratory disease (Basu and Ostro 2008; Hajat et al. 2005). Individuals with limited mobility due to physical disability, or those who are medically fragile will be subject to greater risks (Díaz et al. 2006). The influence of high temperature on mortality has been found in China during the past. Additionally, the frequency and intensity of high temperature are projected to increase along with the global warming in the future in China. Therefore, the risks that high temperature imposes to human health have recently been a hot topic (Kovats and Hajat 2008). Using five models data, we assessed risks of mortality in high temperature under four future emission scenarios (RCP2.6, RCP4.5, RCP6.0, and RCP8.5) (Fig. 3.1).

3.2 Method

3.2.1 High Temperature Days and Heat Waves

The high temperature day is defined as the day with temperature above 35 °C. The heat wave is defined as the weather that continues for at least three days, during which the maximum daily temperature is above the temperature of 95th percentile in the period of 1981–2010. Another criterion for heat waves is that the daily temperature does not fall below 32 °C. In this chapter, the mean number of days with temperature exceeding 35 and 40 °C, heat waves, and the maximum temperature of a year was calculated and mapped.

3.2.2 Population Vulnerability

This study uses the mortality increment (M_i) in each degree centigrade above the threshold temperature to represent the population vulnerability on high temperature. The threshold temperatures and M_i were obtained from relationships between mortality and daily highest temperature in summer of China. The model between M_i and threshold temperature was made through following steps (Fig. 3.1).

Firstly, we searched literatures on the mortality in high temperature by the keywords such as "weather," "temperature," "mortality," and "heat" in the major Chinese and English databases. Secondly, we selected the literatures, which studied the relationship between population mortality and the daily maximum temperature in summer, presenting a mortality threshold temperature or a curve on mortality and high temperature to judge the mortality threshold temperature and its increment value for each 1 °C above the threshold temperature. The relation as reported in the literatures was summarized in Table 3.1.

Based on Table 3.1, the relationship curves between M_i and threshold temperatures is built, and the model equation is given in Eq. 3.1

$$M_i = 5179210 - 794104.82539 \times T_v + 47413.08018 \times T_v^2 - 1431.3192 \times T_v^3 + 21.571523 \times T_v^4 - 0.1298413 \times T_v^5 \quad (3.1)$$

where M_i is the mortality increment for each 1 °C above the threshold temperature, T_v refers to the threshold temperature

X. Xu (✉) · Q. Ge
Key Laboratory of Land Surface Pattern and Simulation, Institute of Geographic Sciences and Natural Resources Research, Chinese Academy of Sciences, Beijing, 100101, China
e-mail: xxch8009@163.com

X. Xu
The School of Resources Environment Science and Technology, Hubei University of Science and Technology, Xianning, 437100, China

X. Liu
Key Laboratory of Water Cycle and Related Land Surface Processes, Institute of Geographic Sciences and Natural Resources Research, Chinese Academy of Sciences, Beijing, 100101, China

Q. Tang and Q. Ge (eds.), *Atlas of Environmental Risks Facing China Under Climate Change*, IHDP/Future Earth-Integrated Risk Governance Project Series, DOI 10.1007/978-981-10-4199-0_3

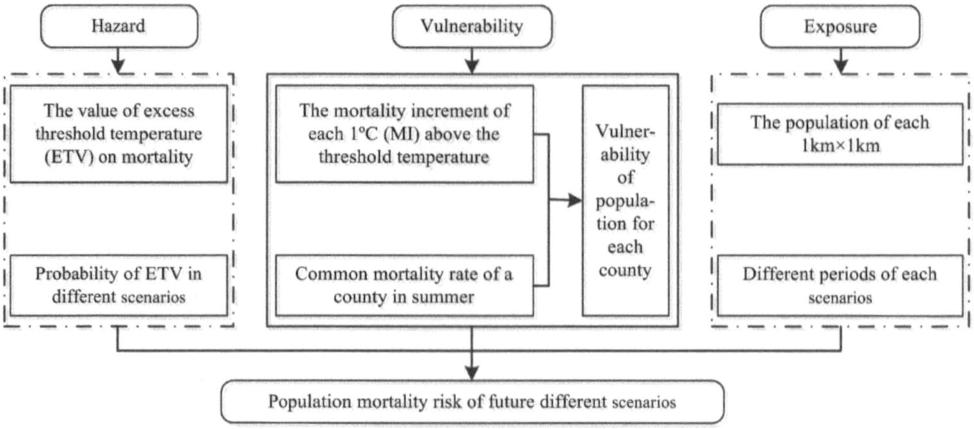

Fig. 3.1 Flow chart of assessing the risk on mortality in high temperature in the future scenarios

Table 3.1 The relation between mortality and maximum daily temperature in summer

City	Period	Threshold temperature (°C)	M_i (%)	Data characters	Literatures
Guangzhou	2000–2004	34	3	T[a]	Qu et al. (2009)
Guangzhou	1979–1989	34	7.5	C/T[a]	Wang and Lei (1997), Tan (1994)
Guangzhou	2006–2010	34	4.8	T	Zeng (2013)
Zhuhai	2006–2010	34	4.6	T	Zeng (2013)
Nanxiong	2006–2010	34	15.4	T	Zeng (2013)
Haerbin	2002–2010	30	7.5	C/T	Li et al. (2014)
Suzhou	2005–2008	36	23.1	T	Lim et al. (2015)
Nanjing	1994–2003	36	25	C[a]	Li et al. (2005)
Shenzhen	2004–2010	33	4	T	Li et al. (2014)
Chongqing	2011–2012	34	5.5	C/T	Li et al. (2014)
Wuhan	1998–2008	36	30	T	Yang et al. (2013)
Shanghai	1979–1989	34	23	C/T	Tan (1994)
Changsha	2006–2009	37	24.89	T	Sun et al. (2012)
Wulumuqi	2006–2007	32	6.6	T	Lim et al. (2015)
Xi'an	2004–2008	35	0.5	T	Lim et al. (2015)
Lanzhou	2004–2008	33	2.1	T	Lim et al. (2015)
Tianjing	2005–2008	35	7	T	Lim et al. (2015)

[a]C refers to the curve, T refers to the threshold, C/T refers that both curve and threshold were given in the literature. M_i is the percent of mortality increment

on population mortality, and the range value T_v is between 30 and 38 °C. The R^2 of the fitting curve is 0.53, and the p is less than 0.01.

Based on Table 3.1, the relationship curves between the threshold temperature and mean temperature was also fitted for cities in summer during the period of 1961–2010. The threshold temperature was calculated in China according to the relation model of Eq. 3.2.

$$T_v = 0.5664\bar{t} + 19.52 \tag{3.2}$$

where T_v refers to the threshold temperature, $\bar{t}$ is the mean temperature of the city in summer. The R^2 of the fitting curve is 0.60, and the p is less than 0.01.

3.2.3 Population Mortality Hazard

This study employed the value of exceeding threshold temperature to represent the hazard intensity for population in high temperature. The hazard was calculated according to Eq. 3.3.

$$H_d = \sum_{i=1}^{7} (T_i - T_v) \times P_i \quad (3.3)$$

where H_d refers to the heat hazard index, T_i represents the temperature above the threshold temperature on mortality (°C), T_v refers to the threshold temperature of population mortality (°C), and P_i refers to probability caused by the daily maximum temperature above the threshold. The hazard indices were divided into ten ranks from one (low) to ten (high) by the natural break method.

3.2.4 Population Exposure

In this study, the population of every 1 × 1 km grid cell in 2000s, 2030s, 2060s, and 2090s was used to represent the exposure of population suffered from high temperature. This data was obtained from the Greenhouse Gas Initiative (GGI) of the International Institute for Applied Systems Analysis (IIASA: http://www.iiasa.ac.at). The data include projections under three scenarios, namely, A2, B1, and B2. We selected the data of A2 and B1 to represent the scenarios of RCP8.5 and RCP2.6.

3.2.5 Population Mortality Risk

Based on the risk theory, the mortality risk for each county was estimated using Eq. 3.4.

$$POP_d = 365 \times H_d \times M_i \times \frac{E \times D_r}{92} \quad (3.4)$$

where POP_d refers to the population mortality risk of a year in different periods (2011–2040, 2041–2070, and 2071–2099). M_i is the mortality increment for each 1.0 °C above the threshold temperature (%). D_r represents the population mortality rate during the summer of each county. E is the population exposure to high temperature. The mortality risk is divided into ten ranks one (low) to ten (high) by the natural break method.

3.3 Results

3.3.1 Spatiotemporal Distribution of High Temperature and Heat Waves

The number of days exceeding 35 °C, 40 °C, and heat waves are all increasing during the different periods under the RCP2.6, RCP4.5, RCP6.0, and RCP8.5 scenarios. When compared with the pre-2040s increase, a more speedy increase is projected after the 2040s.

The values of all variables on high temperature are almost all higher under the RCP8.5 than the RCP2.6 during the different periods. The areas of the north, center, and northwest of China are severely suffered from high temperature.

3.3.2 Spatiotemporal Distribution of Population Mortality Hazard

The mortality hazard on high temperature is the highest in the RCP8.5, and the lowest in the RCP2.6 in China. The increment trend of mortality hazard will be developed from north to south in China during the year 2011–2100. Under all the four scenarios, the highest hazard occurs in the northwest followed by the north and the mid and lower reaches of the Yangtze River.

3.3.3 Spatiotemporal Distribution of Population Mortality Risk

The population mortality risk is the highest in the middle of North China under the four emission scenarios. The area of the mid-lower reaches of the Yangtze River is the area with the second highest risk. The risks of other areas from high to low are southwest, southeast, northeast, and northwest of China.

3.3.4 Uncertainty Analysis in High Temperature Over 2071–2099 for RCP8.5

The various uncertainty of high temperature projection is analyzed by intermodal spread of temperature changes. The projected numbers of days with temperature above 35 °C have high uncertainty in southeast China. The uncertainty of projected number of days with temperature above 40 °C is largest in the middle reaches of the Yangtze River, while the uncertainty is lowest in northeast China. The uncertainty of the maximum temperature in a year is higher in northwest China than those in other regions. For the heat wave projection, the high uncertainty mainly distributes in the southeast and north of China. The higher uncertainty of population mortality hazard is found in the northwest and the middle reaches of the Yangtze River, while the northeast of China has lower uncertainty. The uncertainty of population mortality risk is high in the north China and the middle reaches of the Yangtze River.

3.4 Maps

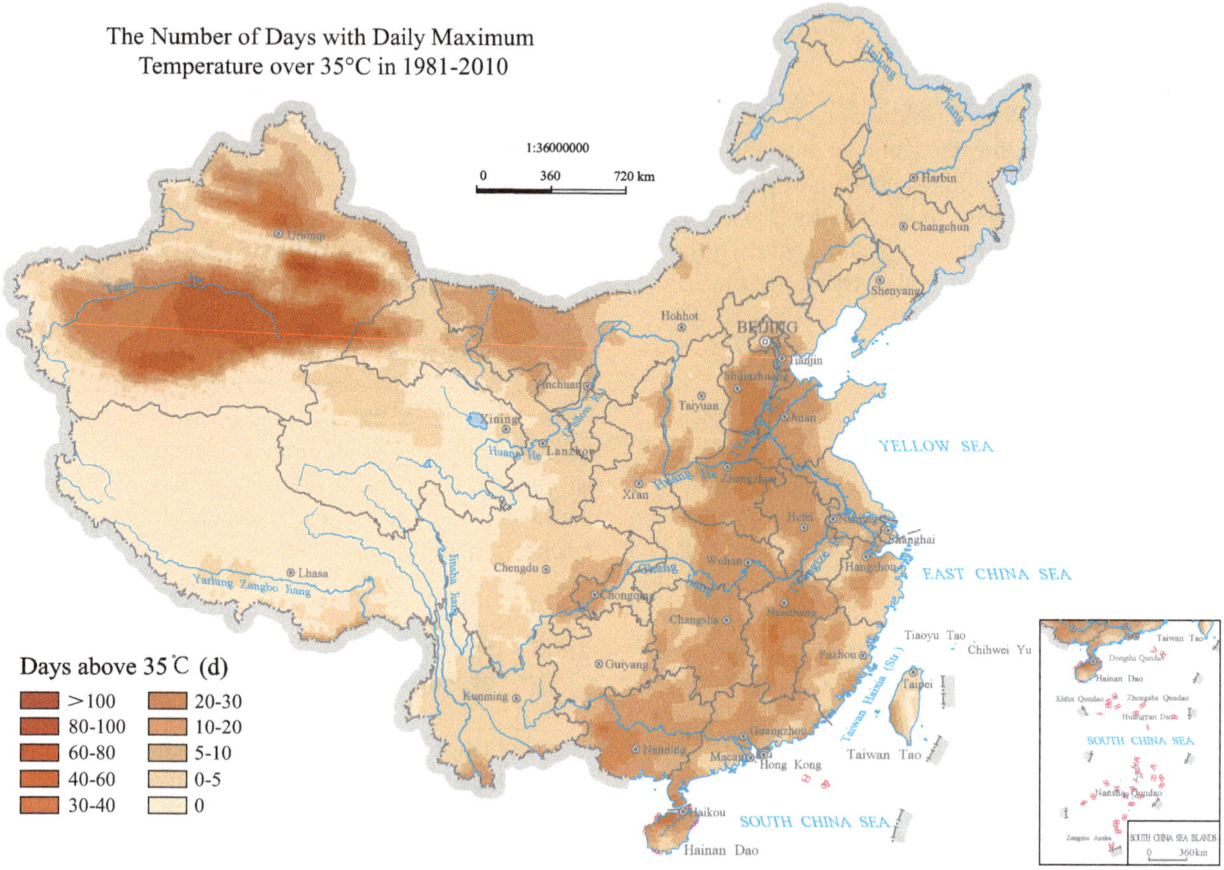
The Number of Days with Daily Maximum Temperature over 35°C in 1981-2010
1:36000000
0 360 720 km
Days above 35℃ (d)
>100
80-100
60-80
40-60
30-40
20-30
10-20
5-10
0-5
0
Harbin
Changchun
Shenyang
Hohhot
BEIJING
Tianjin
Shijiazhuang
Taiyuan
Jinan
Xi'an
Zhengzhou
Hefei
Nanjing
Shanghai
Hangzhou
Wuhan
Nanchang
Changsha
Fuzhou
Taipei
Guangzhou
Hong Kong
Macao
Nanning
Haikou
Hainan Dao
Guiyang
Kunming
Chengdu
Chongqing
Lanzhou
Xining
Yinchuan
Urumqi
Lhasa
Yarlung Zangbo Jiang
Jinsha Jiang
Huang He
YELLOW SEA
EAST CHINA SEA
SOUTH CHINA SEA
Taiwan Tao
Tiaoyu Tao
Chihwei Yu

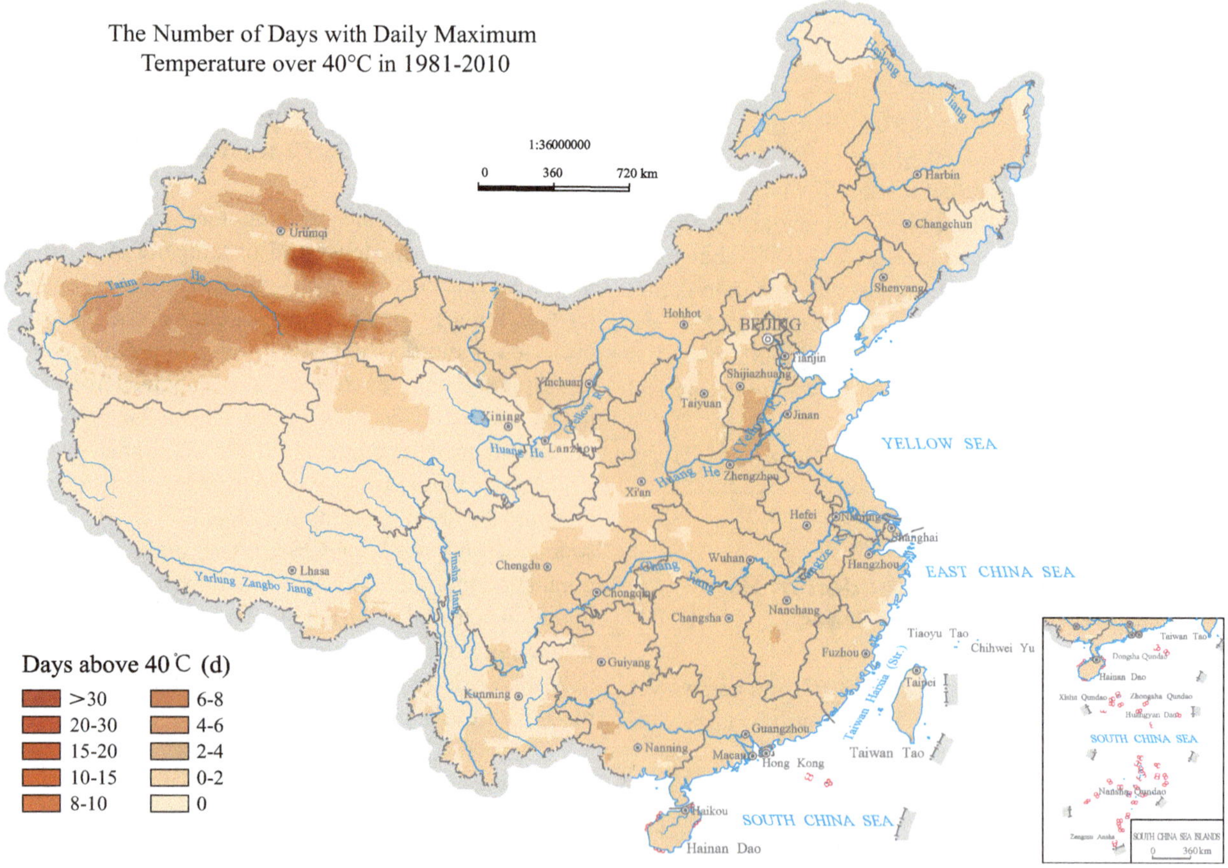
The Number of Days with Daily Maximum Temperature over 40°C in 1981-2010
1:36000000
0 360 720 km
Days above 40℃ (d)
>30
20-30
15-20
10-15
8-10
6-8
4-6
2-4
0-2
0
Harbin
Changchun
Shenyang
Hohhot
BEIJING
Tianjin
Shijiazhuang
Taiyuan
Jinan
Xi'an
Zhengzhou
Hefei
Nanjing
Shanghai
Hangzhou
Wuhan
Nanchang
Changsha
Fuzhou
Taipei
Guangzhou
Hong Kong
Macao
Nanning
Haikou
Hainan Dao
Guiyang
Kunming
Chengdu
Chongqing
Lanzhou
Xining
Yinchuan
Urumqi
Lhasa
Yarlung Zangbo Jiang
Jinsha Jiang
Huang He
YELLOW SEA
EAST CHINA SEA
SOUTH CHINA SEA
Taiwan Tao
Tiaoyu Tao
Chihwei Yu

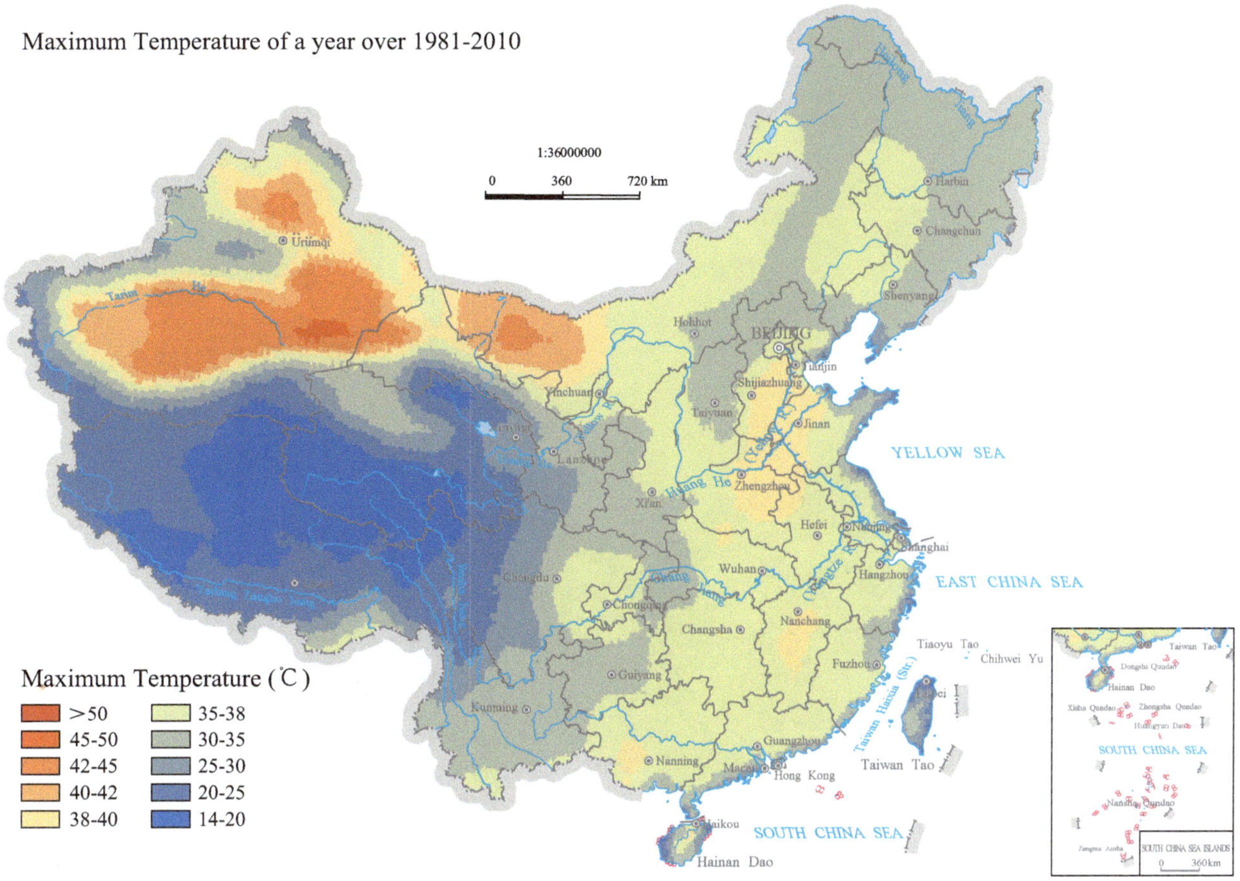
Maximum Temperature of a year over 1981-2010
1:36000000
0 360 720 km
Maximum Temperature (℃)
>50
45-50
42-45
40-42
38-40
35-38
30-35
25-30
20-25
14-20
Harbin
Changchun
Shenyang
Urümqi
Hohhot
BEIJING
Tianjin
Shijiazhuang
Taiyuan
Jinan
Yinchuan
Lanzhou
Xi'an
Zhengzhou
Hefei
Nanjing
Shanghai
Wuhan
Hangzhou
Chengdu
Chongqing
Changsha
Nanchang
Fuzhou
Guiyang
Kunming
Nanning
Guangzhou
Macau
Hong Kong
Haikou
Hainan Dao
Taipei
Taiwan Tao
Tiaoyu Tao
Chihwei Yu
YELLOW SEA
EAST CHINA SEA
SOUTH CHINA SEA

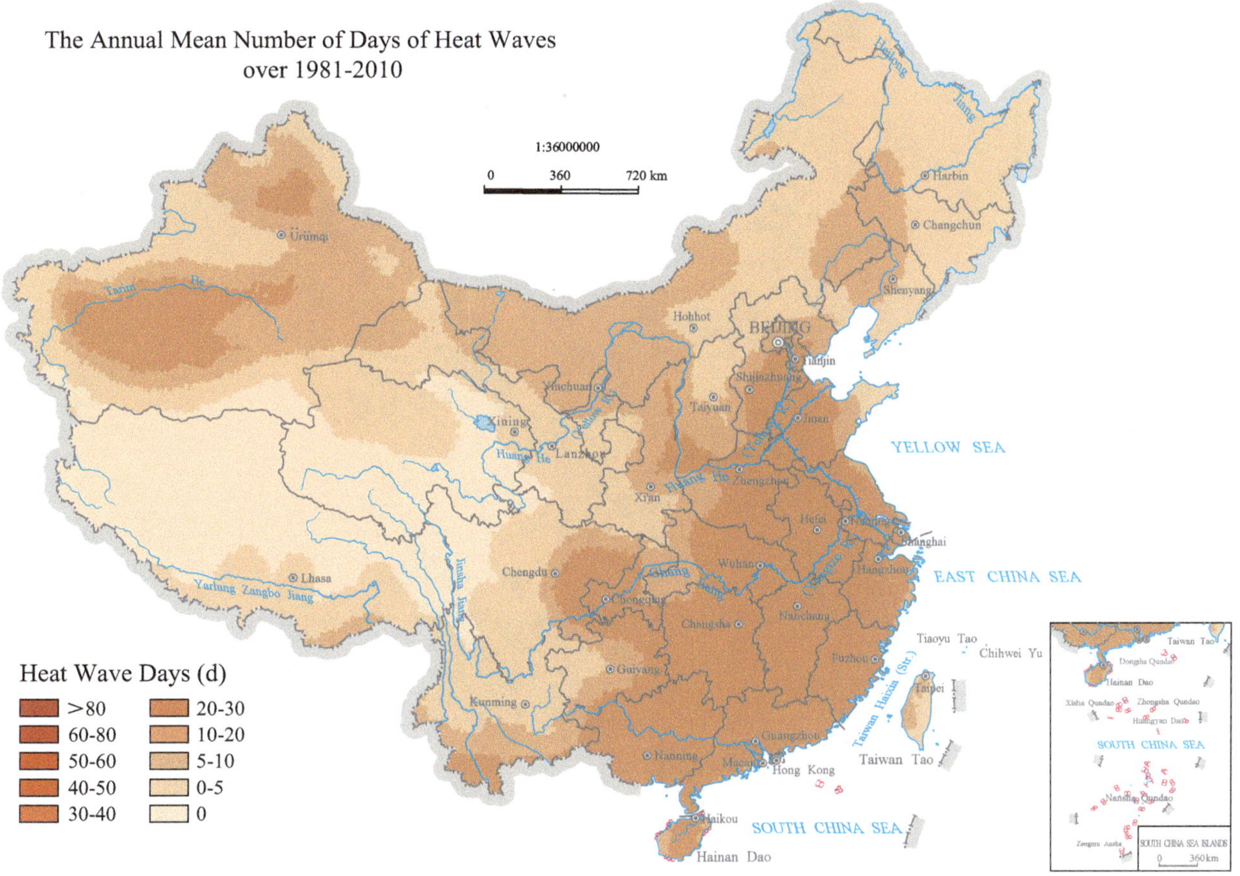
The Annual Mean Number of Days of Heat Waves over 1981-2010
1:36000000
0 360 720 km
Heat Wave Days (d)
>80
60-80
50-60
40-50
30-40
20-30
10-20
5-10
0-5
0
Harbin
Changchun
Shenyang
Urümqi
Hohhot
BEIJING
Tianjin
Shijiazhuang
Taiyuan
Jinan
Yinchuan
Xining
Lanzhou
Xi'an
Zhengzhou
Hefei
Shanghai
Wuhan
Hangzhou
Lhasa
Yarlung Zangbo Jiang
Chengdu
Chongqing
Changsha
Nanchang
Fuzhou
Guiyang
Kunming
Nanning
Guangzhou
Macau
Hong Kong
Haikou
Hainan Dao
Taipei
Taiwan Tao
Tiaoyu Tao
Chihwei Yu
YELLOW SEA
EAST CHINA SEA
SOUTH CHINA SEA

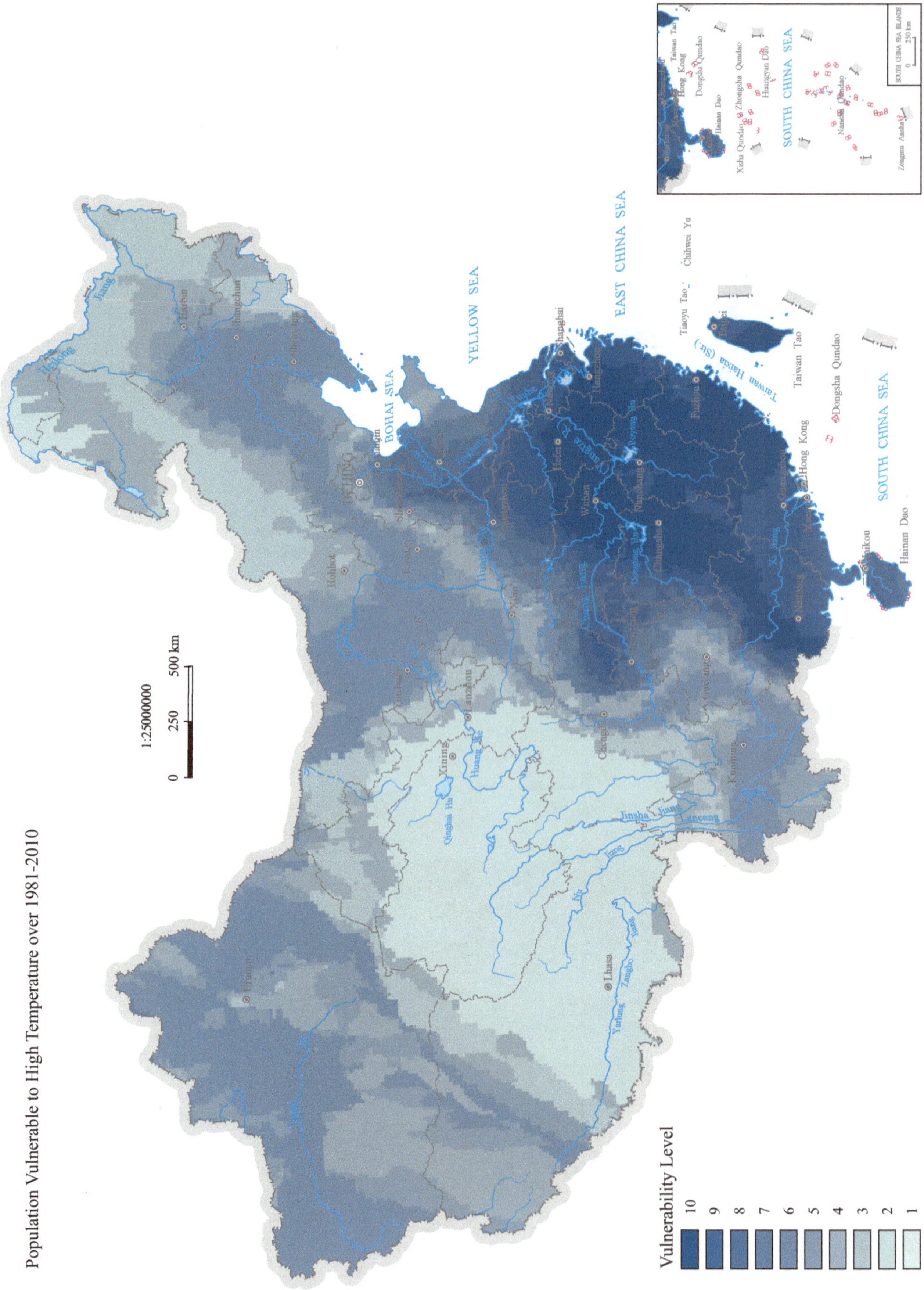
Population Vulnerable to High Temperature over 1981-2010
Vulnerability Level
10
9
8
7
6
5
4
3
2
1
1:25000000
0 250 500 km
BOHAI SEA
YELLOW SEA
EAST CHINA SEA
SOUTH CHINA SEA
BEIJING
Shanghai
Hangzhou
Fuzhou
Hefei
Nanchang
Wuhan
Changsha
Nanning
Haikou
Hong Kong
Taibei
Taiwan Tao
Hainan Dao
Dongsha Qundao
Taiwan Haixia (Str.)
Diaoyu Dao
Chiwei Yu
Hohhot
Lanzhou
Xining
Qinghai Hu
Chengdu
Kunming
Lhasa
Urumqi
Harbin
Changchun
Shenyang
Heilong Jiang
Jinsha Jiang
Lancang Jiang
Nu Jiang
Yarlung Zangbo Jiang
Huang He
Xi Jiang
Yangtze Jiang
Xisha Qundao
Zhongsha Qundao
Huangyan Dao
Nansha Qundao
Zengmu Ansha
SOUTH CHINA SEA ISLANDS
0 250 km

The Number of Days with Daily Maximum Temperature over 35°C in 2011-2040 for RCP2.6

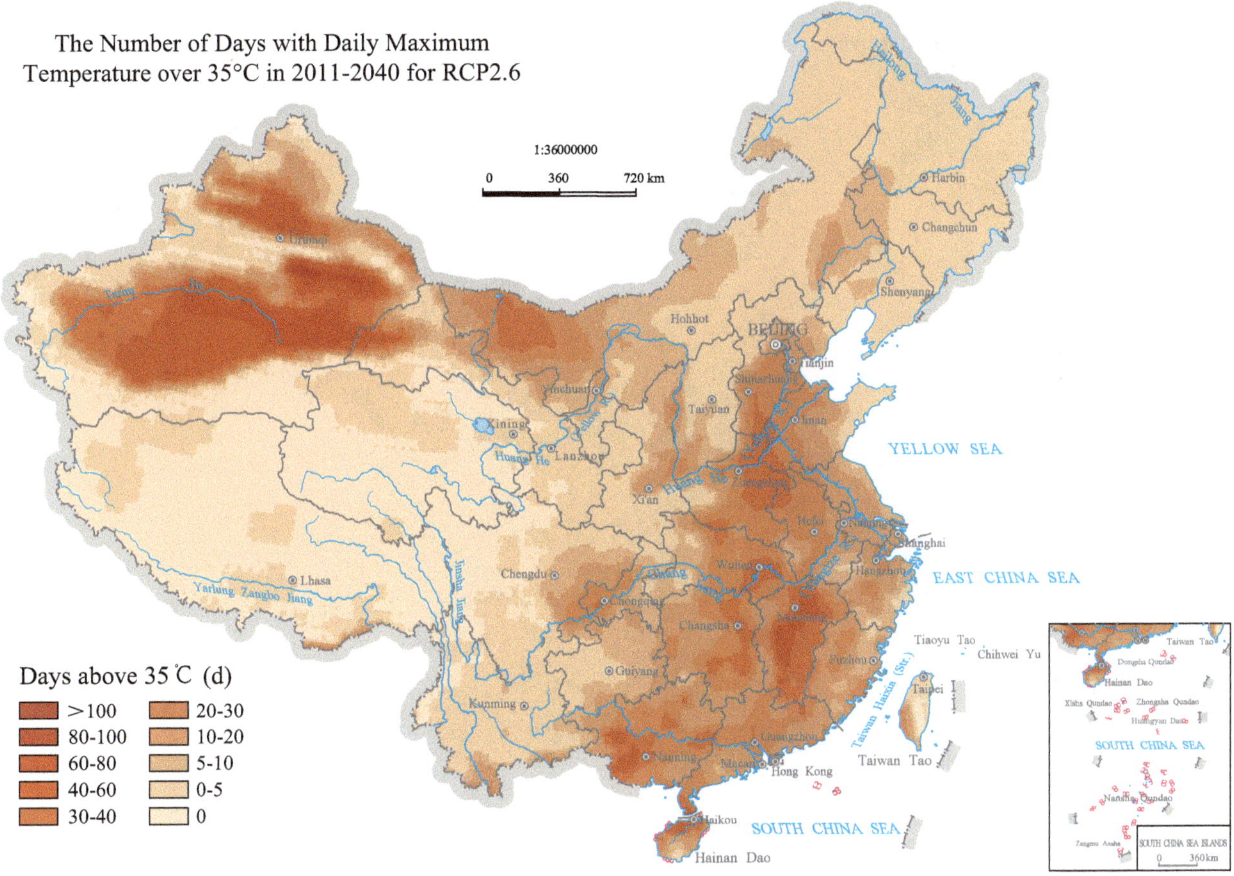

The Number of Days with Daily Maximum Temperature over 40°C in 2011-2040 for RCP2.6

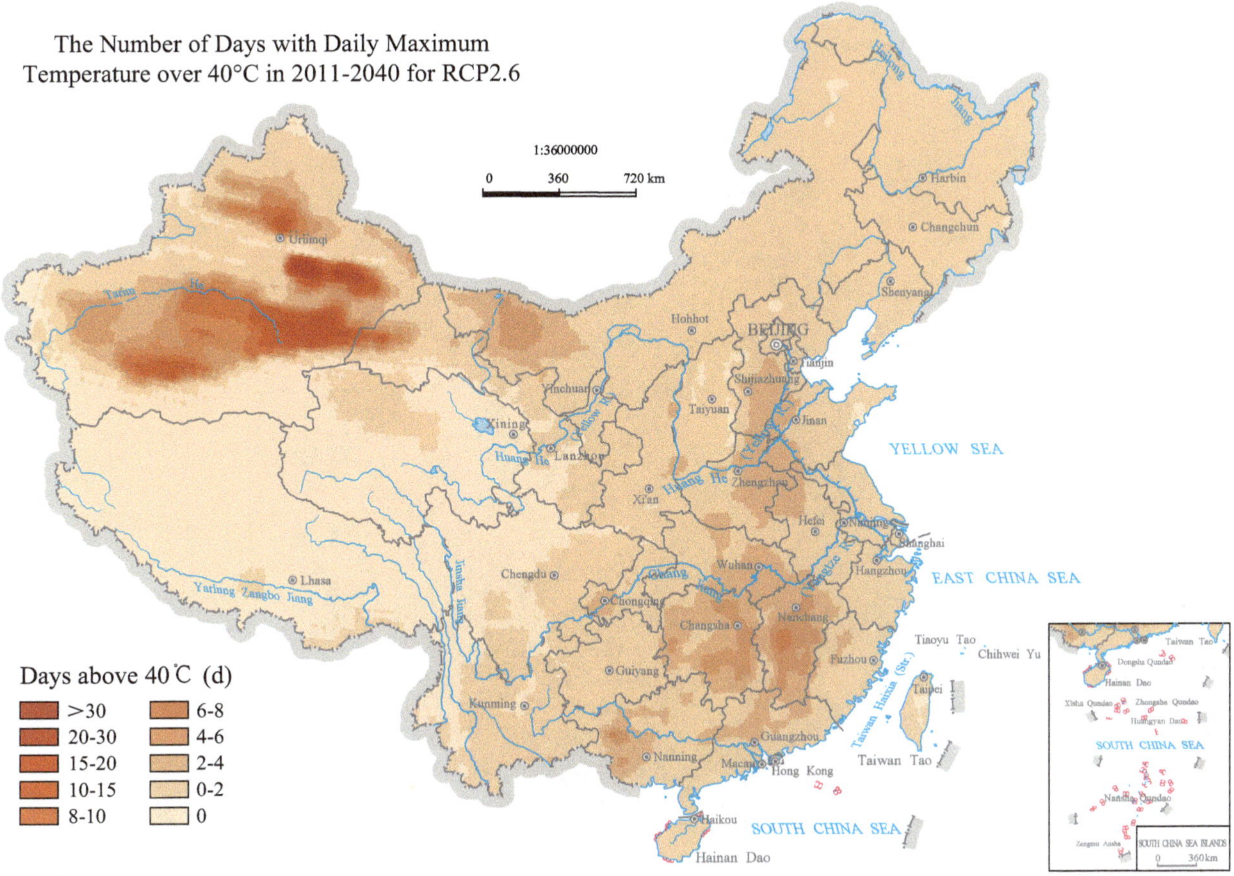

Maximum Temperature of a year over 2011-2040 for RCP2.6

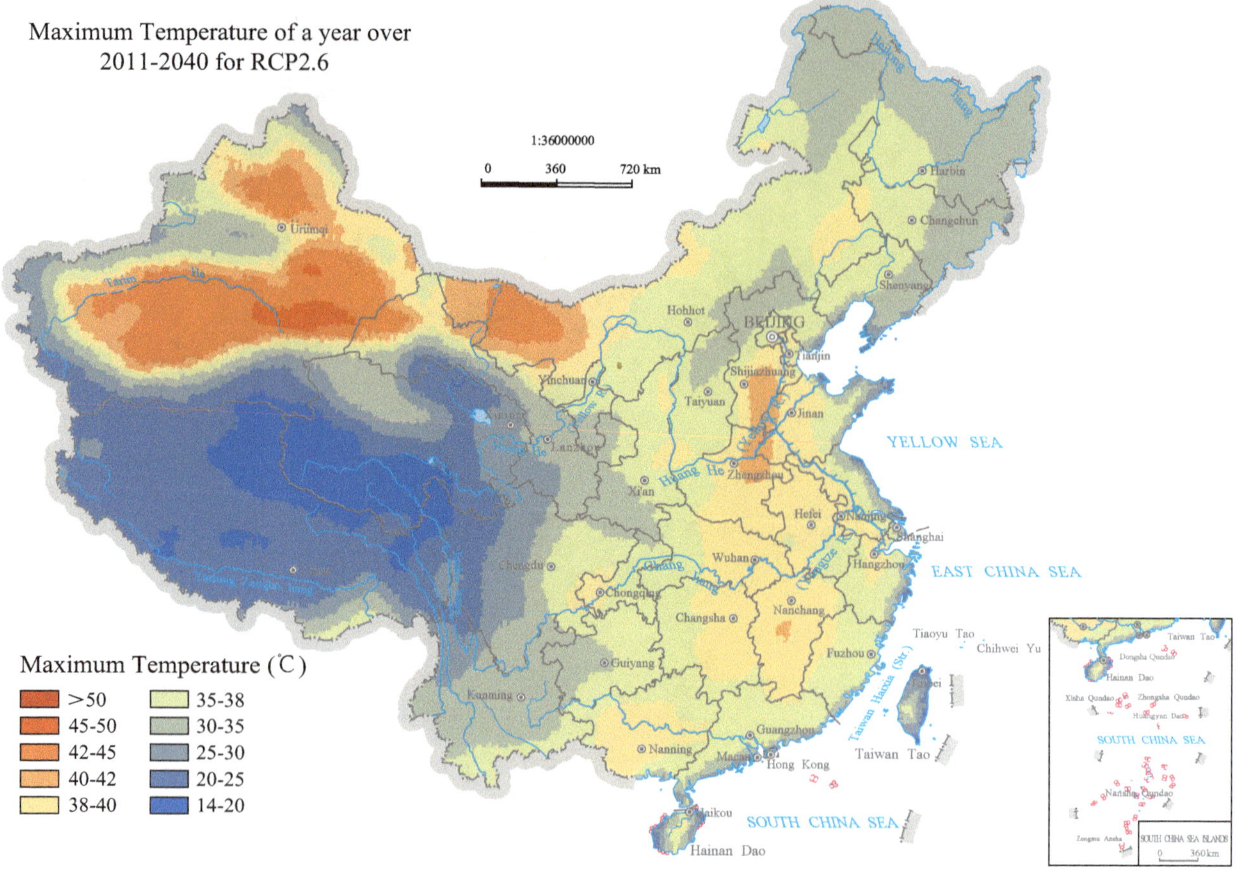

The Number of Days of Heat Waves over 2011-2040 for RCP2.6

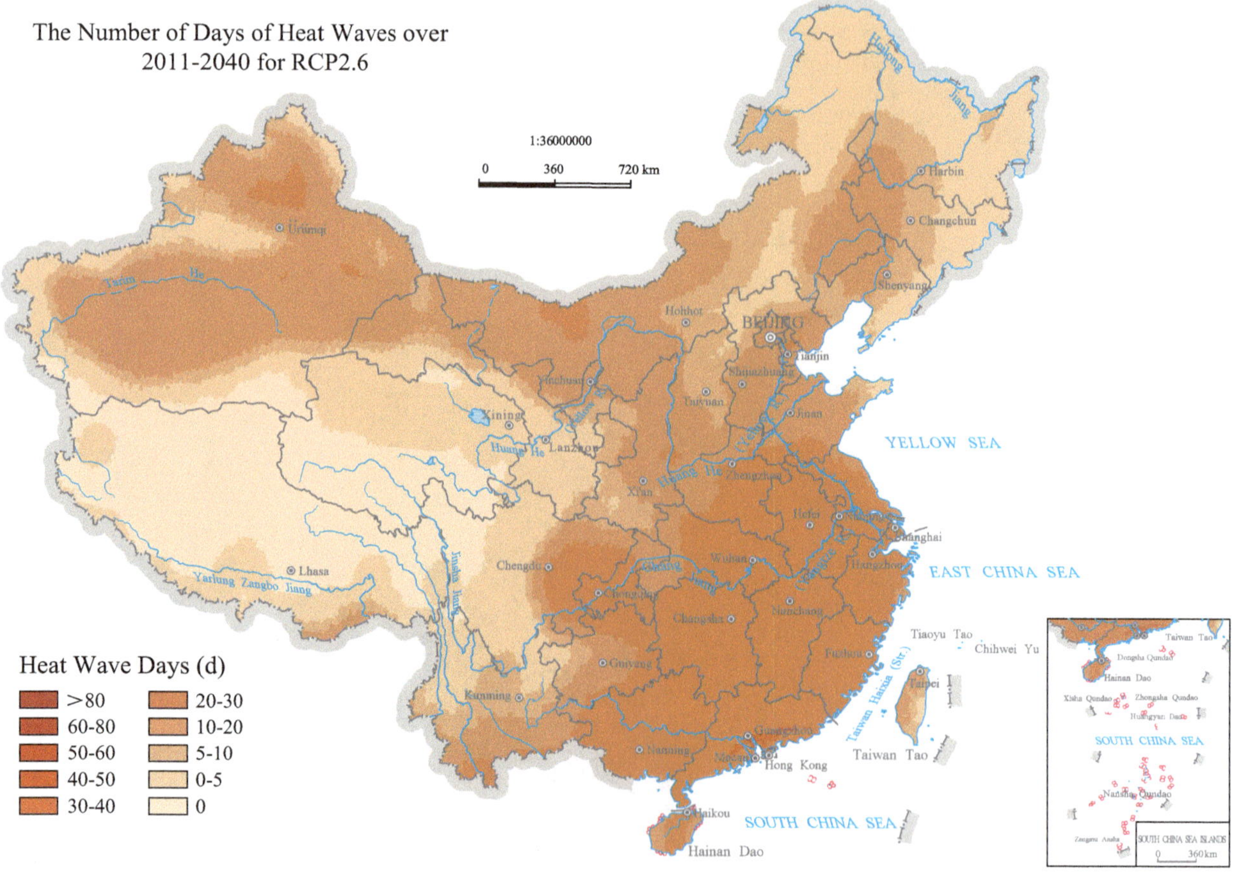

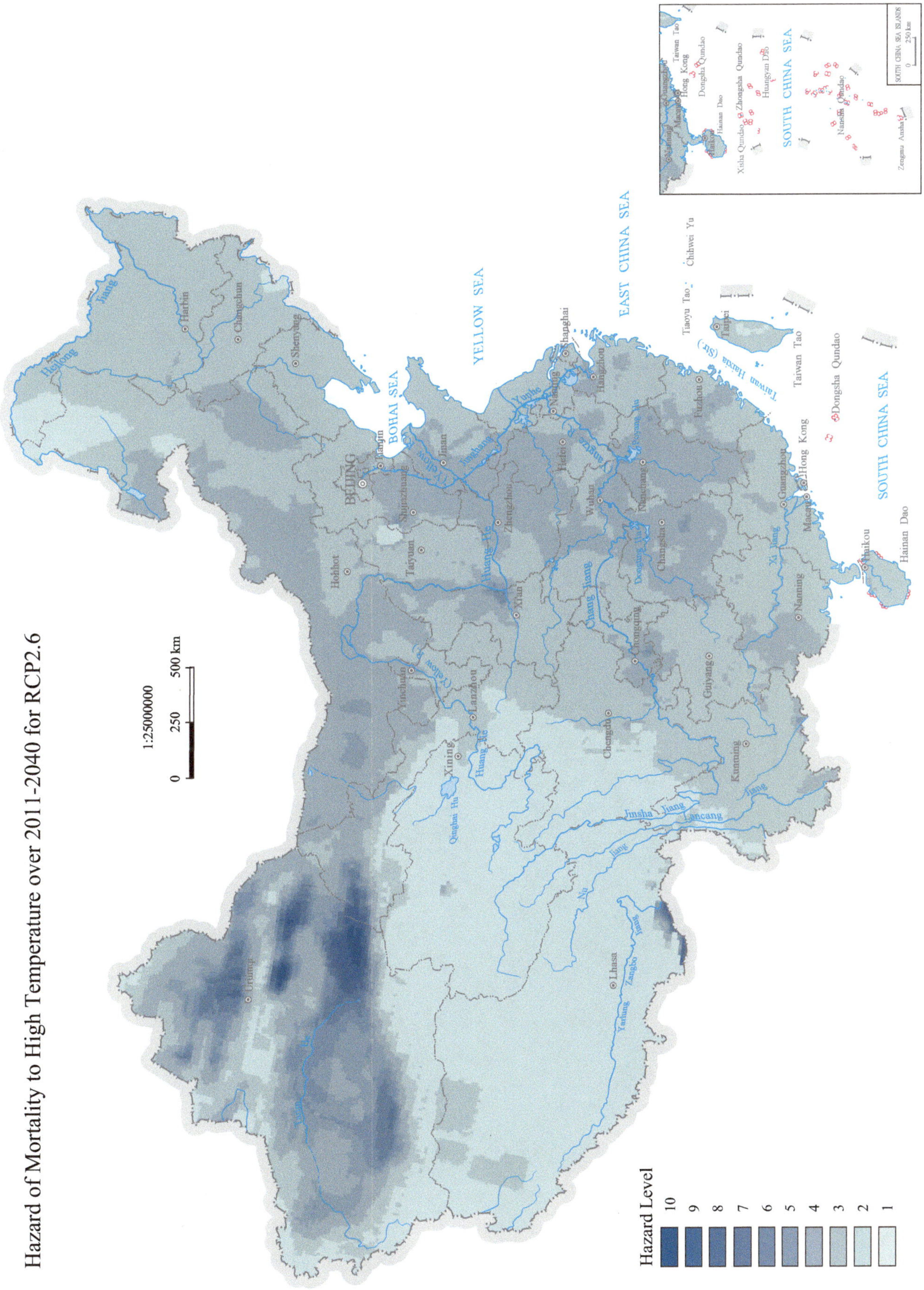
Hazard of Mortality to High Temperature over 2011-2040 for RCP2.6
1:25000000
0 250 500 km
Hazard Level
10
9
8
7
6
5
4
3
2
1
BEIJING
Harbin
Changchun
Shenyang
Tianjin
Jinan
Shijiazhuang
Taiyuan
Hohhot
Zhengzhou
Xi'an
Yinchuan
Lanzhou
Xining
Urumqi
Lhasa
Chengdu
Chongqing
Guiyang
Kunming
Nanning
Changsha
Wuhan
Nanchang
Hefei
Nanjing
Shanghai
Hangzhou
Fuzhou
Taipei
Guangzhou
Hong Kong
Macao
Haikou
Hainan Dao
BOHAI SEA
YELLOW SEA
EAST CHINA SEA
SOUTH CHINA SEA
Taiwan Tao
Dongsha Qundao
Diaoyu Dao
Chiwei Yu
Heilong Jiang
Huang He
Chang Jiang
Jinsha Jiang
Lancang Jiang
Nu Jiang
Yarlung Zangbo Jiang
Qinghai Hu
Xi Jiang
Dongting Hu
Poyang Hu
Taiwan Haixia
Xisha Qundao
Zhongsha Qundao
Nansha Qundao
Huangyan Dao
Zengmu Ansha
SOUTH CHINA SEA ISLANDS

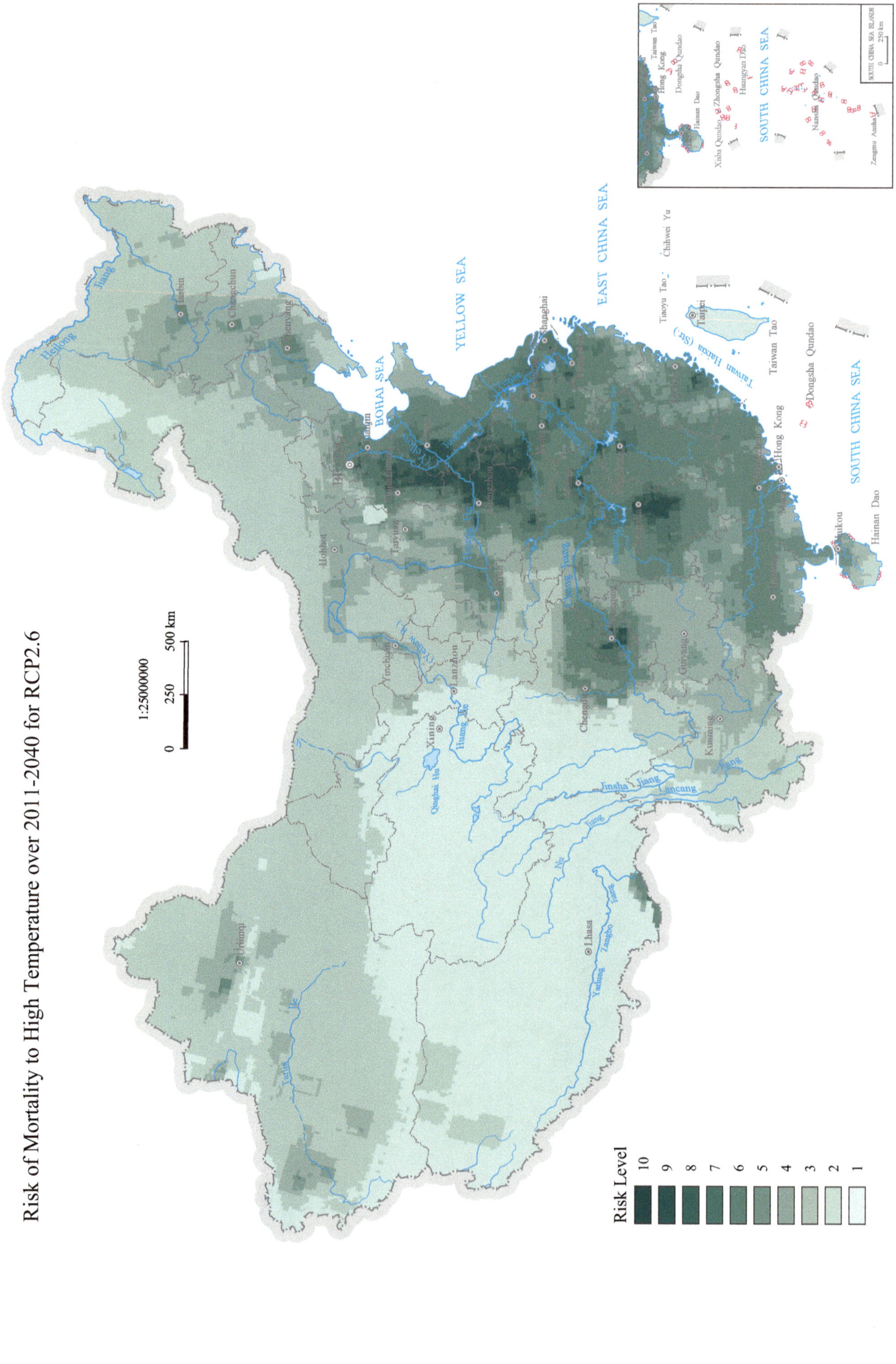

Risk of Mortality to High Temperature over 2011-2040 for RCP2.6
0 250 500 km
1:25000000
Risk Level
10
9
8
7
6
5
4
3
2
1
YELLOW SEA
EAST CHINA SEA
SOUTH CHINA SEA
BOHAI SEA
Shanghai
Harbin
Changchun
Beijing
Hohhot
Taiyuan
Yinchuan
Lanzhou
Xining
Chengdu
Guiyang
Kunming
Lhasa
Urumqi
Taipei
Hong Kong
Haikou
Hainan Dao
Taiwan Tao
Dongsha Qundao
Diaoyu Tao
Chihwei Yu
Heilong Jiang
Huang He
Qinghai Hu
Jinsha Jiang
Lancang Jiang
Nu Jiang
Yarlung Zangbo Jiang
Tarim He
Xisha Qundao
Zhongsha Qundao
Nansha Qundao
Huangyan Dao
SOUTH CHINA SEA ISLANDS

The Number of Days with Daily Maximum Temperature over 35°C in 2041-2070 for RCP2.6

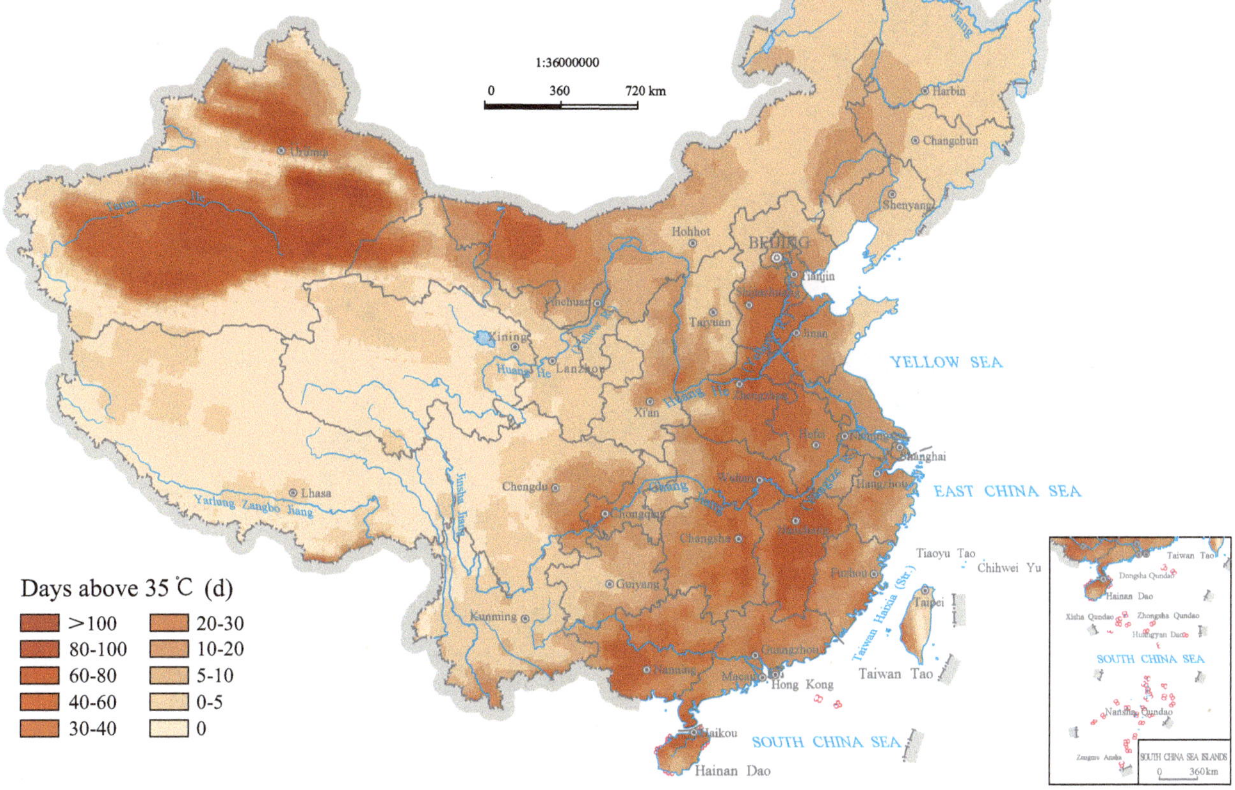

The Number of Days with Daily Maximum Temperature over 40°C in 2041-2070 for RCP2.6

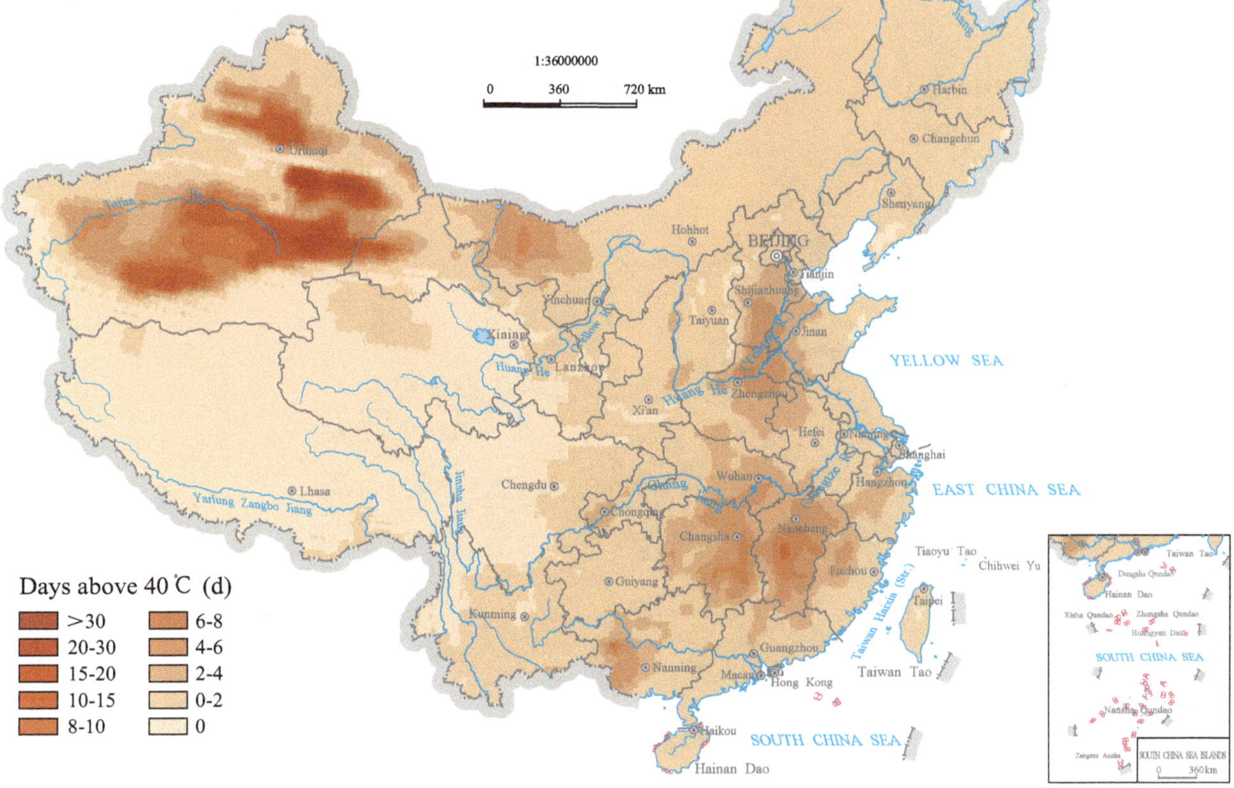

Maximum Temperature of a year over 2041-2070 for RCP2.6

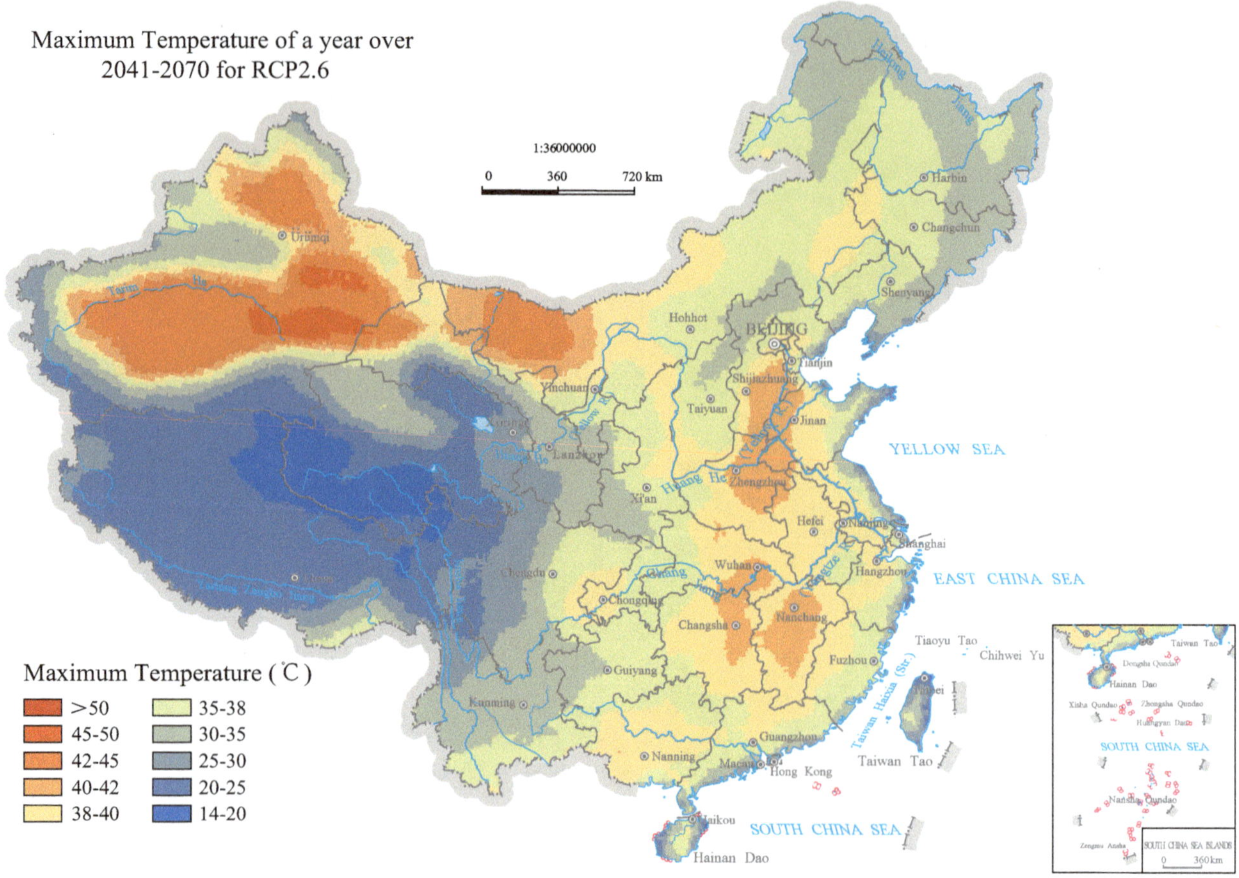

The Number of Days of Heat Waves over 2041-2070 for RCP2.6

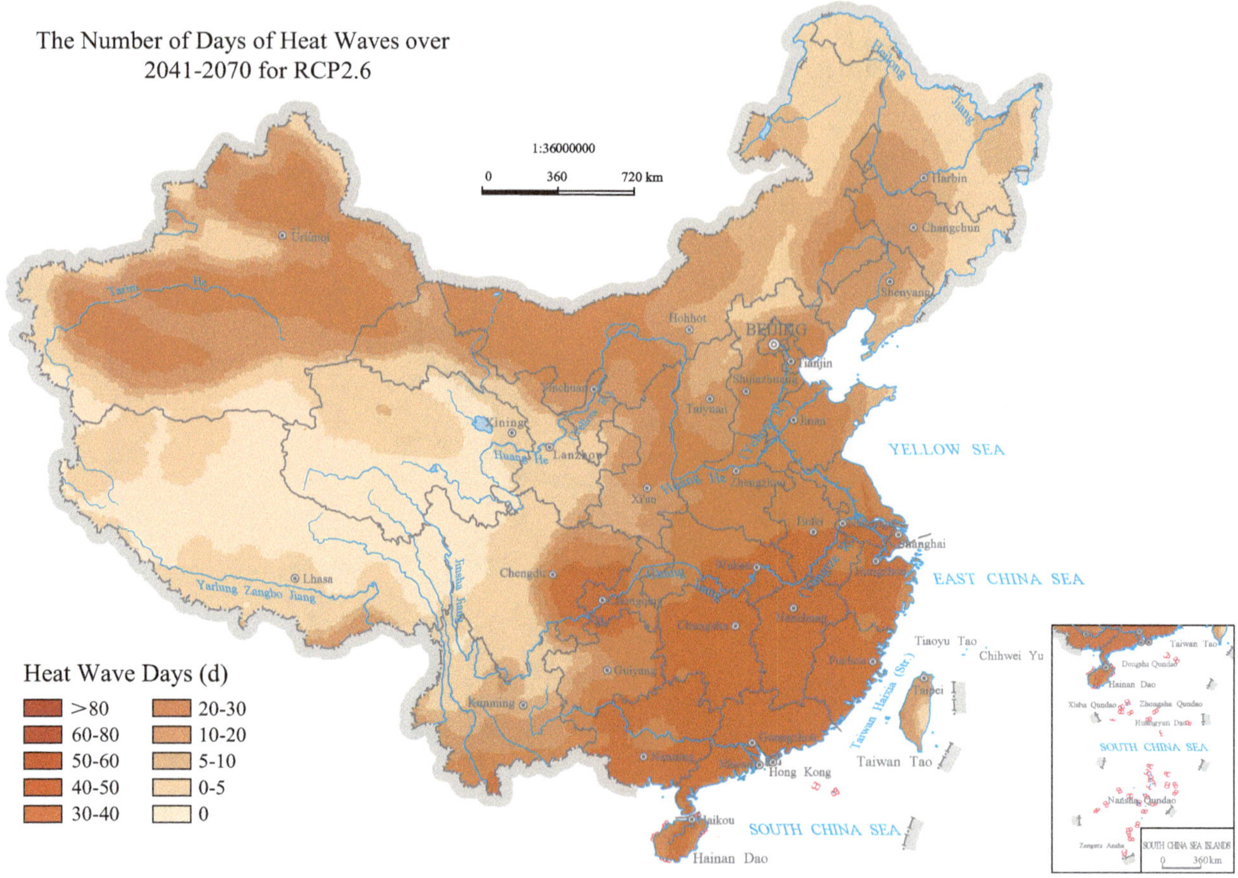

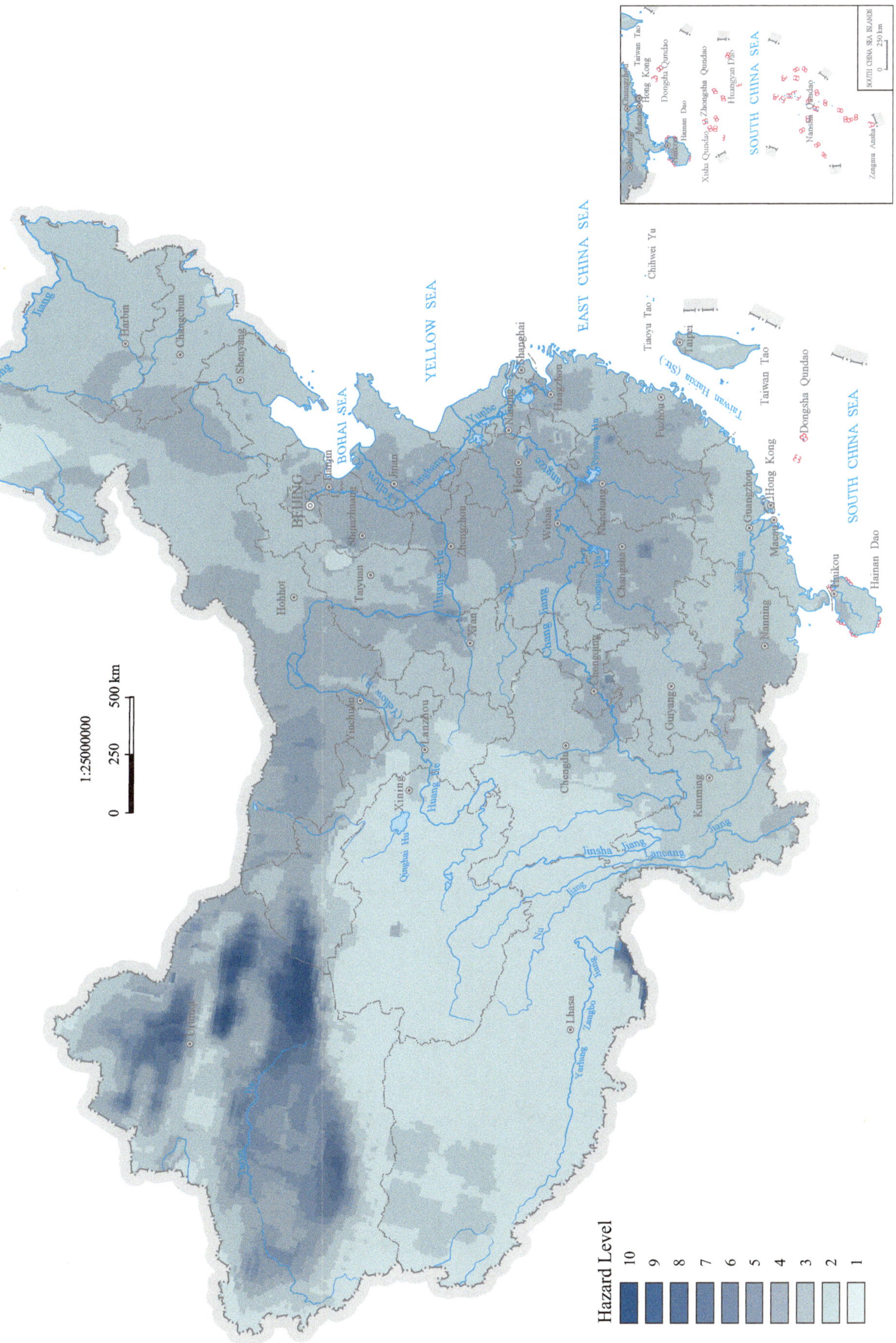

Hazard of Mortality to High Temperature over 2041-2070 for RCP2.6

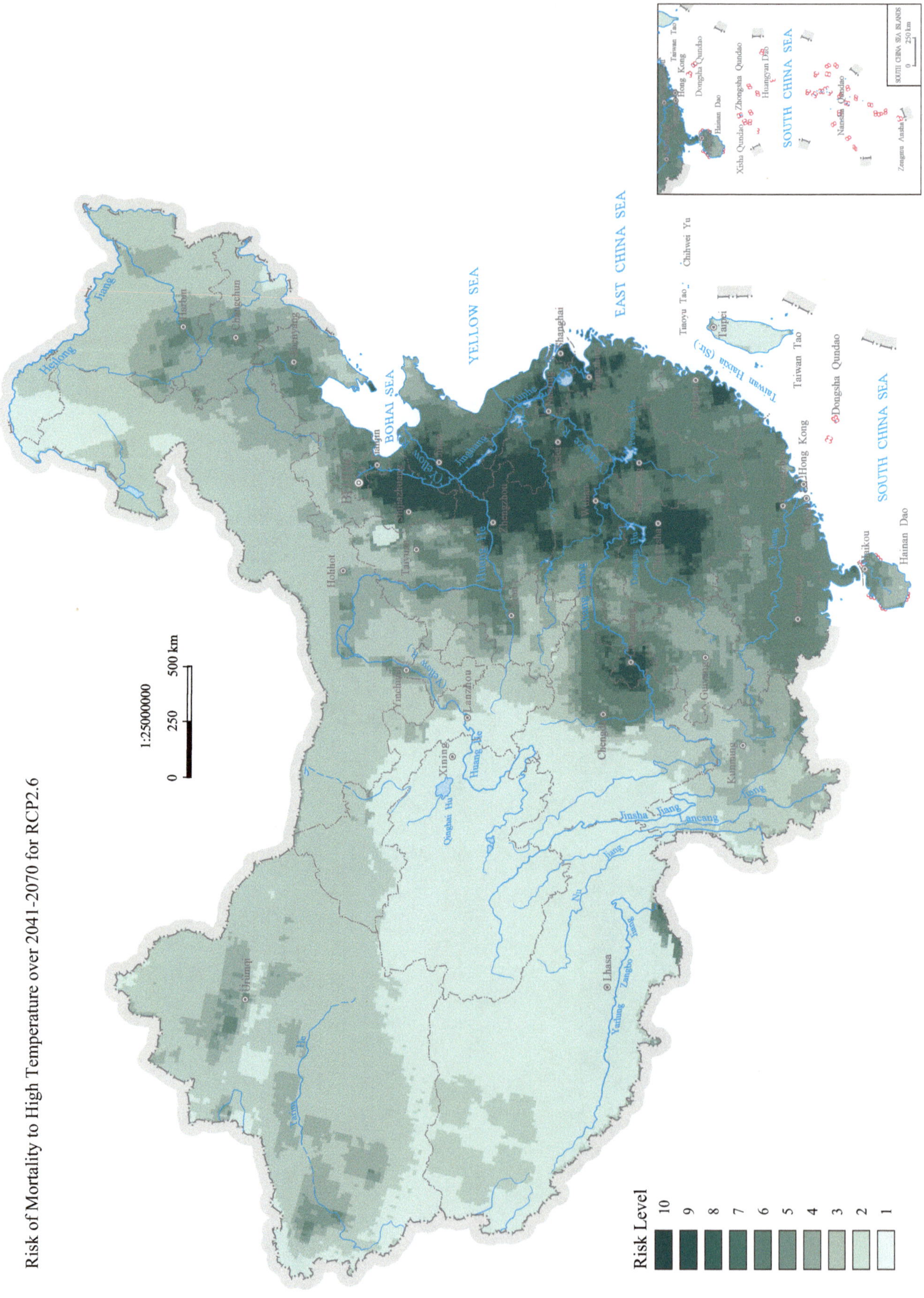

Risk of Mortality to High Temperature over 2041-2070 for RCP2.6
1:25000000
0 250 500 km
Risk Level
10
9
8
7
6
5
4
3
2
1
YELLOW SEA
EAST CHINA SEA
BOHAI SEA
SOUTH CHINA SEA
Shanghai
Hong Kong
Hainan Dao
Taiwan Tao
Taipei
Dongsha Qundao
Chiwei Yu
Taoyu Tao
Harbin
Changchun
Shenyang
Hohhot
Zhengzhou
Lanzhou
Xining
Qinghai Hu
Chengdu
Guiyang
Kunming
Lhasa
Urumqi
Heilong Jiang
Jinsha Jiang
Lancang Jiang
Yarlung Zangbo Jiang
Huang He
Haikou
Taiwan Haixia (Str.)
Xisha Qundao
Zhongsha Qundao
Nansha Qundao
Huangyan Dao
Dongsha Qundao
Zengmu Ansha
SOUTH CHINA SEA ISLANDS
0 250 km

The Number of Days with Daily Maximum Temperature over 35°C in 2071-2099 for RCP2.6

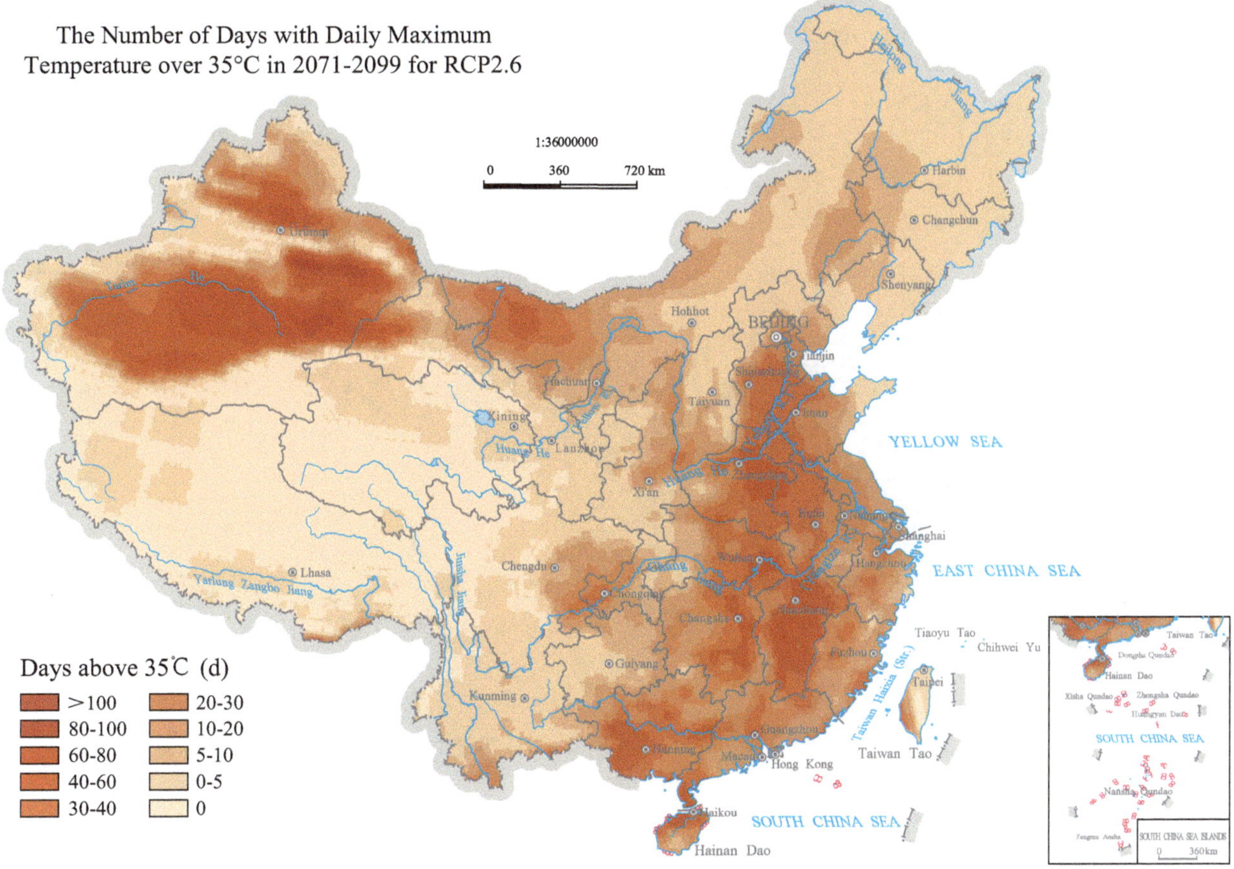

The Number of Days with Daily Maximum Temperature over 40°C in 2071-2099 for RCP2.6

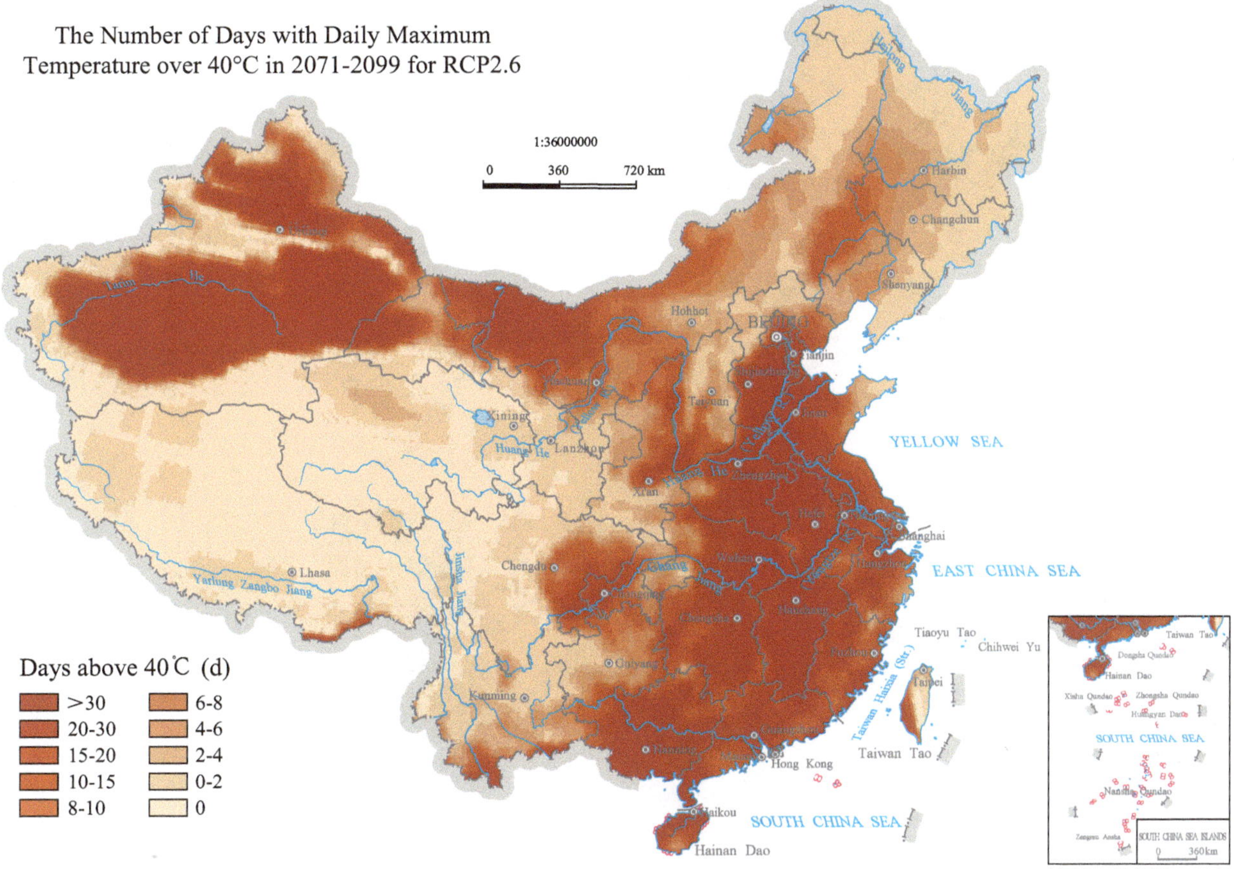

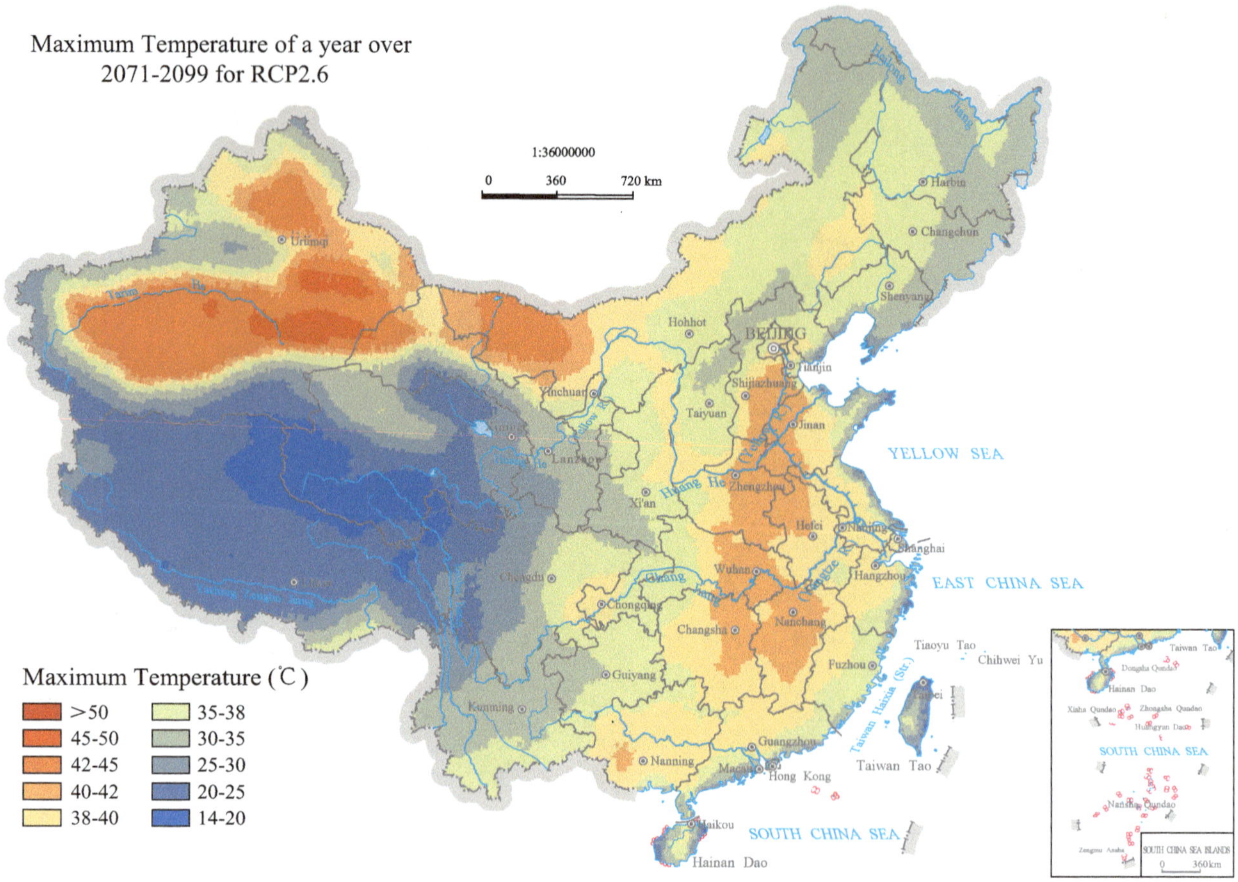
Maximum Temperature of a year over
2071-2099 for RCP2.6
1:36000000
0 360 720 km
Maximum Temperature (℃)
>50
45-50
42-45
40-42
38-40
35-38
30-35
25-30
20-25
14-20
Harbin
Changchun
Shenyang
Urümqi
Hohhot
BEIJING
Tianjin
Shijiazhuang
Taiyuan
Jinan
Xi'an
Zhengzhou
Hefei
Nanjing
Shanghai
Hangzhou
Wuhan
Chengdu
Changsha
Nanchang
Fuzhou
Guiyang
Kunming
Nanning
Guangzhou
Macao
Hong Kong
Haikou
Hainan Dao
YELLOW SEA
EAST CHINA SEA
SOUTH CHINA SEA
Tiaoyu Tao
Chihwei Yu
Taiwan Tao
SOUTH CHINA SEA

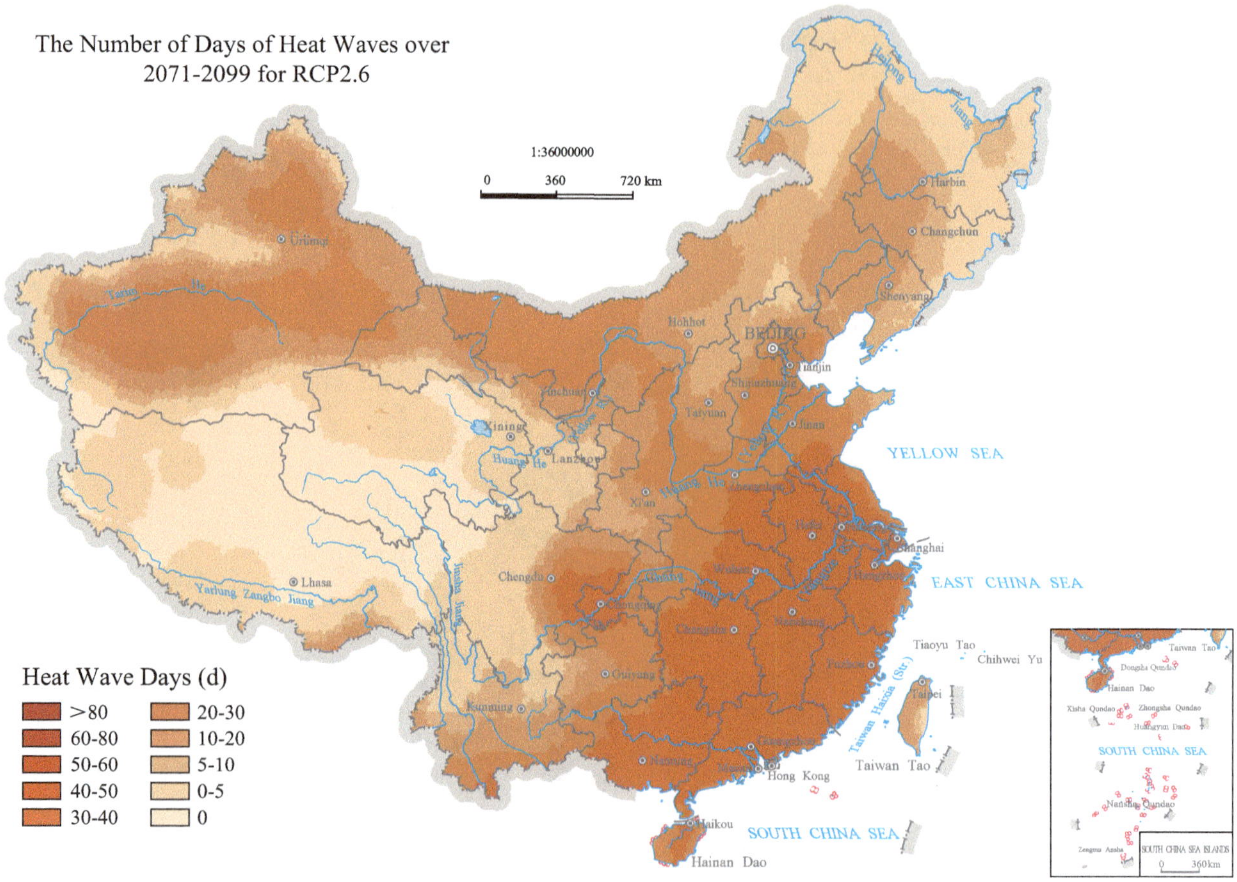
The Number of Days of Heat Waves over
2071-2099 for RCP2.6
1:36000000
0 360 720 km
Heat Wave Days (d)
>80
60-80
50-60
40-50
30-40
20-30
10-20
5-10
0-5
0
Harbin
Changchun
Shenyang
Urümqi
Hohhot
BEIJING
Tianjin
Taiyuan
Jinan
Xining
Lanzhou
Xi'an
Shanghai
Lhasa
Yarlung Zangbo Jiang
Chengdu
Guiyang
Kunming
Hong Kong
Haikou
Hainan Dao
YELLOW SEA
EAST CHINA SEA
SOUTH CHINA SEA
Tiaoyu Tao
Chihwei Yu
Taiwan Tao
SOUTH CHINA SEA

Hazard of Mortality to High Temperature over 2071-2099 for RCP2.6

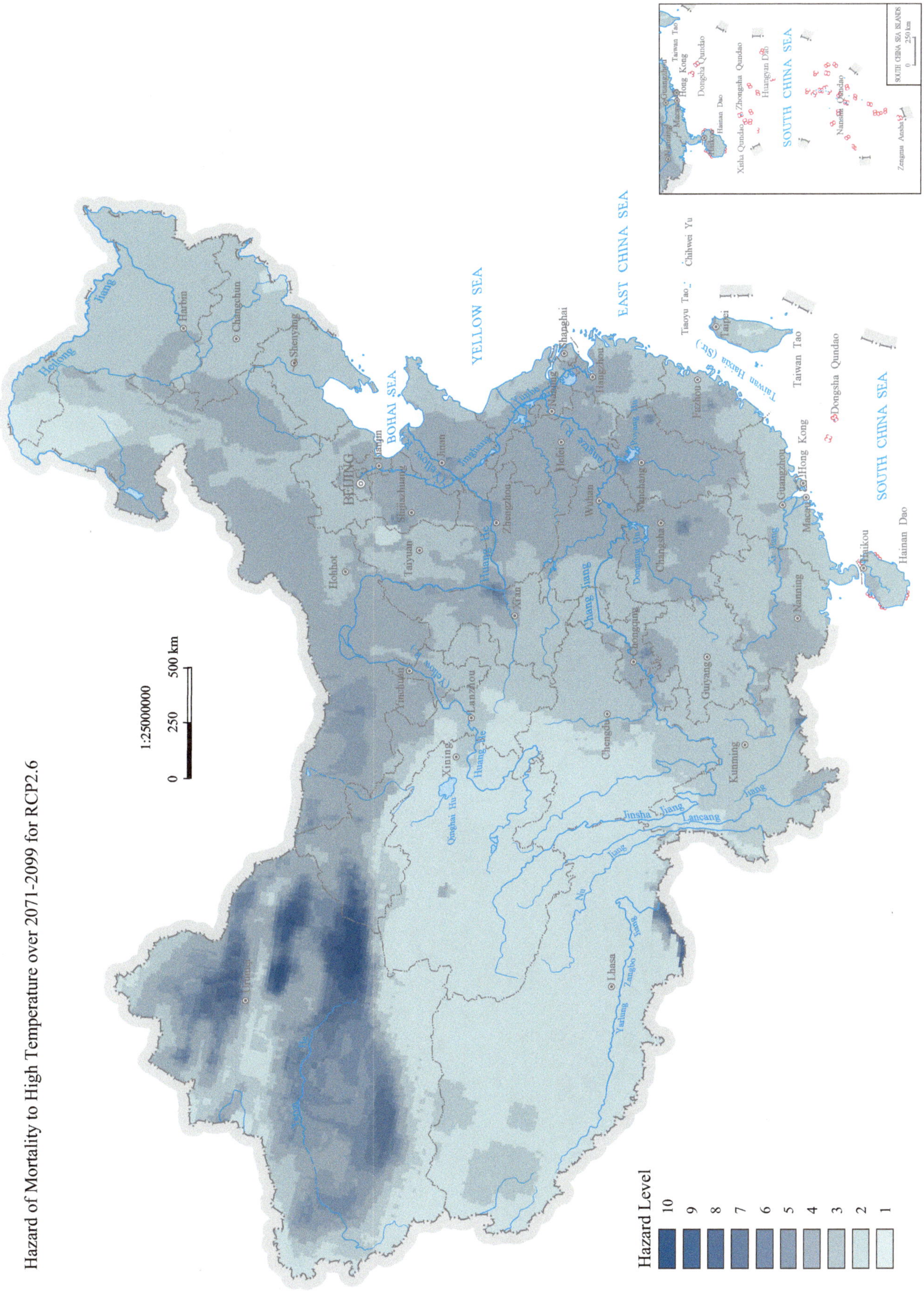

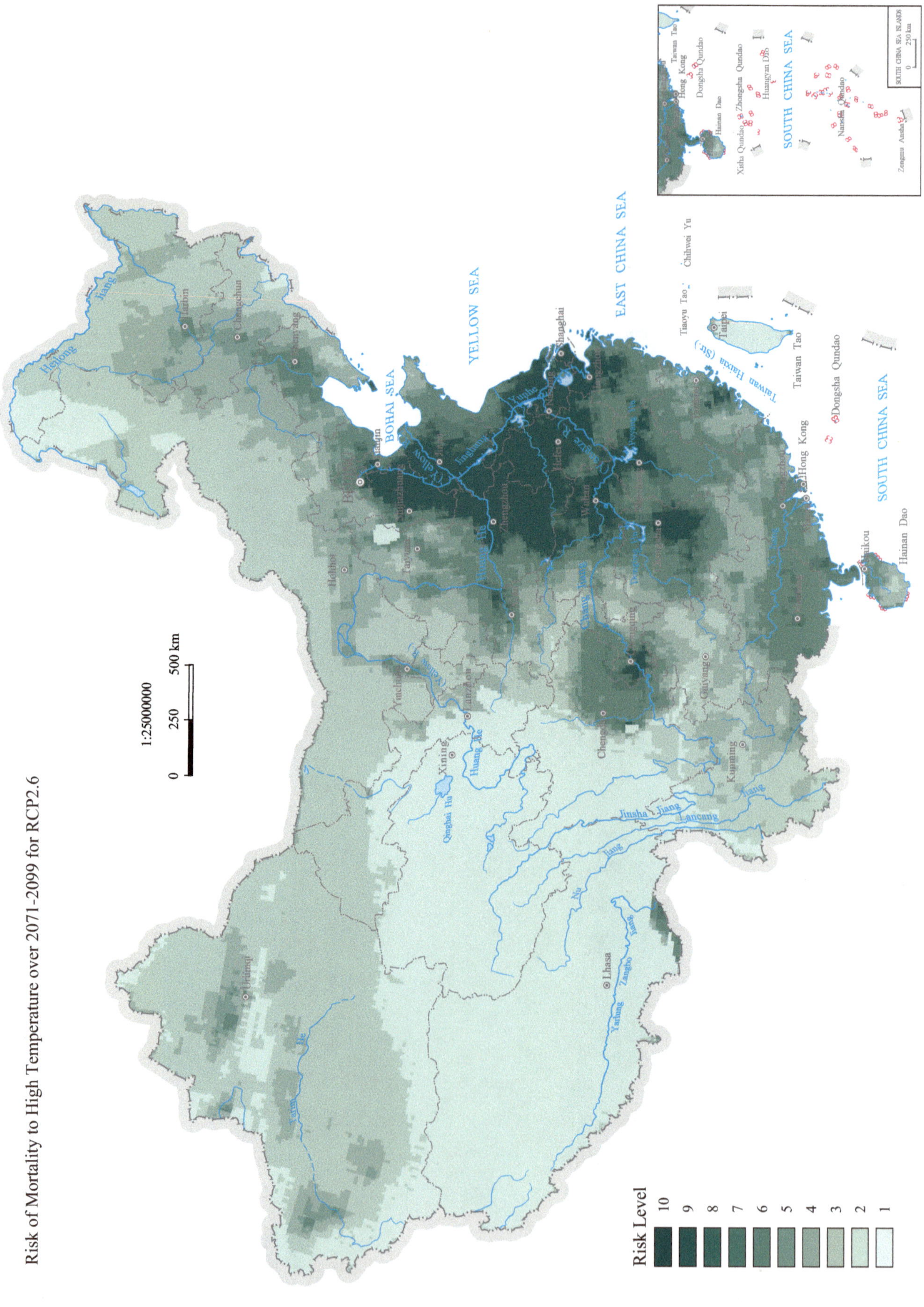
Risk of Mortality to High Temperature over 2071-2099 for RCP2.6
1:25000000
0 250 500 km
Risk Level
10
9
8
7
6
5
4
3
2
1
BOHAI SEA
YELLOW SEA
EAST CHINA SEA
SOUTH CHINA SEA
Taiwan Haixia (Str.)
Taiwan Tao
Dongsha Qundao
Hainan Dao
Hong Kong
Shanghai
Lhasa
Urumqi
Xining
Qinghai Hu
Chengdu
Kunming
Guiyang
Hefei
Wuhan
Nanjing
Zhengzhou
Taiyuan
Hohhot
Harbin
Changchun
Taipei
Haikou
Jinsha Jiang
Lancang
Nu Jiang
Yarlung Zangbo Jiang
Tarim He
Heilong Jiang
Huang He

The Number of Days with Daily Maximum Temperature over 35°C in 2011-2040 for RCP4.5

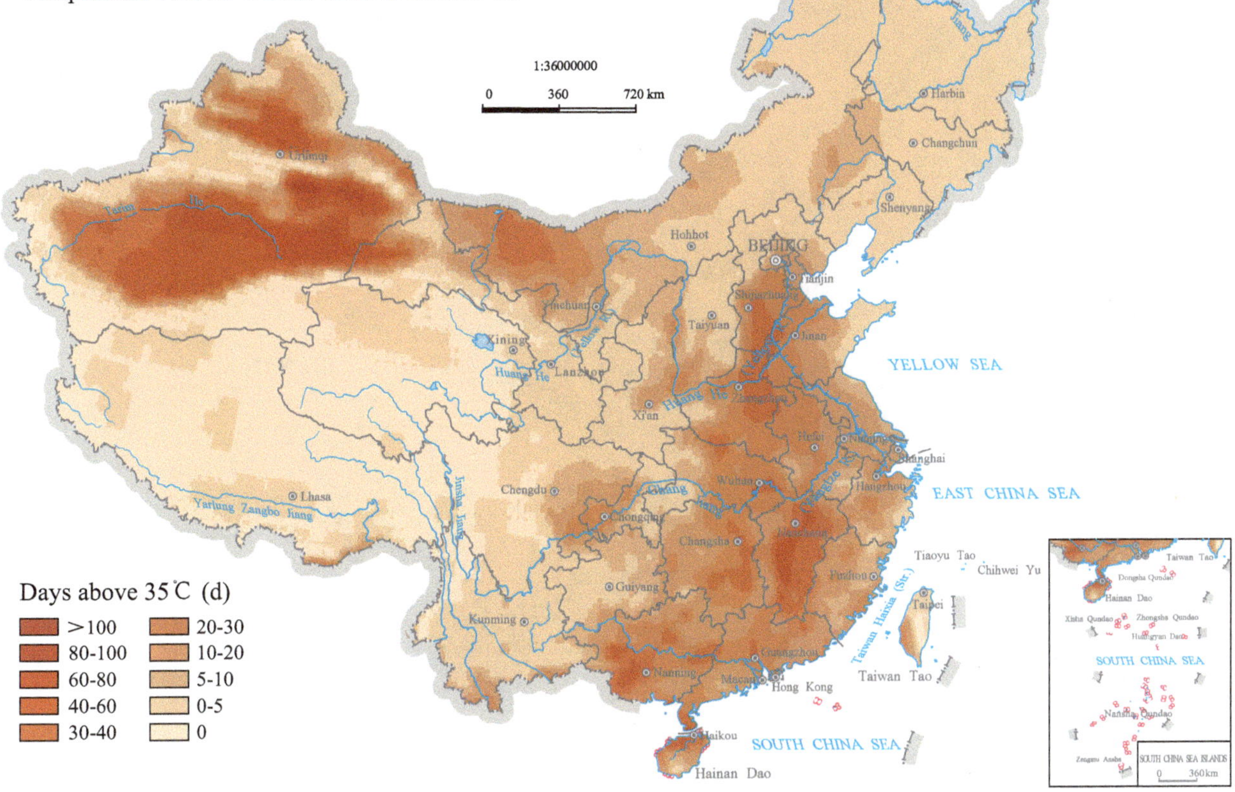

The Number of Days with Daily Maximum Temperature over 40°C in 2011-2040 for RCP4.5

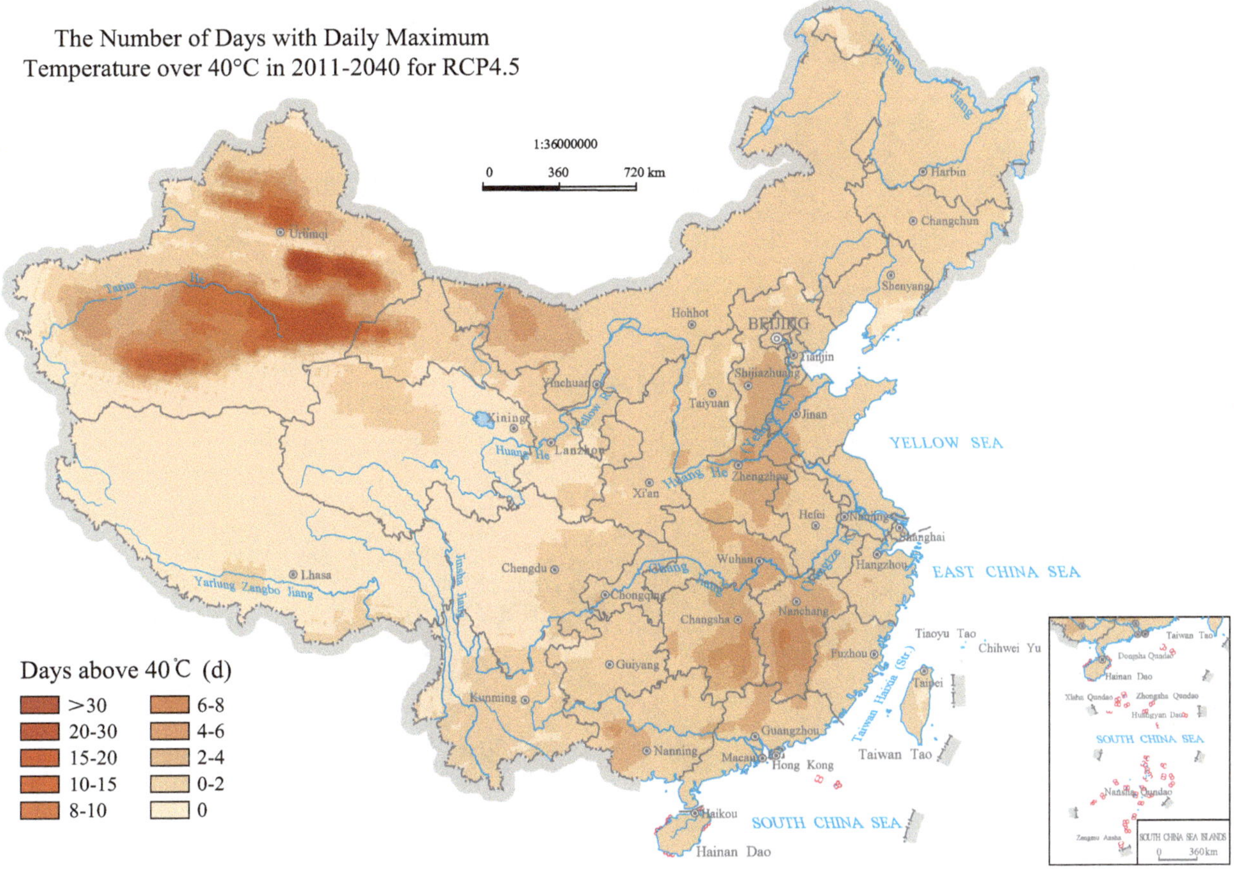

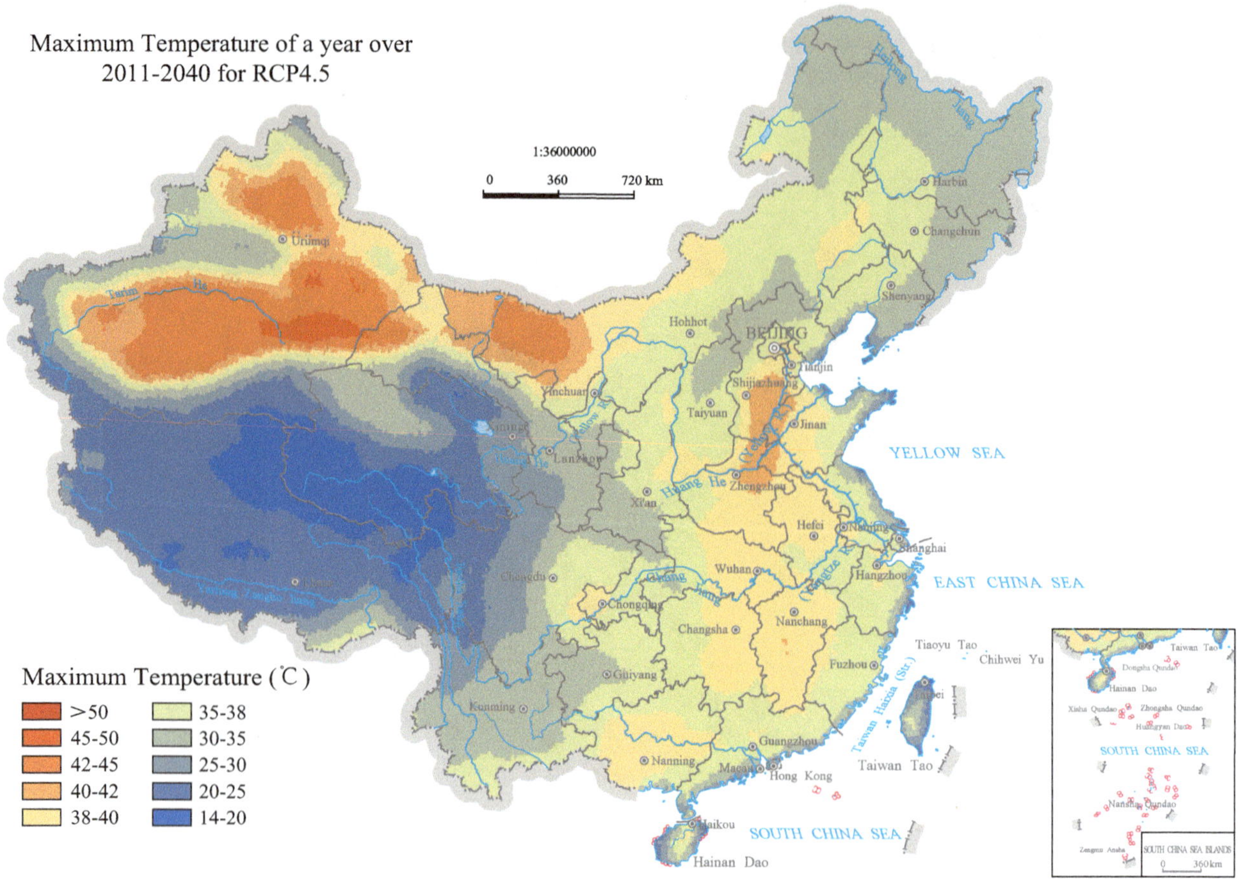
Maximum Temperature of a year over
2011-2040 for RCP4.5
1:36000000
0 360 720 km
Maximum Temperature (℃)
>50
45-50
42-45
40-42
38-40
35-38
30-35
25-30
20-25
14-20
BEIJING
Harbin
Changchun
Shenyang
Urümqi
Hohhot
Tianjin
Shijiazhuang
Taiyuan
Jinan
Yinchuan
Lanzhou
Xi'an
Zhengzhou
Hefei
Nanjing
Shanghai
Wuhan
Hangzhou
Chengdu
Chongqing
Changsha
Nanchang
Fuzhou
Guiyang
Kunming
Nanning
Guangzhou
Macao
Hong Kong
Haikou
Hainan Dao
Taiwan Tao
Tiaoyu Tao
Chihwei Yu
YELLOW SEA
EAST CHINA SEA
SOUTH CHINA SEA
SOUTH CHINA SEA ISLANDS

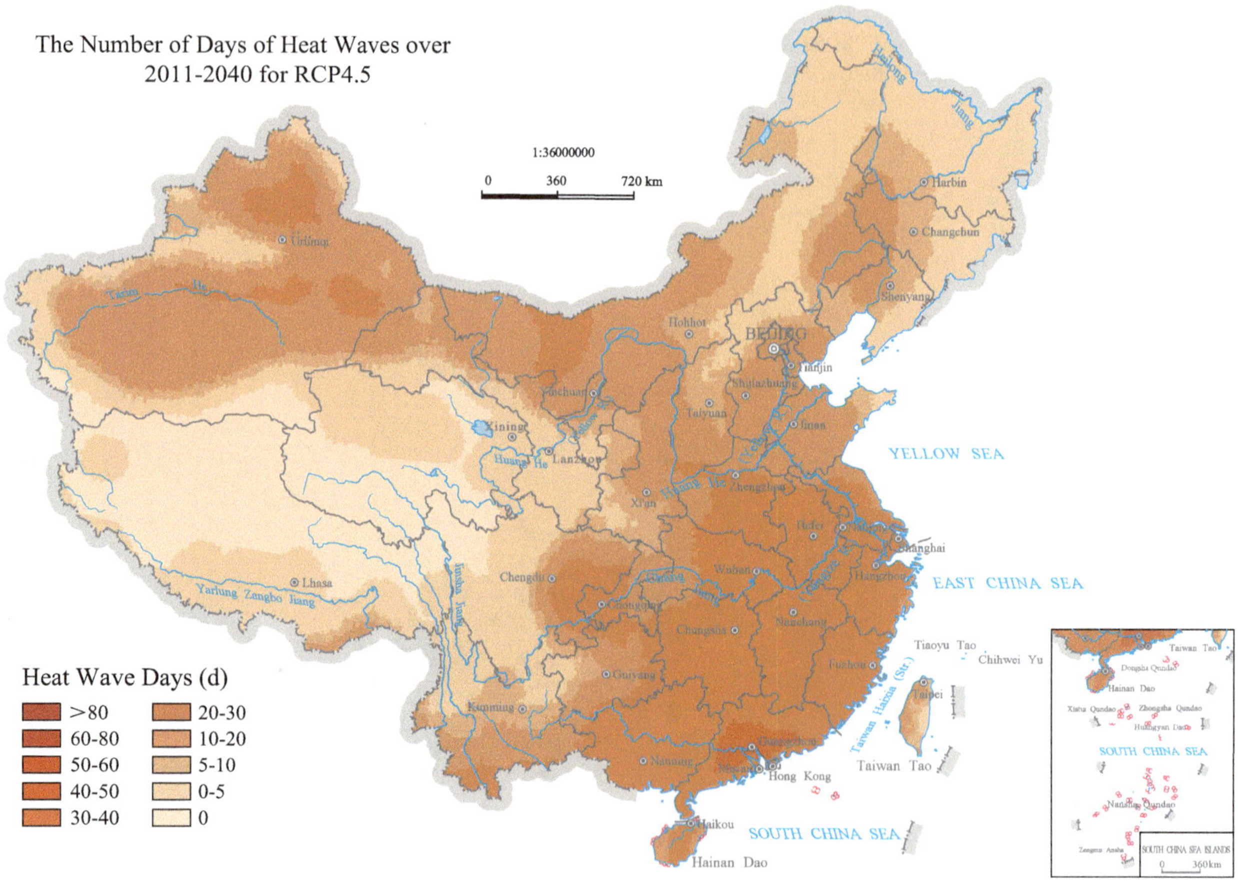
The Number of Days of Heat Waves over
2011-2040 for RCP4.5
1:36000000
0 360 720 km
Heat Wave Days (d)
>80
60-80
50-60
40-50
30-40
20-30
10-20
5-10
0-5
0
BEIJING
Harbin
Changchun
Shenyang
Urümqi
Hohhot
Tianjin
Shijiazhuang
Taiyuan
Jinan
Xining
Lanzhou
Xi'an
Zhengzhou
Hefei
Shanghai
Wuhan
Hangzhou
Lhasa
Yarlung Zangbo Jiang
Chengdu
Chongqing
Changsha
Nanchang
Fuzhou
Guiyang
Kunming
Nanning
Guangzhou
Hong Kong
Haikou
Hainan Dao
Taiwan Tao
Tiaoyu Tao
Chihwei Yu
YELLOW SEA
EAST CHINA SEA
SOUTH CHINA SEA
SOUTH CHINA SEA ISLANDS

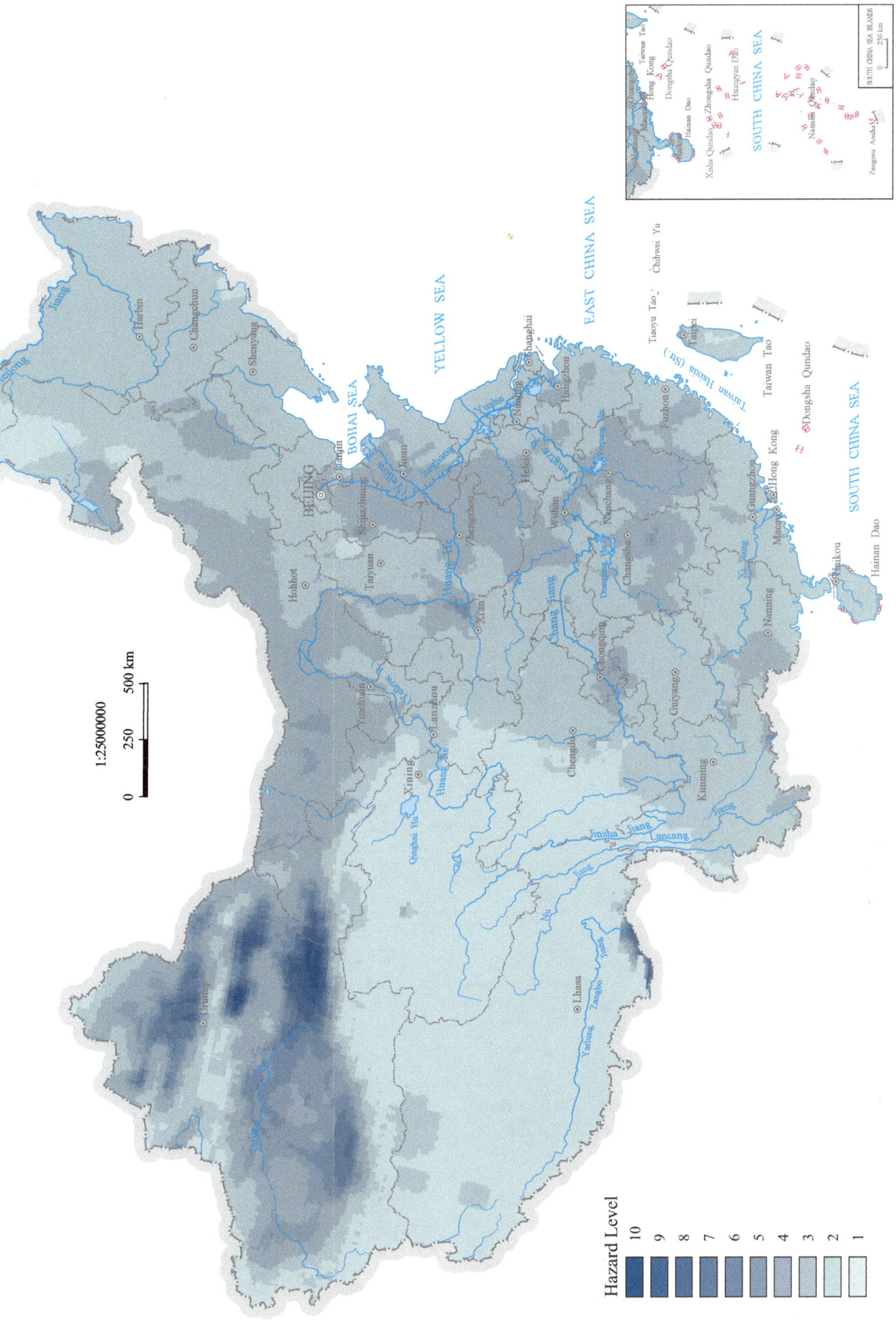

Hazard of Mortality to High Temperature over 2011-2040 for RCP4.5

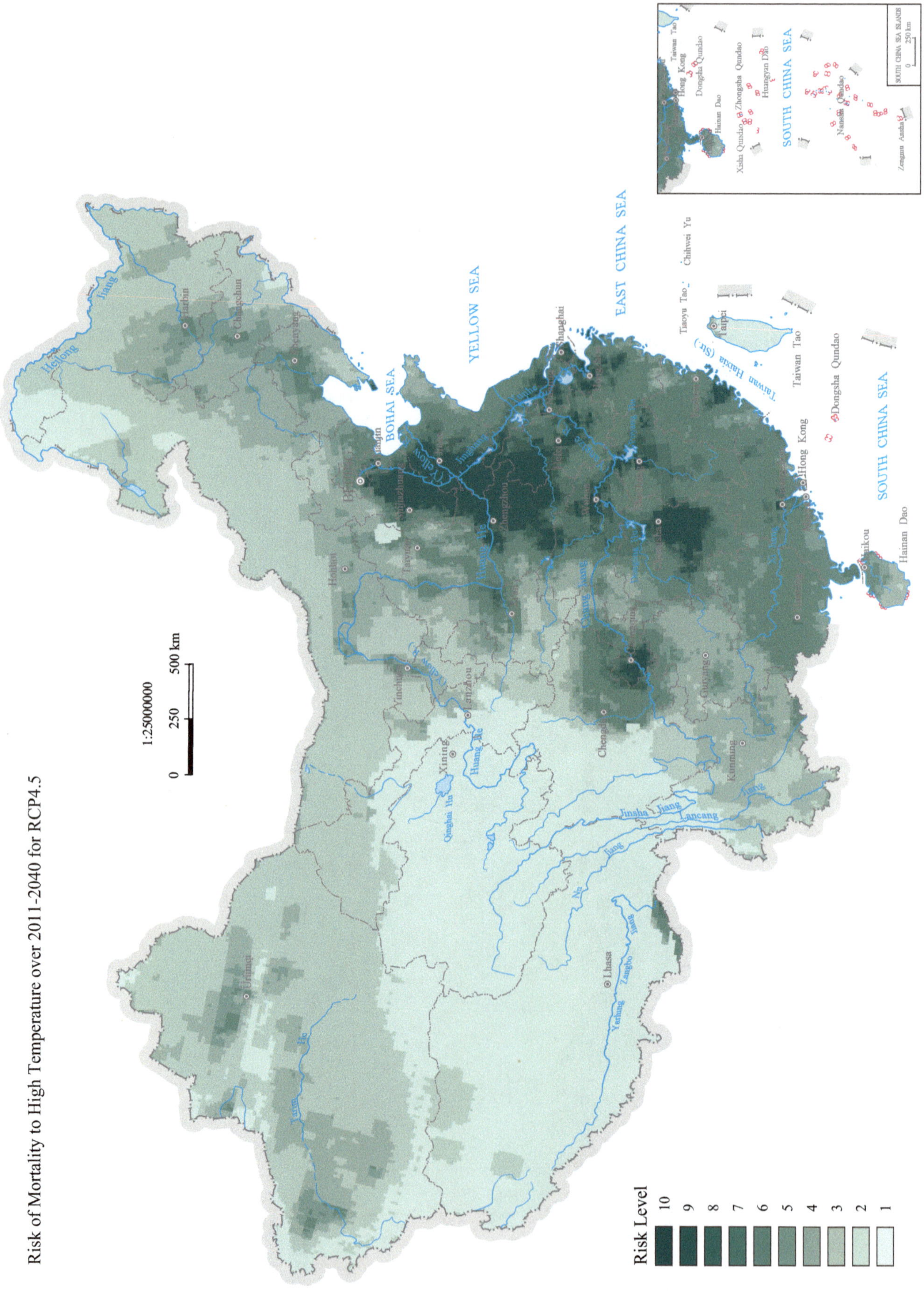
Risk of Mortality to High Temperature over 2011-2040 for RCP4.5
1:25000000
0 250 500 km
Risk Level
10
9
8
7
6
5
4
3
2
1
BOHAI SEA
YELLOW SEA
EAST CHINA SEA
SOUTH CHINA SEA
Taiwan Haixia (Str.)
Taiwan Tao
Diaoyu Tao
Chiwei Yu
Dongsha Qundao
Hainan Dao
Hong Kong
Taipei
Shanghai
Harbin
Changchun
Shenyang
Tianjin
Hohhot
Yinchuan
Lanzhou
Xining
Qinghai Hu
Huang He
Chengdu
Guiyang
Kunming
Haikou
Lhasa
Yarlung Zangbo Jiang
Jinsha Jiang
Lancang Jiang
Heilong Jiang
Urümqi
Tarim He
Nanhai Zhudao
Xisha Qundao
Zhongsha Qundao
Huangyan Dao
SOUTH CHINA SEA ISLANDS
0 250 km

The Number of Days with Daily Maximum Temperature over 35°C in 2041-2070 for RCP4.5

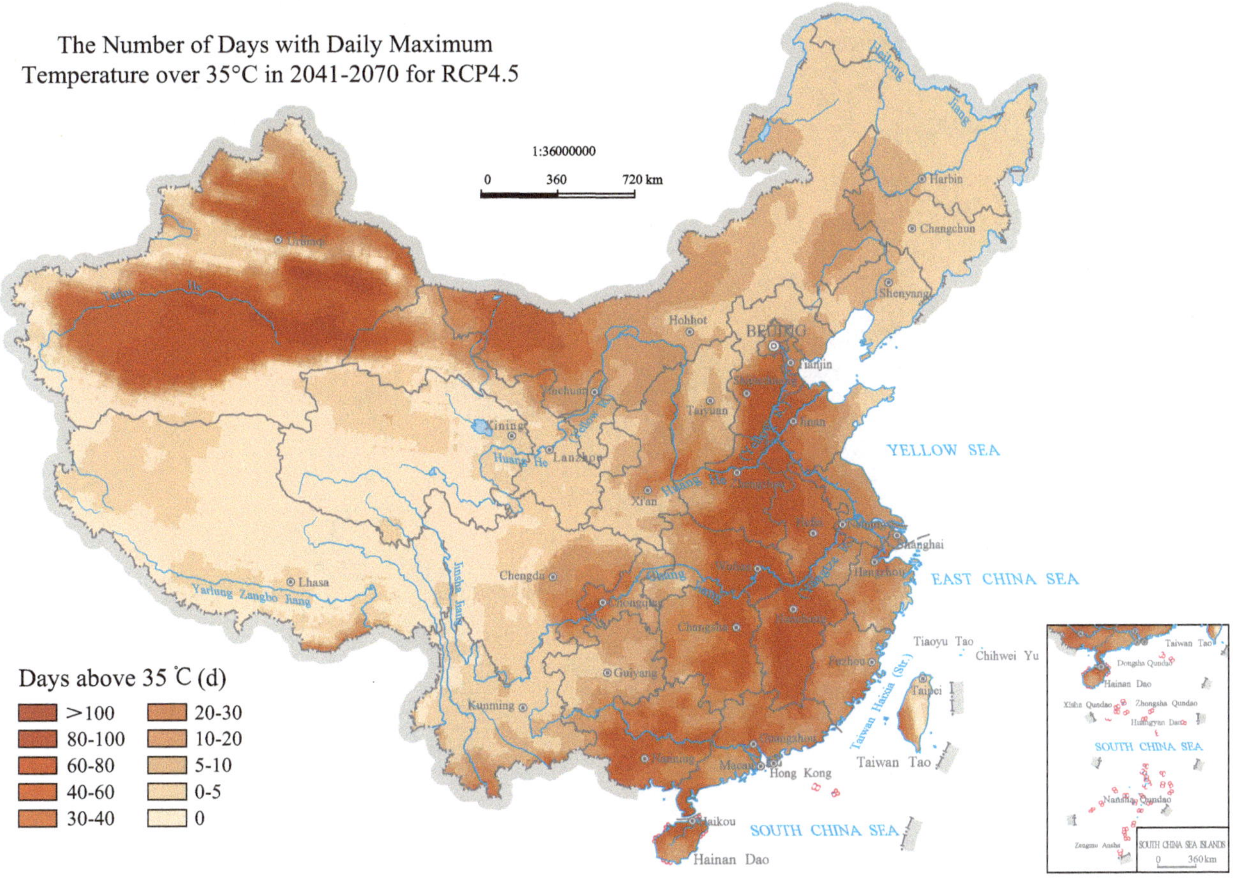

The Number of Days with Daily Maximum Temperature over 40°C in 2041-2070 for RCP4.5

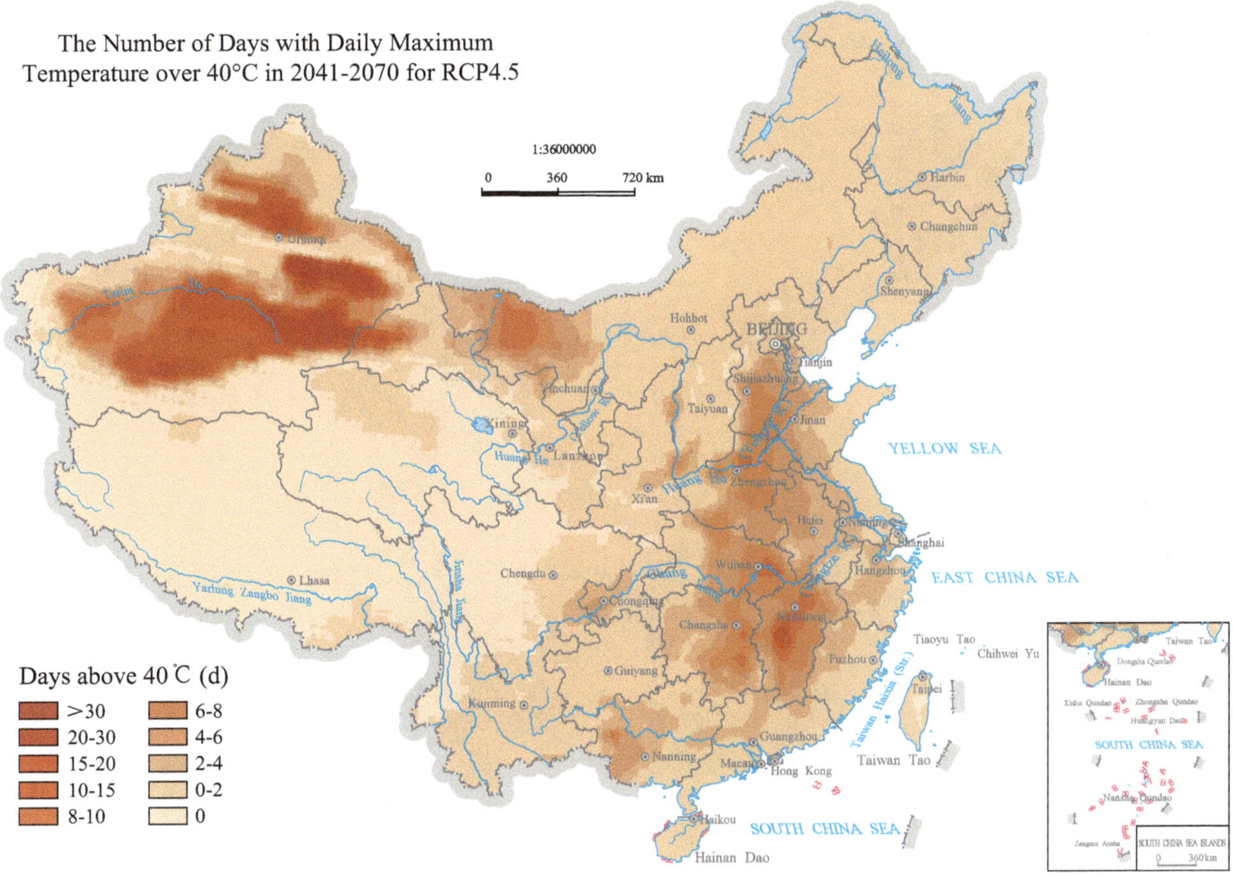

Maximum Temperature of a year over 2041-2070 for RCP4.5

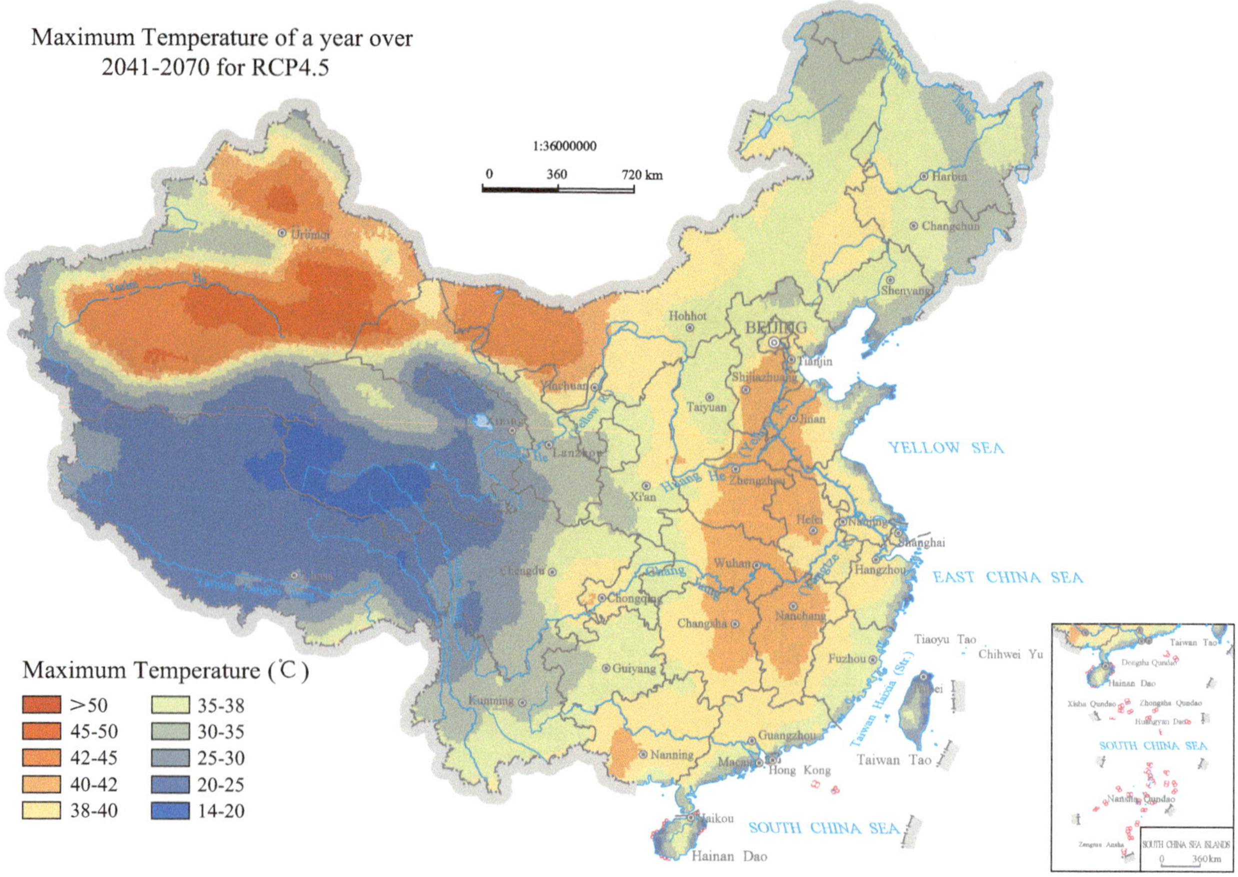

The Number of Days of Heat Waves over 2041-2070 for RCP4.5

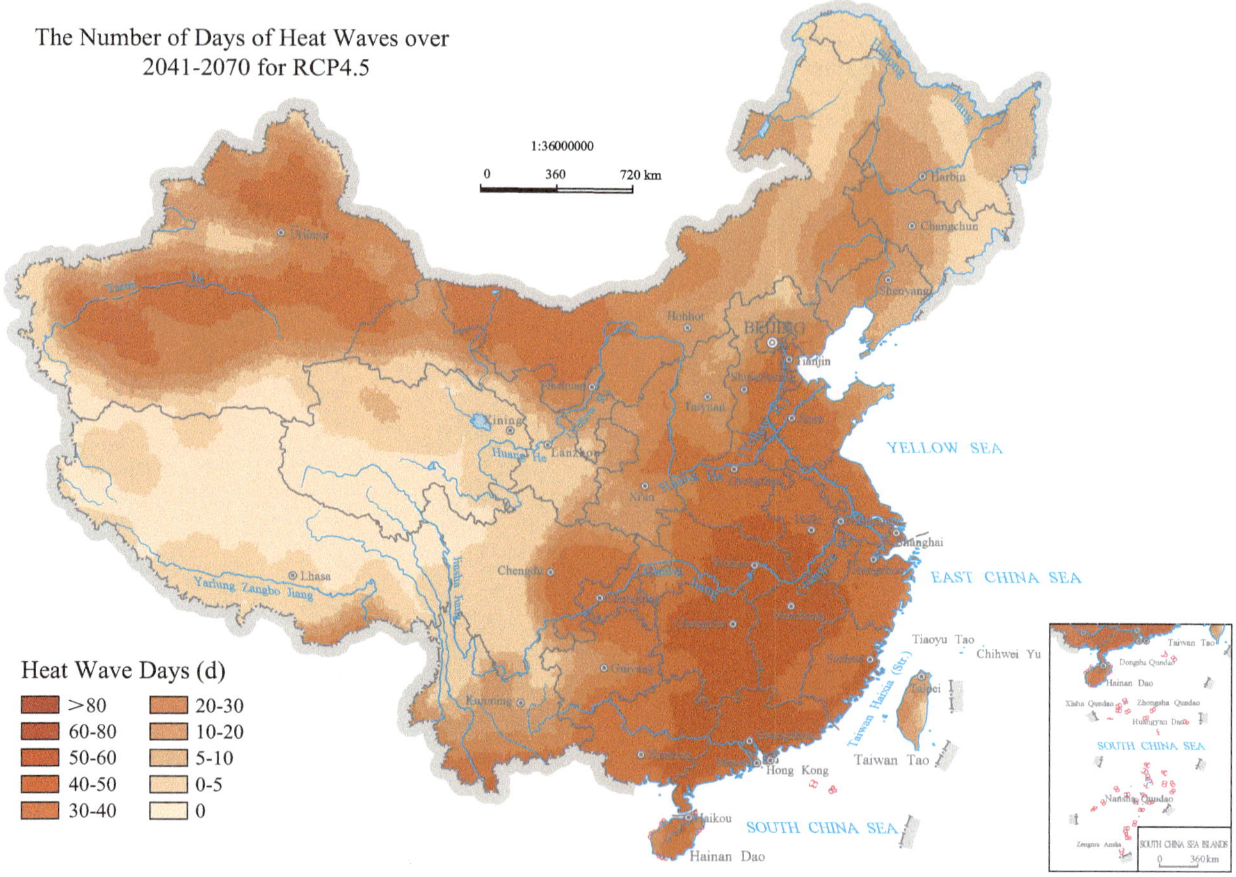

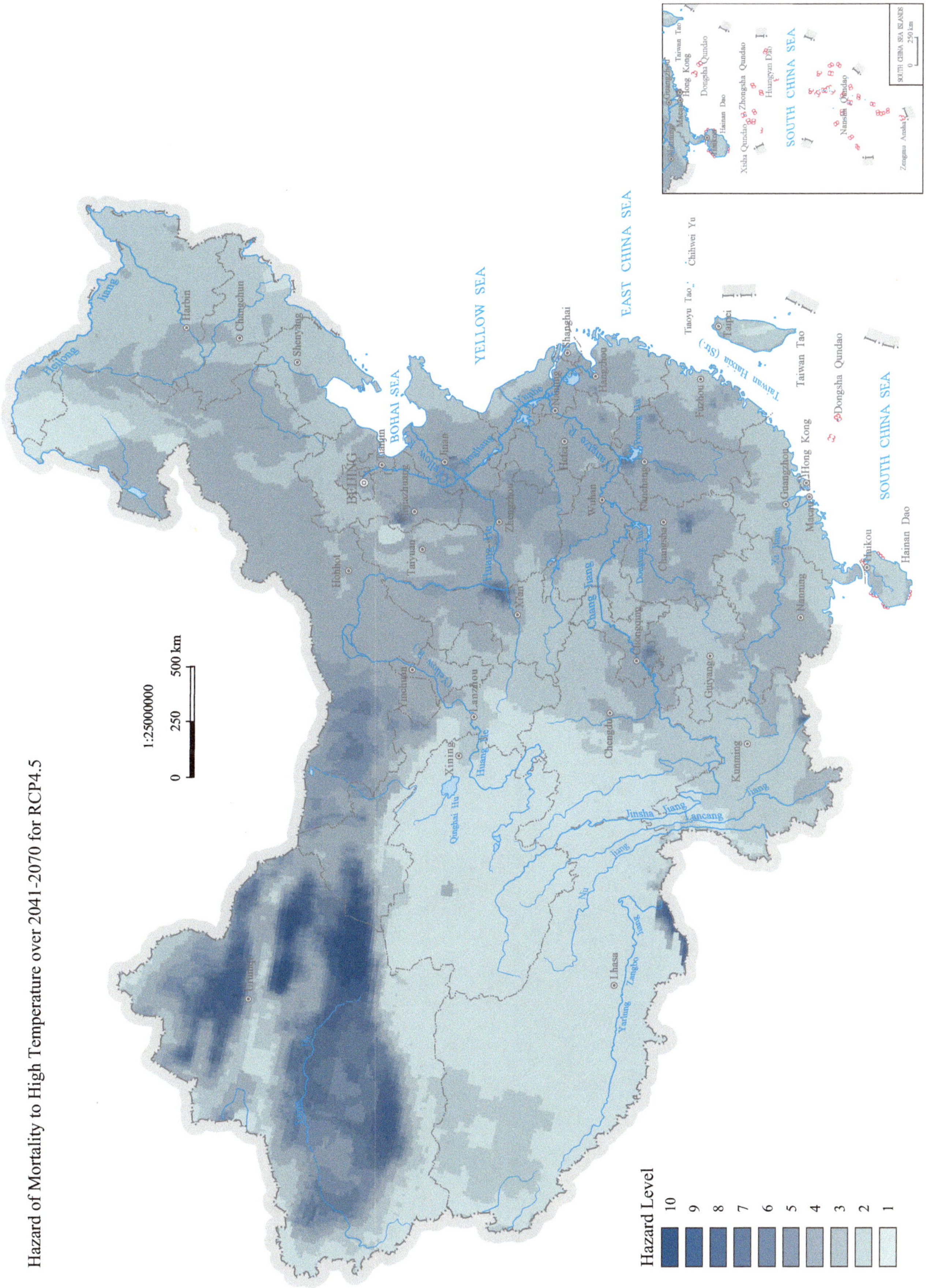
Hazard of Mortality to High Temperature over 2041-2070 for RCP4.5
1:25000000
0 250 500 km
Hazard Level
10
9
8
7
6
5
4
3
2
1
BEIJING
YELLOW SEA
BOHAI SEA
EAST CHINA SEA
SOUTH CHINA SEA
Harbin
Changchun
Shenyang
Tianjin
Jinan
Shanghai
Hangzhou
Nanjing
Hefei
Wuhan
Nanchang
Changsha
Fuzhou
Taipei
Guangzhou
Hong Kong
Macau
Haikou
Hainan Dao
Nanning
Guiyang
Kunming
Chengdu
Chongqing
Xi'an
Zhengzhou
Taiyuan
Shijiazhuang
Hohhot
Yinchuan
Lanzhou
Xining
Qinghai Hu
Lhasa
Urumqi
Heilong
Jiang
Chang Jiang
Huang He
Jinsha Jiang
Lancang
Nu
Yarlung Zangbo Jiang
Taiwan Tao
Diaoyu Tao
Chiwei Yu
Dongsha Qundao
Taiwan Haixia (Str.)
Xisha Qundao
Zhongsha Qundao
Huangyan Dao
Nansha Qundao
Zengmu Ansha
SOUTH CHINA SEA ISLANDS
0 250 km

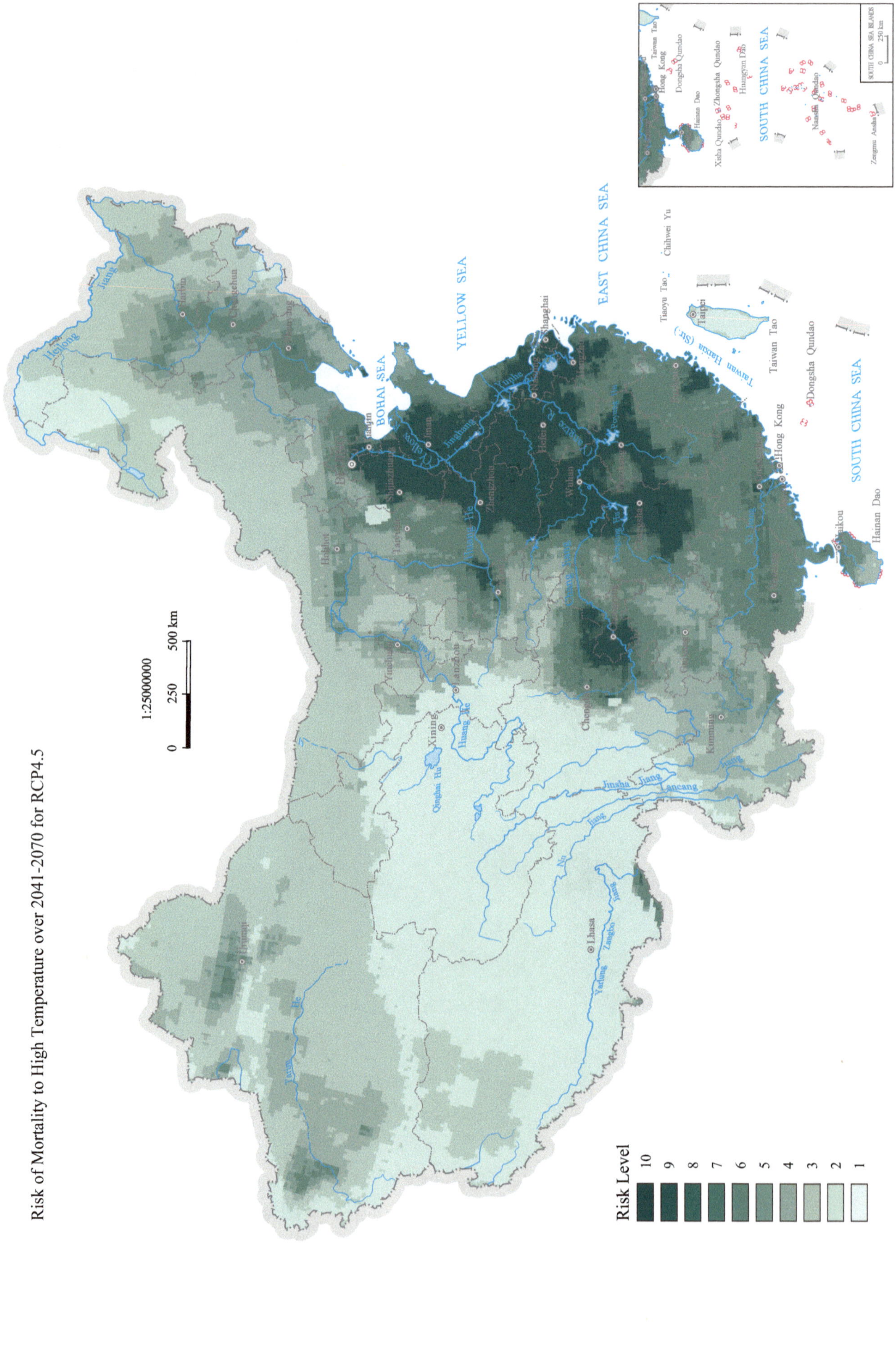
Risk of Mortality to High Temperature over 2041-2070 for RCP4.5
1:25000000
0 250 500 km
Risk Level
10
9
8
7
6
5
4
3
2
1
YELLOW SEA
EAST CHINA SEA
BOHAI SEA
SOUTH CHINA SEA
Taiwan Haixia (Str.)
Taiwan Tao
Dongsha Qundao
Hong Kong
Hainan Dao
Shanghai
Jinan
Zhengzhou
Hefei
Wuhan
Chengdu
Kunming
Lanzhou
Xining
Qinghai Hu
Hohhot
Lhasa
Yarlung Zangbo
Jinsha Jiang
Lancang
Heilong Jiang
Tiaoyu Tao
Chihwei Yu
Taipei
SOUTH CHINA SEA ISLANDS

The Number of Days with Daily Maximum Temperature over 35°C in 2071-2099 for RCP4.5

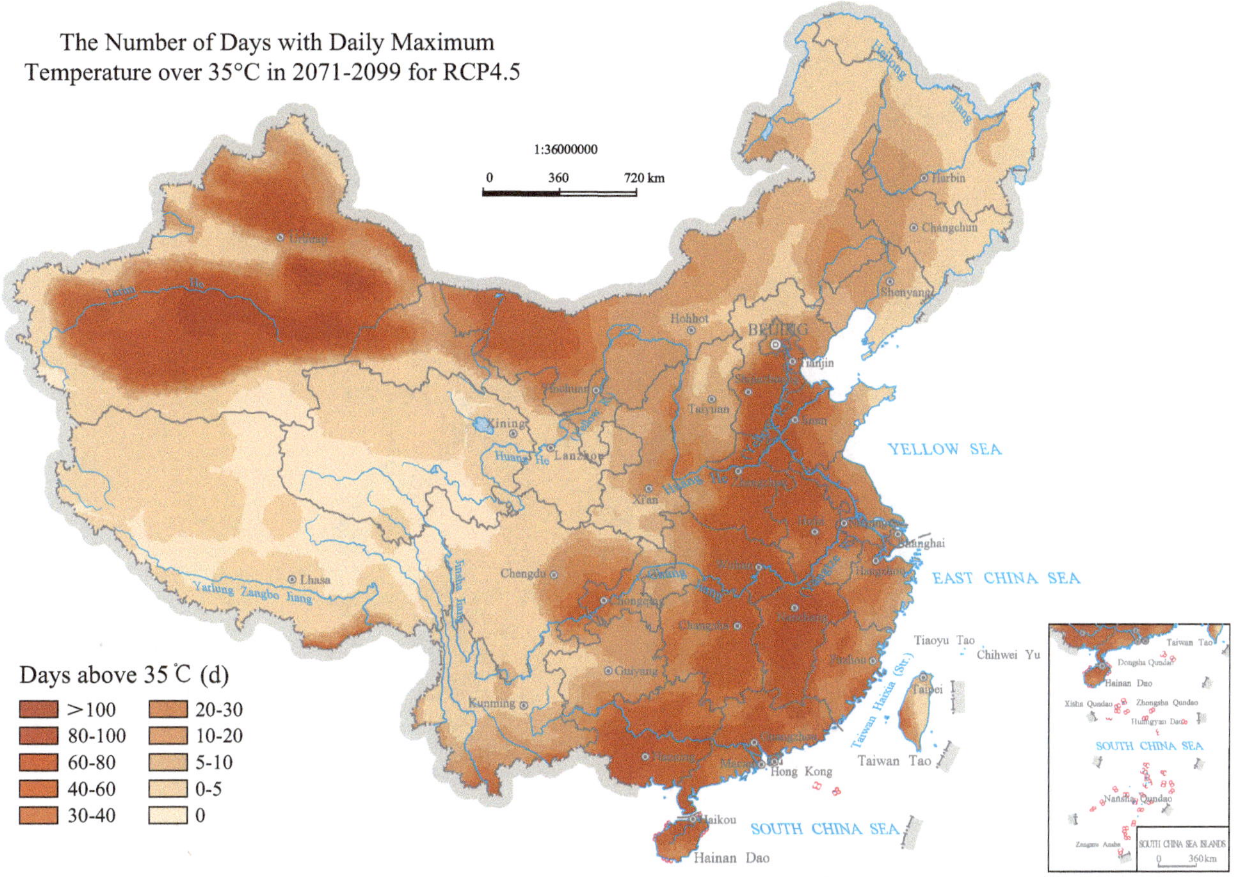

The Number of Days with Daily Maximum Temperature over 40°C in 2071-2099 for RCP4.5

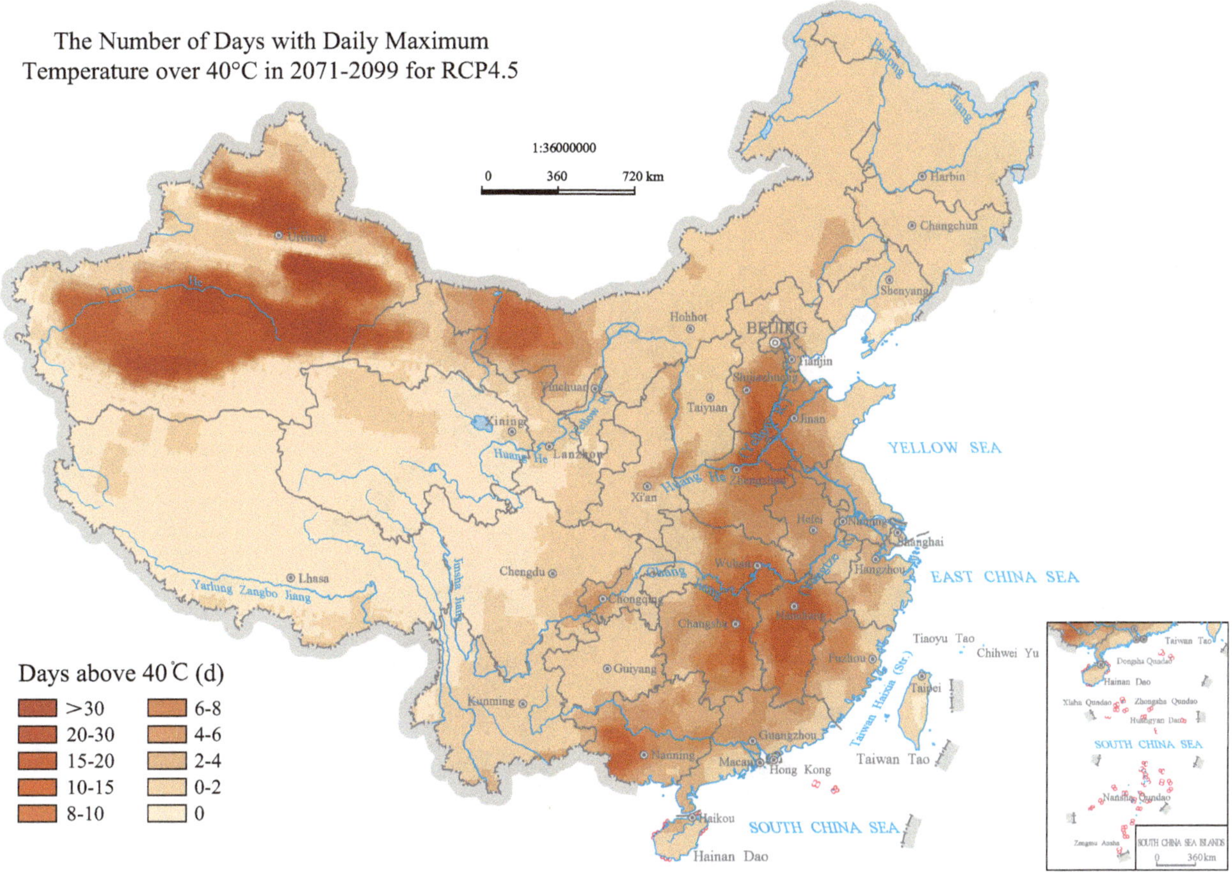

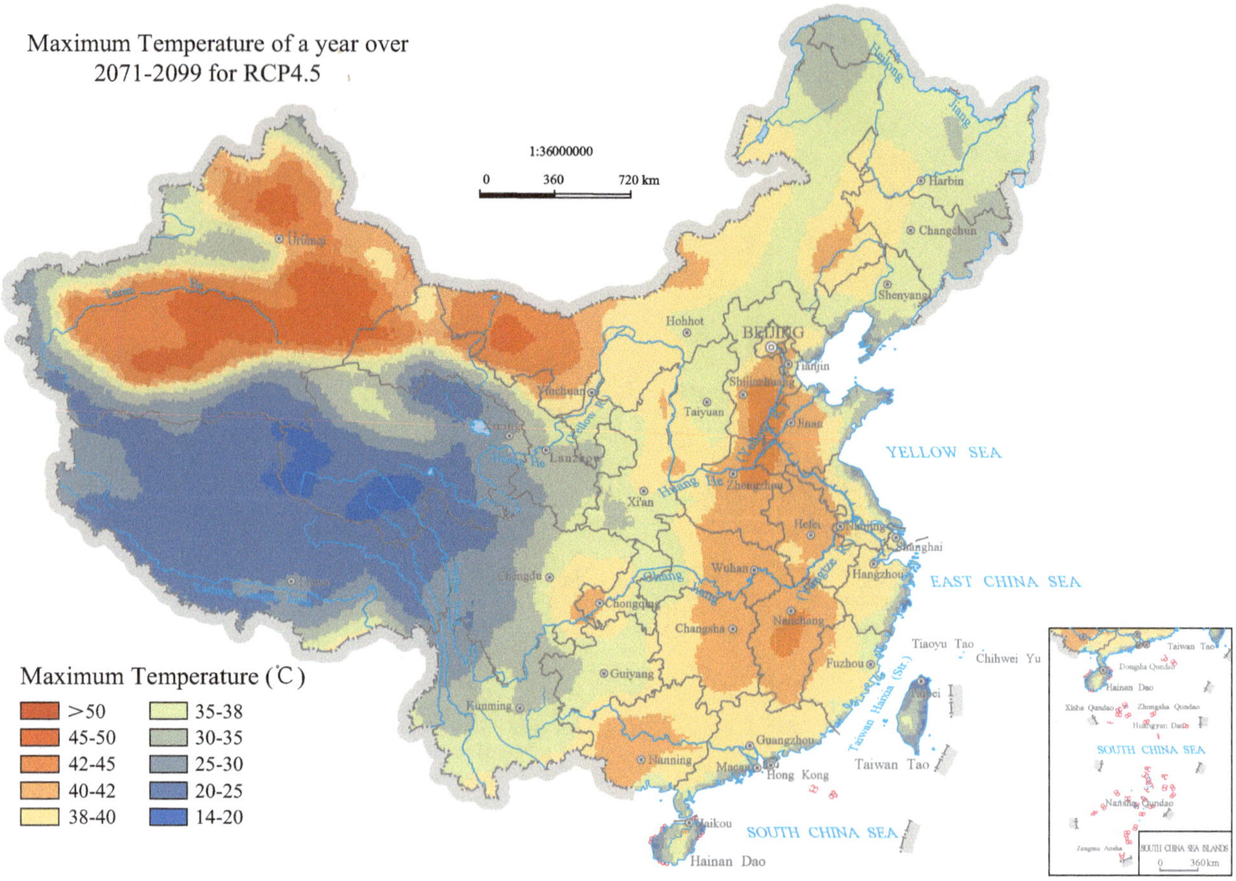
Maximum Temperature of a year over
2071-2099 for RCP4.5
1:36000000
0 360 720 km
Maximum Temperature (℃)
>50
45-50
42-45
40-42
38-40
35-38
30-35
25-30
20-25
14-20
YELLOW SEA
EAST CHINA SEA
SOUTH CHINA SEA
Harbin
Changchun
Shenyang
Hohhot
BEIJING
Tianjin
Taiyuan
Jinan
Xi'an
Lanzhou
Yinchuan
Urumqi
Wuhan
Hefei
Shanghai
Hangzhou
Changsha
Nanchang
Chengdu
Chongqing
Guiyang
Kunming
Nanning
Guangzhou
Fuzhou
Hong Kong
Macao
Haikou
Hainan Dao
Taiwan Tao
Taipei
Tiaoyu Tao
Chihwei Yu

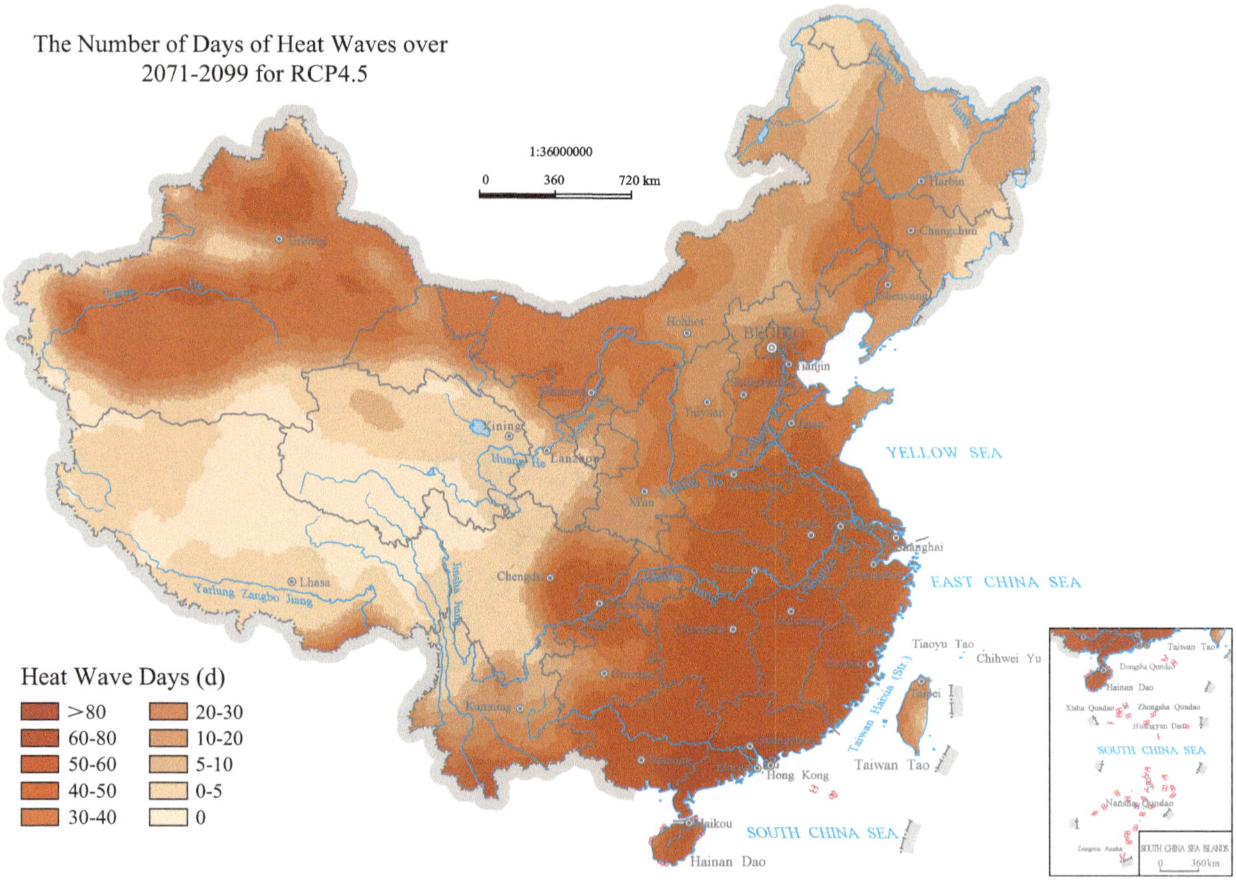
The Number of Days of Heat Waves over
2071-2099 for RCP4.5
1:36000000
0 360 720 km
Heat Wave Days (d)
>80
60-80
50-60
40-50
30-40
20-30
10-20
5-10
0-5
0
YELLOW SEA
EAST CHINA SEA
SOUTH CHINA SEA
Harbin
Changchun
BEIJING
Tianjin
Xining
Lanzhou
Xi'an
Lhasa
Yarlung Zangbo Jiang
Chengdu
Shanghai
Hong Kong
Haikou
Hainan Dao
Taiwan Tao
Tiaoyu Tao
Chihwei Yu

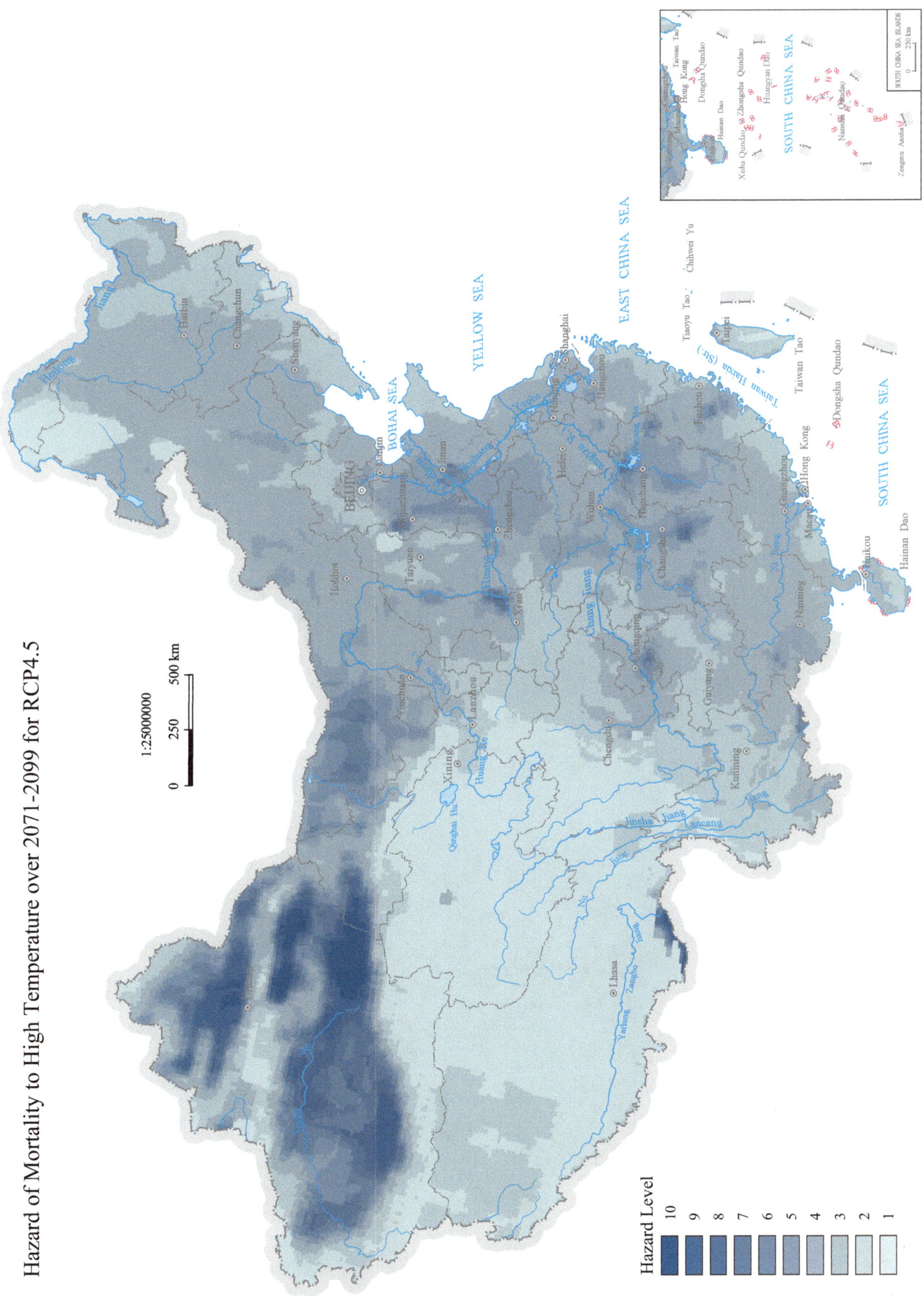
Hazard of Mortality to High Temperature over 2071-2099 for RCP4.5
1:25000000
0 250 500 km
Hazard Level
10
9
8
7
6
5
4
3
2
1
BEIJING
BOHAI SEA
YELLOW SEA
EAST CHINA SEA
SOUTH CHINA SEA
Harbin
Changchun
Shenyang
Tianjin
Shijiazhuang
Jinan
Taiyuan
Hohhot
Zhengzhou
Xi'an
Yinchuan
Lanzhou
Xining
Qinghai Hu
Chengdu
Chongqing
Guiyang
Kunming
Nanning
Haikou
Hainan Dao
Guangzhou
Hong Kong
Macao
Changsha
Nanchang
Wuhan
Hefei
Nanjing
Shanghai
Hangzhou
Fuzhou
Taipei
Taiwan Tao
Taiwan Haixia (Str.)
Dongsha Qundao
Diaoyu Dao
Chiwei Yu
Lhasa
Yarlung Zangbo Jiang
Jinsha Jiang
Lancang Jiang
Chang Jiang
Huang He
Xi Jiang
Nansha Qundao
Xisha Qundao
Zhongsha Qundao
Huangyan Dao
SOUTH CHINA SEA ISLANDS
0 250 km

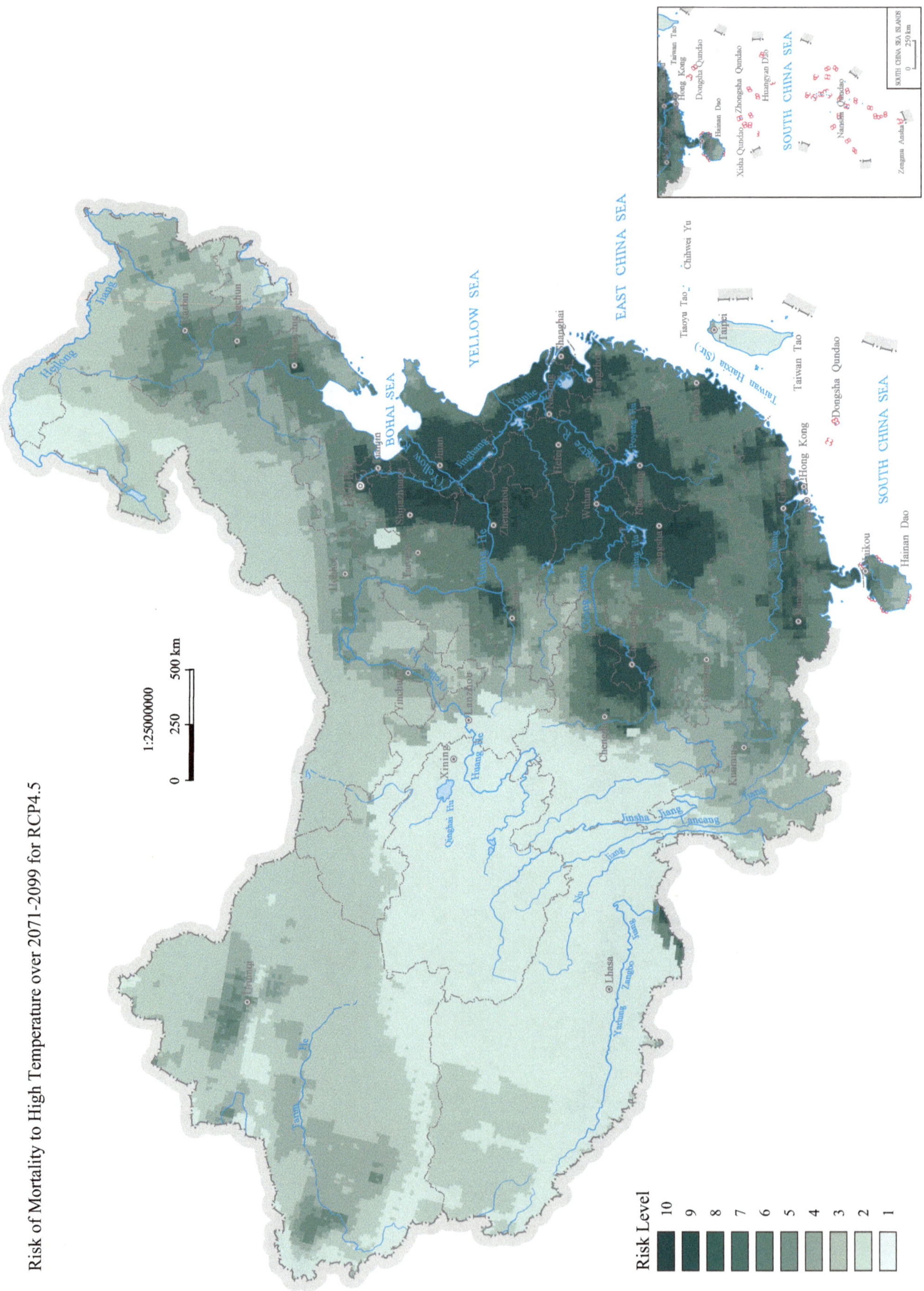
Risk of Mortality to High Temperature over 2071-2099 for RCP4.5
1:25000000
0 250 500 km
Risk Level
10
9
8
7
6
5
4
3
2
1
BOHAI SEA
YELLOW SEA
EAST CHINA SEA
SOUTH CHINA SEA
Taiwan Haixia (Su.)
Taiwan Tao
Dongsha Qundao
Hong Kong
Hainan Dao
Shanghai
Lhasa
Xining
Lanzhou
Qinghai Hu
Jinsha Jiang
Lancang
Yarlung Zangbo
Heilong Jiang
Tarim He
Huang He

The Number of Days with Daily Maximum Temperature over 35°C in 2011-2040 for RCP6.0

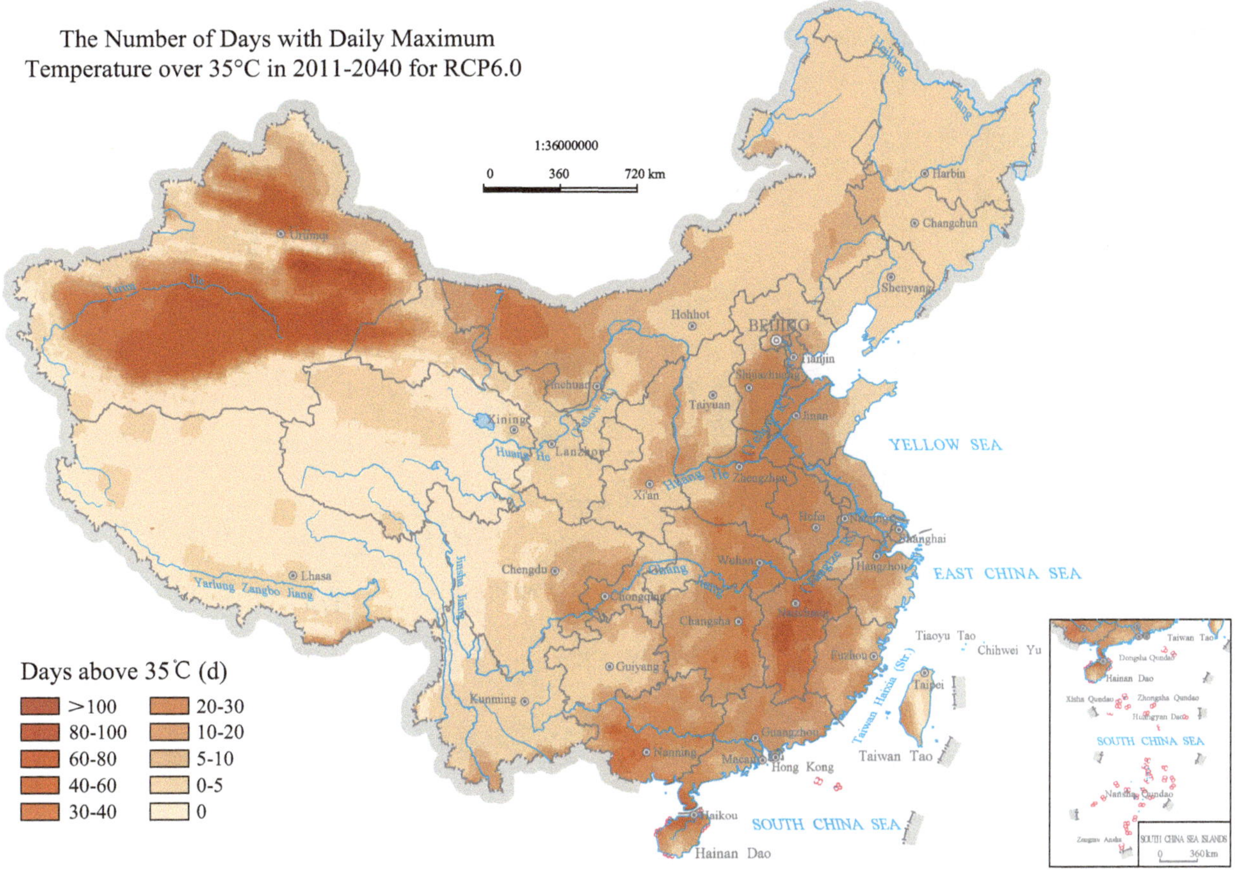

The Number of Days with Daily Maximum Temperature over 40°C in 2011-2040 for RCP6.0

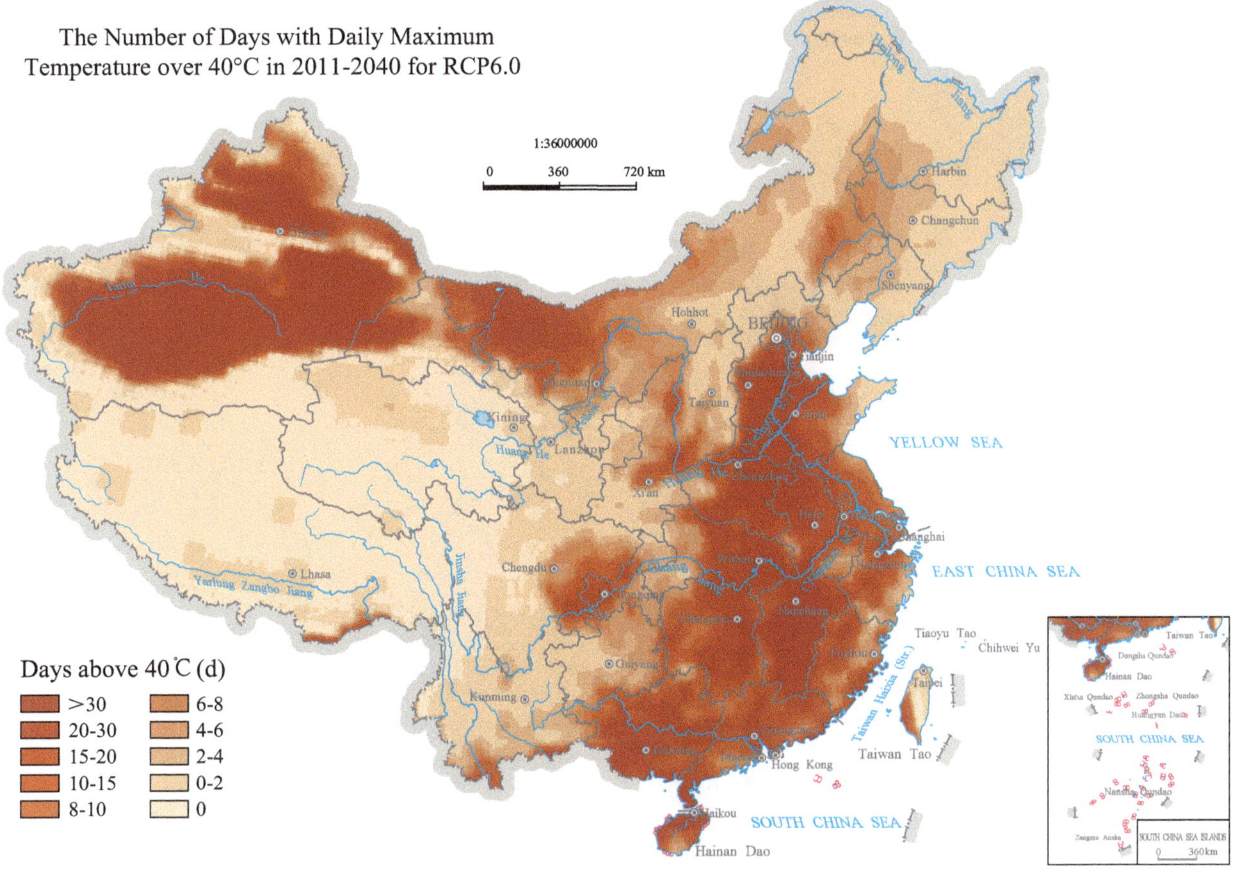

Maximum Temperature of a year over 2011-2040 for RCP6.0

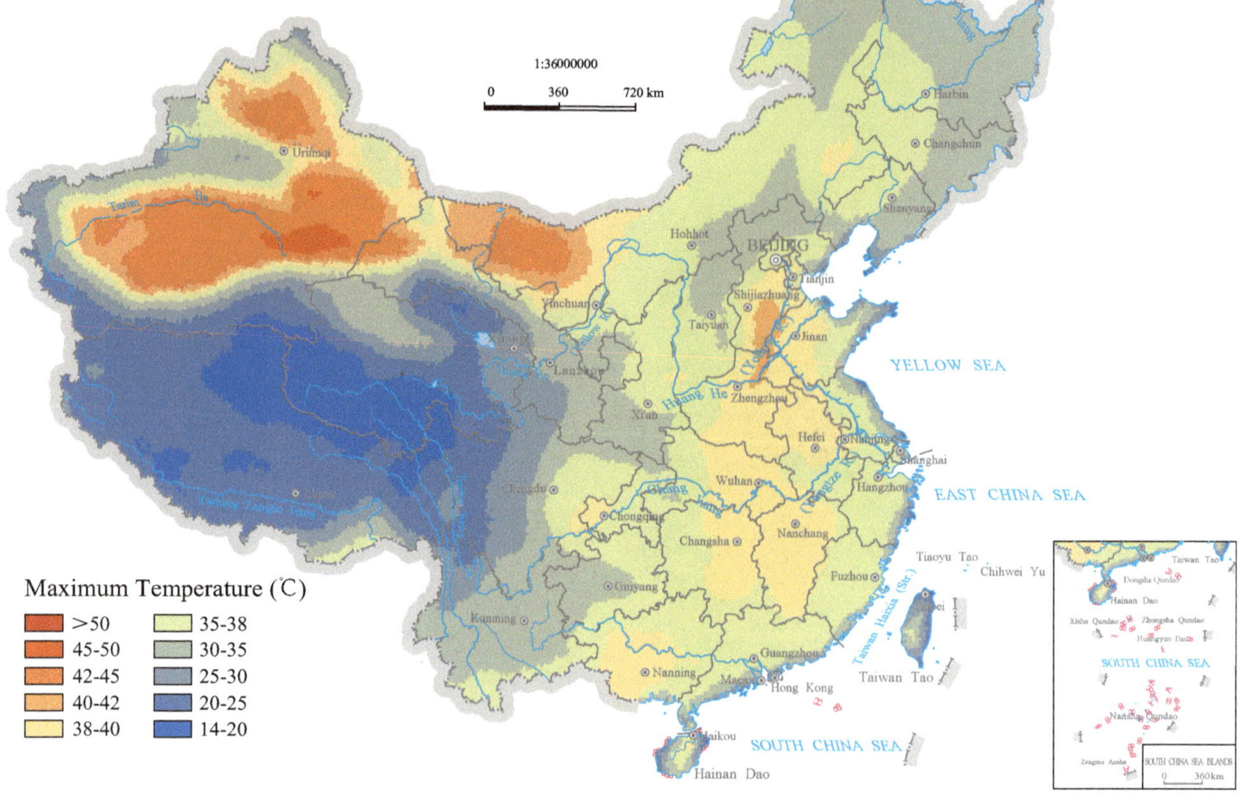

The Number of Days of Heat Waves over 2011-2040 for RCP6.0

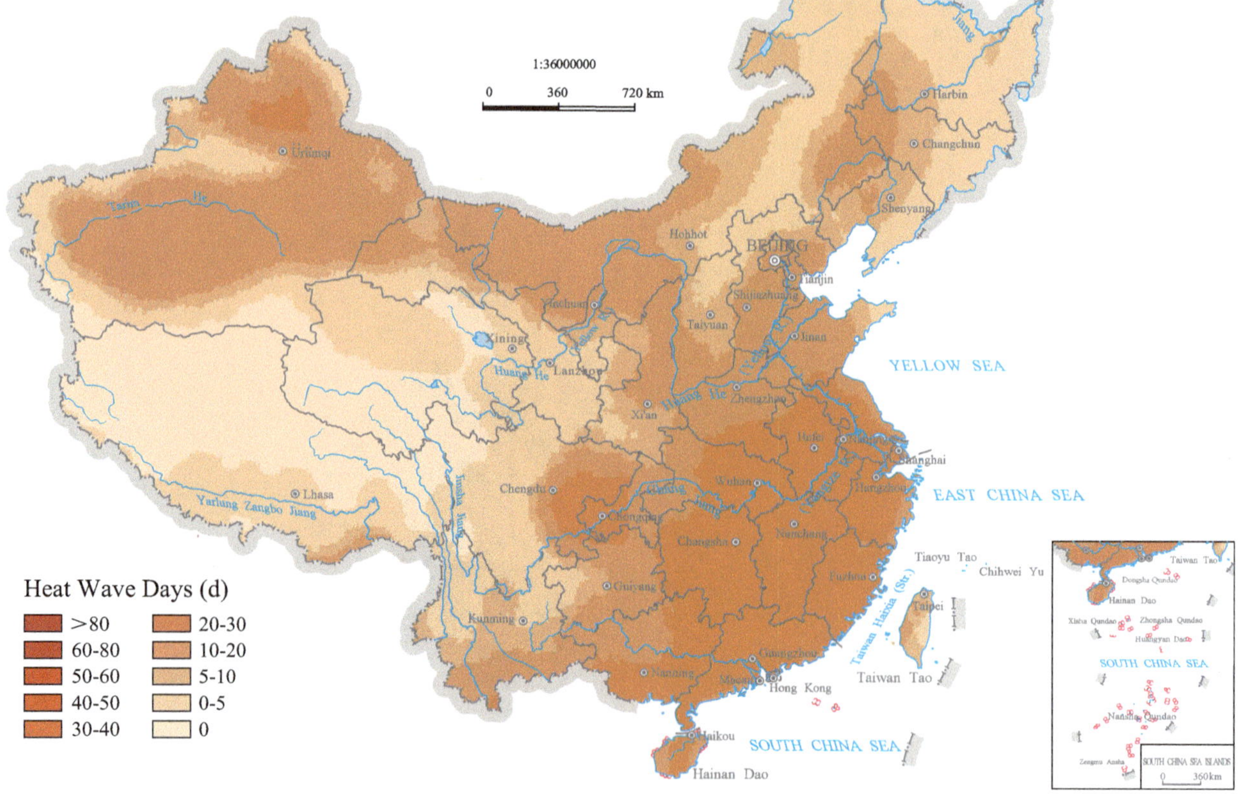

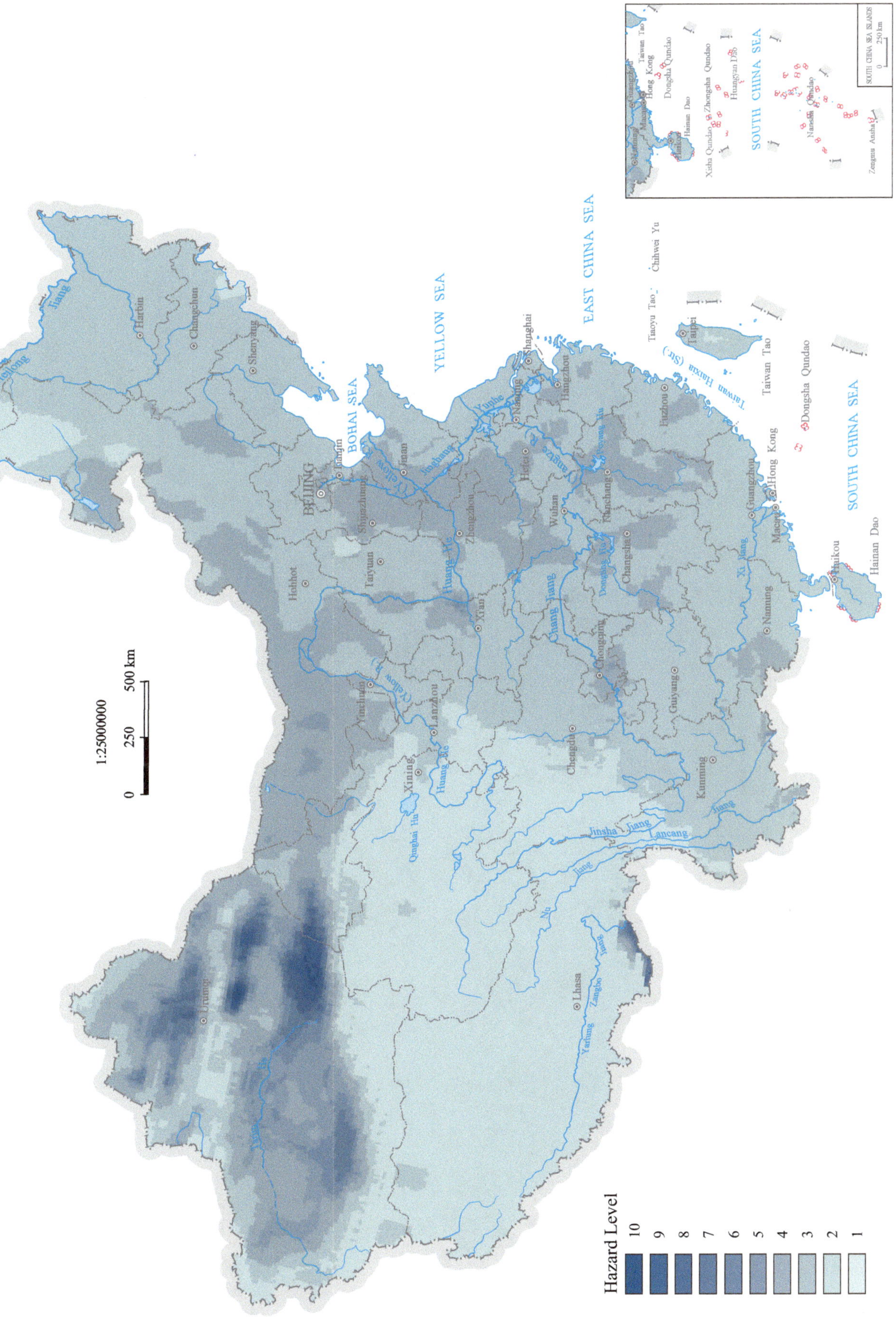

Hazard of Mortality to High Temperature over 2011-2040 for RCP6.0

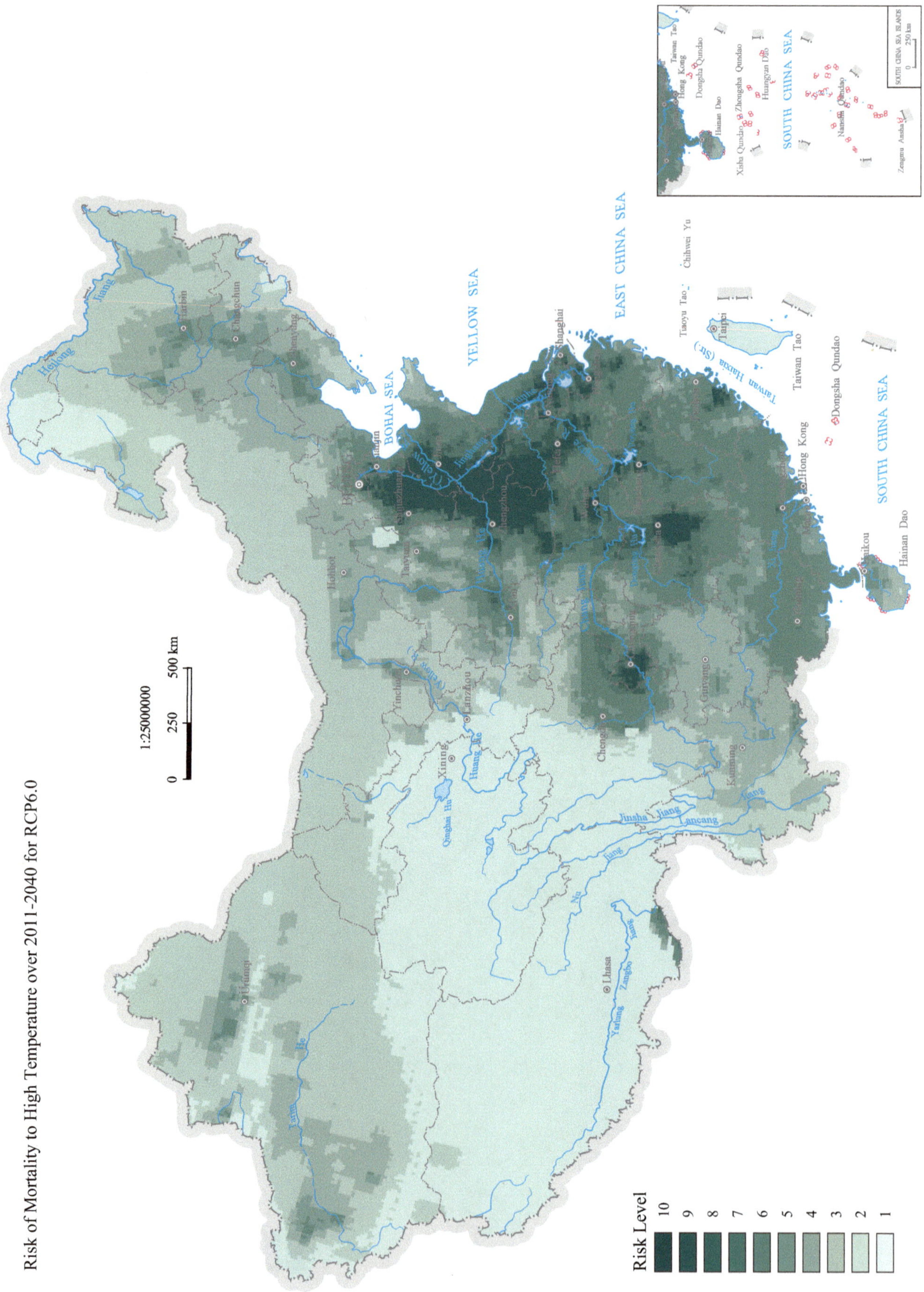
Risk of Mortality to High Temperature over 2011-2040 for RCP6.0
1:25000000
0 250 500 km
Risk Level
10
9
8
7
6
5
4
3
2
1
BOHAI SEA
YELLOW SEA
EAST CHINA SEA
SOUTH CHINA SEA
Taiwan Tao
Dongsha Qundao
Hainan Dao
Hong Kong
Taibei
Shanghai
Harbin
Changchun
Shenyang
Hohhot
Taiyuan
Lanzhou
Xining
Chengdu
Guiyang
Kunming
Lhasa
Urumqi
Haikou
Qinghai Hu
Huang He
Heilong Jiang
Jinsha Jiang
Lancang Jiang
Yarlung Zangbo Jiang
Nu Jiang
Chihwei Yu
Tiaoyu Tao
Taiwan Haixia (Str.)
Xisha Qundao
Zhongsha Qundao
Huangyan Dao
Dongsha Qundao
Nansha Qundao
Zengmu Ansha
SOUTH CHINA SEA ISLANDS
0 250 km

The Number of Days with Daily Maximum Temperature over 35°C in 2041-2070 for RCP6.0

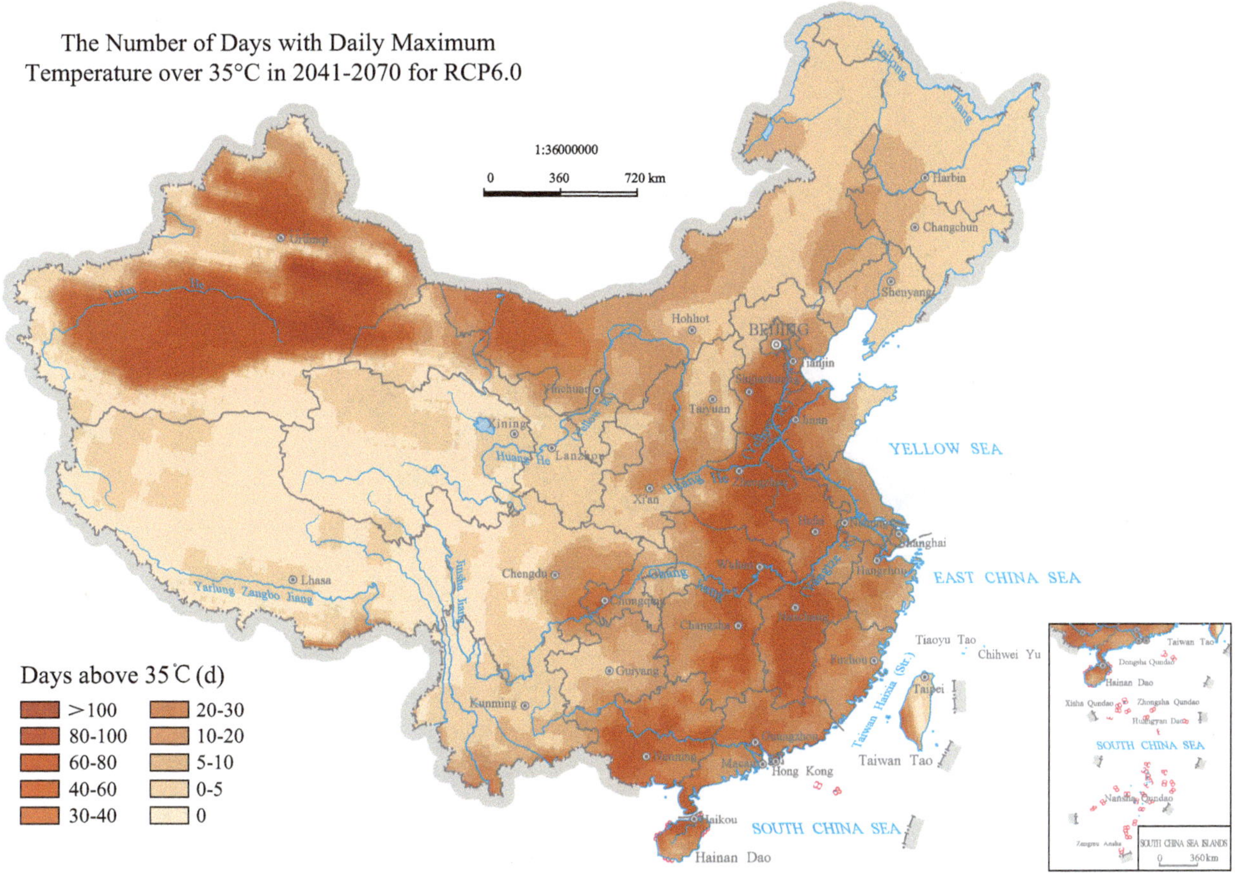

The Number of Days with Daily Maximum Temperature over 40°C in 2041-2070 for RCP6.0

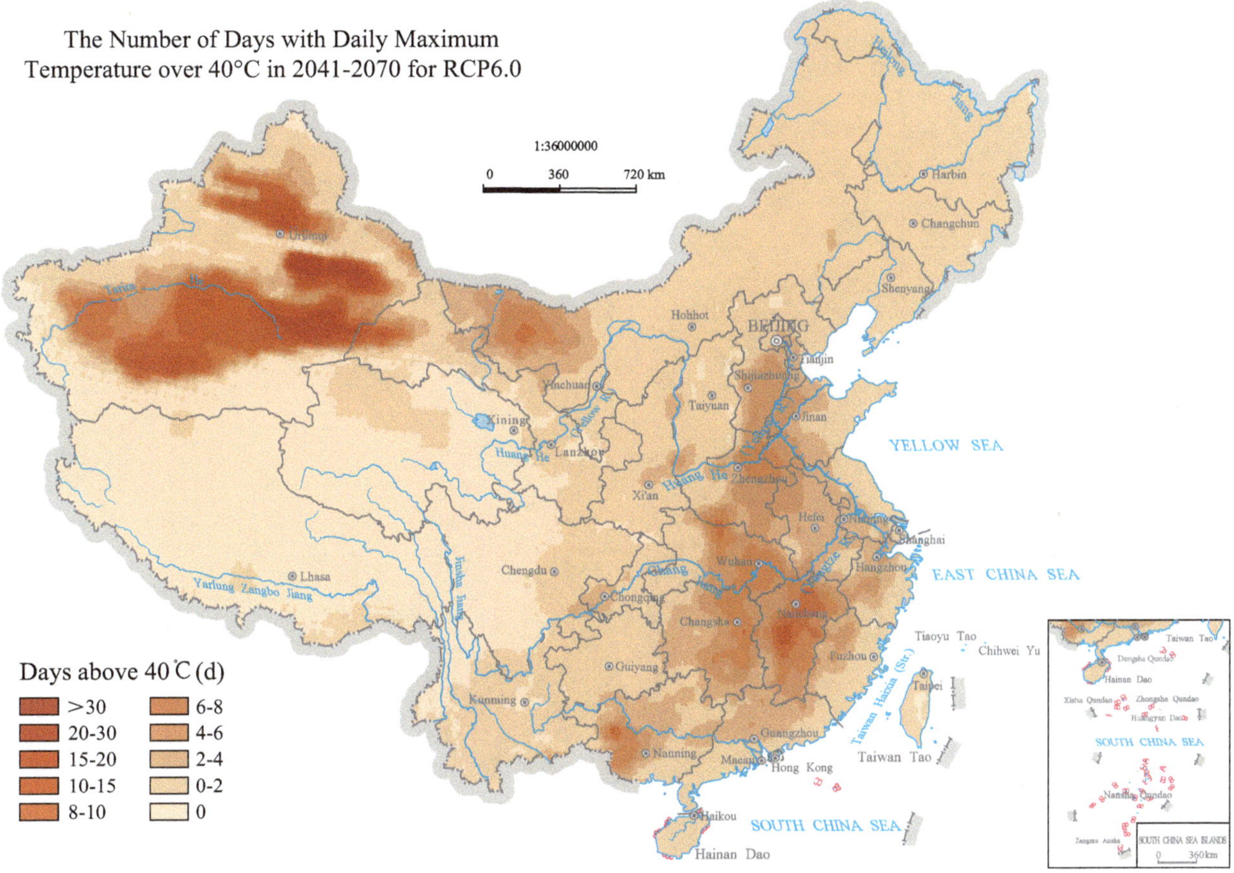

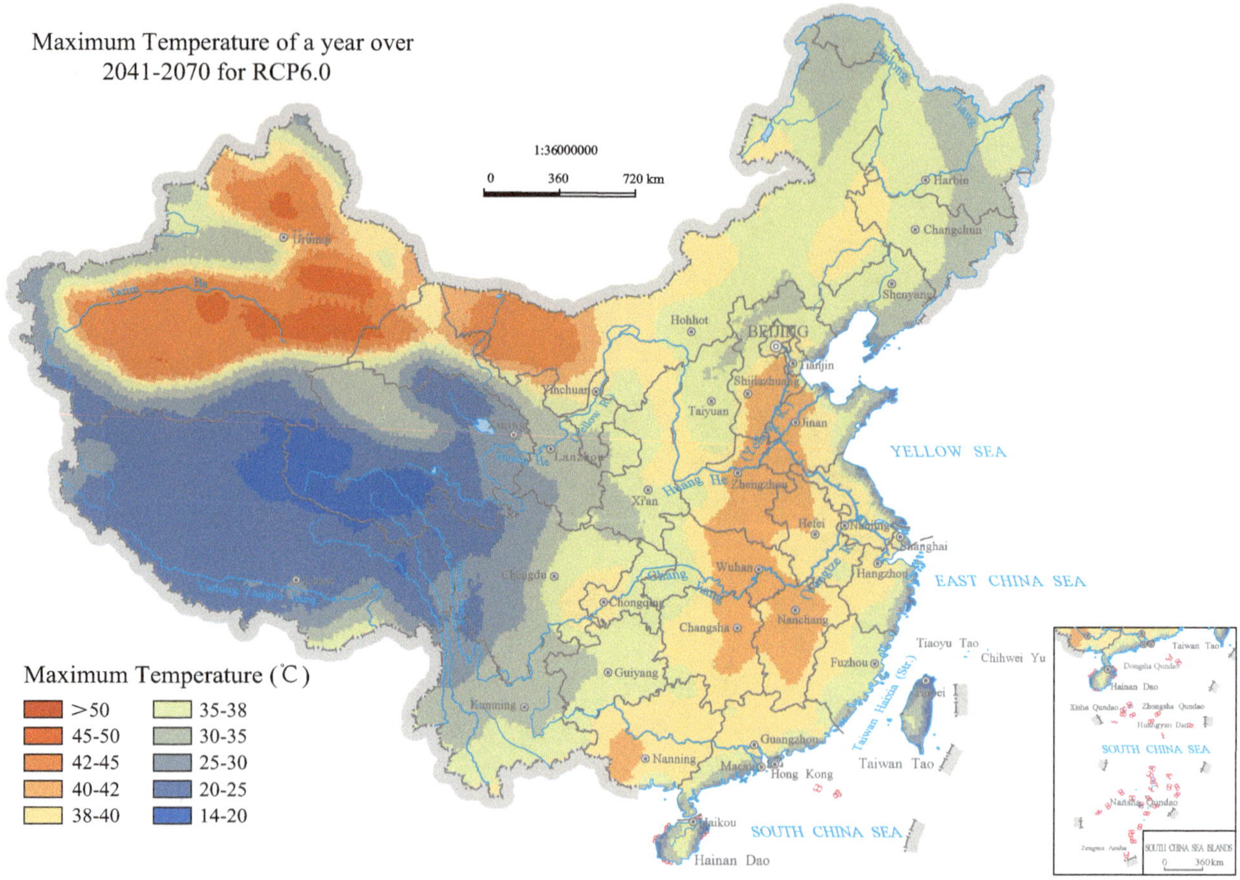

Maximum Temperature of a year over 2041-2070 for RCP6.0
1:36000000
0 360 720 km
Maximum Temperature (℃)
>50
45-50
42-45
40-42
38-40
35-38
30-35
25-30
20-25
14-20
Harbin
Changchun
Shenyang
Hohhot
BEIJING
Tianjin
Shijiazhuang
Taiyuan
Yinchuan
Jinan
Lanzhou
Xi'an
Zhengzhou
Hefei
Nanjing
Shanghai
Wuhan
Hangzhou
Chengdu
Chongqing
Changsha
Nanchang
Guiyang
Fuzhou
Kunming
Nanning
Guangzhou
Macao
Hong Kong
Haikou
Hainan Dao
YELLOW SEA
EAST CHINA SEA
SOUTH CHINA SEA
Tiaoyu Tao
Chihwei Yu
Taiwan Tao

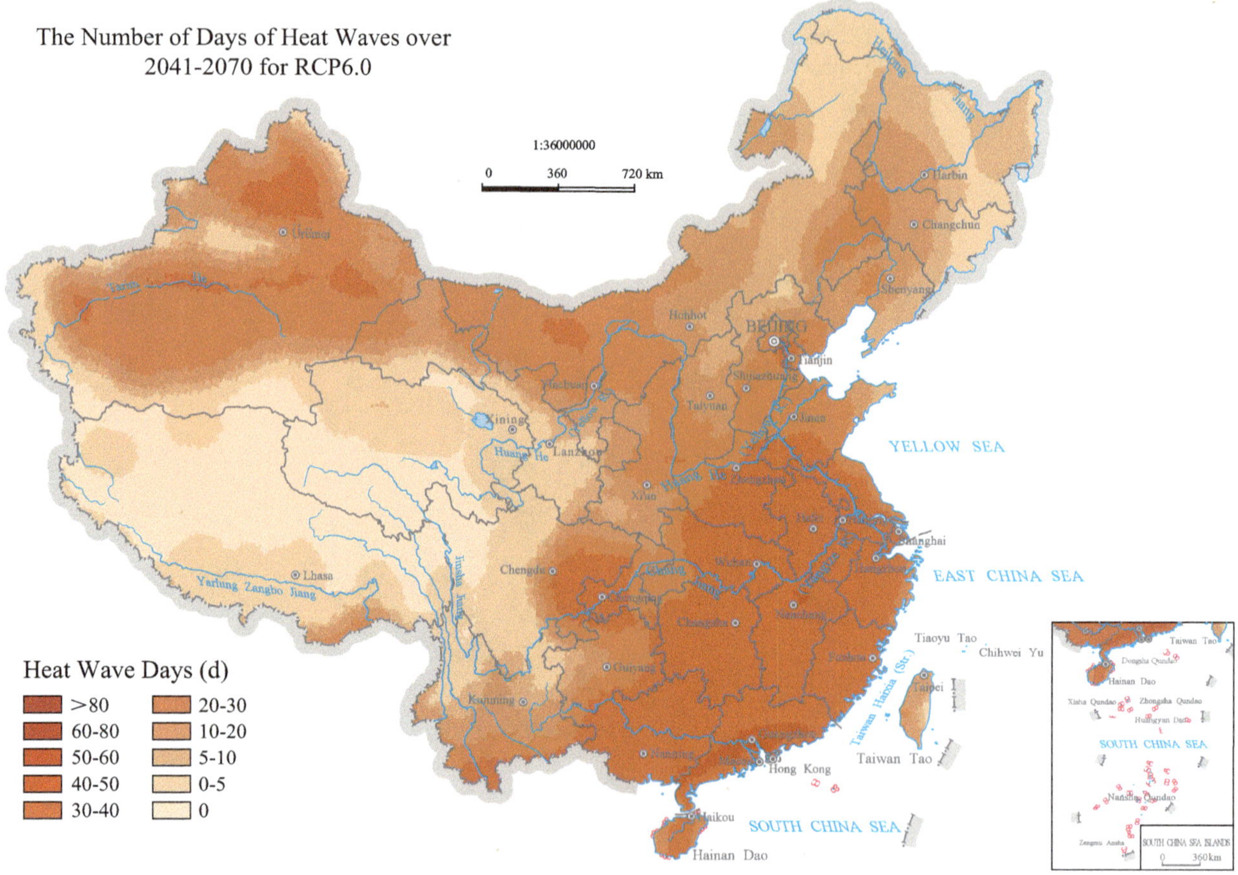

The Number of Days of Heat Waves over 2041-2070 for RCP6.0
1:36000000
0 360 720 km
Heat Wave Days (d)
>80
60-80
50-60
40-50
30-40
20-30
10-20
5-10
0-5
0
Harbin
Changchun
Shenyang
Hohhot
BEIJING
Tianjin
Xining
Lanzhou
Xi'an
Lhasa
Yarlung Zangbo Jiang
Chengdu
Kunming
Shanghai
Hong Kong
Haikou
Hainan Dao
YELLOW SEA
EAST CHINA SEA
SOUTH CHINA SEA
Tiaoyu Tao
Chihwei Yu
Taiwan Tao

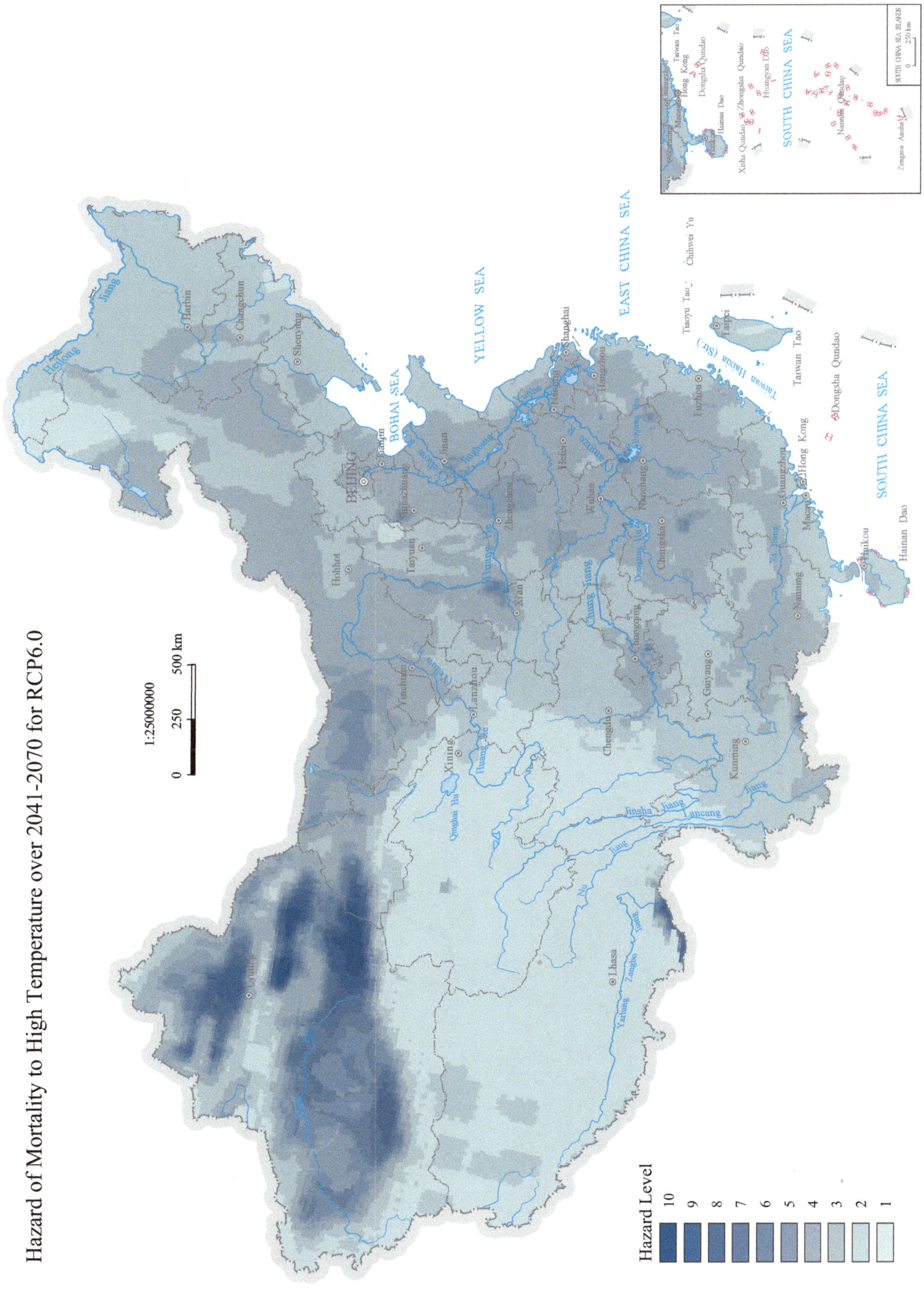
Hazard of Mortality to High Temperature over 2041-2070 for RCP6.0
1:25000000
0
250
500 km
Hazard Level
10
9
8
7
6
5
4
3
2
1
BOHAI SEA
YELLOW SEA
EAST CHINA SEA
SOUTH CHINA SEA
BEIJING
Harbin
Changchun
Shenyang
Tianjin
Shijiazhuang
Jinan
Hohhot
Taiyuan
Zhengzhou
Xi'an
Yinchuan
Lanzhou
Xining
Qinghai Hu
Chengdu
Chongqing
Guiyang
Kunming
Nanning
Guangzhou
Macao
Hong Kong
Haikou
Hainan Dao
Changsha
Wuhan
Nanchang
Hefei
Nanjing
Shanghai
Hangzhou
Fuzhou
Taipei
Taiwan Tao
Diaoyu Dao
Chiwei Yu
Dongsha Qundao
Lhasa
Urumqi
Heilong Jiang
Huang He
Chang Jiang
Jinsha Jiang
Lancang Jiang
Nu Jiang
Yarlung Zangbo Jiang
SOUTH CHINA SEA ISLANDS
Xisha Qundao
Zhongsha Qundao
Nansha Qundao
Huangyan Dao
Zengmu Ansha

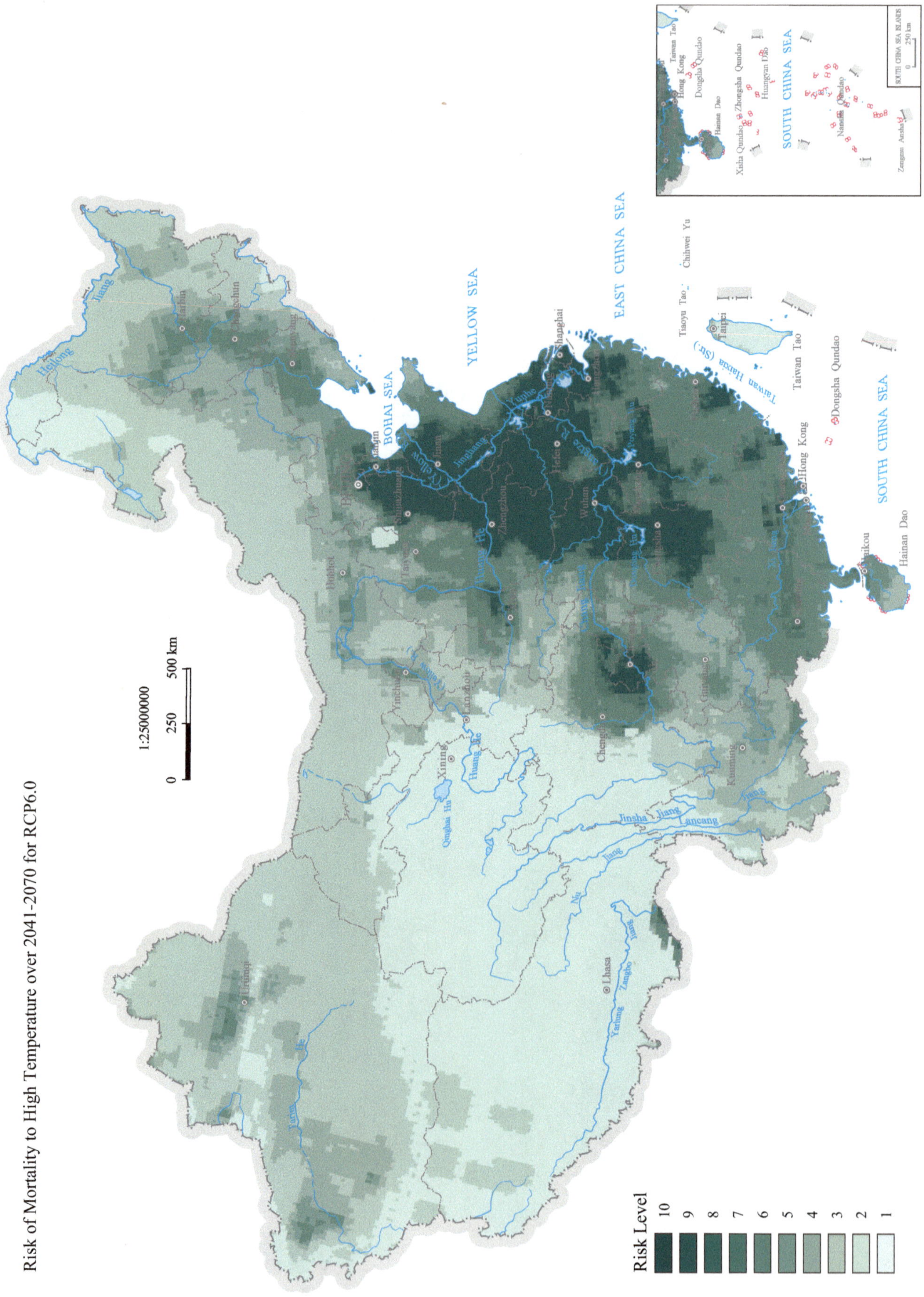
Risk of Mortality to High Temperature over 2041-2070 for RCP6.0
1:25000000
0 250 500 km
Risk Level
10
9
8
7
6
5
4
3
2
1
BOHAI SEA
YELLOW SEA
EAST CHINA SEA
SOUTH CHINA SEA
Taiwan Tao
Dongsha Qundao
Hainan Dao
Hong Kong
Shanghai
Taipei
Lhasa
Xining
Lanzhou
Chengdu
Kunming
Harbin
Changchun
Hohhot
Haikou
Qinghai Hu
Jinsha Jiang
Lancang Jiang
Yarlung Zangbo Jiang
Heilong Jiang
Taiwan Haixia (Str.)
Chiwei Yu
Tiaoyu Tao
SOUTH CHINA SEA ISLANDS

The Number of Days with Daily Maximum Temperature over 35°C in 2071-2099 for RCP6.0

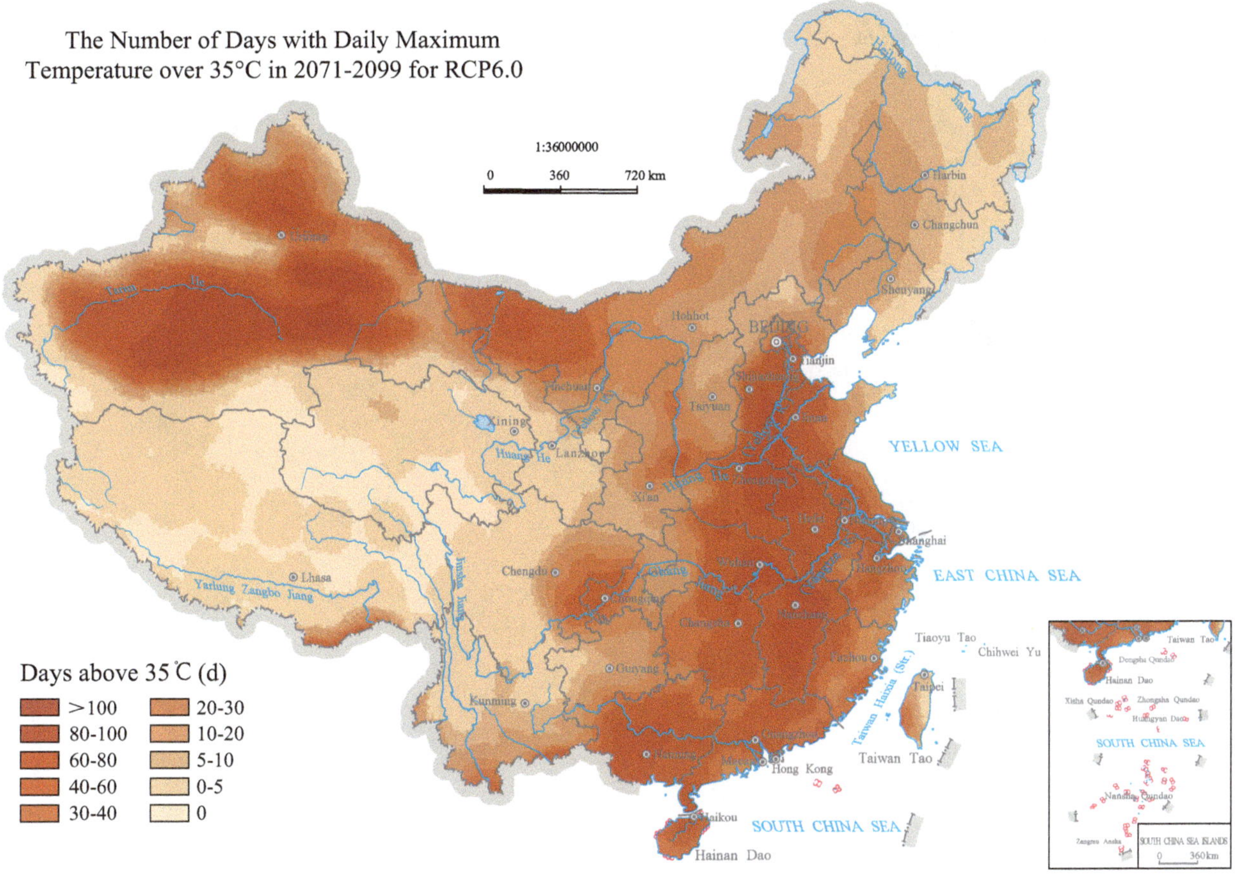

The Number of Days with Daily Maximum Temperature over 40°C in 2071-2099 for RCP6.0

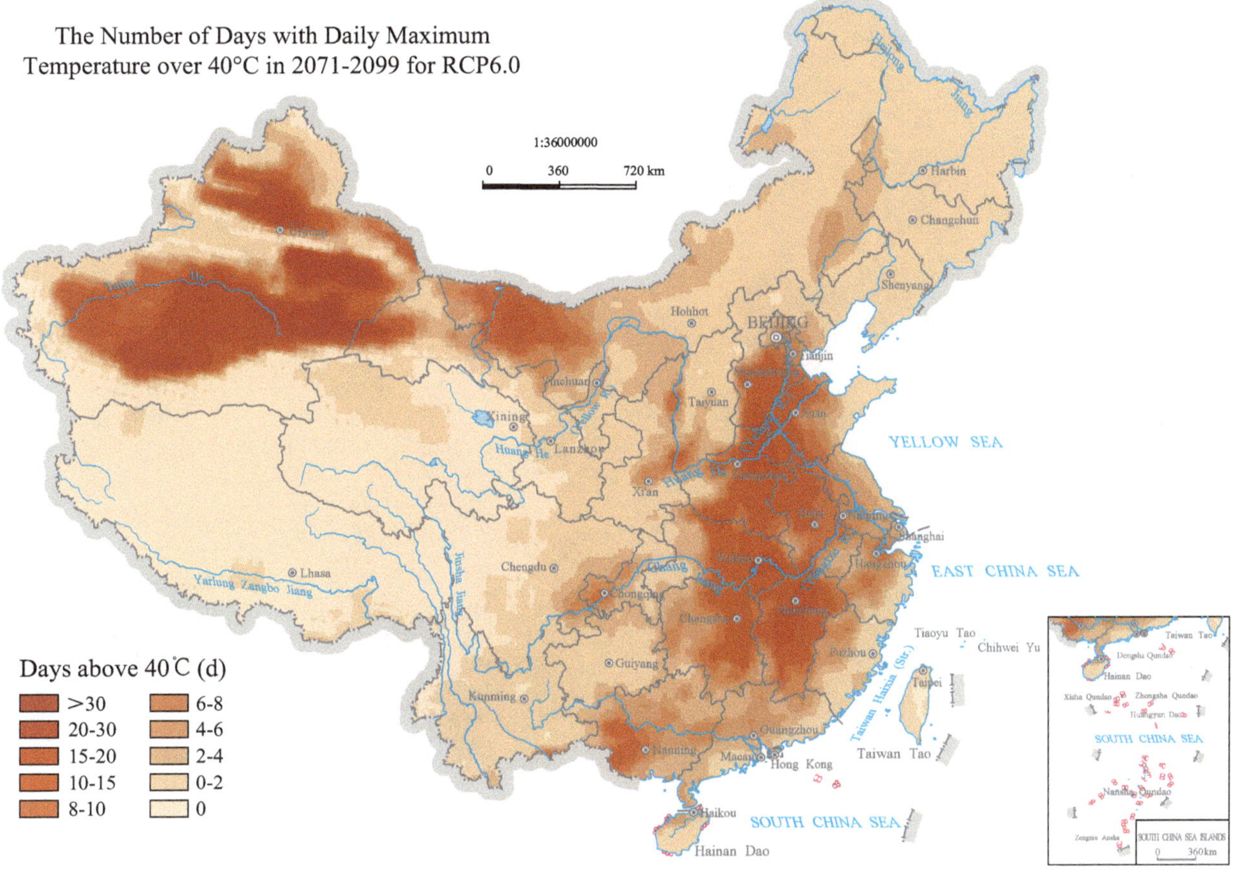

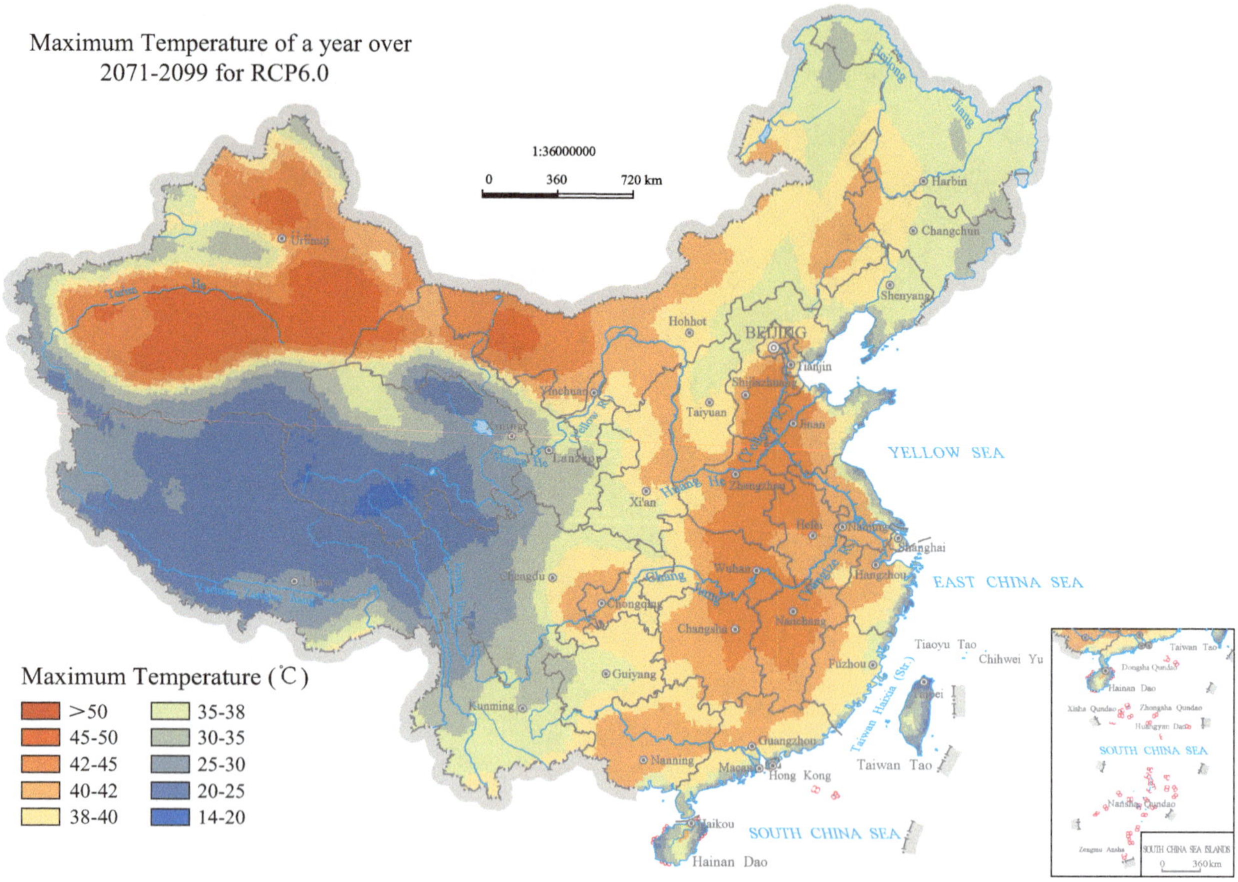
Maximum Temperature of a year over
2071-2099 for RCP6.0
1:36000000
0 360 720 km
Maximum Temperature (℃)
>50
45-50
42-45
40-42
38-40
35-38
30-35
25-30
20-25
14-20
Harbin
Changchun
Shenyang
Hohhot
BEIJING
Tianjin
Taiyuan
Jinan
Xi'an
Zhengzhou
Hefei
Nanjing
Shanghai
Hangzhou
Wuhan
Chengdu
Chongqing
Changsha
Nanchang
Fuzhou
Guiyang
Kunming
Nanning
Guangzhou
Macao
Hong Kong
Taipei
Haikou
Hainan Dao
YELLOW SEA
EAST CHINA SEA
SOUTH CHINA SEA
Tiaoyu Tao
Chihwei Yu
Taiwan Tao
SOUTH CHINA SEA ISLANDS

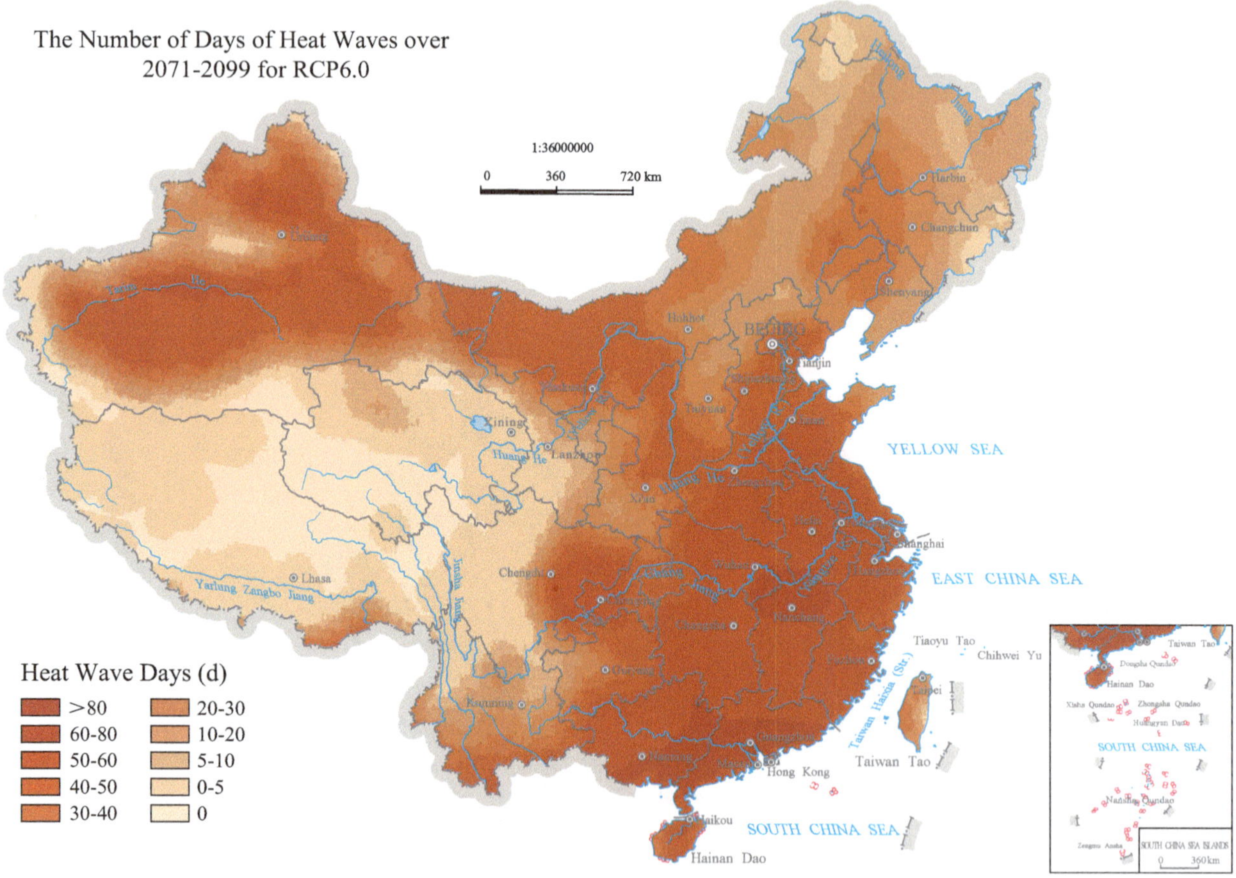
The Number of Days of Heat Waves over
2071-2099 for RCP6.0
1:36000000
0 360 720 km
Heat Wave Days (d)
>80
60-80
50-60
40-50
30-40
20-30
10-20
5-10
0-5
0
Harbin
Changchun
Shenyang
BEIJING
Tianjin
Xining
Lanzhou
Huang He
Lhasa
Yarlung Zangbo Jiang
Chengdu
Kunming
Shanghai
Hong Kong
Haikou
Hainan Dao
YELLOW SEA
EAST CHINA SEA
SOUTH CHINA SEA
Tiaoyu Tao
Chihwei Yu
Taiwan Tao
SOUTH CHINA SEA ISLANDS

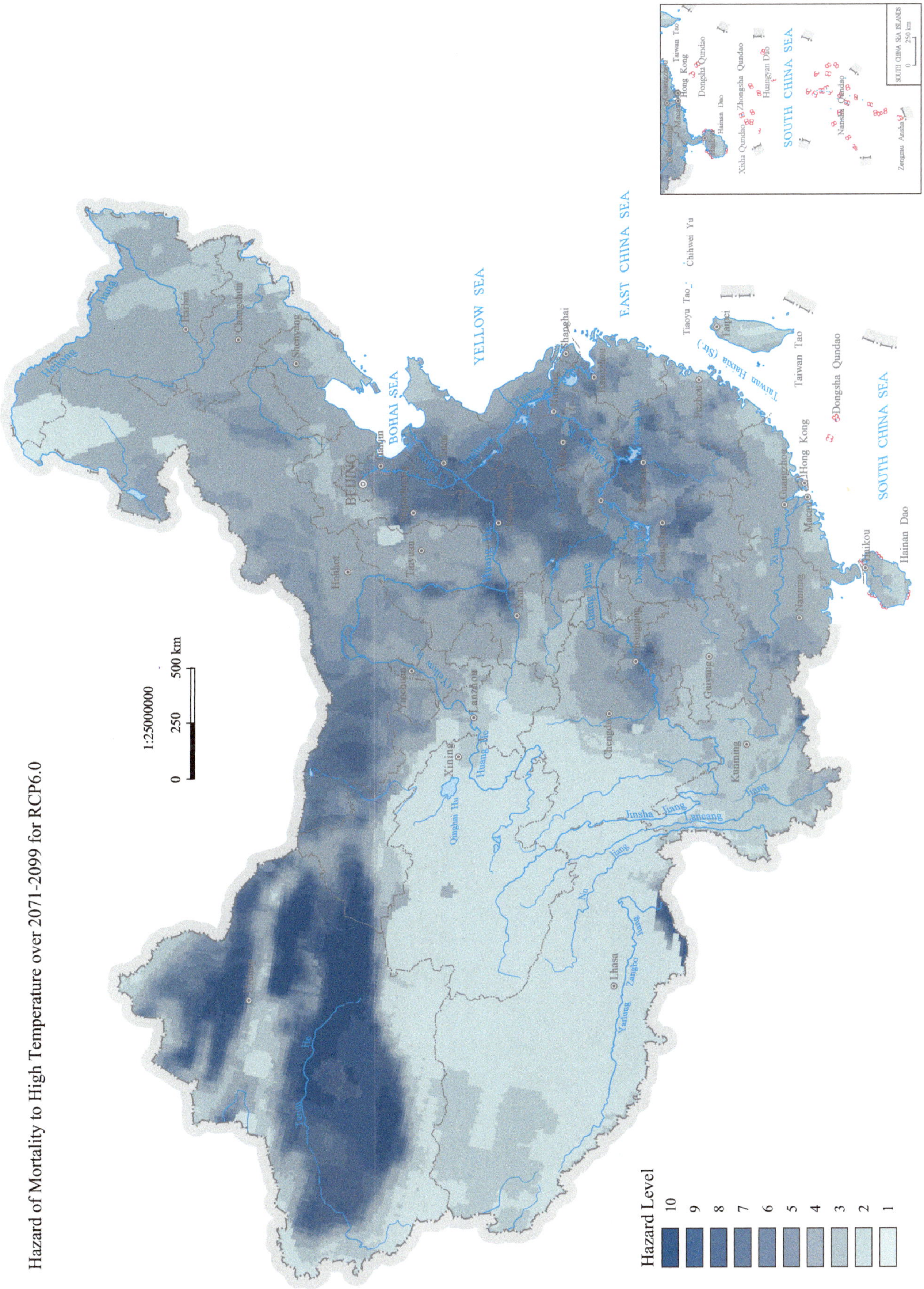

Hazard of Mortality to High Temperature over 2071-2099 for RCP6.0

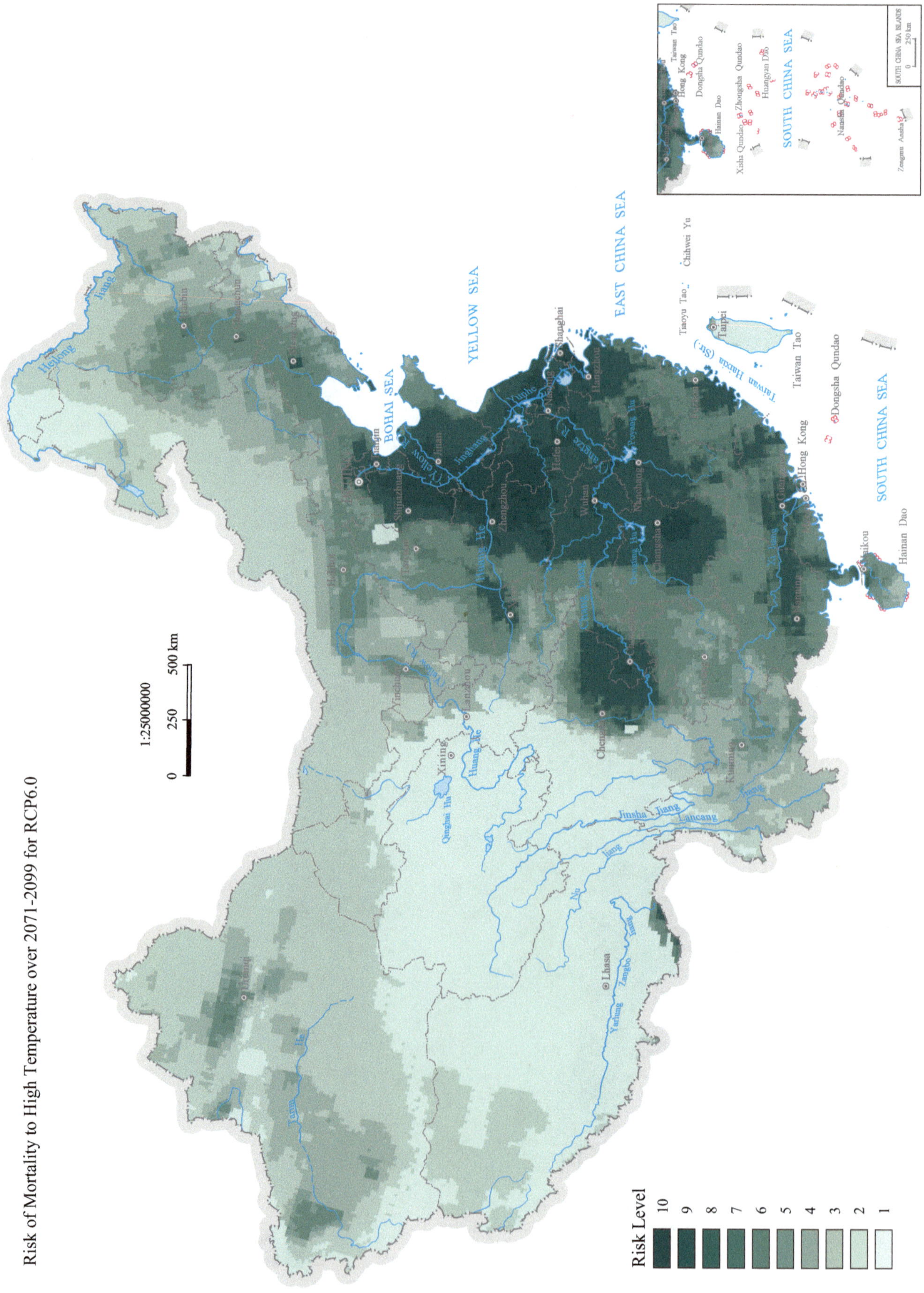
Risk of Mortality to High Temperature over 2071-2099 for RCP6.0
1:25000000
0 250 500 km
Risk Level
10
9
8
7
6
5
4
3
2
1
YELLOW SEA
EAST CHINA SEA
BOHAI SEA
SOUTH CHINA SEA
Taiwan Haixia (Str.)
Taiwan Tao
Dongsha Qundao
Hong Kong
Hainan Dao
Diaoyu Dao
Chihwei Yu
Taipei
Shanghai
Nanjing
Hangzhou
Hefei
Wuhan
Nanchang
Changsha
Zhengzhou
Jinan
Shijiazhuang
Yinchuan
Lanzhou
Xining
Chengdu
Kunming
Lhasa
Qinghai Hu
Huang He
Jinsha Jiang
Lancang
Yarlung Zangbo
Heilong Jiang
Yangtze R.
Poyang Hu
SOUTH CHINA SEA ISLANDS

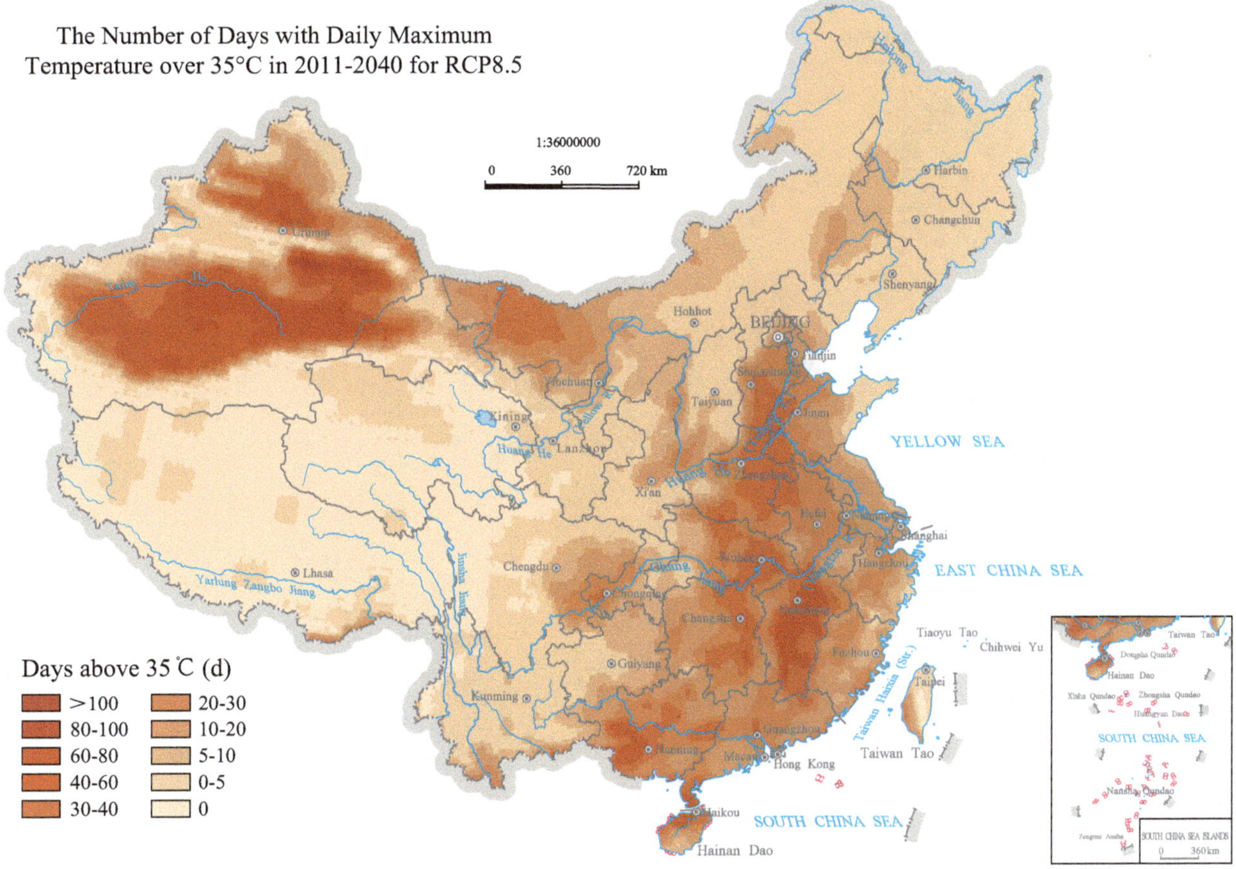
The Number of Days with Daily Maximum
Temperature over 35°C in 2011-2040 for RCP8.5
1:36000000
0 360 720 km
Harbin
Changchun
Shenyang
Urumqi
Hohhot
BEIJING
Tianjin
Taiyuan
Jinan
Xining
Lanzhou
Xi'an
YELLOW SEA
Shanghai
Hefei
EAST CHINA SEA
Lhasa
Yarlung Zangbo Jiang
Chengdu
Changsha
Tiaoyu Tao
Chihwei Yu
Fuzhou
Guiyang
Kunming
Taipei
Taiwan Tao
Hong Kong
SOUTH CHINA SEA
Haikou
Hainan Dao
Days above 35℃ (d)
>100
80-100
60-80
40-60
30-40
20-30
10-20
5-10
0-5
0

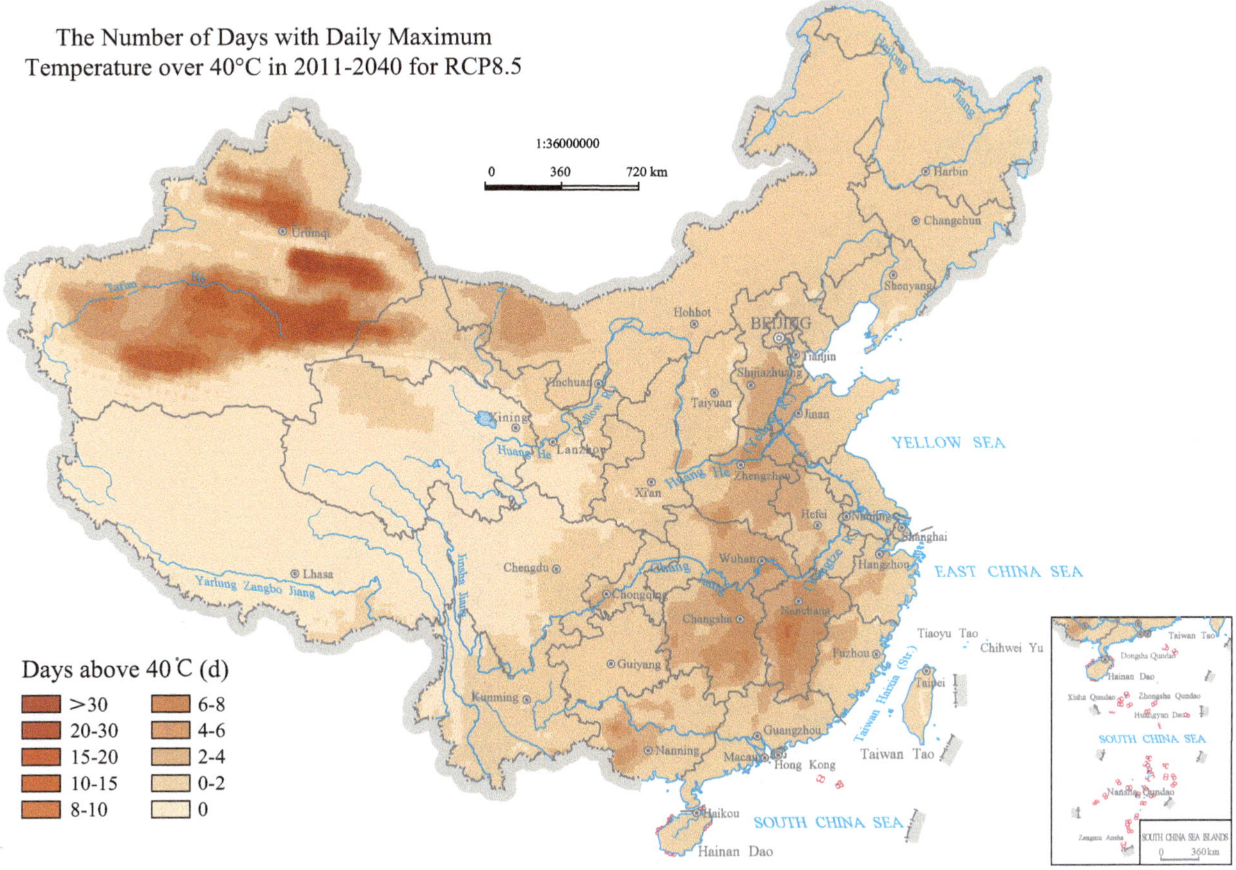
The Number of Days with Daily Maximum
Temperature over 40°C in 2011-2040 for RCP8.5
1:36000000
0 360 720 km
Harbin
Changchun
Shenyang
Urumqi
Hohhot
BEIJING
Tianjin
Shijiazhuang
Taiyuan
Jinan
Xining
Lanzhou
Xi'an
YELLOW SEA
Hefei
Nanjing
Shanghai
Wuhan
Hangzhou
EAST CHINA SEA
Lhasa
Yarlung Zangbo Jiang
Chengdu
Chongqing
Changsha
Nanchang
Tiaoyu Tao
Chihwei Yu
Fuzhou
Guiyang
Kunming
Taipei
Guangzhou
Nanning
Macao
Hong Kong
Taiwan Tao
SOUTH CHINA SEA
Haikou
Hainan Dao
Days above 40℃ (d)
>30
20-30
15-20
10-15
8-10
6-8
4-6
2-4
0-2
0

Maximum Temperature of a year over 2011-2040 for RCP8.5

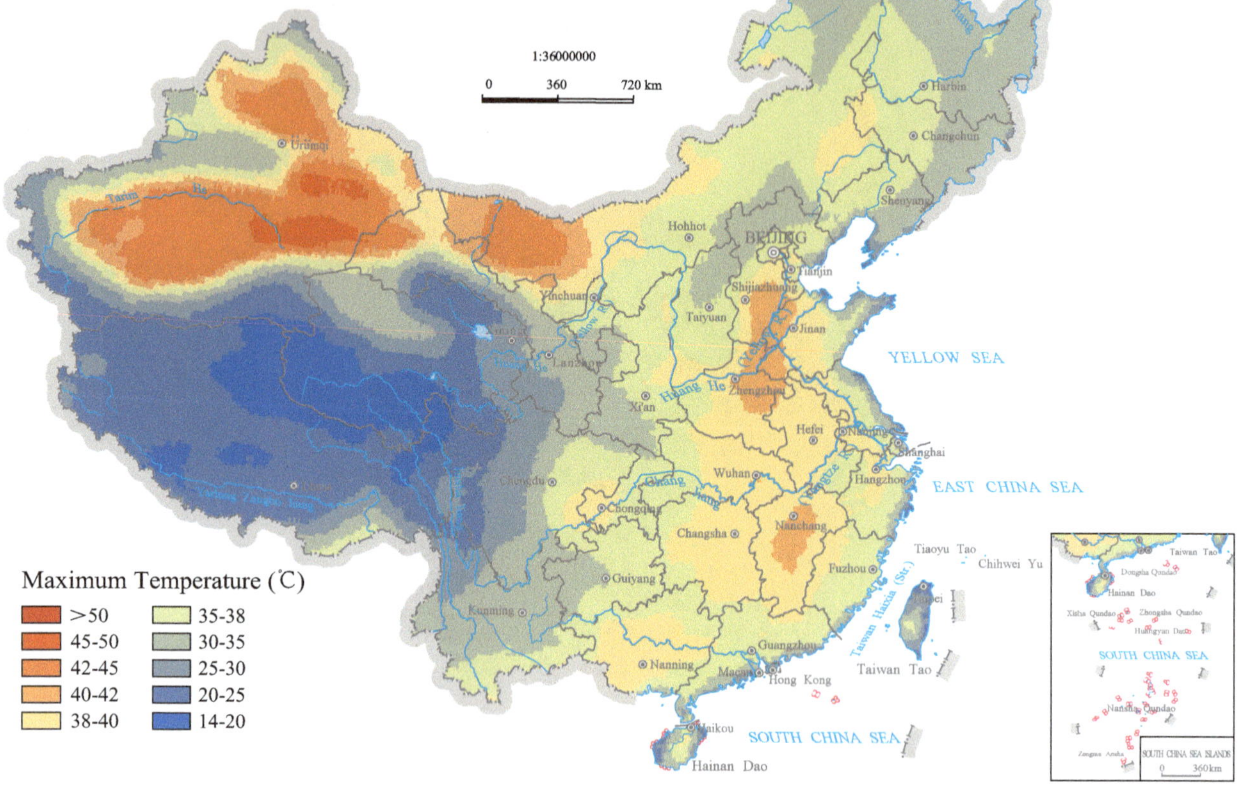

The Number of Days of Heat Waves over 2011-2040 for RCP8.5

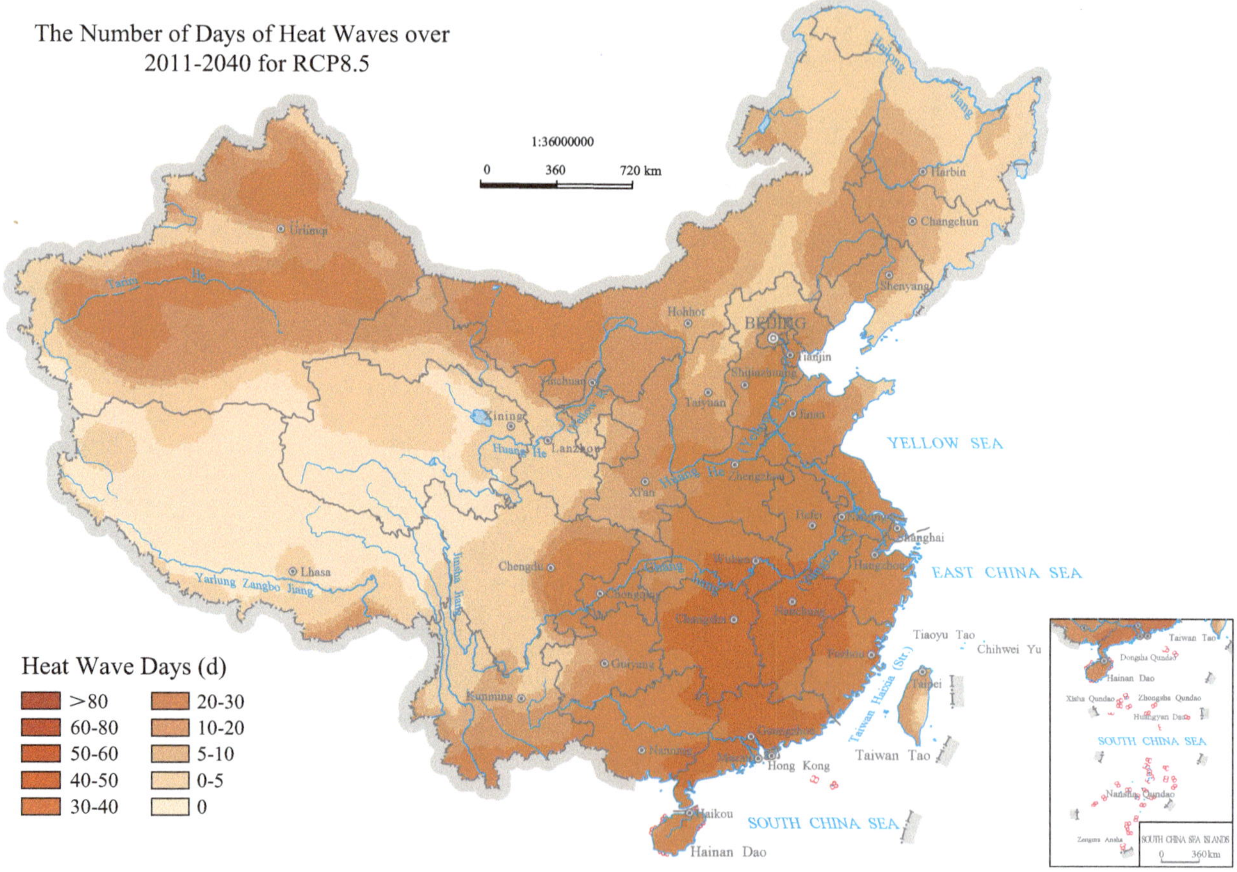

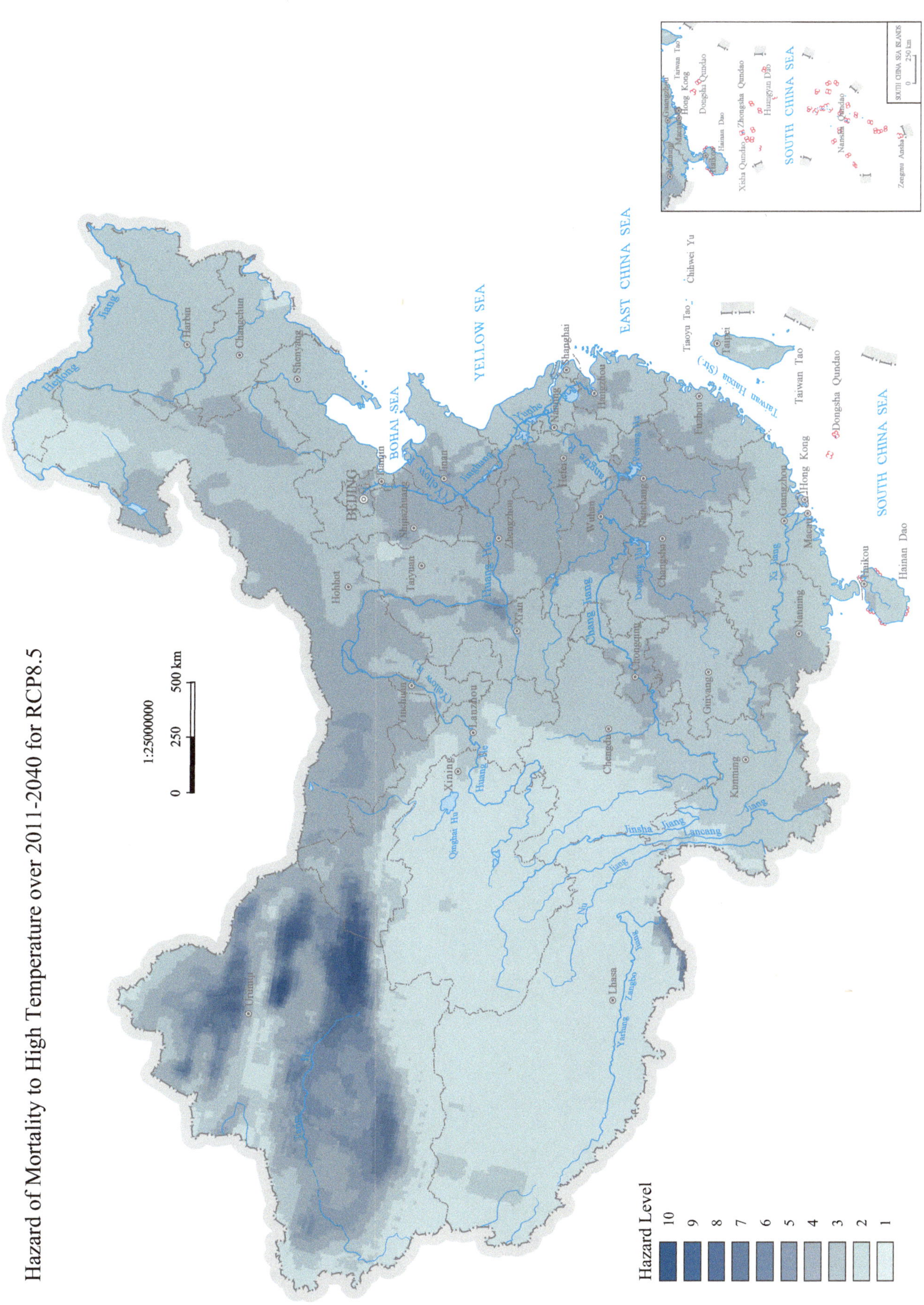
Hazard of Mortality to High Temperature over 2011-2040 for RCP8.5
1:25000000
0 250 500 km
Hazard Level
10
9
8
7
6
5
4
3
2
1
BEIJING
YELLOW SEA
EAST CHINA SEA
BOHAI SEA
SOUTH CHINA SEA
Harbin
Changchun
Shenyang
Tianjin
Jinan
Shijiazhuang
Taiyuan
Hohhot
Zhengzhou
Xi'an
Yinchuan
Lanzhou
Xining
Qinghai Hu
Chengdu
Chongqing
Guiyang
Kunming
Nanning
Lhasa
Urumqi
Wuhan
Hefei
Nanjing
Shanghai
Hangzhou
Fuzhou
Nanchang
Changsha
Guangzhou
Hong Kong
Macau
Haikou
Taipei
Hainan Dao
Taiwan Tao
Dongsha Qundao
Diaoyu Dao
Chiwei Yu
Taiwan Haixia (Su)
Jinsha Jiang
Lancang Jiang
Nu Jiang
Yarlung Zangbo Jiang
Heilong Jiang
Huang He
Chang Jiang
Xi Jiang
Xisha Qundao
Zhongsha Qundao
Huangyan Dao
Nansha Qundao
Zengmu Ansha
SOUTH CHINA SEA ISLANDS

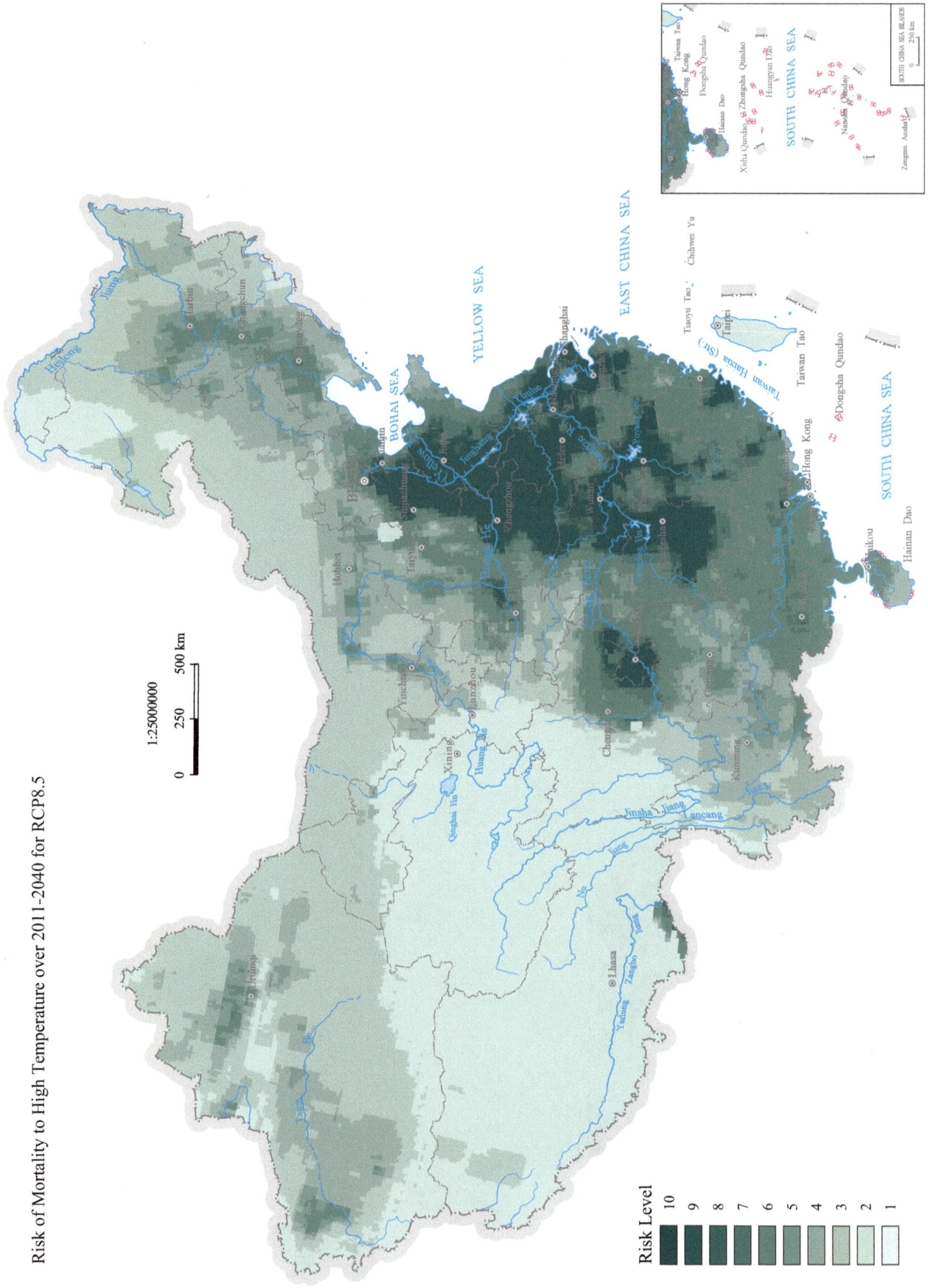
Risk of Mortality to High Temperature over 2011-2040 for RCP8.5
1:25000000
0 250 500 km
Risk Level
10
9
8
7
6
5
4
3
2
1
BOHAI SEA
YELLOW SEA
EAST CHINA SEA
SOUTH CHINA SEA
Taiwan Haixia (Str.)
Taiwan Tao
Dongsha Qundao
Hainan Dao
Hong Kong
Taipei
Shanghai
Lhasa
Xining
Lanzhou
Chengdu
Kunming
Hohhot
Zhengzhou
Harbin
Changchun
Qinghai Hu
Jinsha Jiang
Lancang
Heilong Jiang
Yarlung Zangbo Jiang

The Number of Days with Daily Maximum Temperature over 35°C in 2041-2070 for RCP8.5

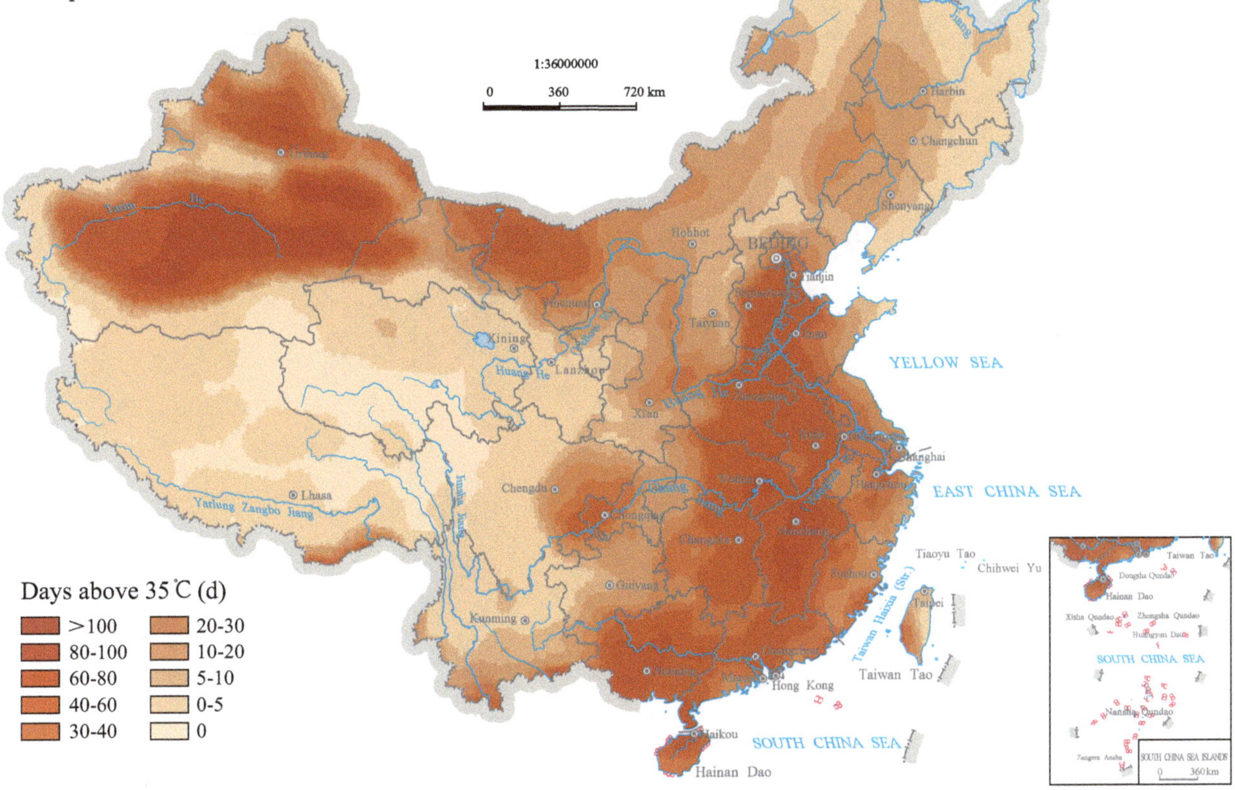

The Number of Days with Daily Maximum Temperature over 40°C in 2041-2070 for RCP8.5

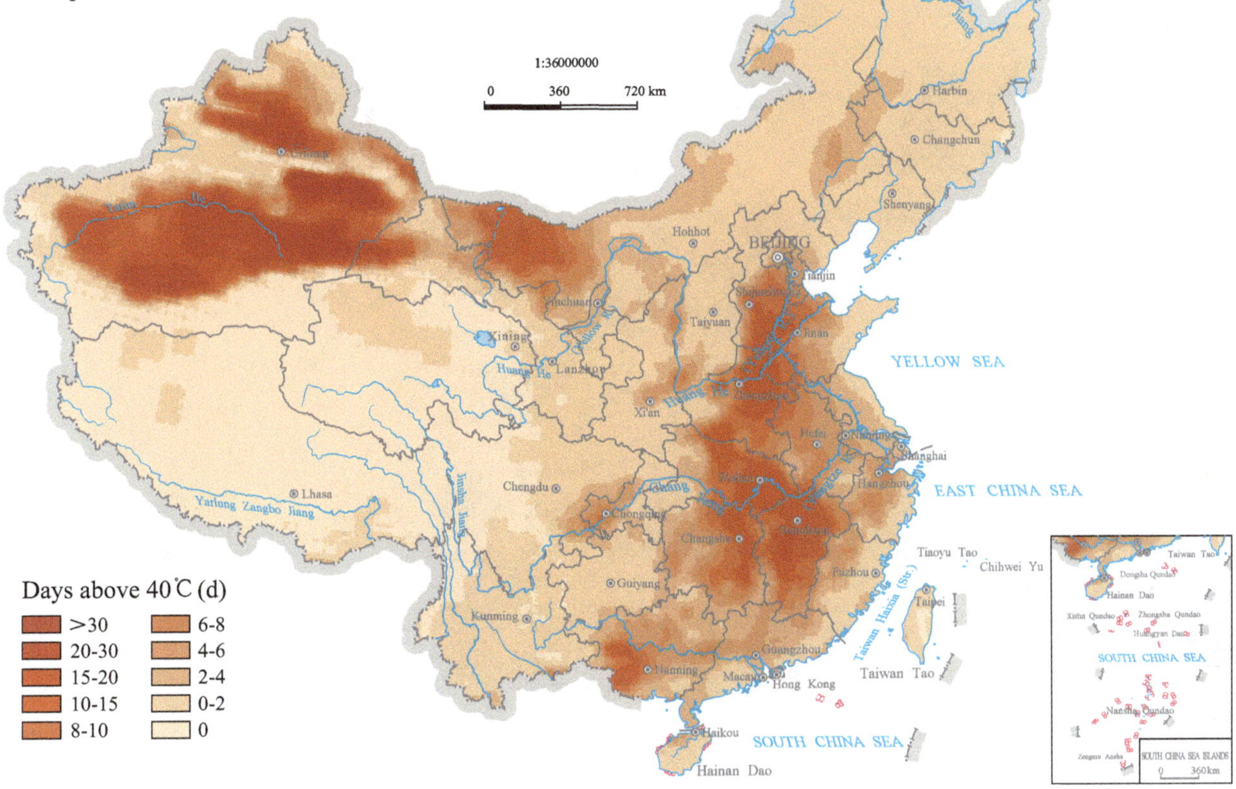

Maximum Temperature of a year over 2041-2070 for RCP8.5

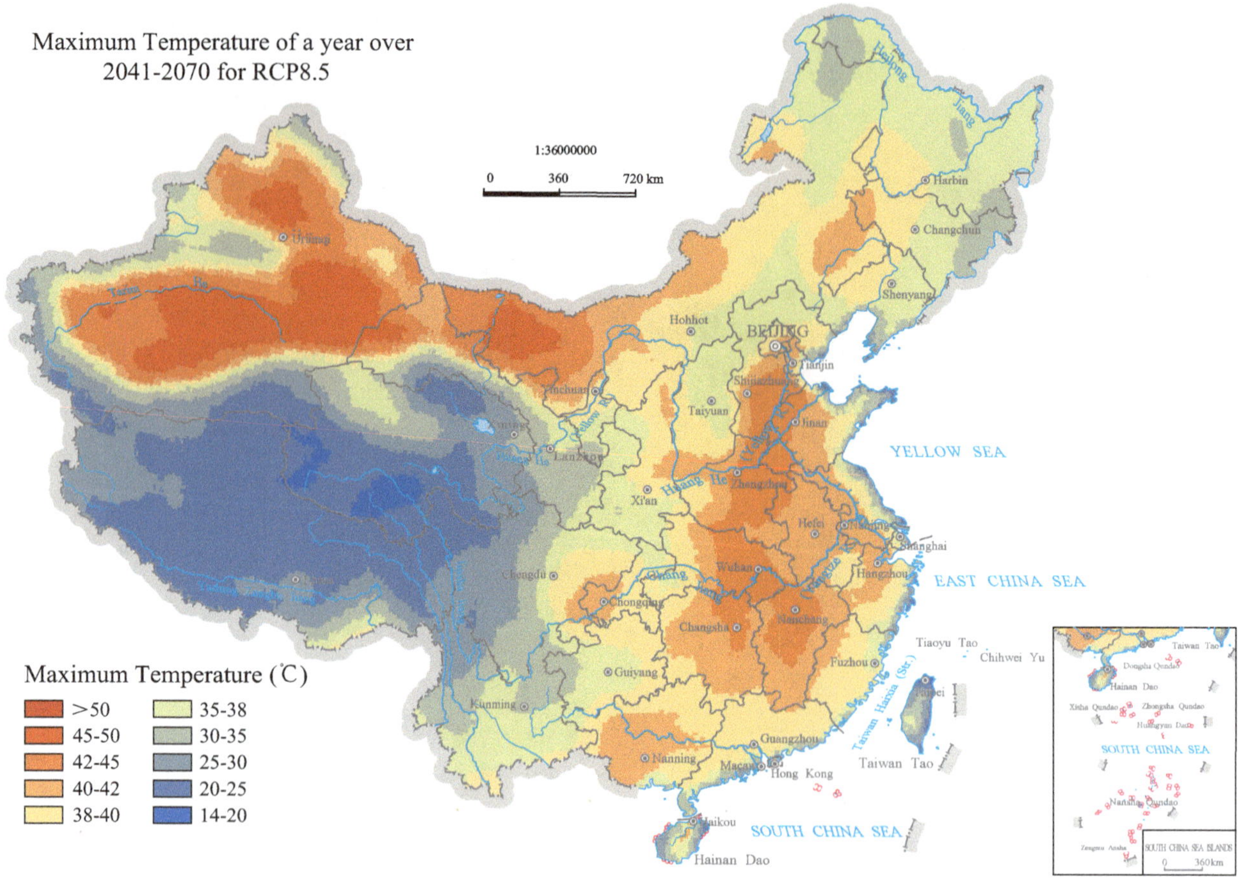

The Number of Days of Heat Waves over 2041-2070 for RCP8.5

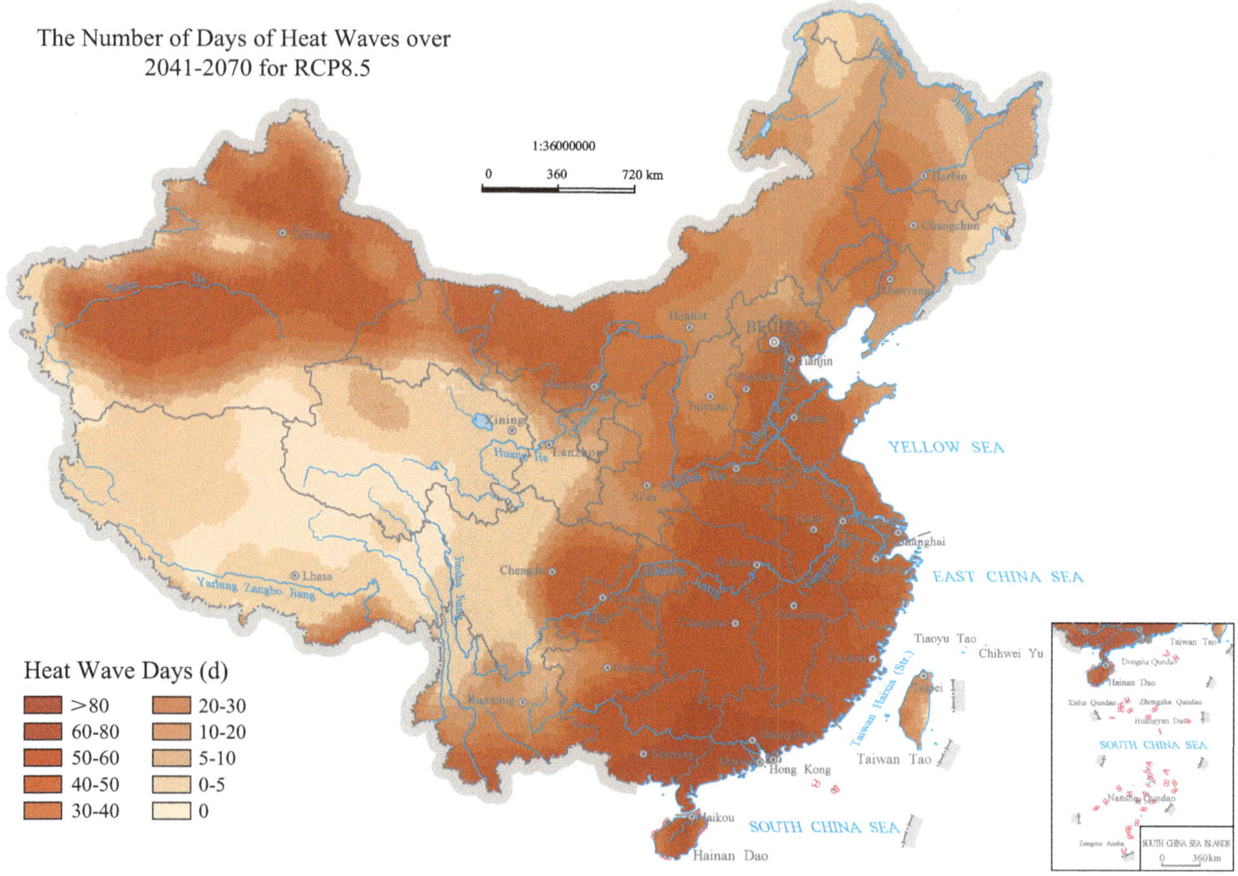

Hazard of Mortality to High Temperature over 2041-2070 for RCP8.5

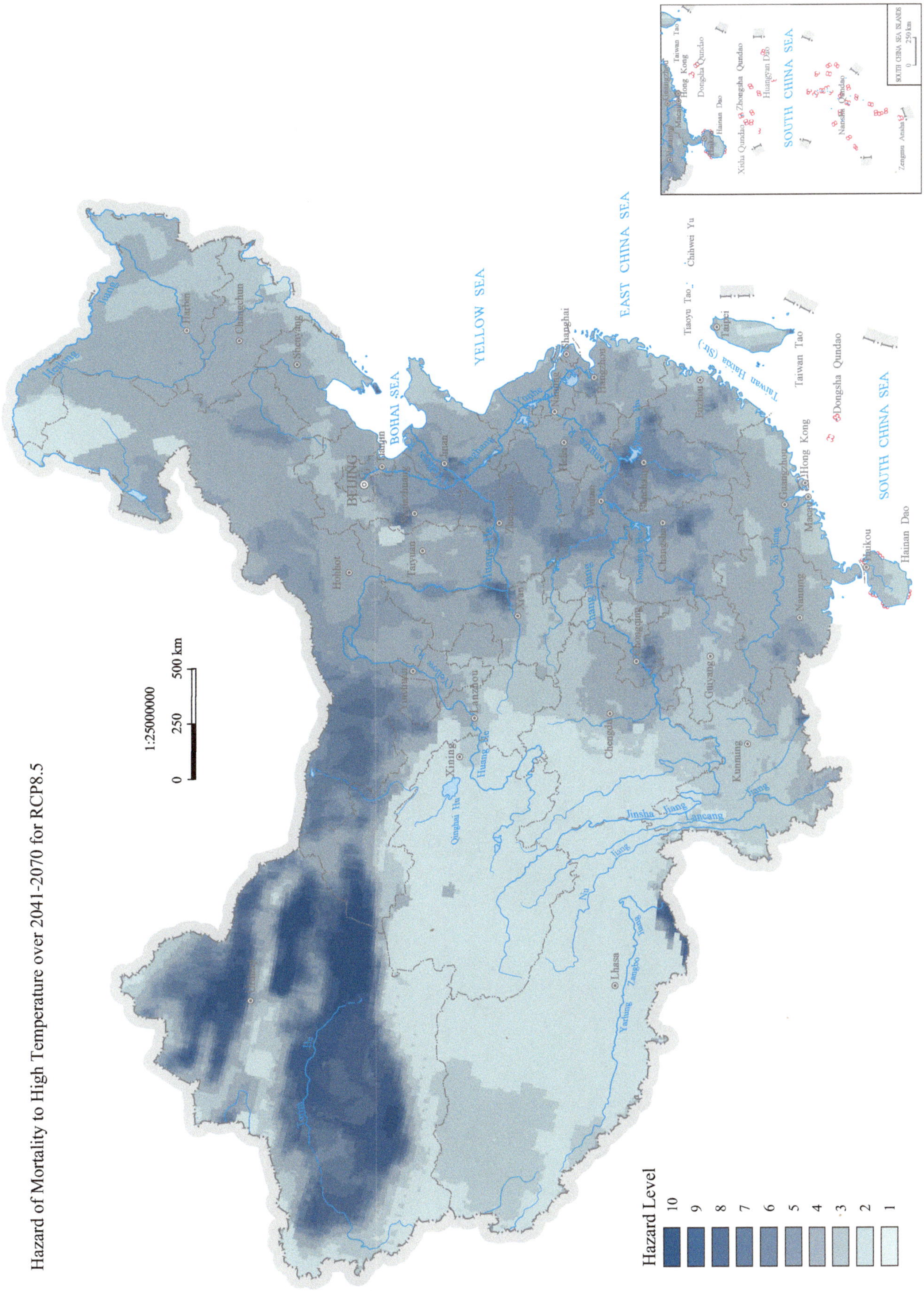

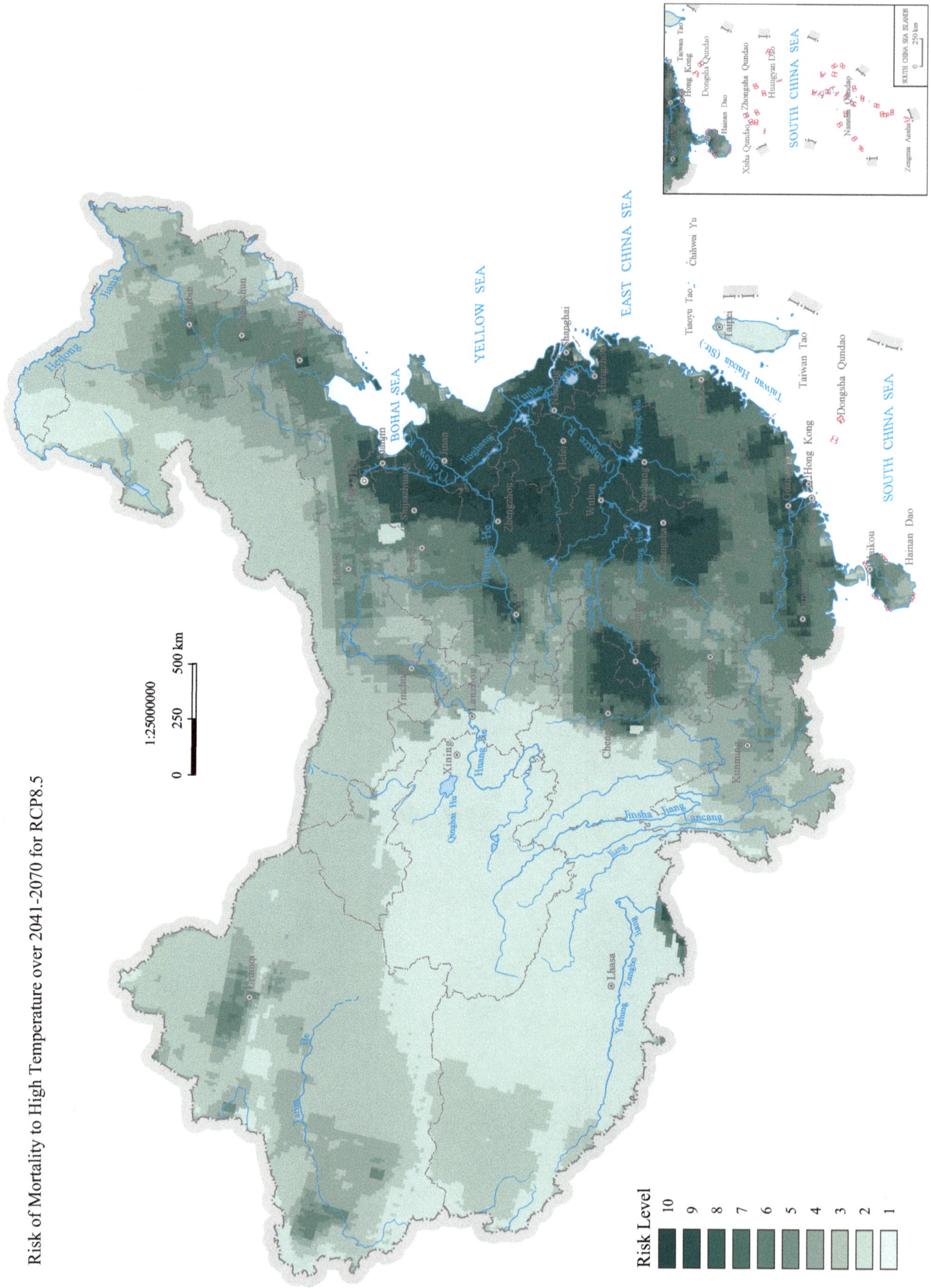
Risk of Mortality to High Temperature over 2041-2070 for RCP8.5
1:25000000
0 250 500 km
Risk Level
10
9
8
7
6
5
4
3
2
1
BOHAI SEA
YELLOW SEA
EAST CHINA SEA
SOUTH CHINA SEA
Taiwan Haixia (Str.)
Taiwan Tao
Taiwan Tao
Tiaoyu Tao
Chihwei Yu
Dongsha Qundao
Hainan Dao
Taipei
Shanghai
Nanjing
Hefei
Wuhan
Zhengzhou
Jinan
Tianjin
Shijiazhuang
Xining
Lanzhou
Yinchuan
Chengdu
Kunming
Lhasa
Haikou
Hong Kong
Hong Kong
Jinsha Jiang
Lancang
Qinghai Hu
Huang He
Heilong
Yarlung Zangbo
Xisha Qundao
Zhongsha Qundao
Huangyan Dao
Nansha Qundao
Zengmu Ansha
SOUTH CHINA SEA ISLANDS
0 250 km

The Number of Days with Daily Maximum Temperature over 35°C in 2071-2099 for RCP8.5

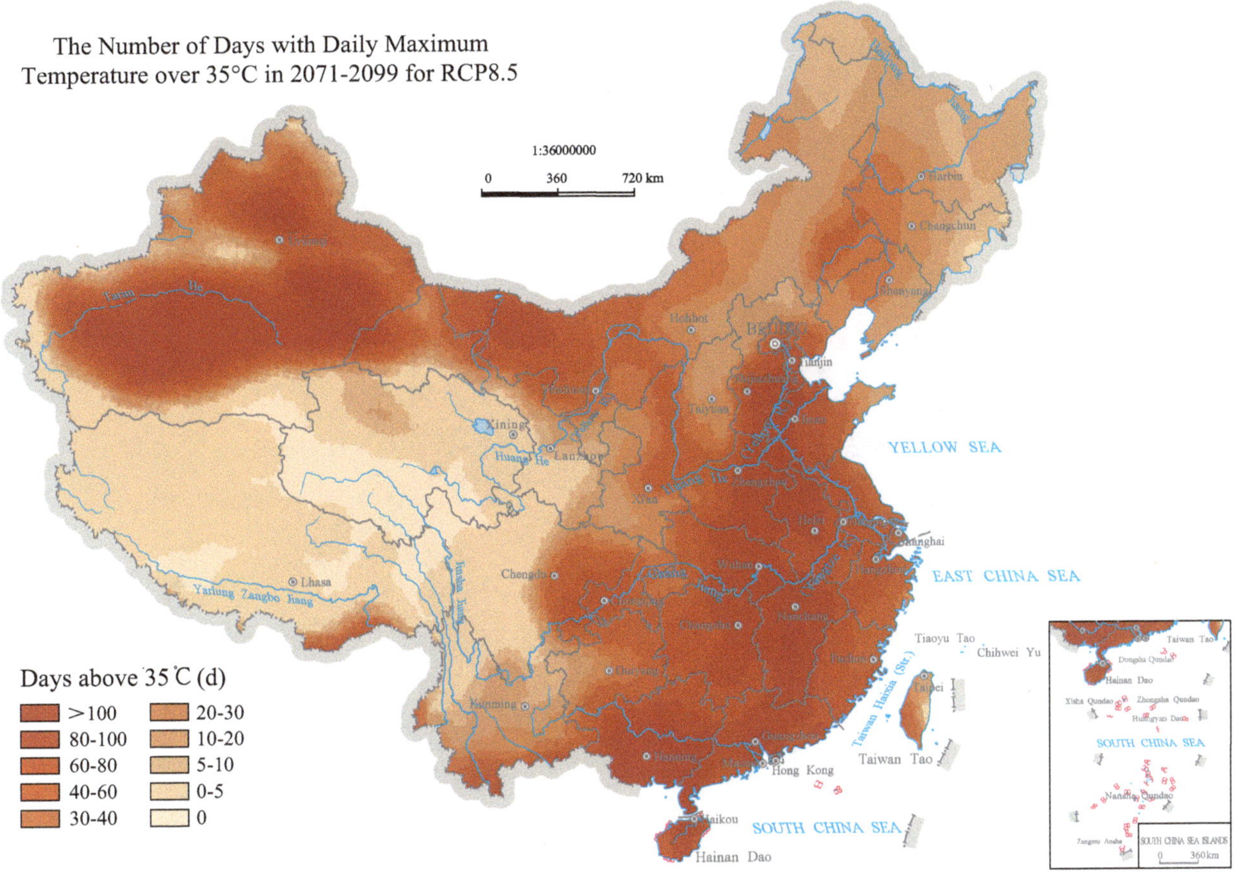

The Number of Days with Daily Maximum Temperature over 40°C in 2071-2099 for RCP8.5

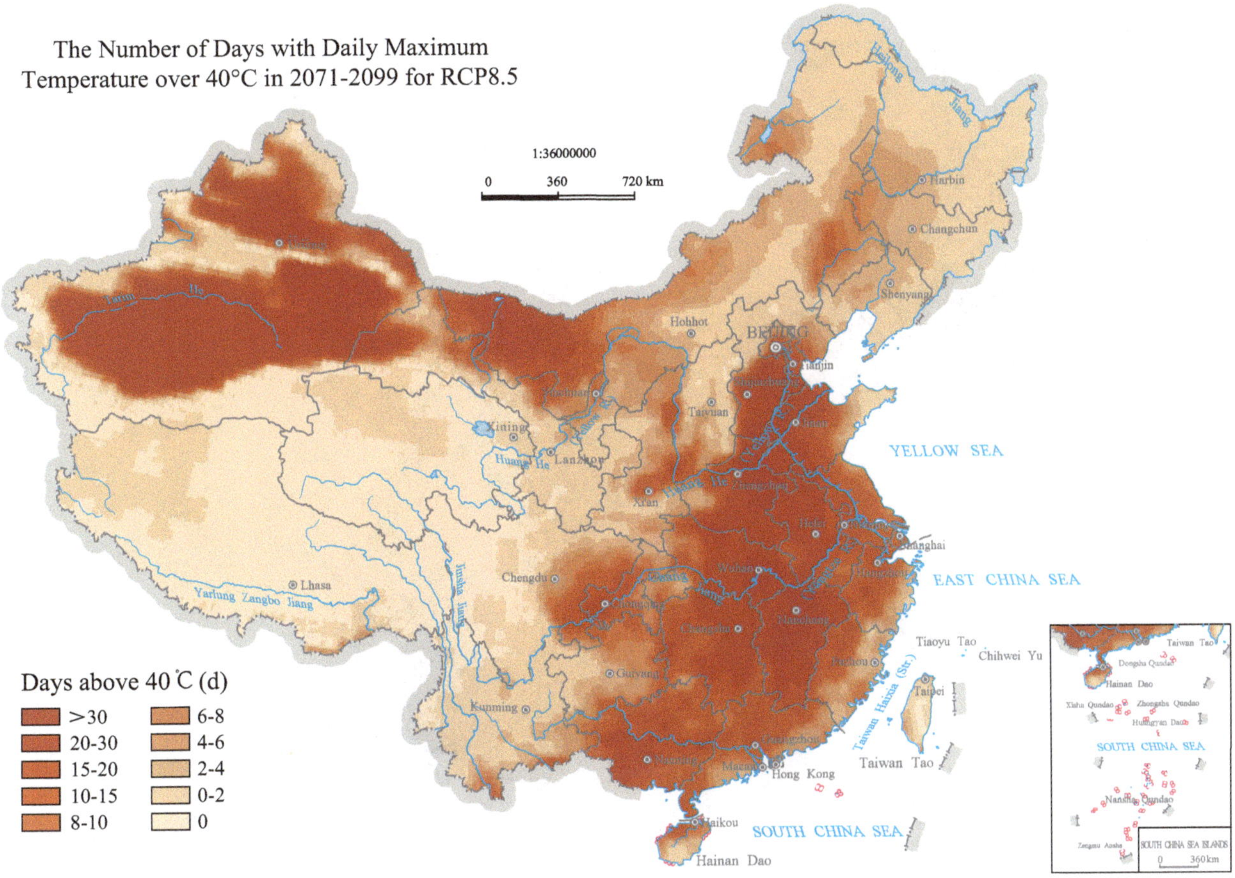

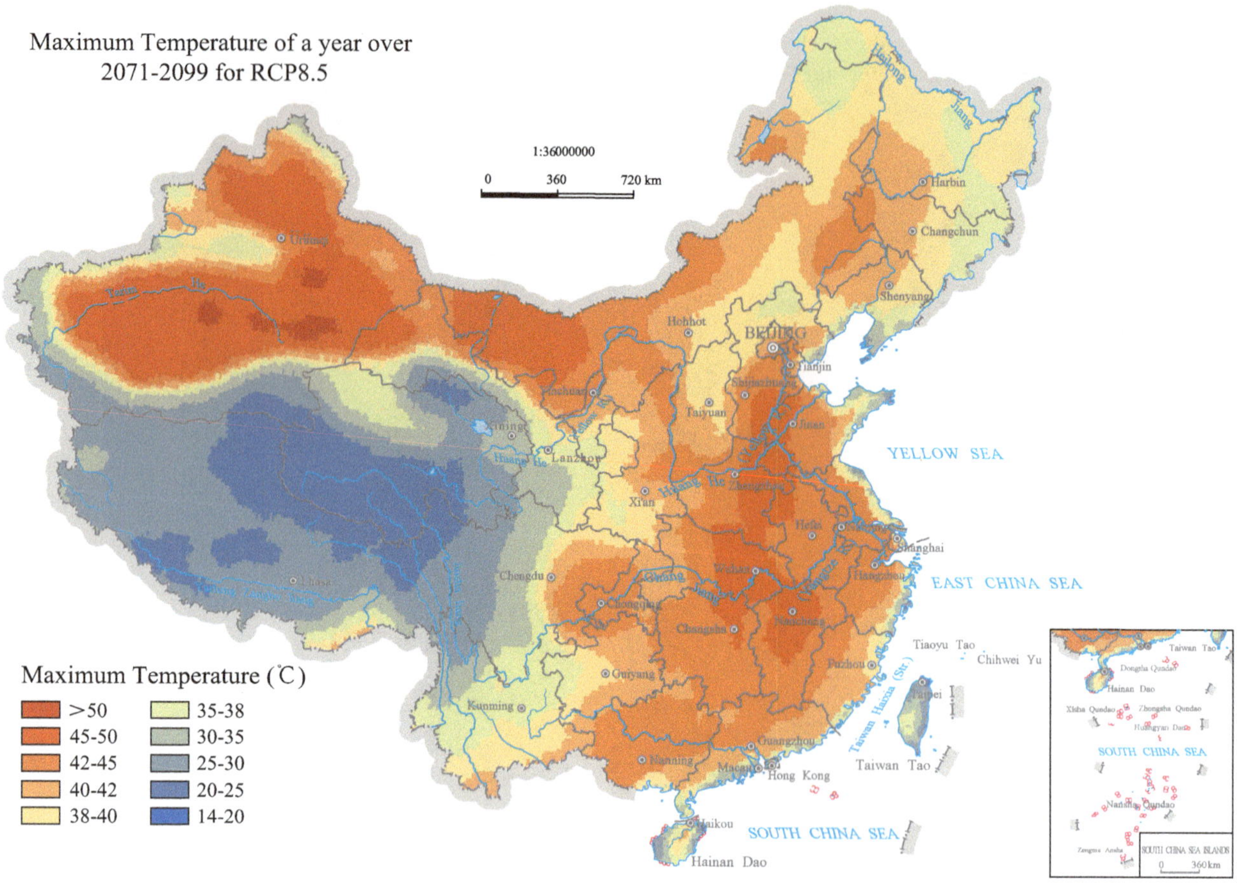
Maximum Temperature of a year over
2071-2099 for RCP8.5
1:36000000
0 360 720 km
Maximum Temperature (℃)
>50
45-50
42-45
40-42
38-40
35-38
30-35
25-30
20-25
14-20
YELLOW SEA
EAST CHINA SEA
SOUTH CHINA SEA
Harbin
Changchun
Shenyang
Hohhot
BEIJING
Tianjin
Taiyuan
Xining
Lanzhou
Xi'an
Chengdu
Guiyang
Kunming
Nanning
Guangzhou
Hong Kong
Fuzhou
Shanghai
Tiaoyu Tao
Chihwei Yu
Taiwan Tao
Hainan Dao

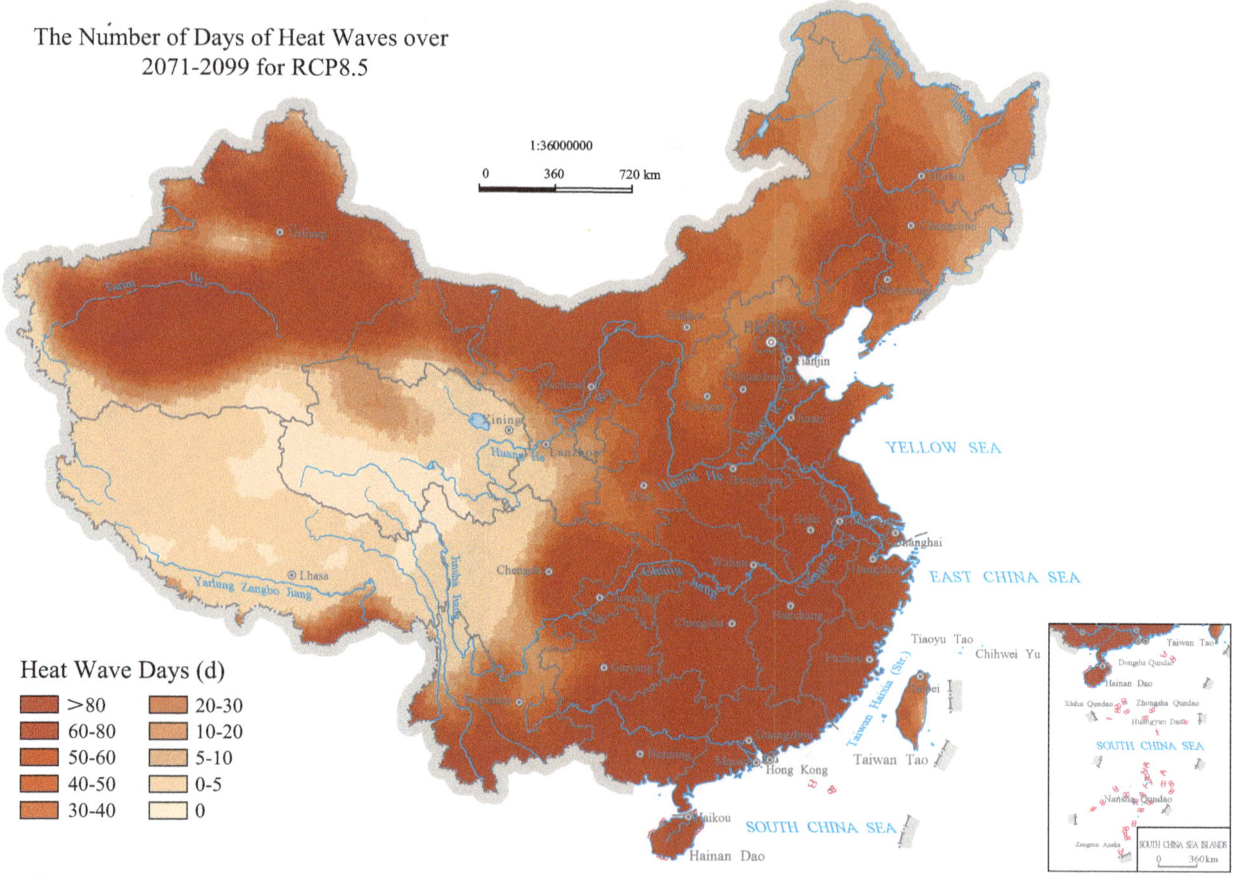
The Number of Days of Heat Waves over
2071-2099 for RCP8.5
1:36000000
0 360 720 km
Heat Wave Days (d)
>80
60-80
50-60
40-50
30-40
20-30
10-20
5-10
0-5
0
YELLOW SEA
EAST CHINA SEA
SOUTH CHINA SEA
Tianjin
Xining
Lhasa
Yarlung Zangbo Jiang
Shanghai
Tiaoyu Tao
Chihwei Yu
Taiwan Tao
Hong Kong
Hainan Dao

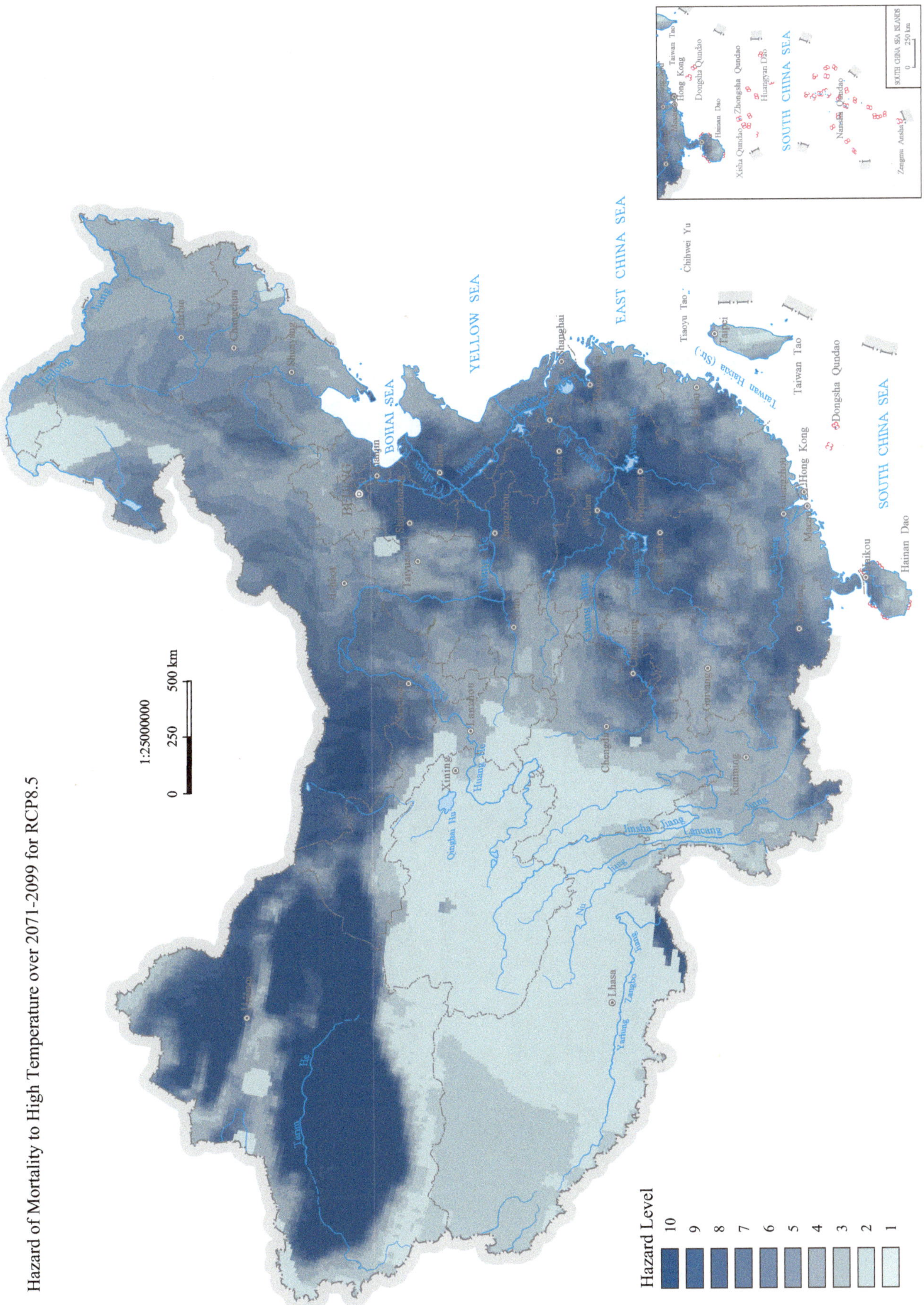

Hazard of Mortality to High Temperature over 2071-2099 for RCP8.5

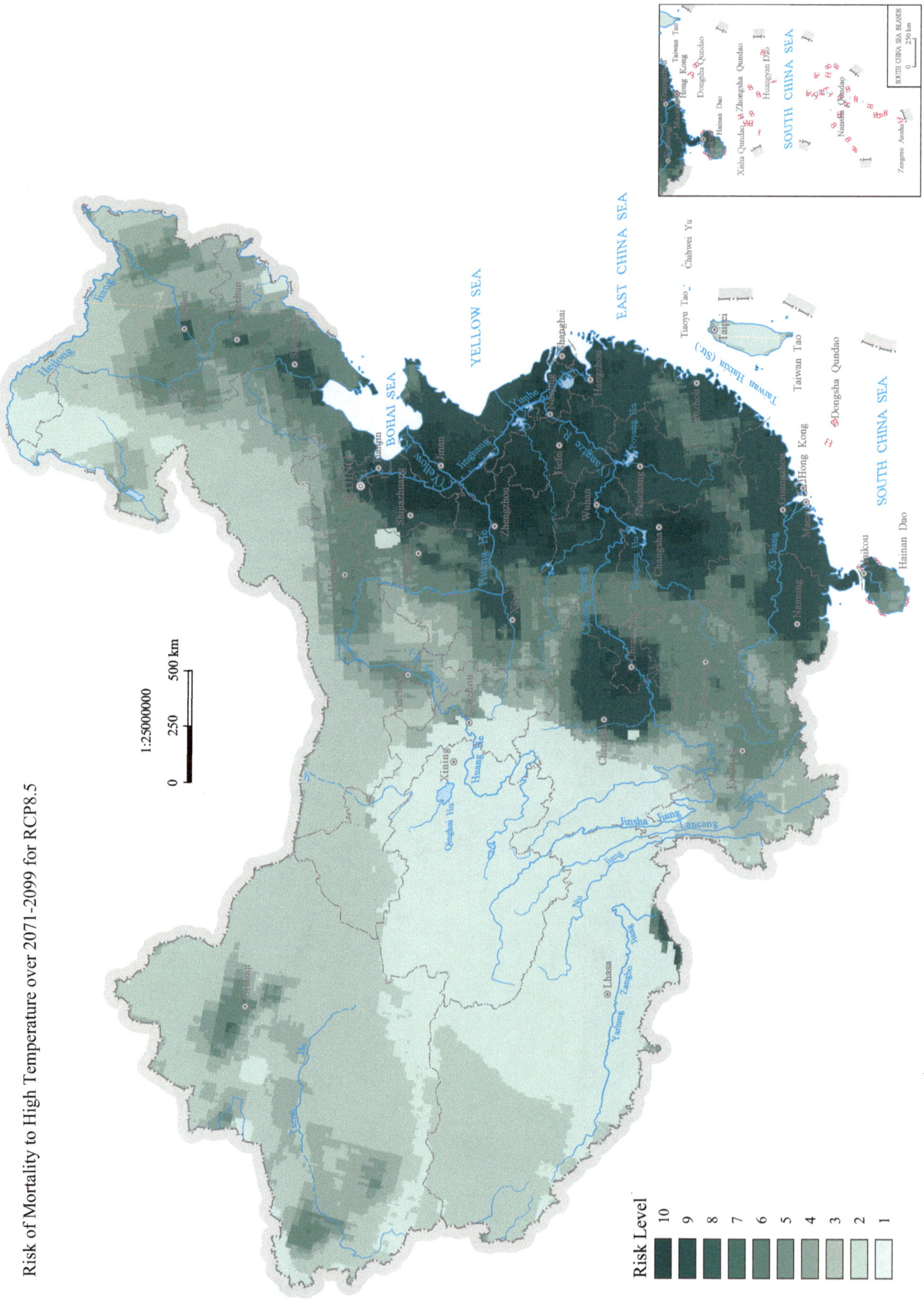

Risk of Mortality to High Temperature over 2071-2099 for RCP8.5
1:25000000
0 250 500 km
Risk Level
10
9
8
7
6
5
4
3
2
1
BOHAI SEA
YELLOW SEA
EAST CHINA SEA
SOUTH CHINA SEA
Taiwan Haixia (Str.)
Taiwan Tao
Hong Kong
Dongsha Qundao
Hainan Dao
Shanghai
Beijing
Tianjin
Jinan
Shijiazhuang
Zhengzhou
Hefei
Nanjing
Hangzhou
Wuhan
Nanchang
Changsha
Guangzhou
Nanning
Haikou
Taipei
Xining
Lhasa
Qinghai Hu
Huang He
Jinsha Jiang
Lancang
Yarlung Zangbo Jiang
Heilong Jiang
Poyang Hu
Tiaoyu Tao
Chihwei Yu
SOUTH CHINA SEA ISLANDS

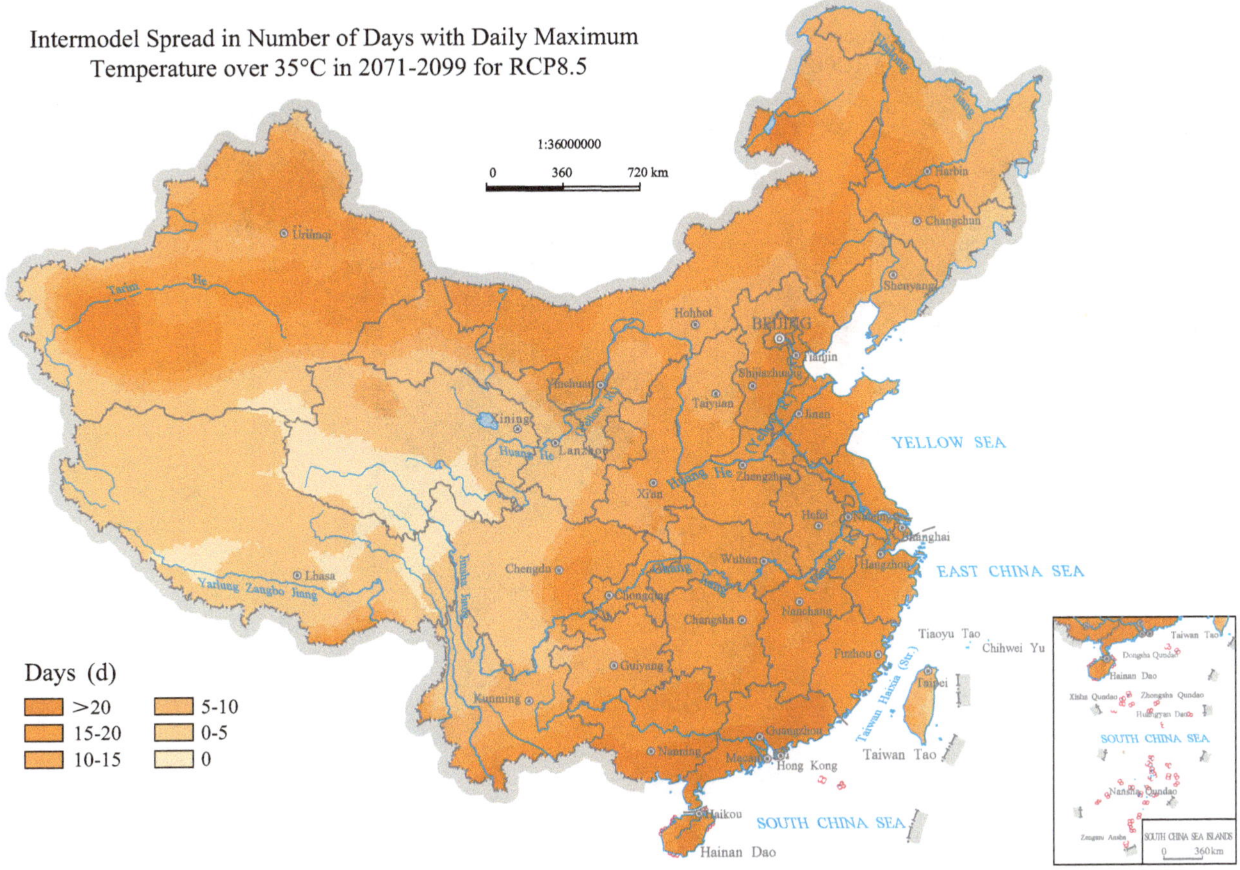
Intermodel Spread in Number of Days with Daily Maximum Temperature over 35°C in 2071-2099 for RCP8.5
1:36000000
0 360 720 km
Days (d)
>20
15-20
10-15
5-10
0-5
0
Ürümqi
Tarim He
Lhasa
Yarlung Zangbo Jiang
Xining
Lanzhou
Huang He
Yinchuan
Hohhot
BEIJING
Tianjin
Shijiazhuang
Taiyuan
Jinan
Xi'an
Zhengzhou
Hefei
Nanjing
Shanghai
Hangzhou
Wuhan
Chengdu
Chongqing
Changsha
Nanchang
Fuzhou
Guiyang
Kunming
Nanning
Guangzhou
Macao
Hong Kong
Haikou
Hainan Dao
Taipei
Taiwan Tao
Harbin
Changchun
Shenyang
YELLOW SEA
EAST CHINA SEA
SOUTH CHINA SEA
Tiaoyu Tao
Chihwei Yu
SOUTH CHINA SEA ISLANDS
360km

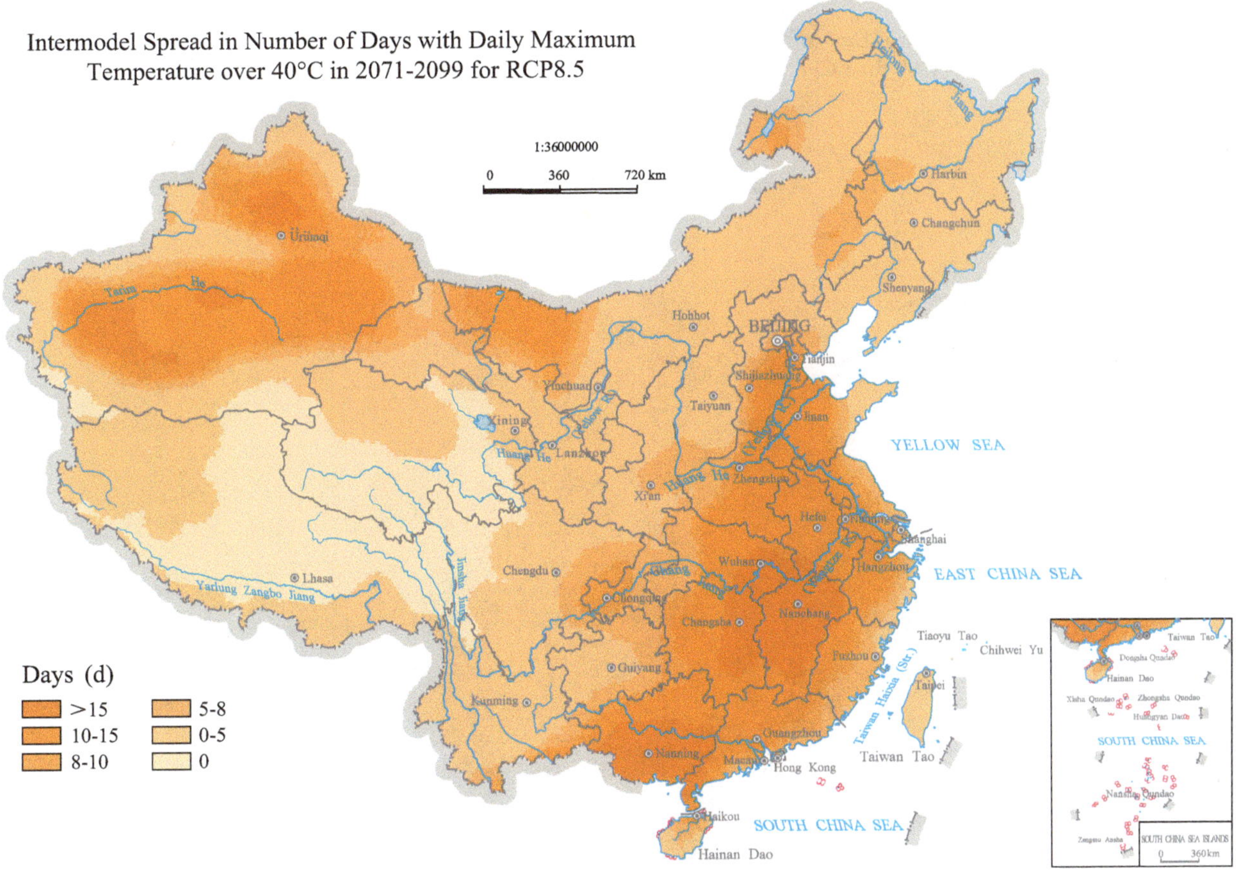
Intermodel Spread in Number of Days with Daily Maximum Temperature over 40°C in 2071-2099 for RCP8.5
1:36000000
0 360 720 km
Days (d)
>15
10-15
8-10
5-8
0-5
0
Ürümqi
Tarim He
Lhasa
Yarlung Zangbo Jiang
Xining
Lanzhou
Huang He
Yinchuan
Hohhot
BEIJING
Tianjin
Shijiazhuang
Taiyuan
Jinan
Xi'an
Zhengzhou
Hefei
Nanjing
Shanghai
Hangzhou
Wuhan
Chengdu
Chongqing
Changsha
Nanchang
Fuzhou
Guiyang
Kunming
Nanning
Guangzhou
Macao
Hong Kong
Haikou
Hainan Dao
Taipei
Taiwan Tao
Harbin
Changchun
Shenyang
YELLOW SEA
EAST CHINA SEA
SOUTH CHINA SEA
Tiaoyu Tao
Chihwei Yu
SOUTH CHINA SEA ISLANDS
360km

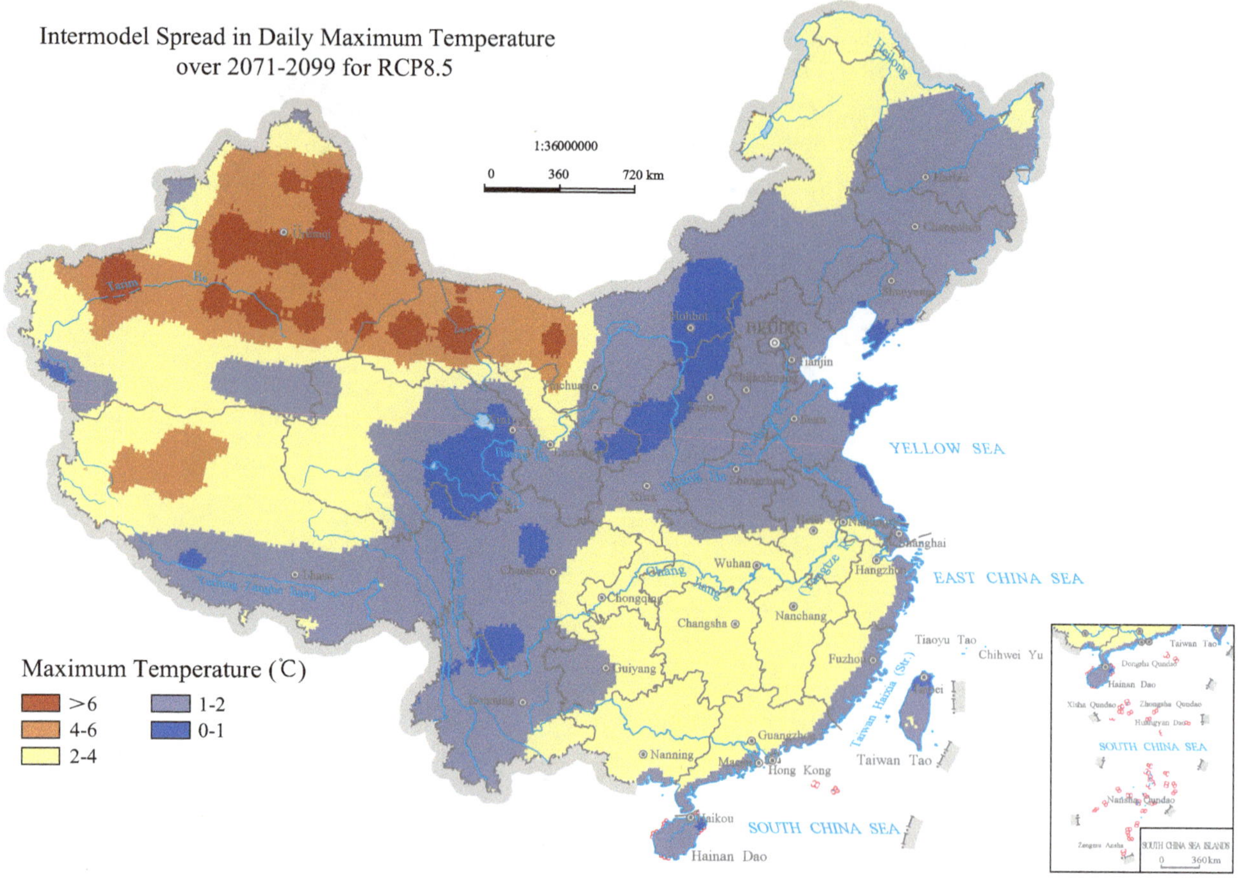
Intermodel Spread in Daily Maximum Temperature
over 2071-2099 for RCP8.5
1:36000000
0 360 720 km
Maximum Temperature (℃)
>6
4-6
2-4
1-2
0-1
YELLOW SEA
EAST CHINA SEA
SOUTH CHINA SEA
Hainan Dao
Taiwan Tao
Hong Kong
Beijing
Tianjin
Shanghai
Hangzhou
Wuhan
Nanchang
Changsha
Chongqing
Guiyang
Nanning
Haikou
Fuzhou
Lhasa
Tiaoyu Tao
Chihwei Yu

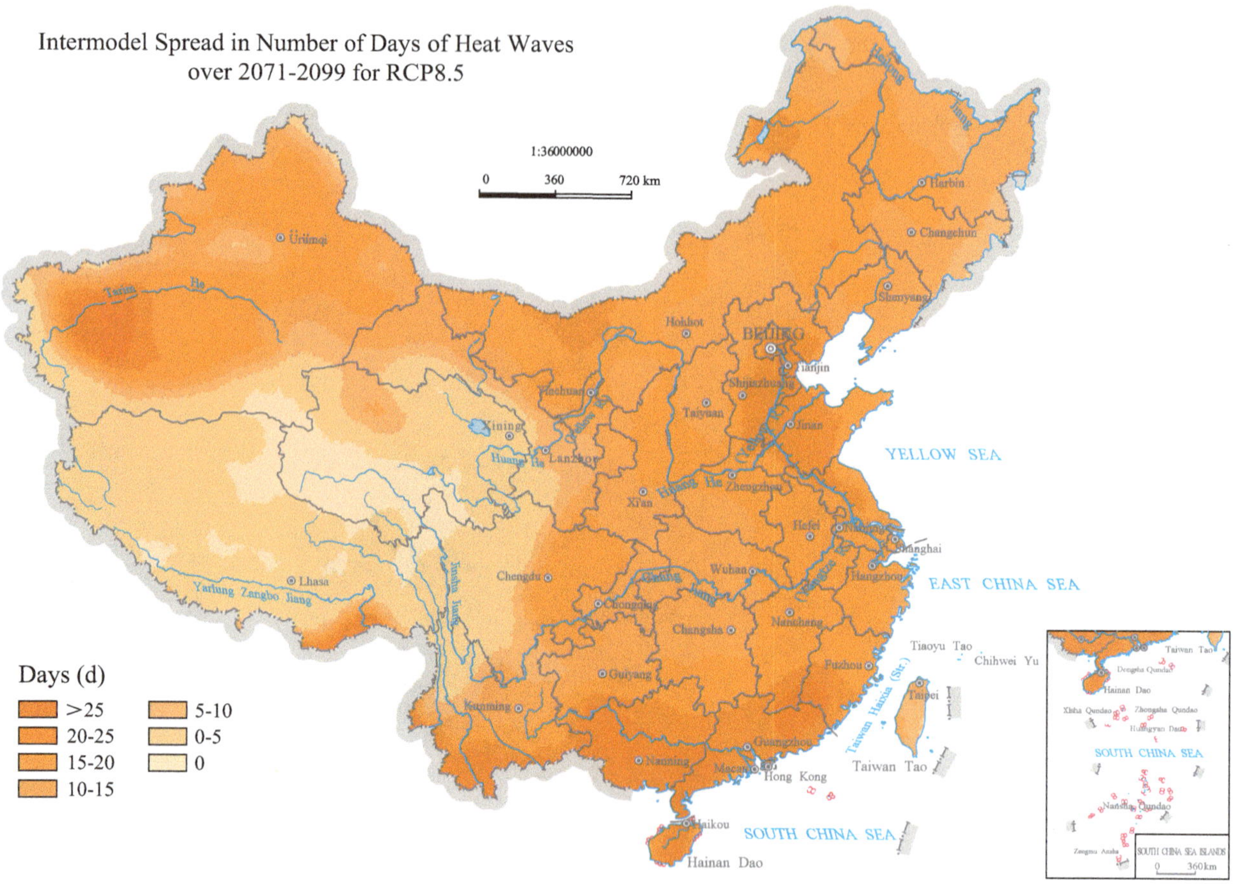
Intermodel Spread in Number of Days of Heat Waves
over 2071-2099 for RCP8.5
1:36000000
0 360 720 km
Days (d)
>25
20-25
15-20
10-15
5-10
0-5
0
YELLOW SEA
EAST CHINA SEA
SOUTH CHINA SEA
Hainan Dao
Taiwan Tao
Hong Kong
Ürümqi
Harbin
Changchun
Shenyang
Hohhot
Tianjin
Shijiazhuang
Taiyuan
Jinan
Xining
Lanzhou
Xi'an
Zhengzhou
Hefei
Shanghai
Hangzhou
Wuhan
Chengdu
Chongqing
Changsha
Nanchang
Fuzhou
Guiyang
Kunming
Nanning
Guangzhou
Macao
Haikou
Lhasa
Yarlung Zangbo Jiang
Tiaoyu Tao
Chihwei Yu

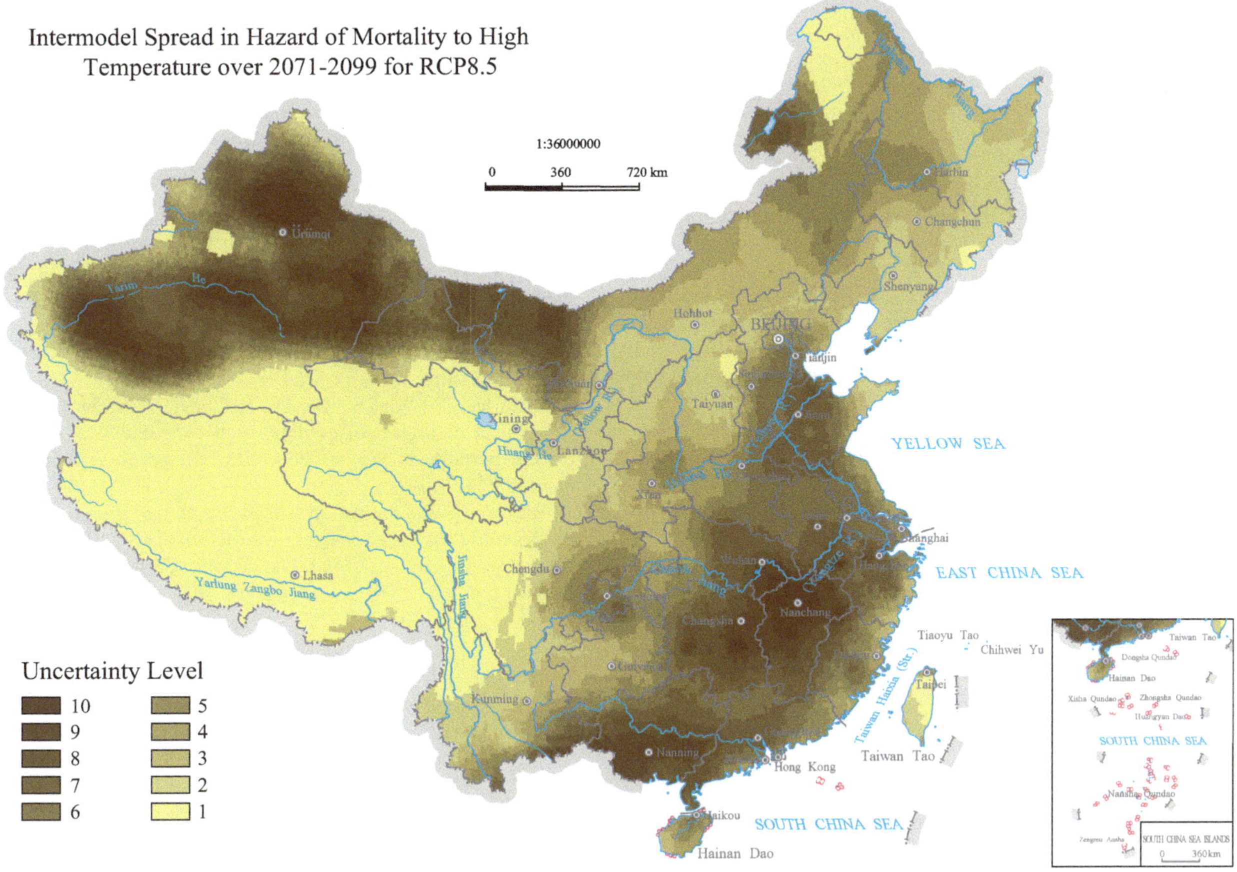
Intermodel Spread in Hazard of Mortality to High Temperature over 2071-2099 for RCP8.5
1:36000000
0 360 720 km
Uncertainty Level
10
9
8
7
6
5
4
3
2
1
YELLOW SEA
EAST CHINA SEA
SOUTH CHINA SEA
Hainan Dao
Taiwan Tao
Hong Kong
Lhasa
Chengdu
Kunming
Nanning
Beijing
Harbin
Changchun
Shenyang
Hohhot
Taiyuan
Xining
Lanzhou
Xi'an
Nanchang
Tiaoyu Tao
Chihwei Yu
SOUTH CHINA SEA ISLANDS

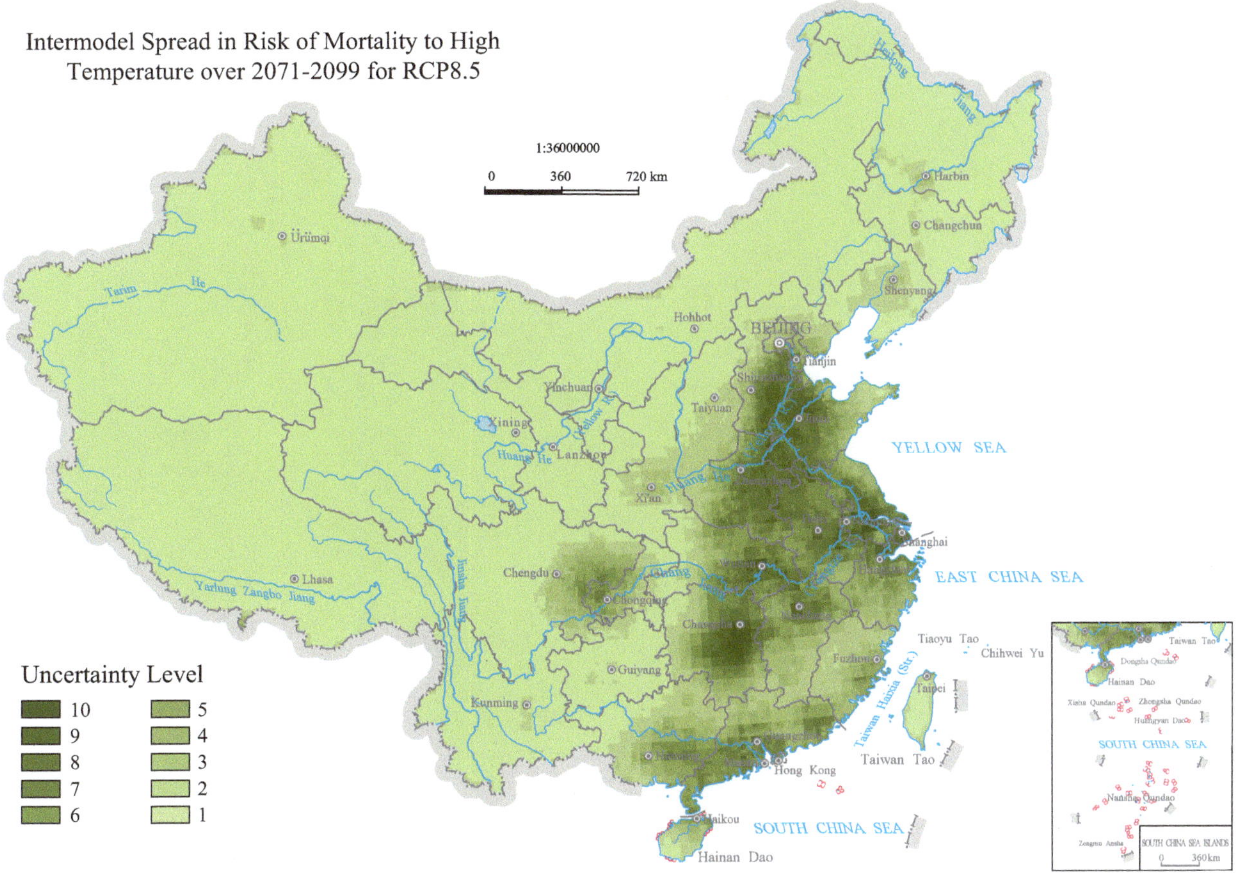
Intermodel Spread in Risk of Mortality to High Temperature over 2071-2099 for RCP8.5
1:36000000
0 360 720 km
Uncertainty Level
10
9
8
7
6
5
4
3
2
1
YELLOW SEA
EAST CHINA SEA
SOUTH CHINA SEA
Hainan Dao
Taiwan Tao
Hong Kong
Lhasa
Chengdu
Chongqing
Guiyang
Kunming
Harbin
Changchun
Shenyang
Hohhot
Taiyuan
Yinchuan
Xining
Lanzhou
Xi'an
Fuzhou
Tiaoyu Tao
Chihwei Yu
SOUTH CHINA SEA ISLANDS

References

Basu, R., & Ostro, B. (2008). Multicounty analysis identifying the populations vulnerable to mortality associated with high ambient temperature in California. *American Journal of Epidemiology, 168,* 632–637.

Díaz, J., Linares, C., & Tobías, A. (2006). Impact of extreme temperatures on daily mortality in Madrid (Spain) among the 45–64 age-group. *International Journal of Biometeorology, 150,* 342–348.

Hajat, S., Armstrong, B., Gouvcia, N., & Wilkinson, P. (2005). Mortality displacement of heat-related deaths: A comparison of Delhi, Sao Paulo, and London. *Epidemology, 16,* 613–620.

Kovats, R., & Hajat, S. (2008). Heat stress and public health: A critical review. *Annual Review Public Health, 29,* 41–55.

Li, Y., Chen, X., & Lin, P. (2005). The impact of high temperature on population death in a district of Nanjing. *Journal of Environmental Health, 22*(1), 6–8.

Li, Y., Lan, L., Cheng, Y., Cui, G., Yang, C., & Liu, Y., et al. (2014). High temperature sensitive diseases and vulnerable population in one district of Harbin. *Journal of Environmental Hygiene, 4*(4), 321–325 (in Chinese).

Lim, Y., Bell, M., Kan, H., Honda, Y., Guo, Y., & Kim, H. (2015). Economic status and temperature-related mortality in Asia. *International Journal of Biometeorology, 59*(10), 1405–1412.

Qu, Y., Zhang, J., Dai, C., Wang, L., Liang, H., He, C., et al. (2009). Study on the relationship between temperature and residential death in a district of Guangzhou city from 2000 to 2004. *Preventive Medicine Tribune, 15*(9), 807–810.

Sun, Y., Li, L., & Zhou, M. (2012). Analysis of the lag-effects of temperature on the five cities' mortality in China. *Chinese Journal of Preventive Medicine, 11,* 1015–1019. (in Chinese).

Tan, G. (1994). Potential impacts of global warming on human mortality in Shanghai and Guangzhou, China. *Acta Scientiae Circumstantiae, 14*(3), 368–373.

Wang, L., & Lei, L. (1997). The impact of the changing weather on the human mortality. *Ecology Science, 16*(2), 81–87.

Yang, H., Chen, Z., Xie, S., Ye, D., & Gong, J. (2013). Quantitative assessment of impact of extreme high temperature in summer on excess mortality in Wuhan. *Journal of Meteorology and Environment, 29*(5), 140–143.

Zeng, W. (2013). *The impact of heat waves on mortality and the effect modification by heat waves characteristics in four areas in Guangdong Province, China*. Guangdong, China: Jinan University.

4 Agricultural Risks

Yuanyuan Yin, Qiuhong Tang, Xingcai Liu and Huijuan Cui

4.1 Background

Climate change may affect crop growth and yield, which consequently casts a shadow of doubt over China's food self-sufficiency efforts. Although the projections of climate change impacts on crop yields may be inherently uncertain (Asseng et al. 2013), the projections of changes in crop yields in China are widely reported using crop models with general circulation model (GCM) outputs generated for the Assessment Report of IPCC. It has been suggested that along with increase in global mean temperature the yields of maize and rice would decline, while wheat yield would increase in some regions in China (Challinor et al. 2010; Ju et al. 2013).

Understanding the effects of climate change on crop yield is important for development of adaptation and mitigation measures in agricultural regions of China. Rice, maize, and wheat are the major crops in China. The total area of the three mentioned crops is about 54% of the total cropland areas in China. Soybean is a globally important crop, providing oil and protein. In recent years, China's rising demand for soybean has brought China to the top of the list of importers. Therefore, the yield changes in the four crops (i.e., maize, rice, soybean, and wheat) are vital for assessing the climate change impact on food security in China. In this chapter, the projections derived from four global gridded crop models (GGCropMs) were used to assess the risk of future climate change on the yields of the major crops (i.e., maize, rice, soybean, and wheat) and the production of agriculture in China (Yin et al. 2015).

Y. Yin · Q. Tang (✉) · X. Liu
Key Laboratory of Water Cycle and Related Land Surface Processes, Institute of Geographic Sciences and Natural Resources Research, Chinese Academy of Sciences, Beijing, 100101, China
e-mail: tangqh@igsnrr.ac.cn

H. Cui
Key Laboratory of Land Surface Pattern and Simulation, Institute of Geographic Sciences and Natural Resources Research, Chinese Academy of Sciences, Beijing, 100101, China

4.2 Method

The agriculture risk assessment was carried out based on three data sets: the global irrigated and rain-fed crop area data (MIRCA2000), the China Land Cover Data 2010, and the simulated crop yield data with CO_2 effect provided by the Inter-Sectoral Impact Model Intercomparison Project (ISI-MIP). The MIRCA2000 data consist of all major food crops including wheat, rice, maize, and soybean (Portmann et al. 2010). The data sets refer to the period of 1998–2002 and are made available with a spatial resolution of 0.5° × 0.5° by ISI-MIP (Warszawski et al. 2014). The China Land Cover Data 2010 with a spatial resolution of 1 km × 1 km is provided by the Data Center for Resources and Environmental Sciences, Chinese Academy of Sciences (RESDC) (http://www.resdc.cn). The simulated crop yield data were taken from four GGCropMs with a spatial resolution of 0.5° × 0.5° (Table 4.1). We assigned the value of a 0.5° × 0.5° grid, to which the central point of a 1 km grid belong, to the 1 km grid (Taylor et al. 2012). The GGCropMs were forced with the bias-corrected climatic data (Hempel et al. 2013) for the historical period of 1971–2005 (except EPIC which was for 1980–2010) and four representative concentration pathways (RCPs), namely, RCP2.6, RCP4.5, RCP6.0, and RCP8.5, for the future period of 2006–2099 (except EPIC which is for 2011–2099) of five GCMs (Table 4.1) from the Fifth Coupled Model Intercomparison Project (CMIP5) (Taylor et al. 2012). All the four GGCropMs provided the simulated yields for maize, rice, wheat, and soybean excluding PEGASUS which offered no simulation for rice yield. The yield simulations of EPIC were missing in 2066, 2067, and 2068. The GGCropMs provided the simulated crop yields in both irrigated and rain-fed croplands.

For each 1 km × 1 km grid, crop yield was calculated as the area-weighted yield in the irrigated and rain-fed portions of the grid according to the crop-specific irrigated and rain-fed areas. Agricultural production is the sum of calories content by the major crop yields. The crop yield and agricultural production of each grid for every year was calculated for each GCM-GGCropM pair. There are 20 model pairs for maize,

Q. Tang and Q. Ge (eds.), *Atlas of Environmental Risks Facing China Under Climate Change*,
IHDP/Future Earth-Integrated Risk Governance Project Series, DOI 10.1007/978-981-10-4199-0_4

Table 4.1 Overview of the GCMs and GGCropMs

	Name	Institute
GCMs	HadGEM2-ES	Met Office Hadley Centre
	IPSL-CM5A-LR	Institute Pierre-Simon Laplace
	MIROC-ESM-CHEM	Japan Agency for Marine-Earth Science and Technology, Atmosphere and Ocean Research Institute, and National Institute for Environmental Studies
	GFDL-ESM2M	Geophysical Fluid Dynamics Laboratory
	NorESM1-M	Norwegian Climate Centre
GGCropMs	EPIC	BOKU, University of Natural Resources and Life Sciences, Vienna
	GEPIC	EAWAG; Swiss Federal Institute of Aquatic Science and Technology
	pDSSAT	University of Chicago Computation Institute
	PEGASUS	Tyndall Centre, University of East Anglia UK/McGill University, Canada

Table 4.2 Agricultural risk levels

Risk level	1	2	3	4	5	6	7	8	9	10
MM of relative yield change	≤ 0	0–2.5%	2.5–5%	5–7.5%	7.5–10%	10–12.5%	12.5–15%	15–17.5%	17.5–20%	>20%

wheat, and soybean. Meanwhile, there are 15 GCM-GGCropM pairs for rice because the rice yield is missing in PEGASUS simulations. The averages of the crop yield and agricultural production were computed for the periods 1981–2010, 2011–2040, 2041–2070, and 2071–2099, respectively. The Multimodel-ensemble Medians (MMs) of the crop yield or agricultural production from all available GCM-GGCropM pairs were computed. The relative change of the crop yield or agricultural production was computed as the difference between a 30-year period in future and the historical period of 1981–2010, divided by the yield or production in the historical period. The MMs of relative yield or production changes were calculated over the periods 2011–2040, 2041–2070, and 2071–2099, respectively. The MMs of relative yield changes were divided into ten levels (Table 4.2). The standard deviation (SD) of the relative changes from all available GCM-GGCropM pairs was used to quantify the model uncertainty.

4.3 Results

4.3.1 Simulated Crop Yield and Agricultural Production in 1981–2010

Nine maps of yield and production during 1981–2010 were developed, with four maps of major crop yields, and five maps of major crop productions and agricultural productions. The yield of maize is higher in North China (NC), and parts of Northwest China (NWC), Xinjiang (XJ), and Sichuan province in Southwest China (SWC), while the production of maize is higher in the NC, and parts of Northeast China (NEC) and Sichuan province in SWC. Wheat yield is higher in certain regions of SWC, Fujian, and Jiangxi province in Southeast China (SEC), while the production of wheat is greater in quantity in the NC and Sichuan province. The yields of rice and soybean are higher in eastern China. The production of rice has greater quantity in the NC, central China, and Sichuan province, whereas the production of soybean is higher only in specific parts of NC and NEC. The agricultural production is higher in NC, NEC, and some parts of central China and Sichuan province.

4.3.2 Risk of Crop Yield and Agricultural Production Under Climate Change

The largest decrease is found to be at the main agricultural areas in the northern and southern China. The simulated rice yield would increase over a large portion of China with the largest increase in the high-altitude and cold regions. Rice would decrease in some of the current main rice planting areas such as the eastern and southern China. The relative change in soybean yield shows a spatial pattern similar to that of rice yield. The soybean yield would increase in the regions not in areas occupied by the traditional agriculture but decrease in the main agricultural areas in the eastern China. The relative change in wheat yield tends to be decreasing across China except for a small area in the Tibetan Plateau and NEC. The agricultural production would increase in the regions not in areas engaged for the traditional agriculture but decrease in the main crops agricultural areas in the eastern China, such as NC and South China.

4.3.3 Model Spread and Uncertainty

Five maps of model spread in the relative change across all the available GCM-GGCropM pairs were developed, including four maps of model spread of major crop yields at the end of the twenty-first century. The SDs from the crop model ensembles are more than 60% in the Tibetan Plateau, suggesting the model uncertainty is large in this region. The model spread for maize is generally less than 40% and the model spread for rice and wheat is generally less than 30% in the eastern China. The model spread for soybean and wheat is more than 50% in many parts of the eastern China, showing the model uncertainty is especially large for these crop types. The model spread for agricultural production is generally large in the areas extended north-eastward from the Tibetan Plateau to northeast China.

4.4 Maps

Maize Growing Areas

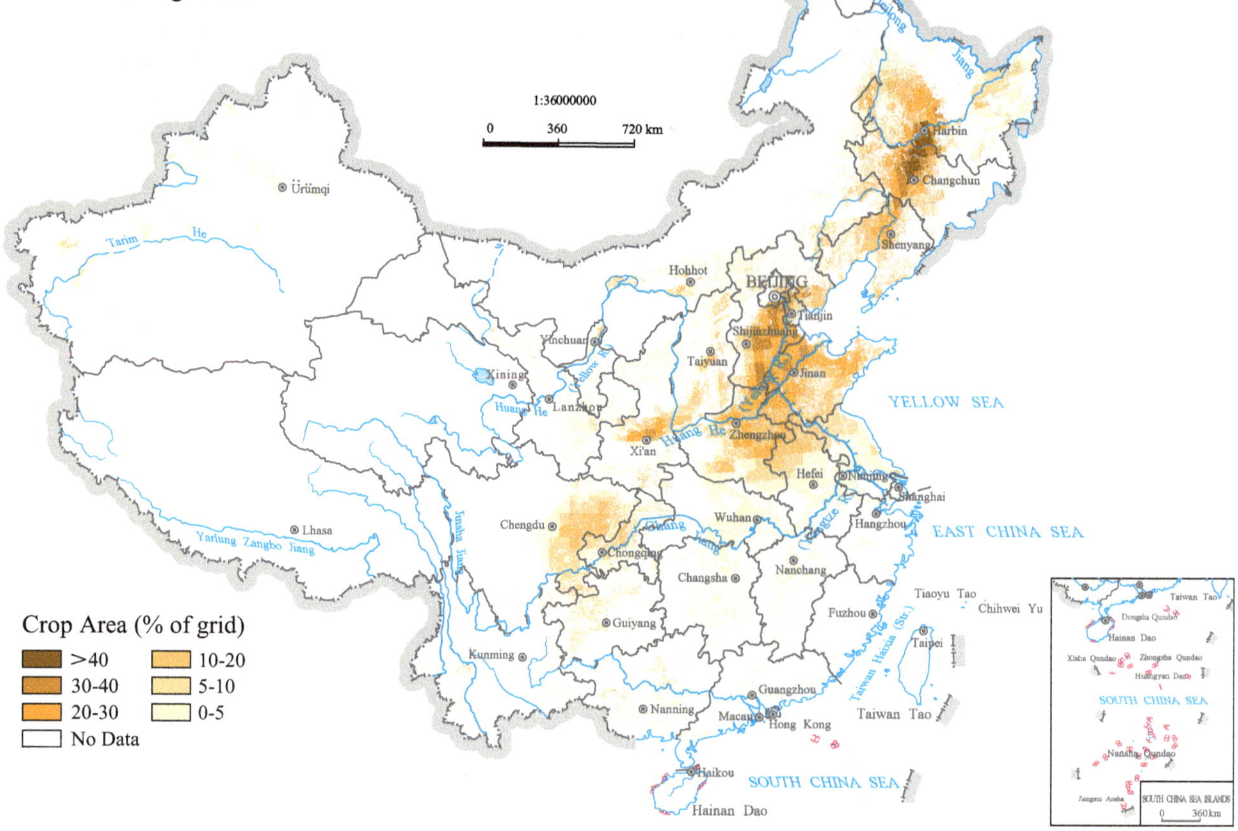

Wheat Growing Areas

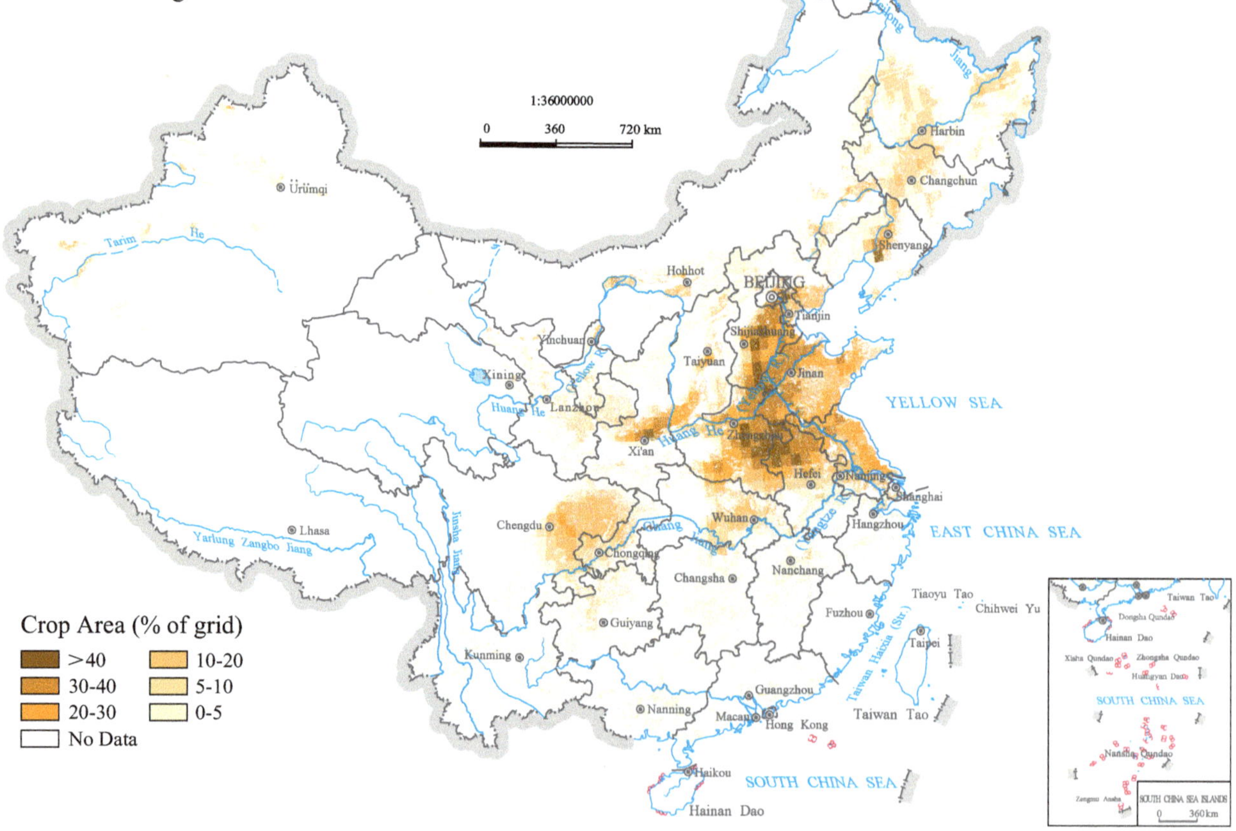

Rice Growing Areas

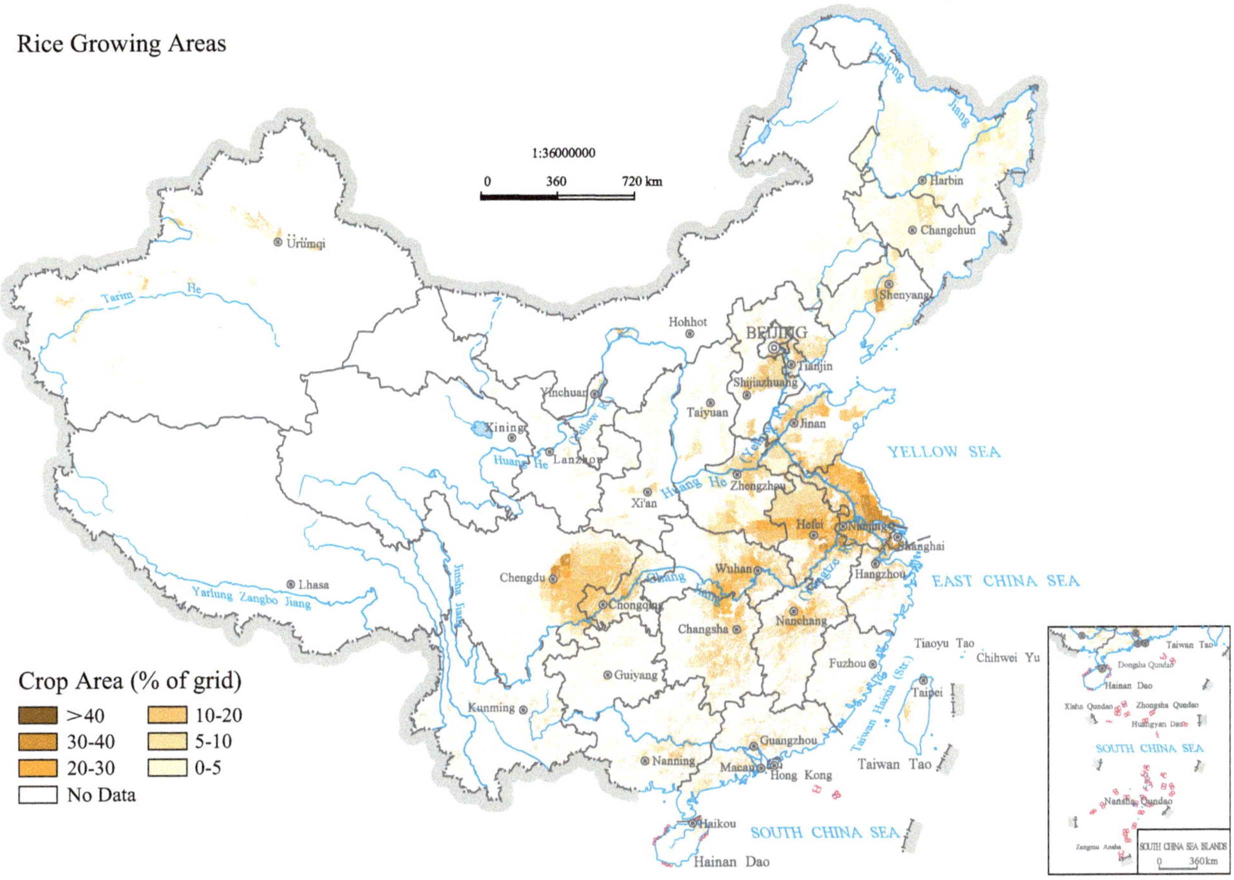

Soybean Growing Areas

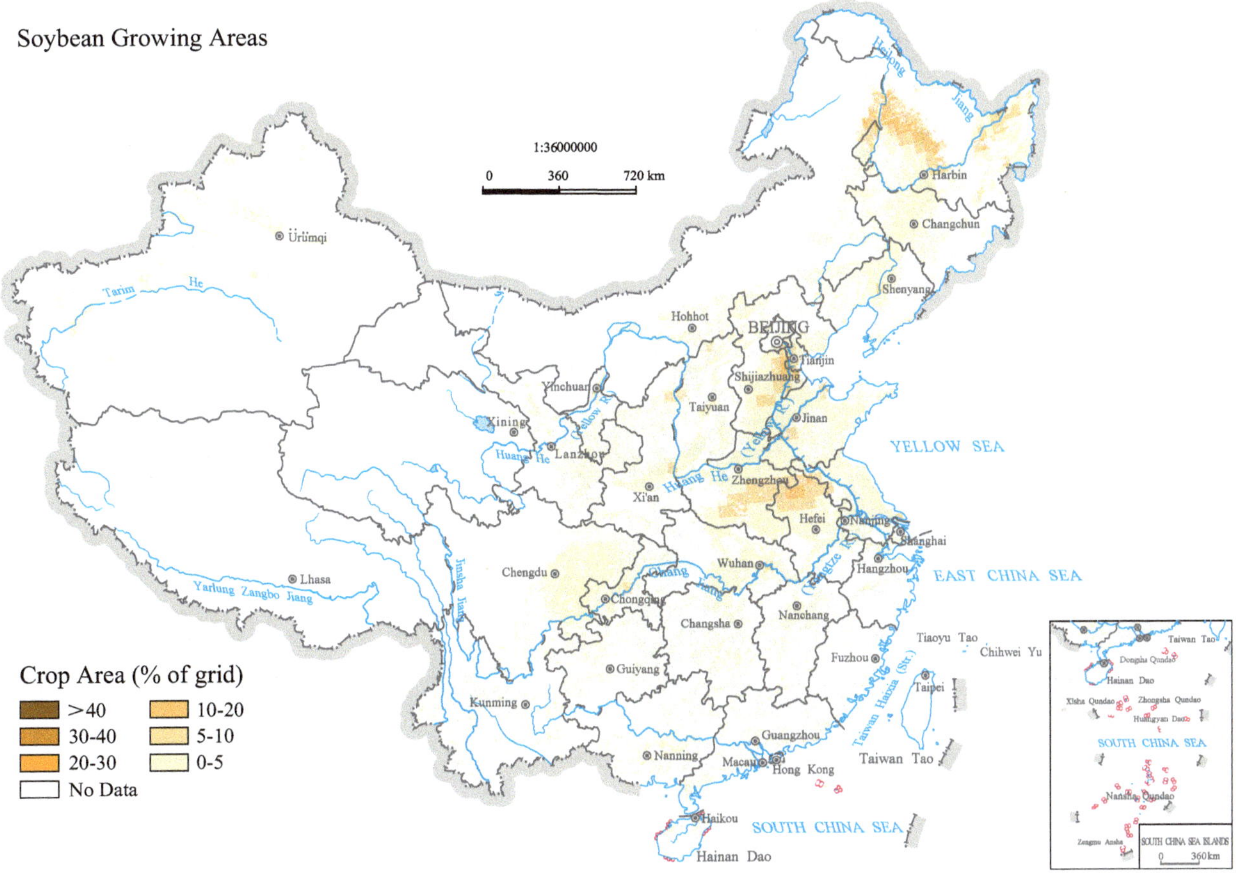

Annual Mean Yield of Maize over 1981-2010

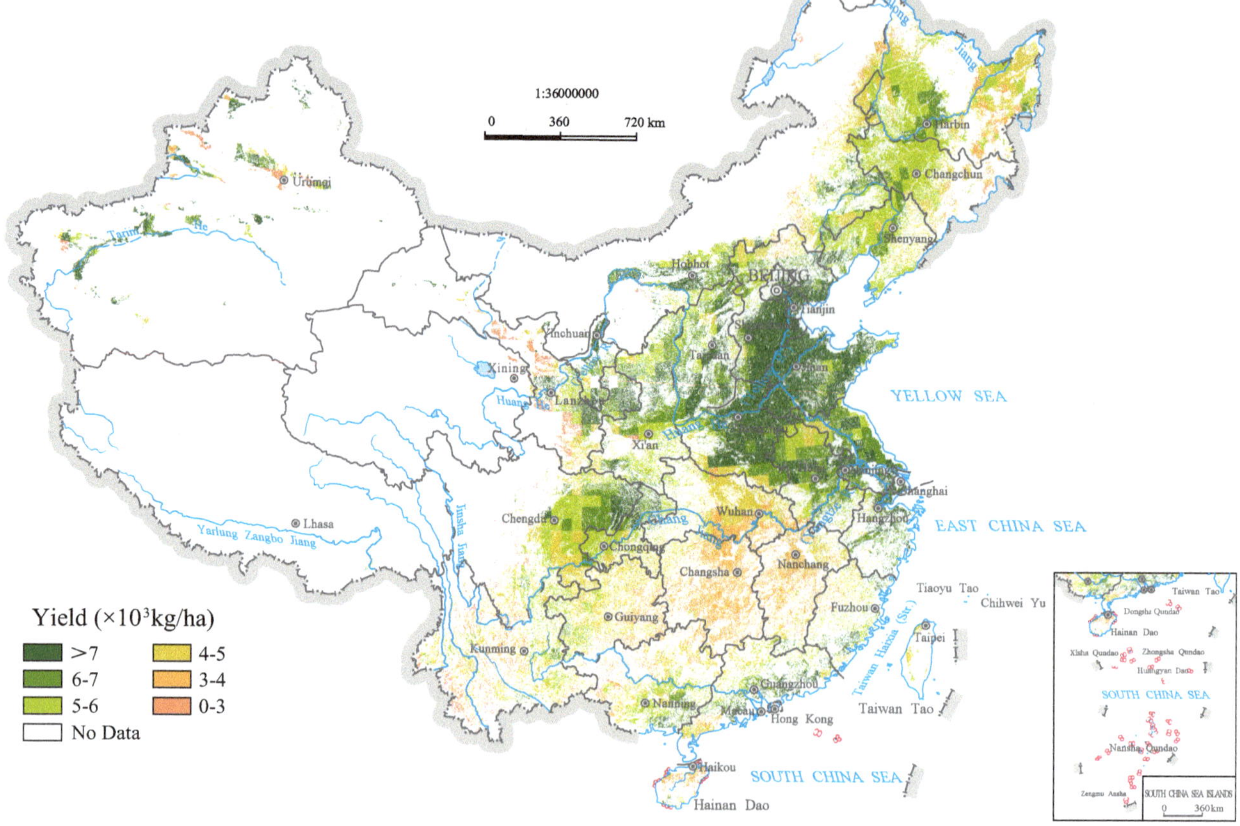

Annual Mean Yield of Wheat over 1981-2010

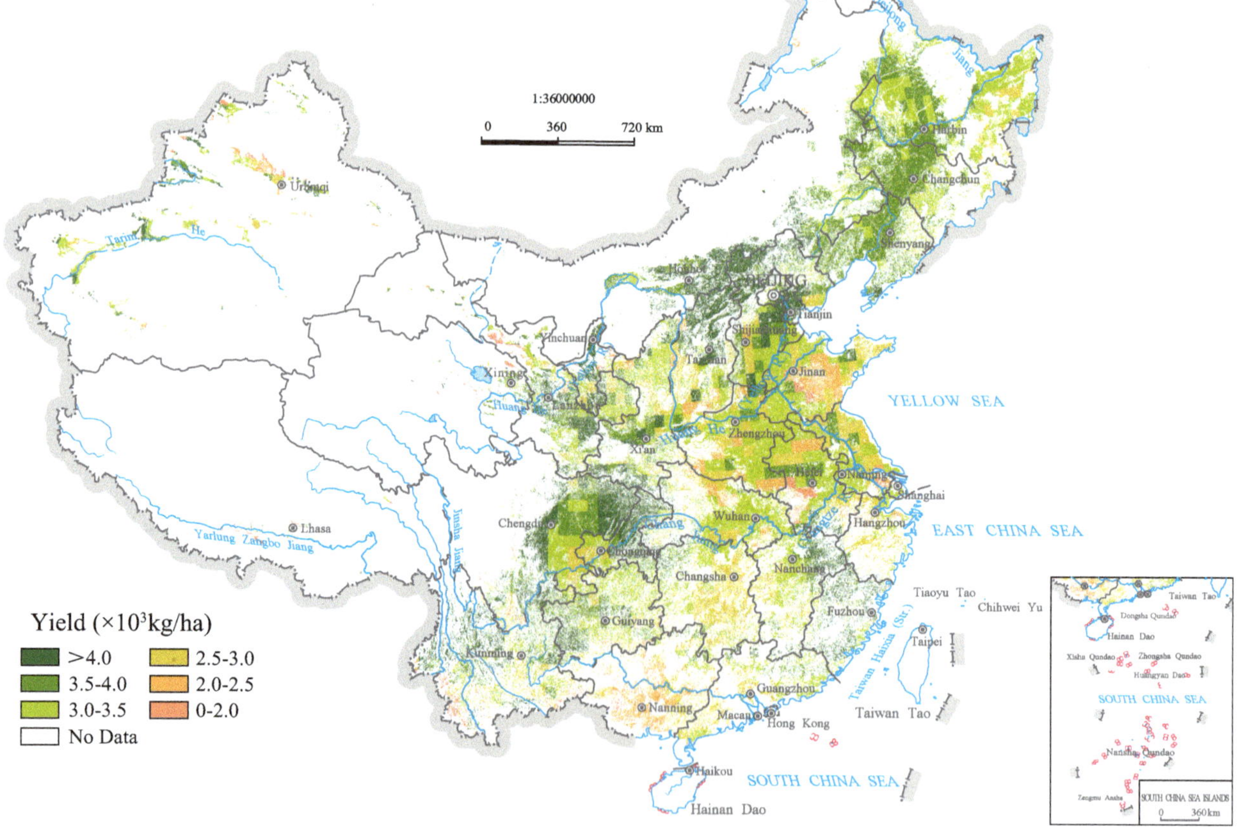

Annual Mean Yield of Rice over 1981-2010

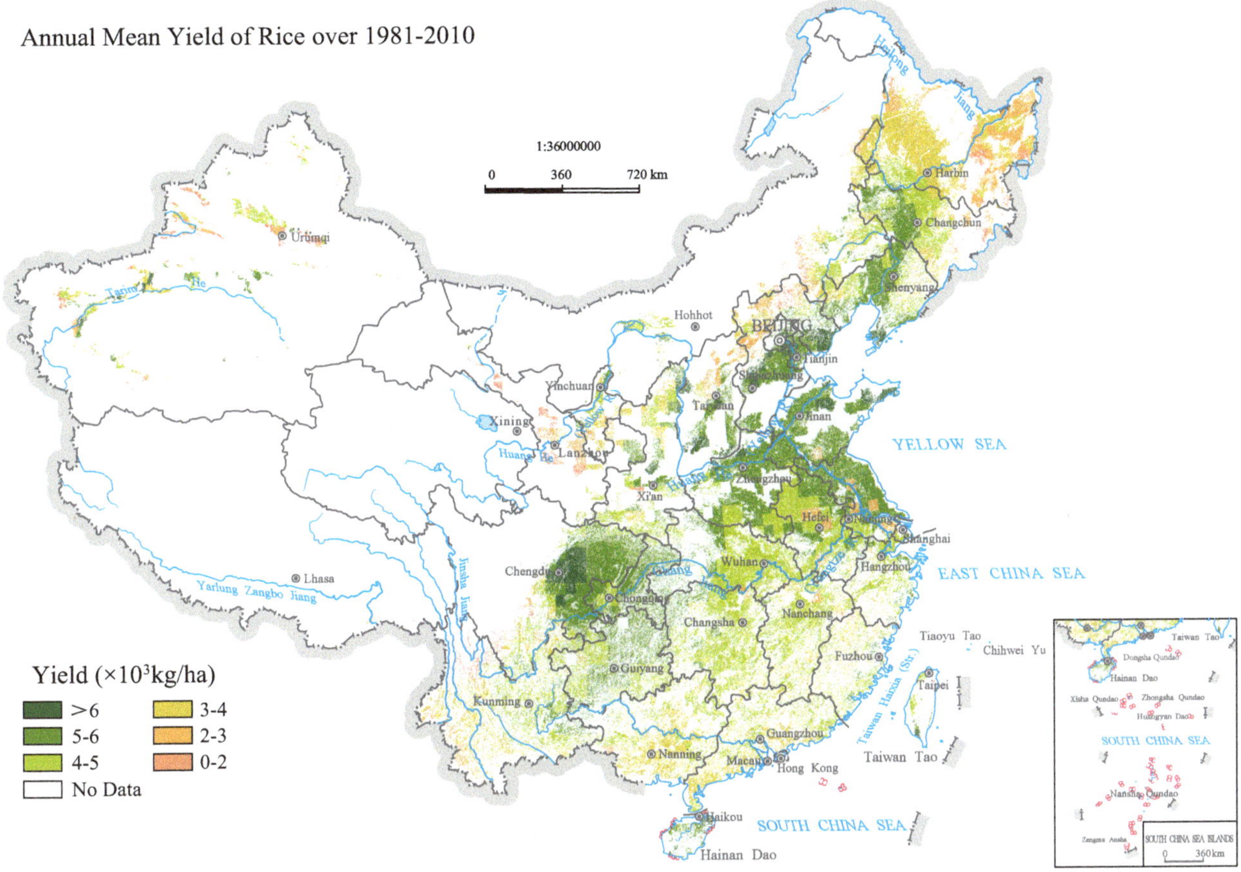

Yield ($\times10^3$kg/ha)

- >6
- 5-6
- 4-5
- 3-4
- 2-3
- 0-2
- No Data

Annual Mean Yield of Soybean over 1981-2010

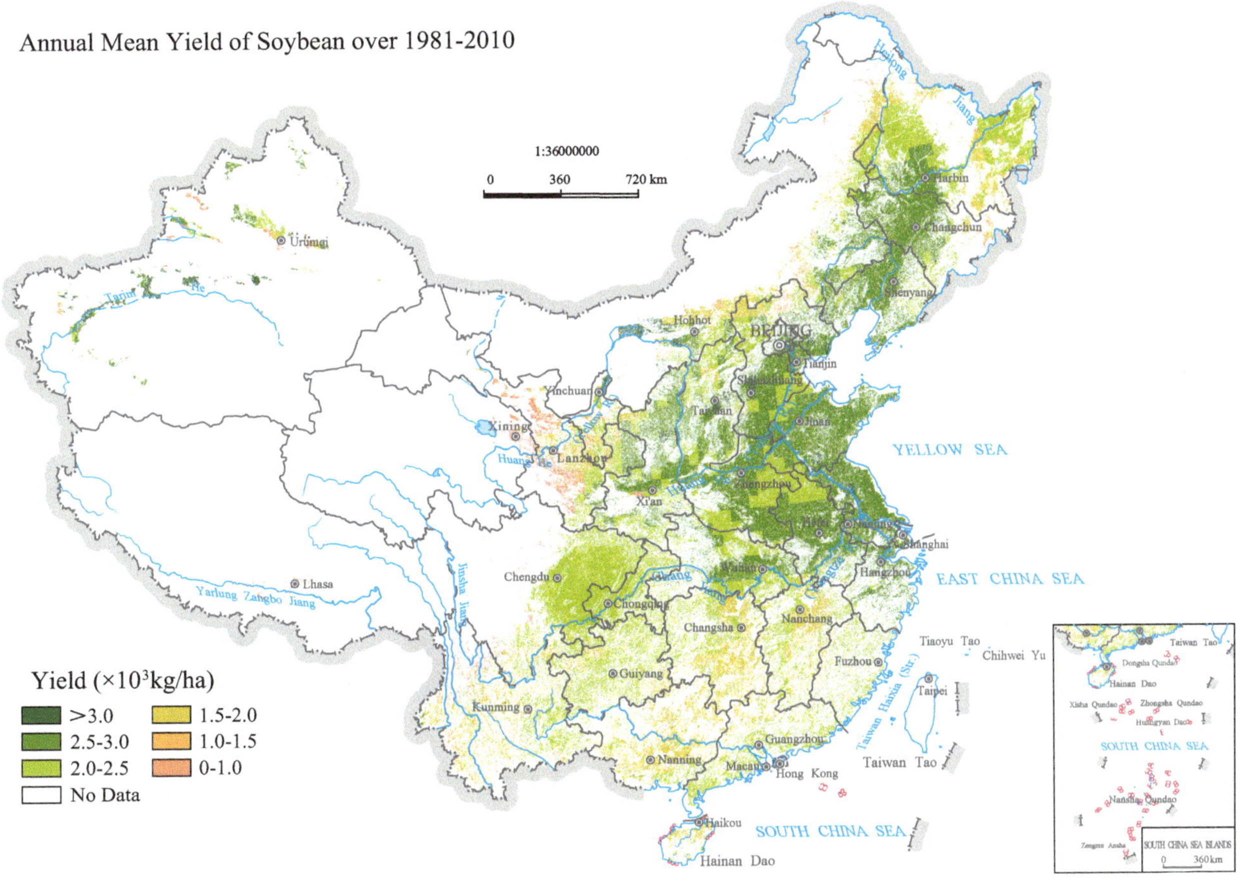

Yield ($\times10^3$kg/ha)

- >3.0
- 2.5-3.0
- 2.0-2.5
- 1.5-2.0
- 1.0-1.5
- 0-1.0
- No Data

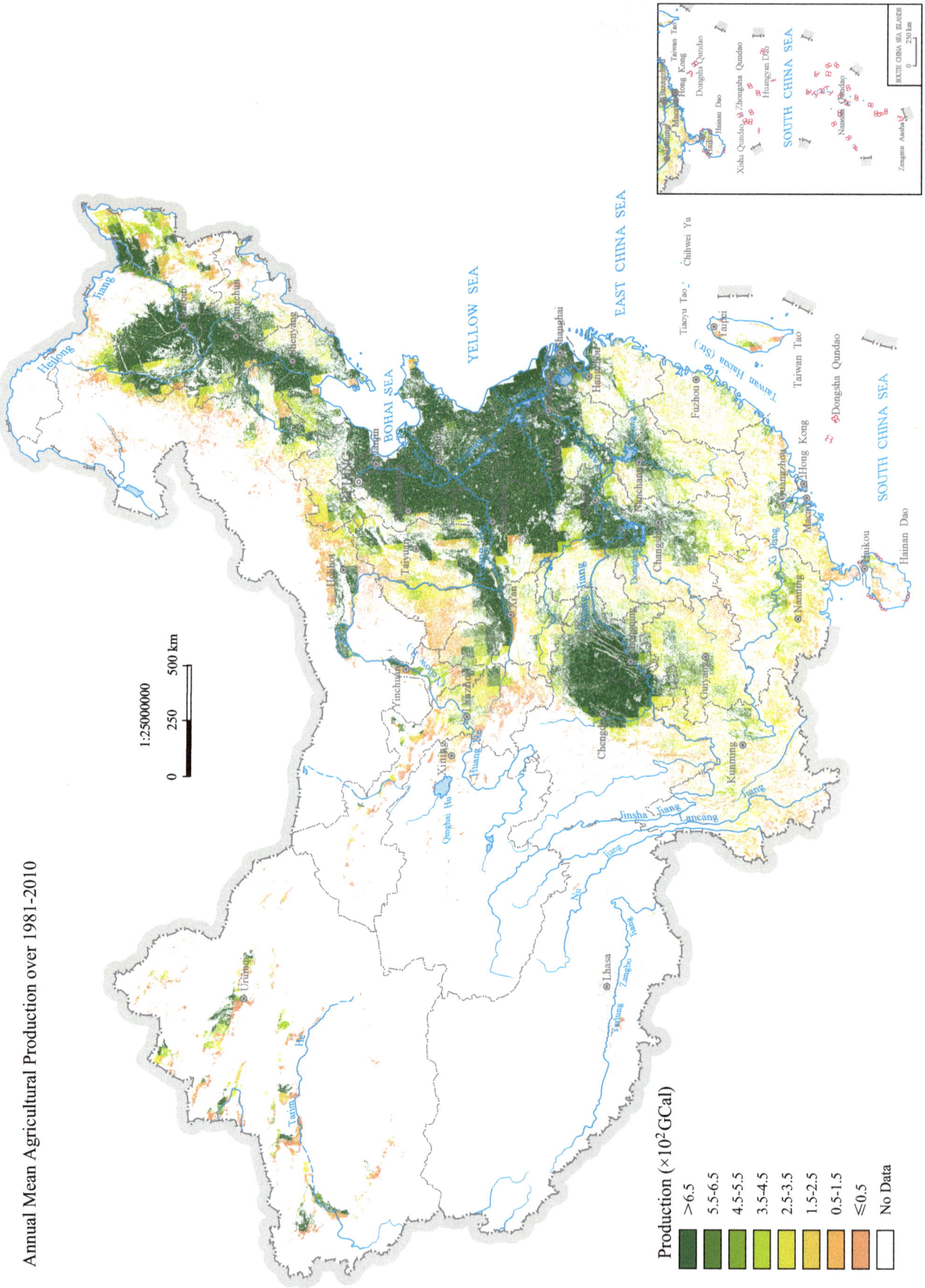
Annual Mean Agricultural Production over 1981-2010
1:25000000
0 250 500 km
Production (×10^2 GCal)
>6.5
5.5-6.5
4.5-5.5
3.5-4.5
2.5-3.5
1.5-2.5
0.5-1.5
≤0.5
No Data
BOHAI SEA
YELLOW SEA
EAST CHINA SEA
SOUTH CHINA SEA
Taiwan Tao
Hainan Dao
Dongsha Qundao
Hong Kong
Shanghai
Lhasa
Urümqi
Kunming
Guiyang
Nanning
Haikou
Fuzhou
Taipei
Hangzhou
Xining
Yinchuan
Taiyuan
Chengdu
Changsha
Nanchang
Shenyang
Changchun
Heilong Jiang
Jinsha Jiang
Lancang Jiang
Nu Jiang
Yarlung Zangbo Jiang
Qinghai Hu
Tarim He
Xi Jiang
Taiwan Haixia (Str.)
Diaoyu Tao
Chiwei Yu

Risk Levels of Maize Yield over 2011-2040 for RCP2.6

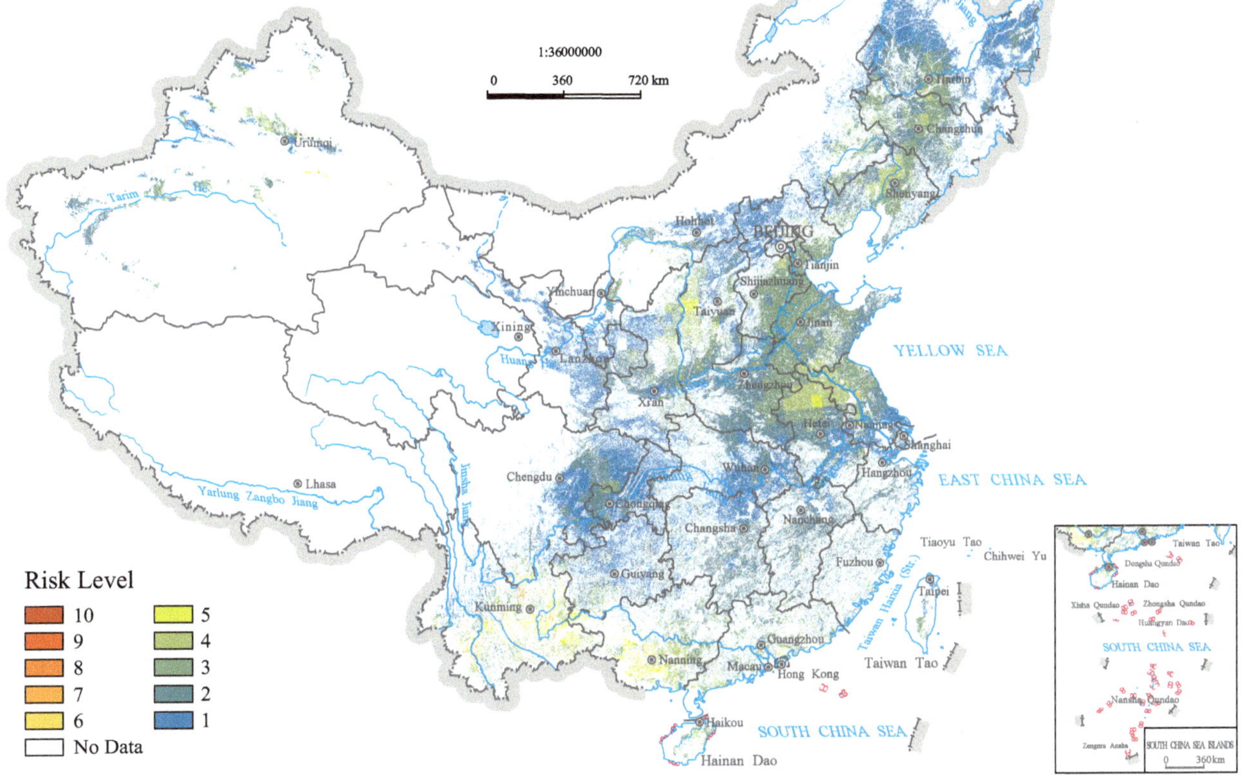

Risk Levels of Wheat Yield over 2011-2040 for RCP2.6

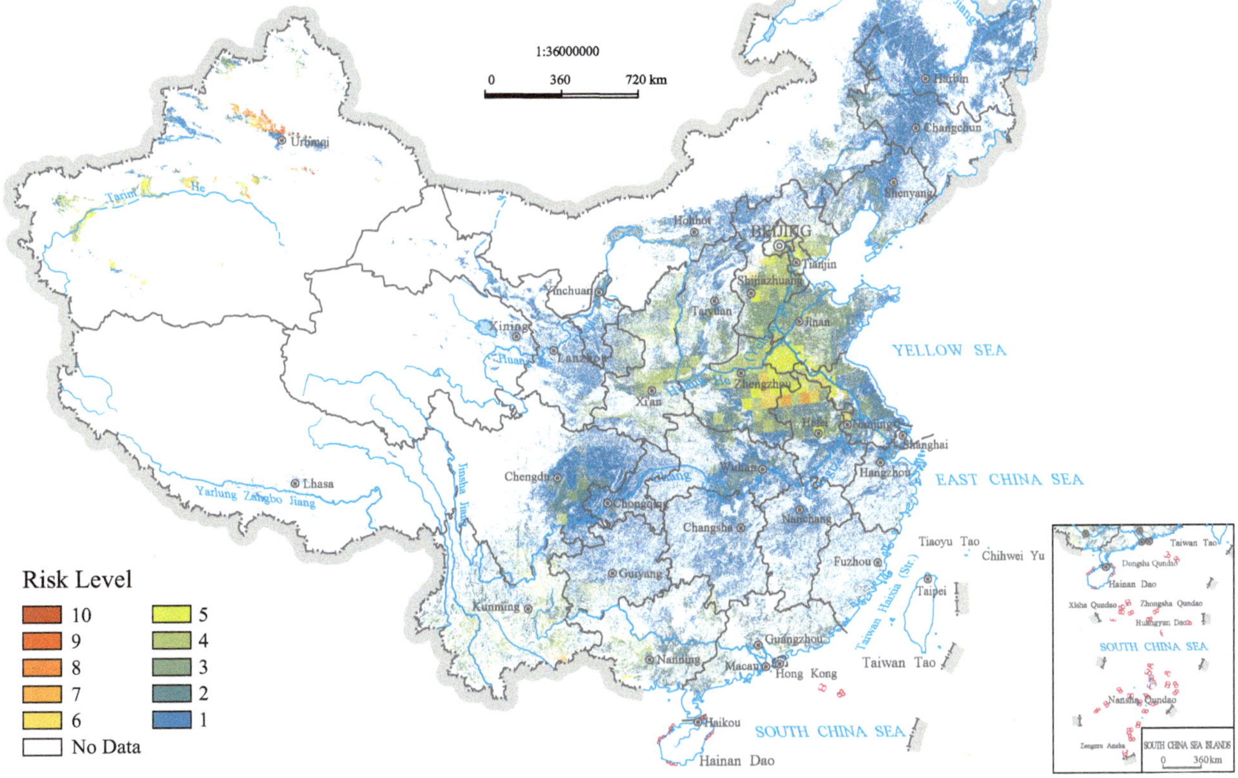

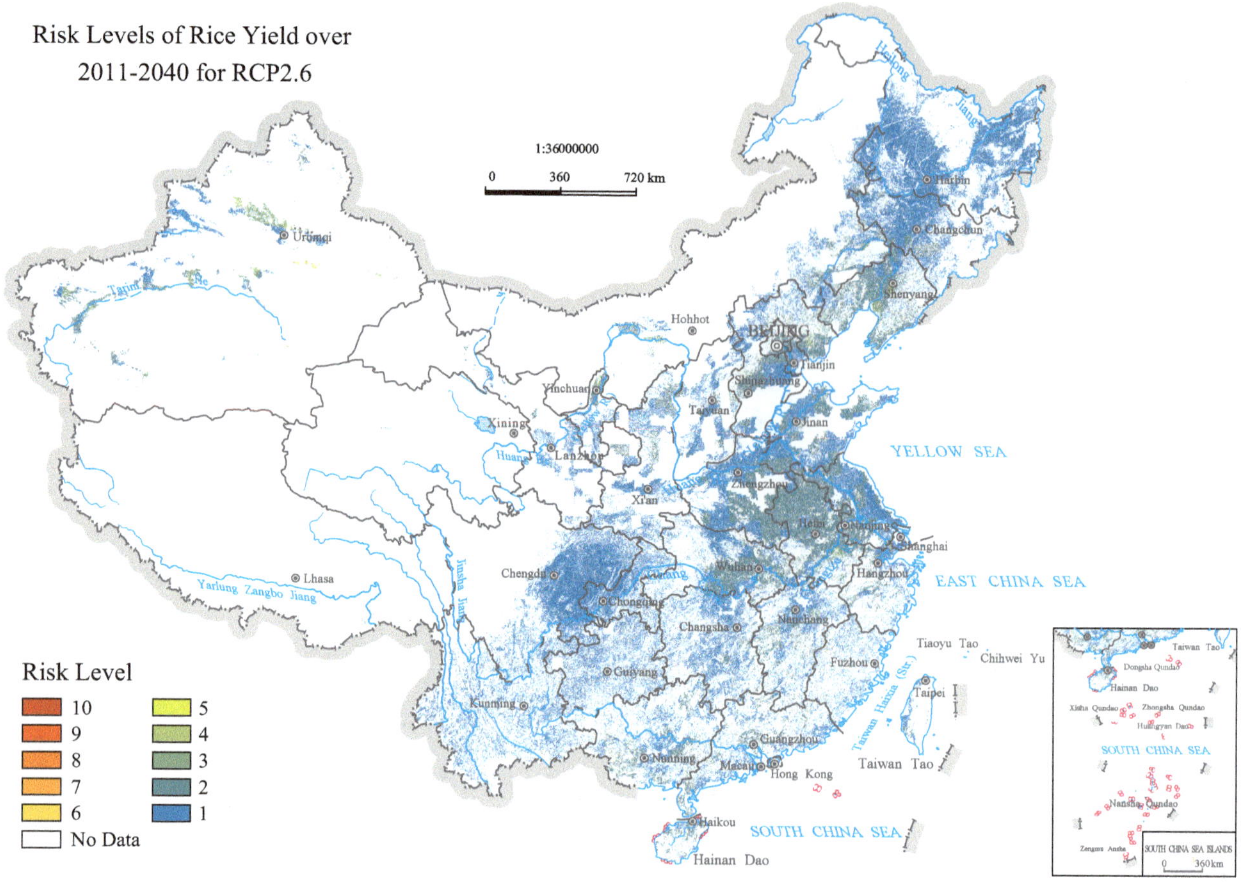
Risk Levels of Rice Yield over
2011-2040 for RCP2.6
1:36000000
0 360 720 km
Risk Level
10
9
8
7
6
5
4
3
2
1
No Data
YELLOW SEA
EAST CHINA SEA
SOUTH CHINA SEA
Hainan Dao
Taiwan Tao
Tiaoyu Tao
Chihwei Yu
Harbin
Changchun
Shenyang
Hohhot
BEIJING
Tianjin
Shijiazhuang
Taiyuan
Jinan
Yinchuan
Xining
Lanzhou
Xi'an
Zhengzhou
Hefei
Nanjing
Shanghai
Hangzhou
Wuhan
Nanchang
Changsha
Chengdu
Chongqing
Guiyang
Kunming
Nanning
Guangzhou
Macau
Hong Kong
Haikou
Fuzhou
Taipei
Lhasa
Urumqi
Yarlung Zangbo Jiang
Jinsha Jiang
SOUTH CHINA SEA
SOUTH CHINA SEA ISLANDS
0 360km

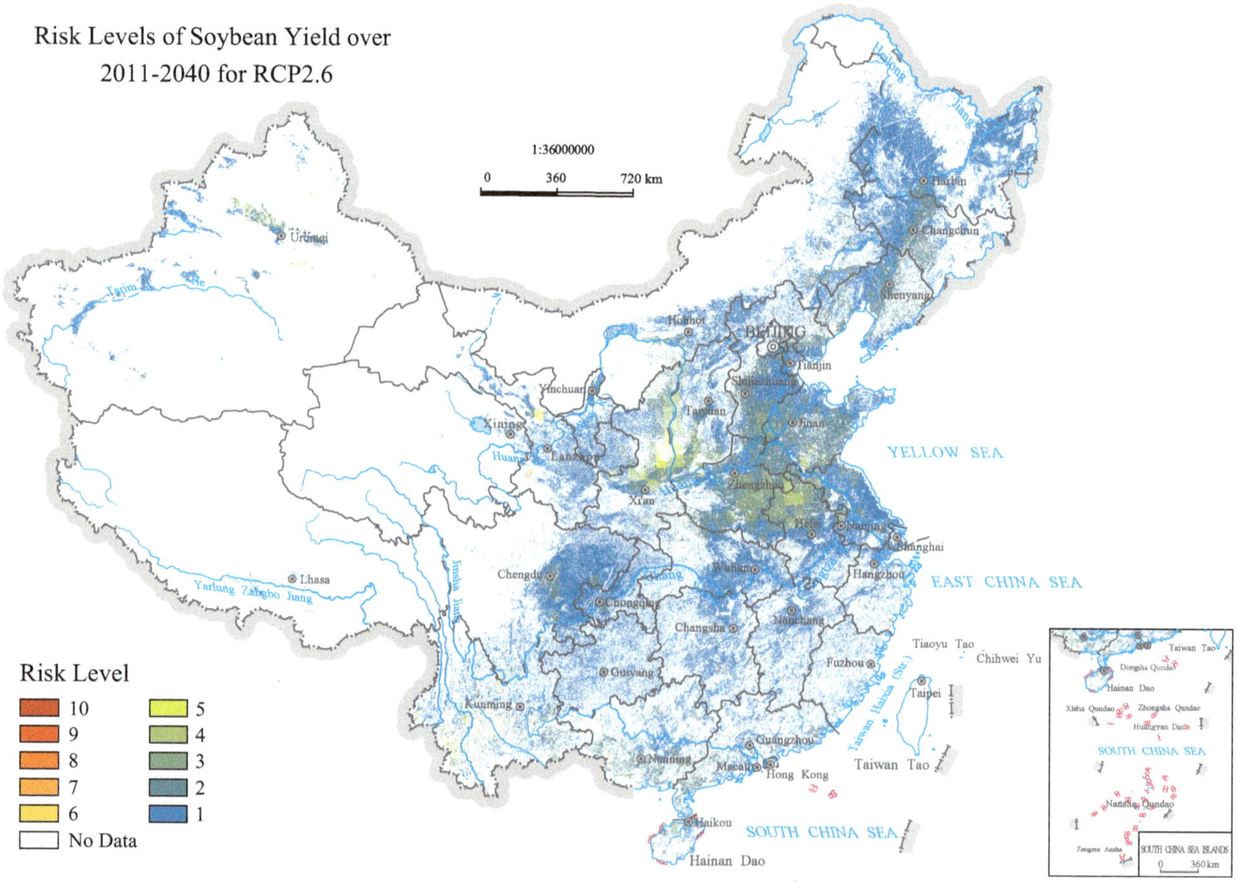
Risk Levels of Soybean Yield over
2011-2040 for RCP2.6
1:36000000
0 360 720 km
Risk Level
10
9
8
7
6
5
4
3
2
1
No Data
YELLOW SEA
EAST CHINA SEA
SOUTH CHINA SEA
Hainan Dao
Taiwan Tao
Tiaoyu Tao
Chihwei Yu
Harbin
Changchun
Shenyang
Hohhot
BEIJING
Tianjin
Shijiazhuang
Taiyuan
Jinan
Yinchuan
Xining
Lanzhou
Xi'an
Zhengzhou
Hefei
Nanjing
Shanghai
Hangzhou
Wuhan
Nanchang
Changsha
Chengdu
Chongqing
Guiyang
Kunming
Nanning
Guangzhou
Macau
Hong Kong
Haikou
Fuzhou
Taipei
Lhasa
Urumqi
Yarlung Zangbo Jiang
Jinsha Jiang
SOUTH CHINA SEA
SOUTH CHINA SEA ISLANDS
0 360km

Risk Levels of Agricultural Production over 2011-2040 for RCP2.6

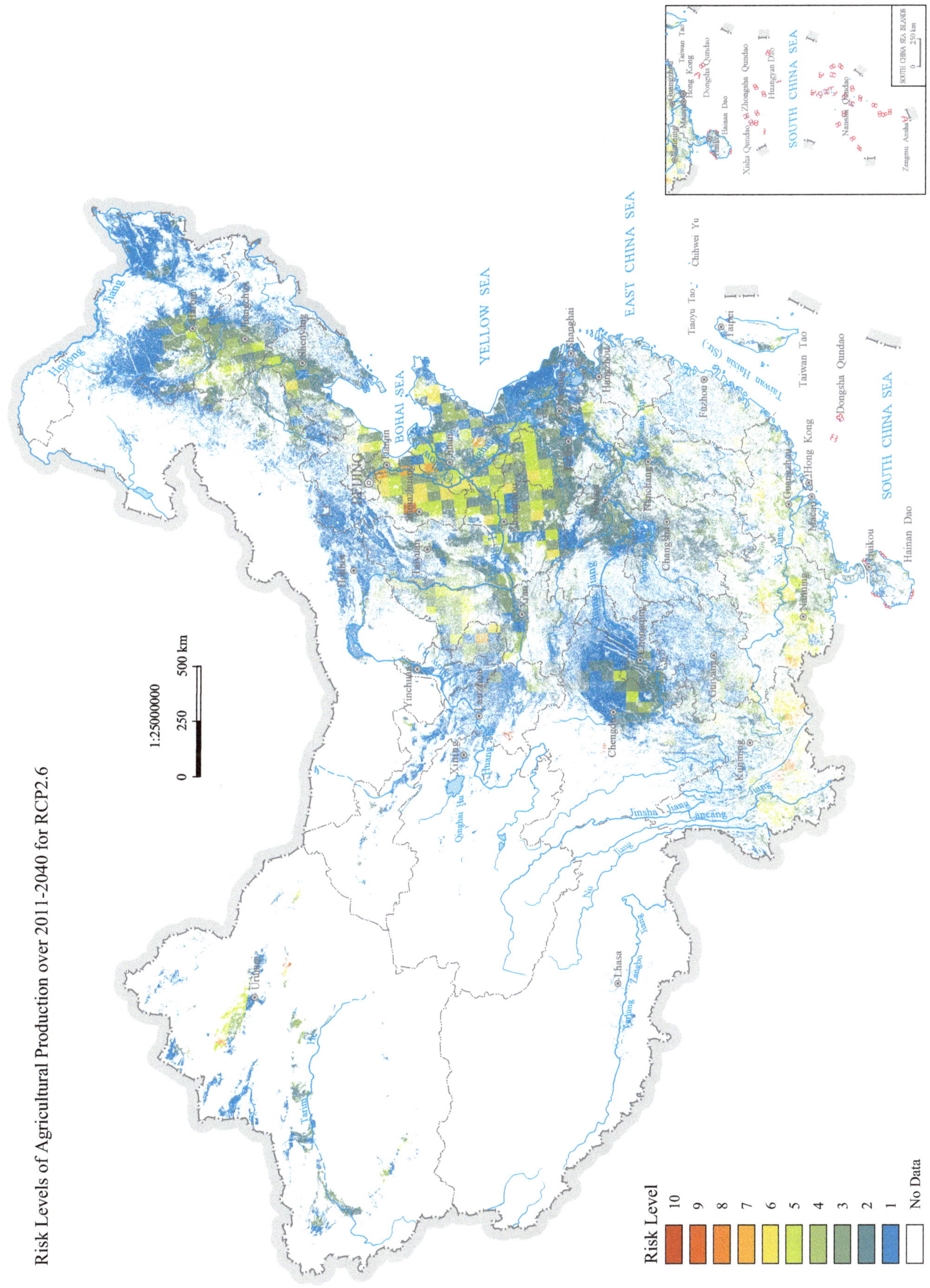

Risk Levels of Maize Yield over 2041-2070 for RCP2.6

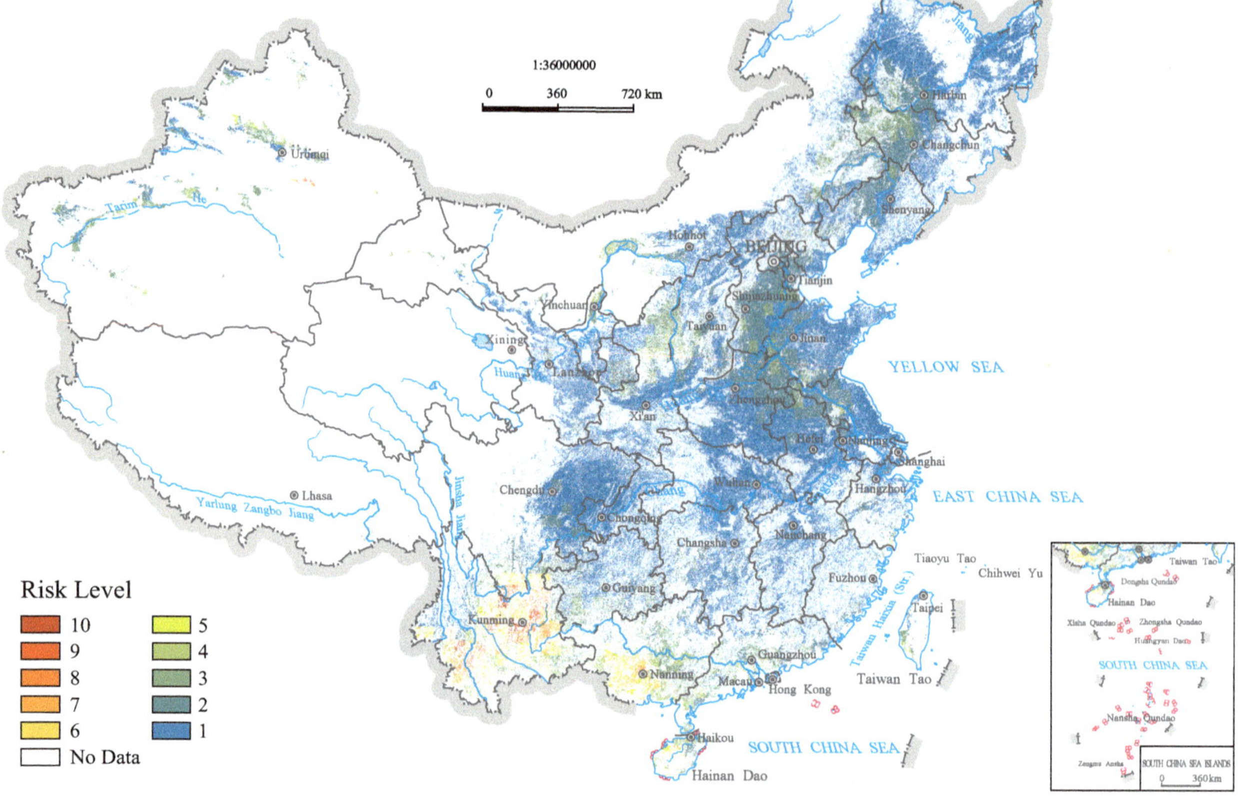

Risk Levels of Wheat Yield over 2041-2070 for RCP2.6

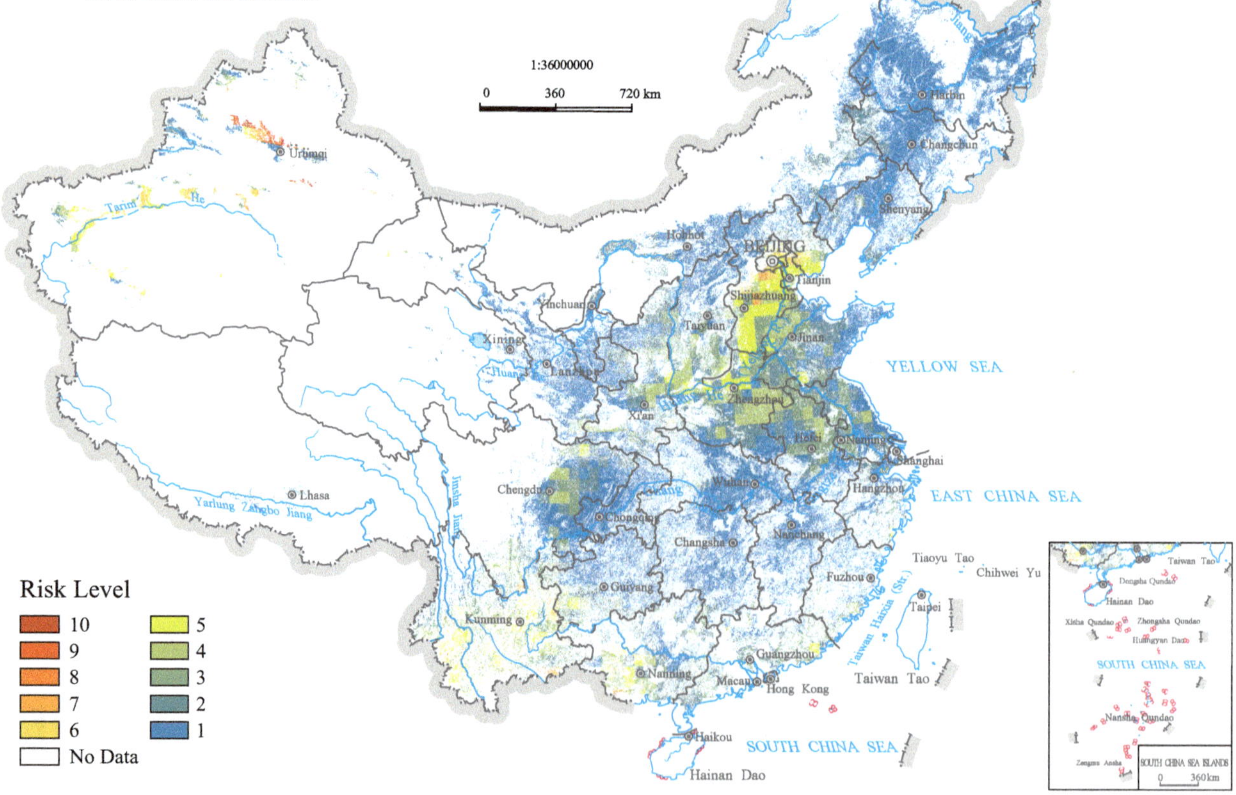

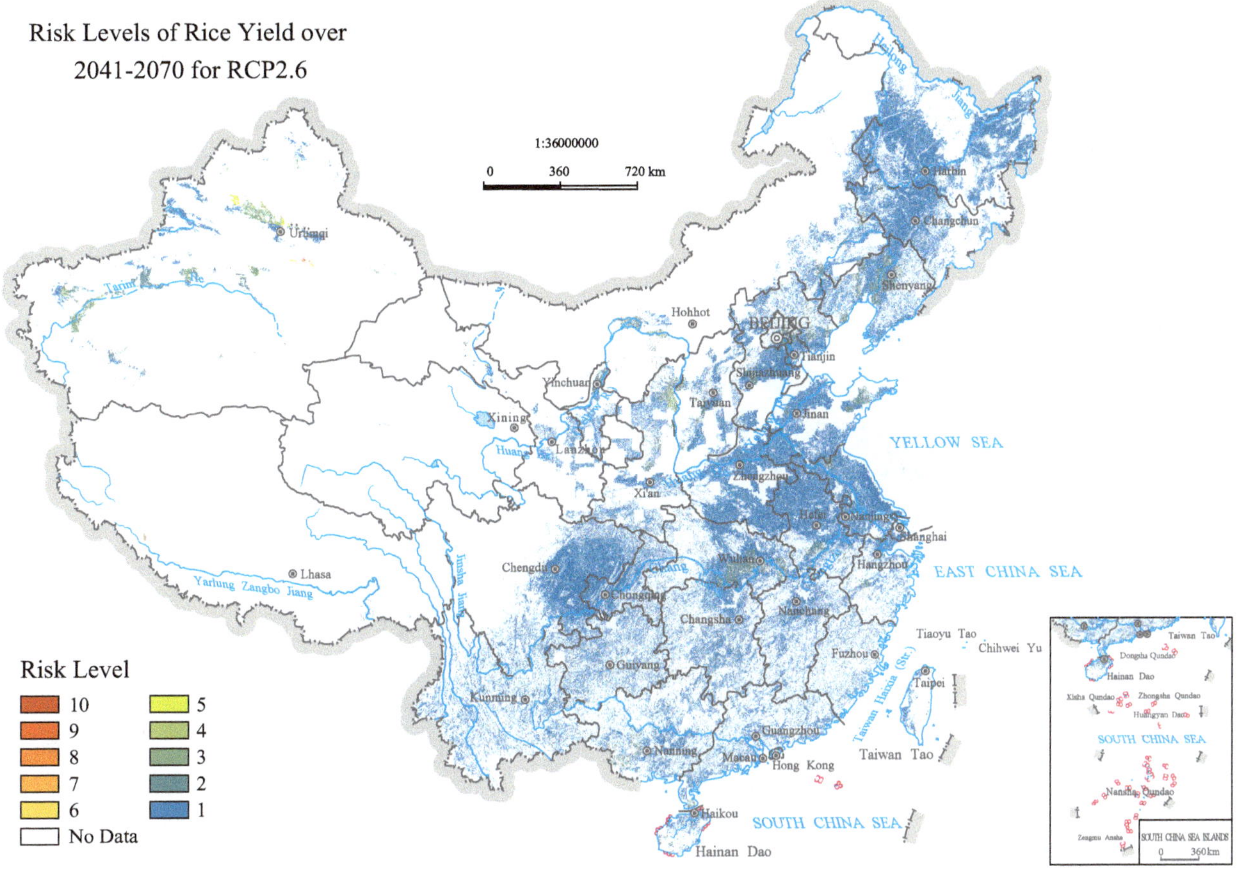

Risk Levels of Rice Yield over 2041-2070 for RCP2.6

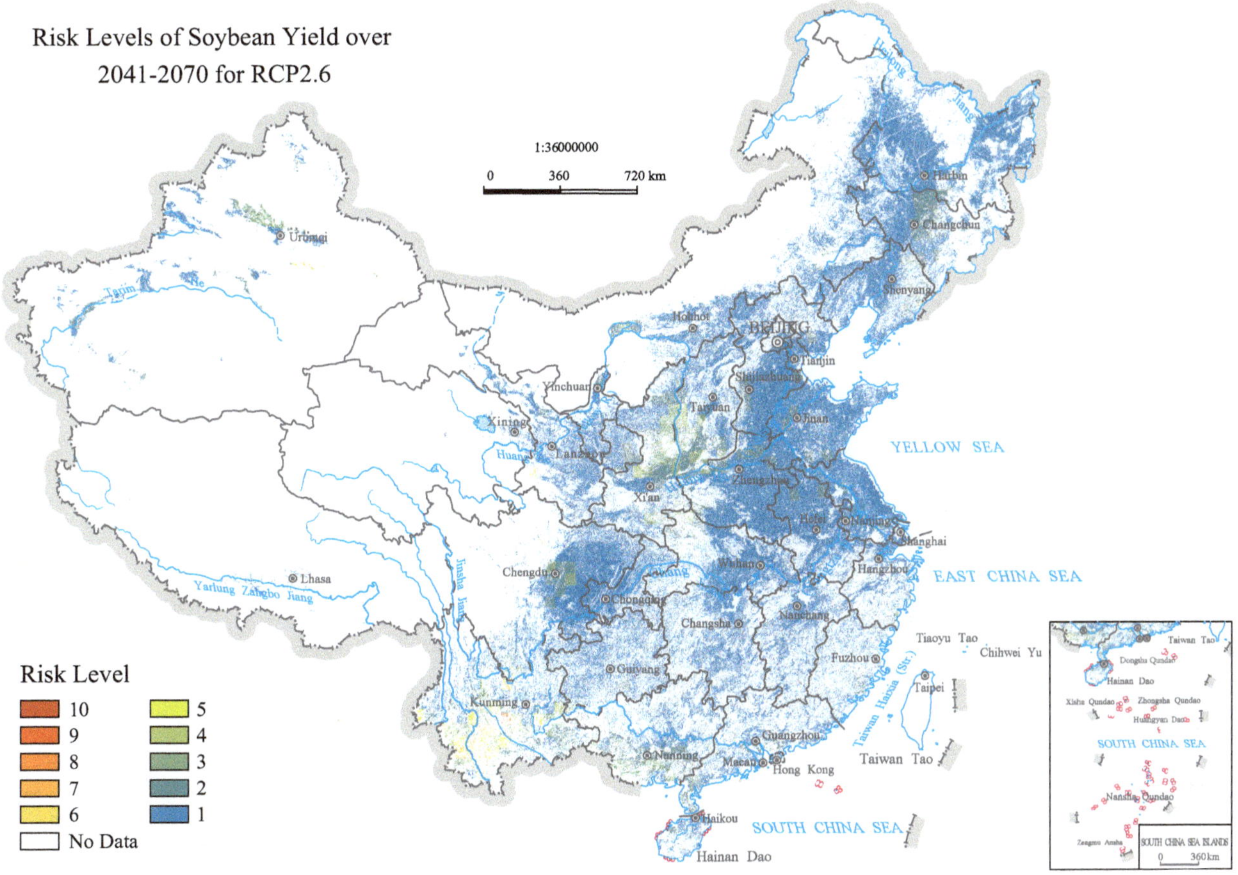

Risk Levels of Soybean Yield over 2041-2070 for RCP2.6

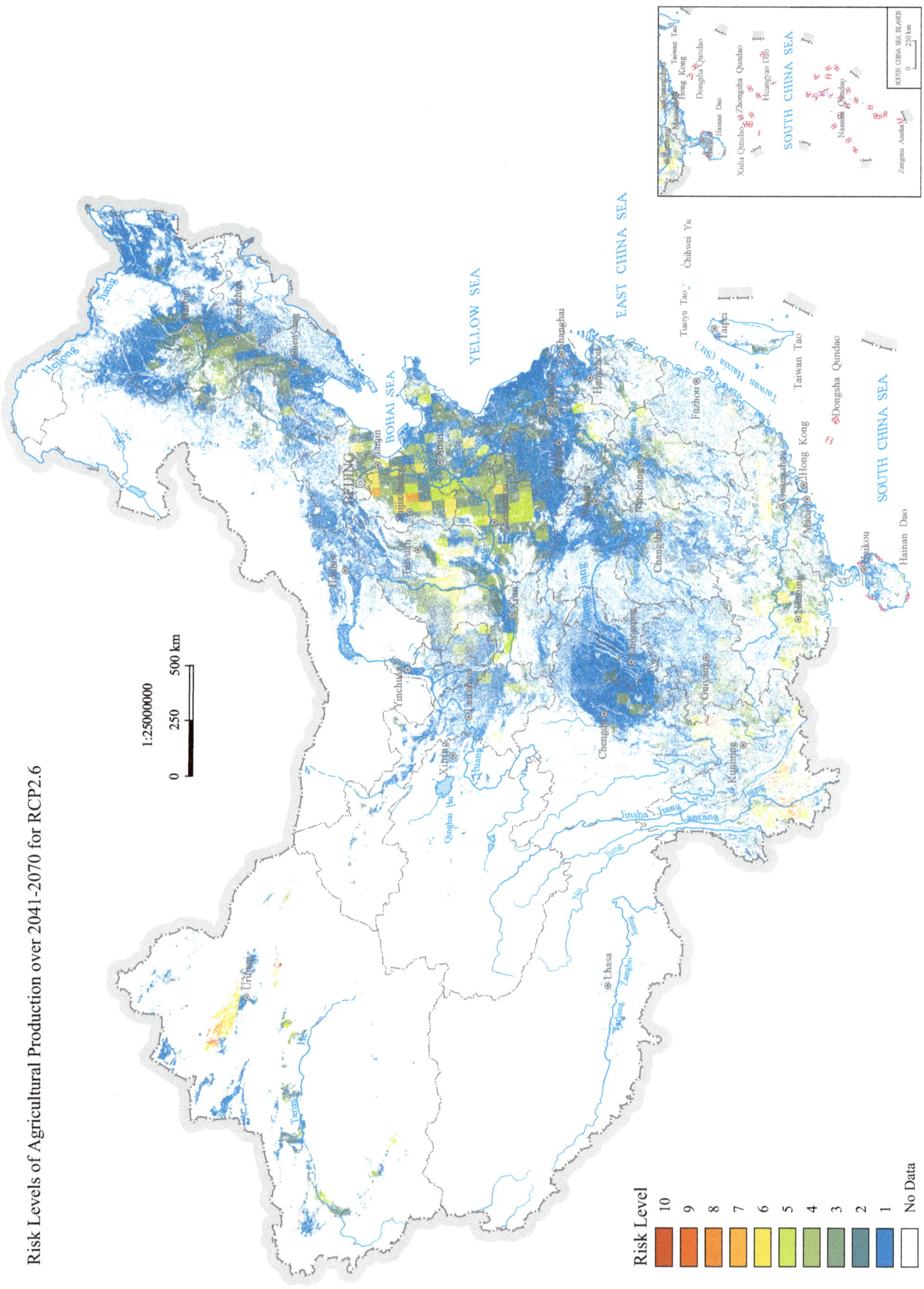

Risk Levels of Agricultural Production over 2041-2070 for RCP2.6
1:25000000
0 250 500 km
Risk Level
10
9
8
7
6
5
4
3
2
1
No Data
BEIJING
Tianjin
Shenyang
Heilong Jiang
BOHAI SEA
YELLOW SEA
EAST CHINA SEA
SOUTH CHINA SEA
Shanghai
Fuzhou
Taipei
Taiwan Tao
Tiaoyu Tao
Chihwei Yu
Dongsha Qundao
Hong Kong
Haikou
Hainan Dao
Nanchang
Changsha
Taiyuan
Yinchuan
Lanzhou
Xining
Qinghai Hu
Chengdu
Guiyang
Kunming
Jinsha Jiang
Lancang Jiang
Nu Jiang
Lhasa
Yarlung Zangbo Jiang
Urumqi
Xisha Qundao
Zhongsha Qundao
Nansha Qundao
SOUTH CHINA SEA ISLANDS

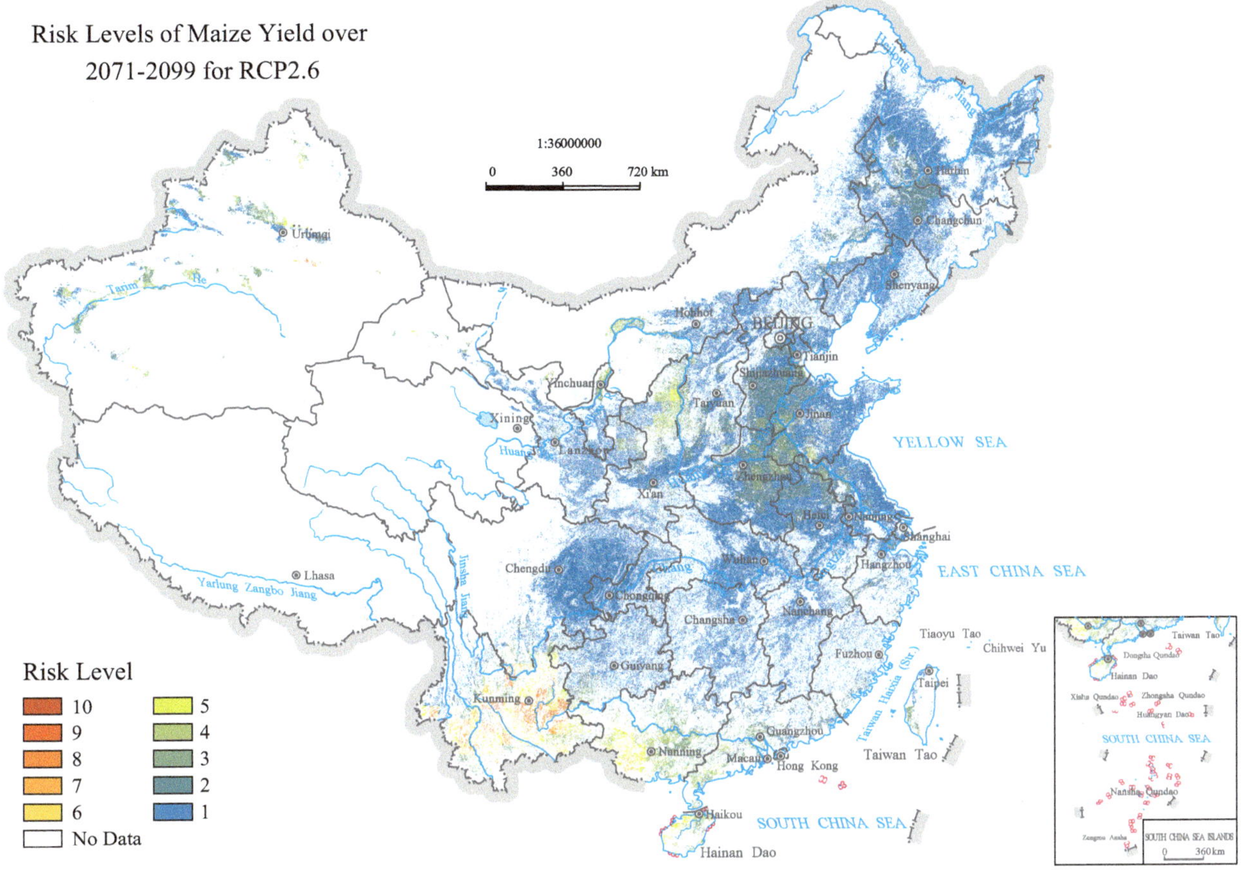

Risk Levels of Maize Yield over 2071-2099 for RCP2.6

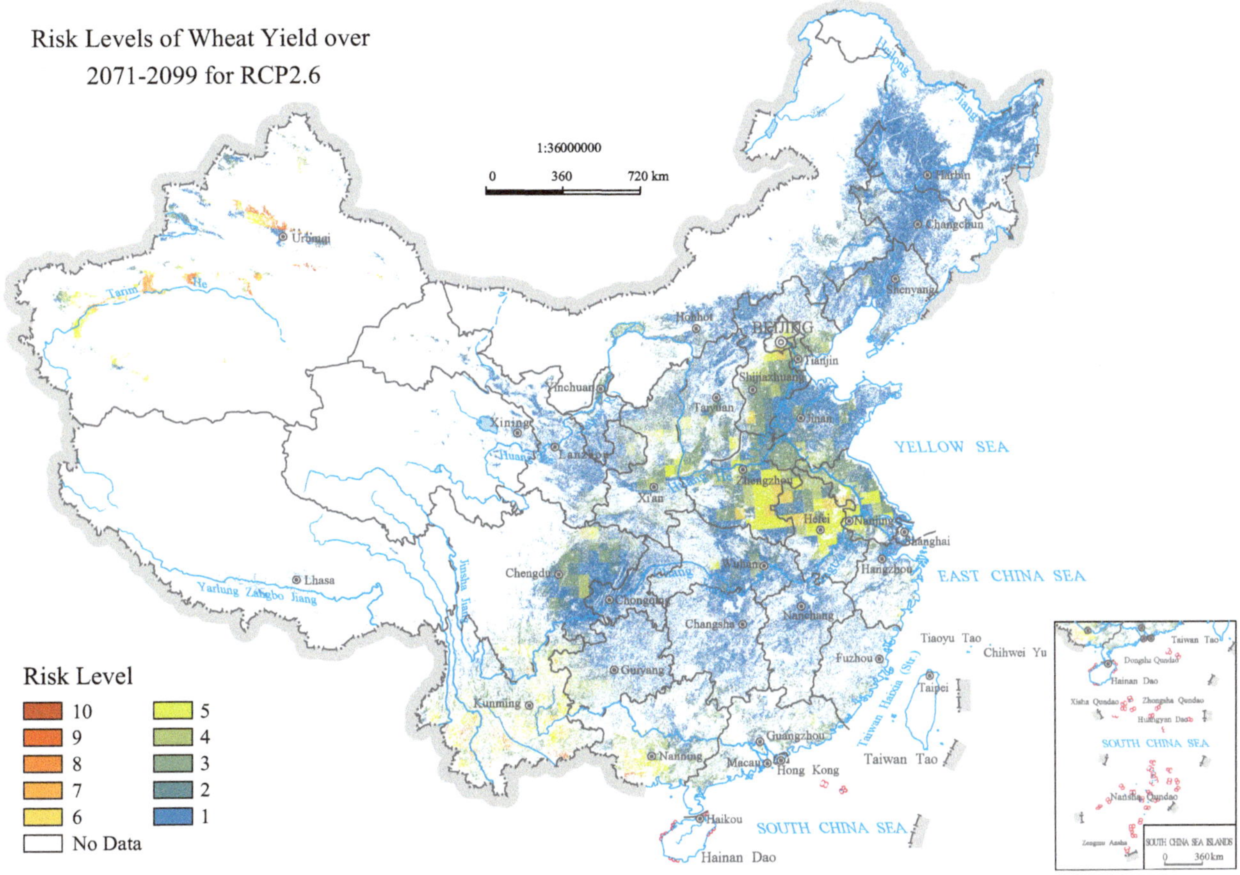

Risk Levels of Wheat Yield over 2071-2099 for RCP2.6

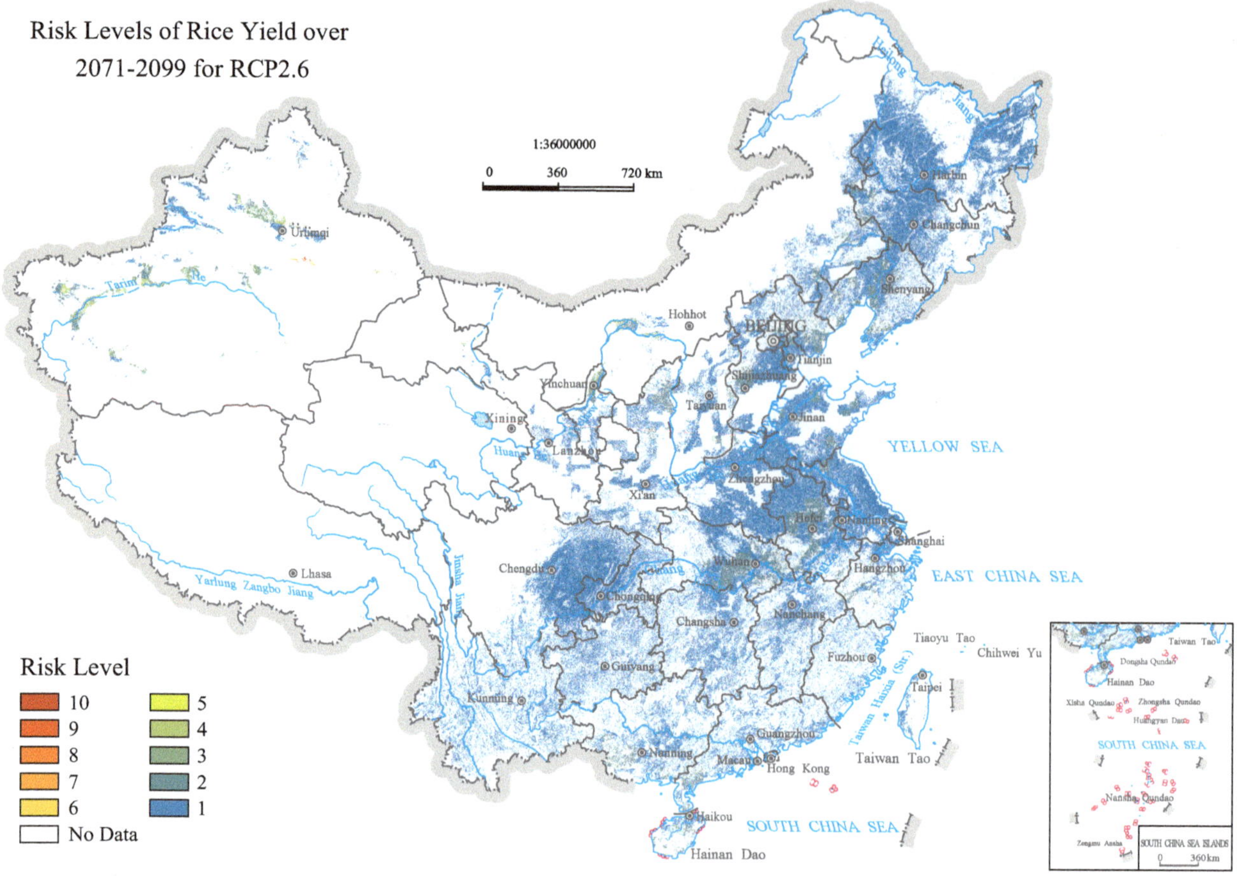
Risk Levels of Rice Yield over 2071-2099 for RCP2.6
1:36000000
0 360 720 km
Risk Level
10
9
8
7
6
5
4
3
2
1
No Data
Urümqi
Hohhot
BEIJING
Tianjin
Shijiazhuang
Taiyuan
Yinchuan
Xining
Lanzhou
Jinan
Zhengzhou
Xi'an
Hefei
Nanjing
Shanghai
Hangzhou
Wuhan
Chengdu
Chongqing
Changsha
Nanchang
Fuzhou
Taipei
Guiyang
Kunming
Nanning
Guangzhou
Macau
Hong Kong
Haikou
Harbin
Changchun
Shenyang
Lhasa
Yarlung Zangbo Jiang
Jinsha Jiang
Tarim
YELLOW SEA
EAST CHINA SEA
SOUTH CHINA SEA
Tiaoyu Tao
Chihwei Yu
Taiwan Tao
Hainan Dao
SOUTH CHINA SEA ISLANDS
0 360km

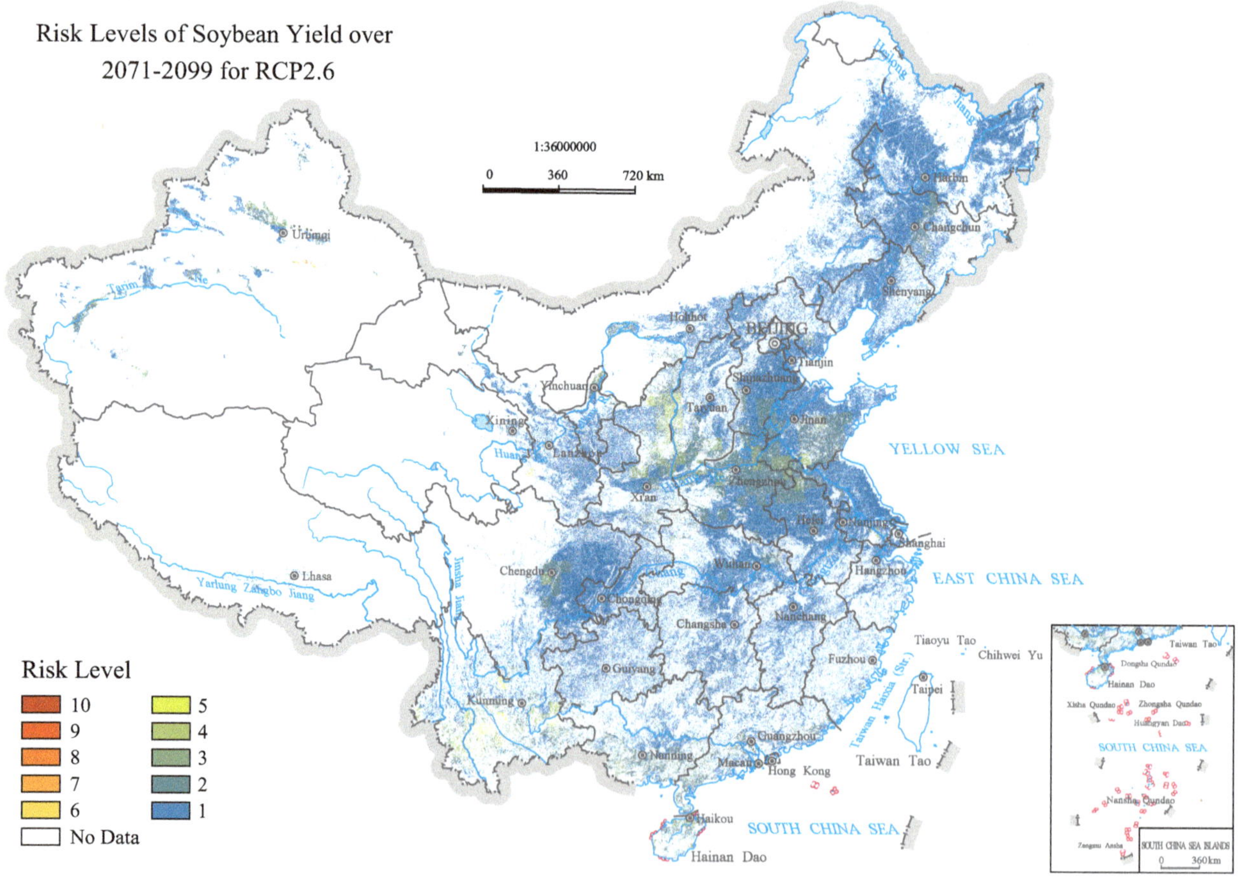
Risk Levels of Soybean Yield over 2071-2099 for RCP2.6
1:36000000
0 360 720 km
Risk Level
10
9
8
7
6
5
4
3
2
1
No Data
Urümqi
Hohhot
BEIJING
Tianjin
Shijiazhuang
Taiyuan
Yinchuan
Xining
Lanzhou
Jinan
Zhengzhou
Xi'an
Hefei
Nanjing
Shanghai
Hangzhou
Wuhan
Chengdu
Chongqing
Changsha
Nanchang
Fuzhou
Taipei
Guiyang
Kunming
Nanning
Guangzhou
Macau
Hong Kong
Haikou
Harbin
Changchun
Shenyang
Lhasa
Yarlung Zangbo Jiang
Jinsha Jiang
Tarim
YELLOW SEA
EAST CHINA SEA
SOUTH CHINA SEA
Tiaoyu Tao
Chihwei Yu
Taiwan Tao
Hainan Dao
SOUTH CHINA SEA ISLANDS
0 360km

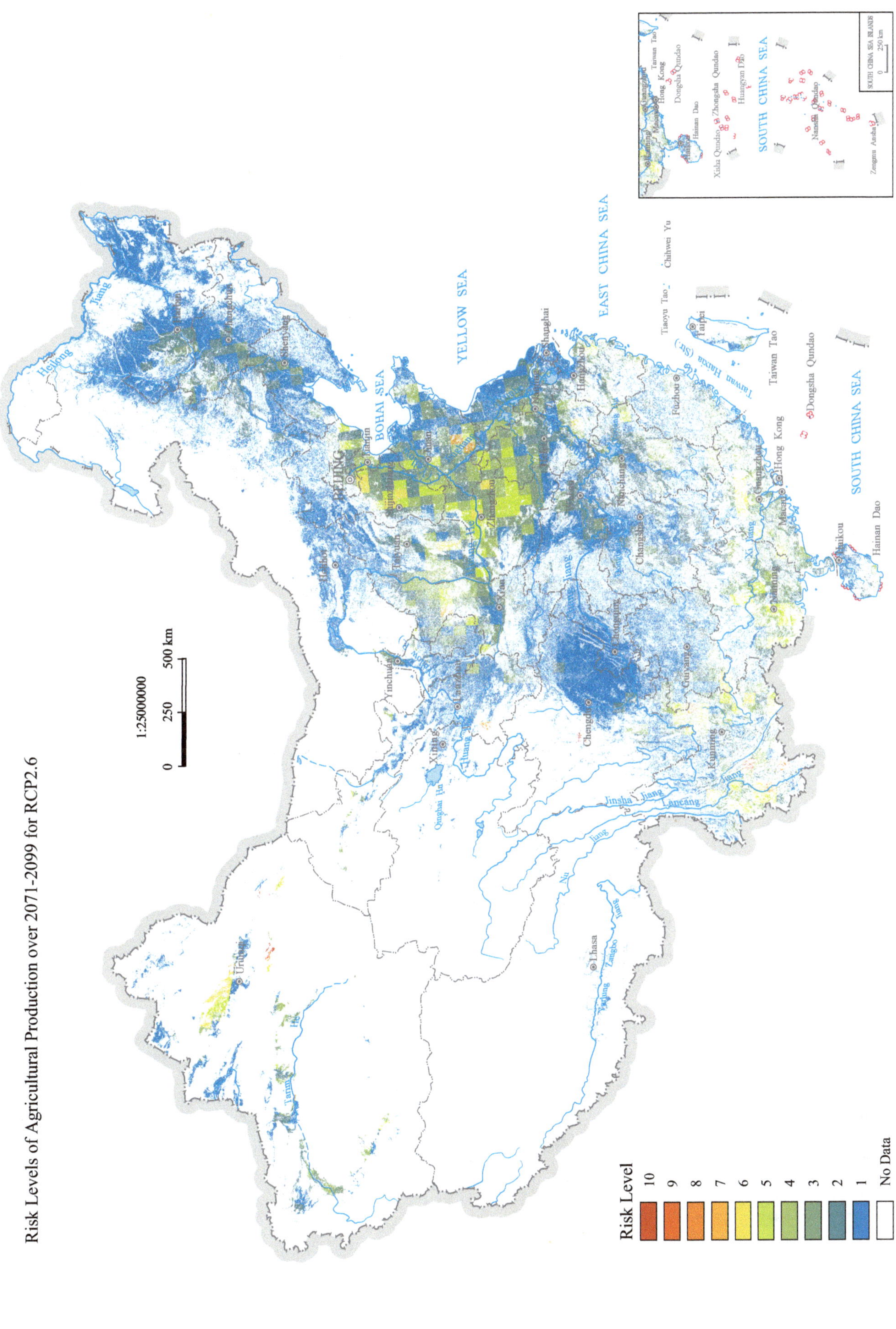
Risk Levels of Agricultural Production over 2071-2099 for RCP2.6
1:25000000
0 250 500 km
Risk Level
10
9
8
7
6
5
4
3
2
1
No Data
BOHAI SEA
YELLOW SEA
EAST CHINA SEA
SOUTH CHINA SEA
BEIJING
Shanghai
Lhasa
Urumqi
Xining
Yinchuan
Chengdu
Kunming
Guangzhou
Hong Kong
Haikou
Hainan Dao
Taibei
Taiwan Tao
Dongsha Qundao
Jinsha Jiang
Lancang Jiang
Heilong Jiang
Qinghai Hu

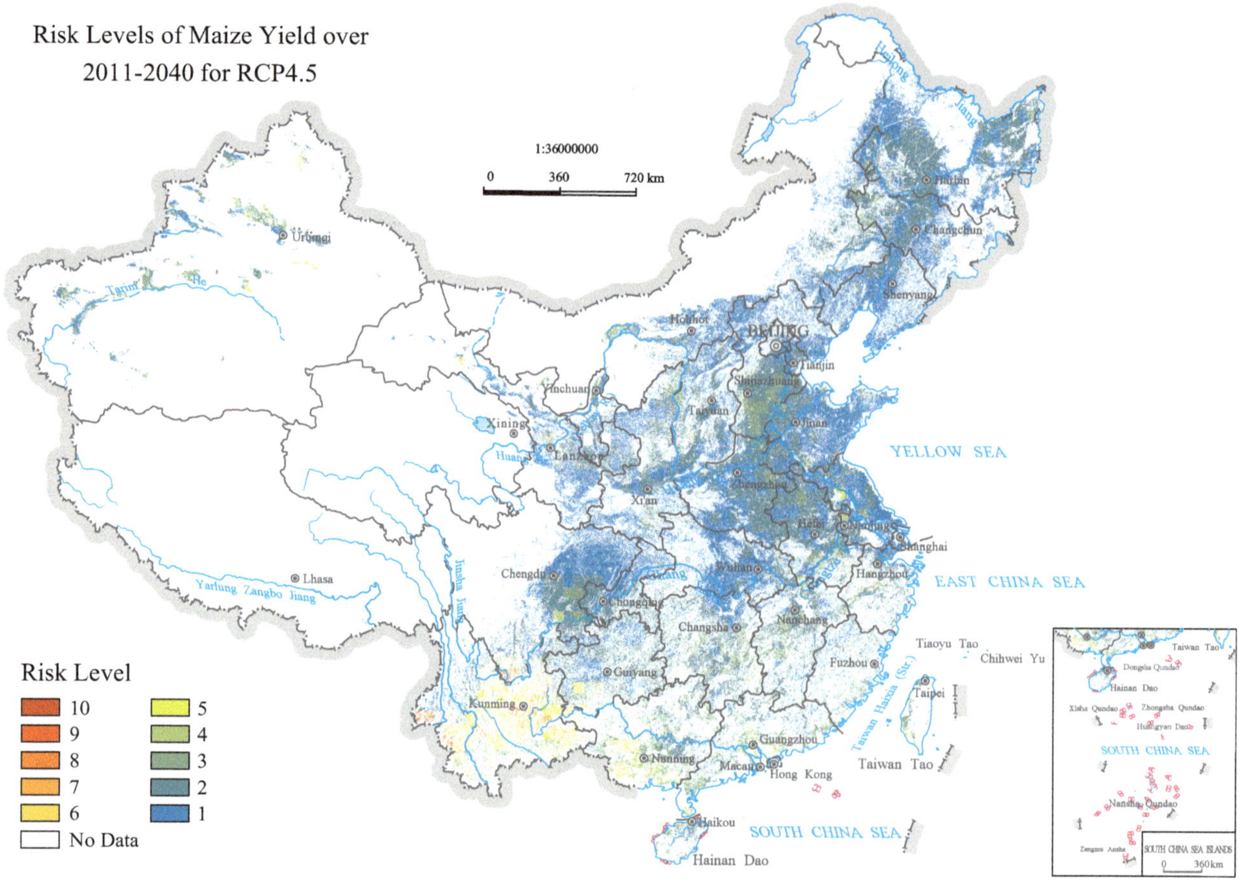
Risk Levels of Maize Yield over 2011-2040 for RCP4.5
1:36000000
0 360 720 km
Risk Level
10
9
8
7
6
5
4
3
2
1
No Data
YELLOW SEA
EAST CHINA SEA
SOUTH CHINA SEA
Hainan Dao
Taiwan Tao
Tiaoyu Tao
Chihwei Yu
BEIJING
Tianjin
Harbin
Changchun
Shenyang
Hohhot
Shijiazhuang
Taiyuan
Jinan
Yinchuan
Xining
Lanzhou
Xi'an
Zhengzhou
Hefei
Nanjing
Shanghai
Hangzhou
Wuhan
Chengdu
Chongqing
Changsha
Nanchang
Fuzhou
Taipei
Guiyang
Kunming
Nanning
Guangzhou
Macau
Hong Kong
Haikou
Lhasa
Urümqi
Yarlung Zangbo Jiang
Jinsha Jiang
Tarim He
Heilong Jiang
SOUTH CHINA SEA ISLANDS
360km

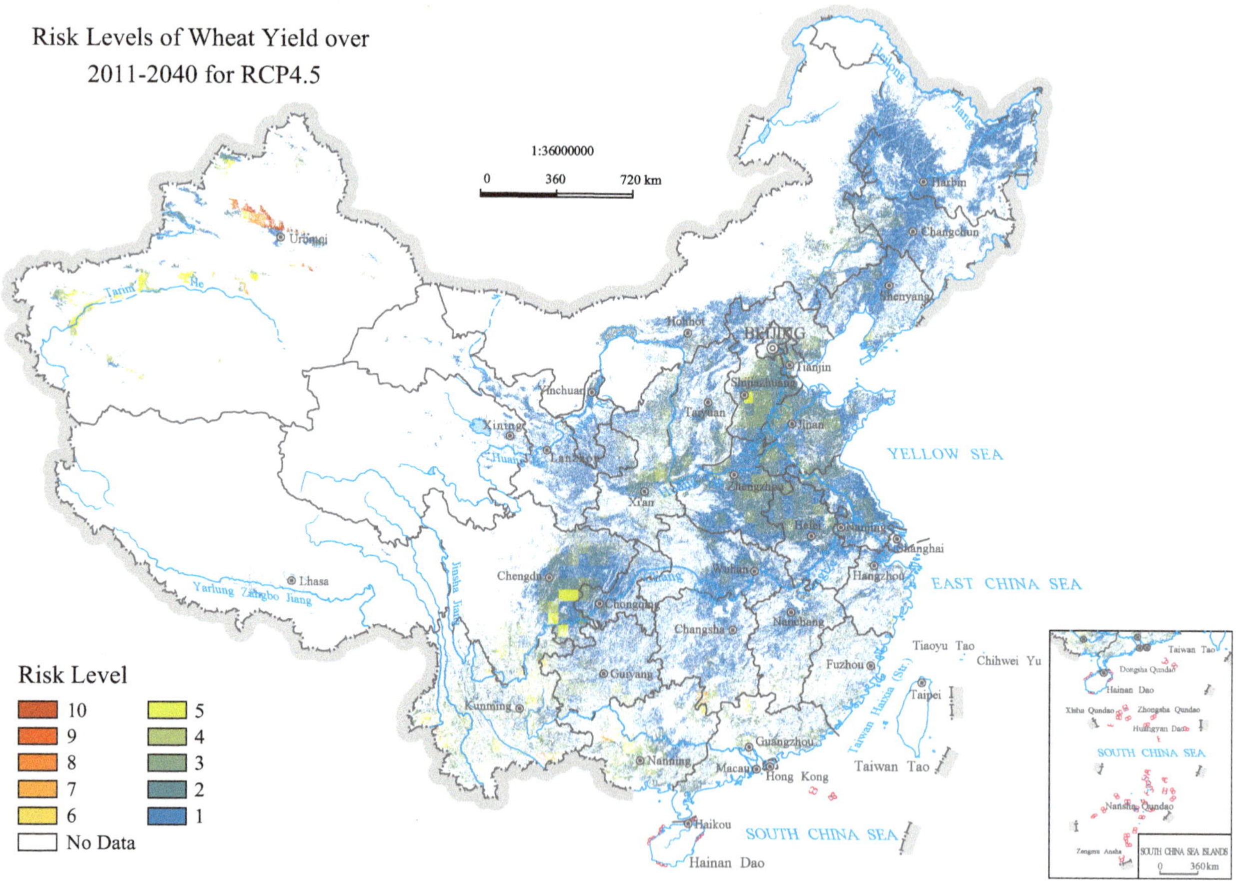
Risk Levels of Wheat Yield over 2011-2040 for RCP4.5
1:36000000
0 360 720 km
Risk Level
10
9
8
7
6
5
4
3
2
1
No Data
YELLOW SEA
EAST CHINA SEA
SOUTH CHINA SEA
Hainan Dao
Taiwan Tao
Tiaoyu Tao
Chihwei Yu
BEIJING
Tianjin
Harbin
Changchun
Shenyang
Hohhot
Shijiazhuang
Taiyuan
Jinan
Yinchuan
Xining
Lanzhou
Xi'an
Zhengzhou
Hefei
Nanjing
Shanghai
Hangzhou
Wuhan
Chengdu
Chongqing
Changsha
Nanchang
Fuzhou
Taipei
Guiyang
Kunming
Nanning
Guangzhou
Macau
Hong Kong
Haikou
Lhasa
Urümqi
Yarlung Zangbo Jiang
Jinsha Jiang
Tarim He
Heilong Jiang
SOUTH CHINA SEA ISLANDS
360km

Risk Levels of Rice Yield over 2011-2040 for RCP4.5

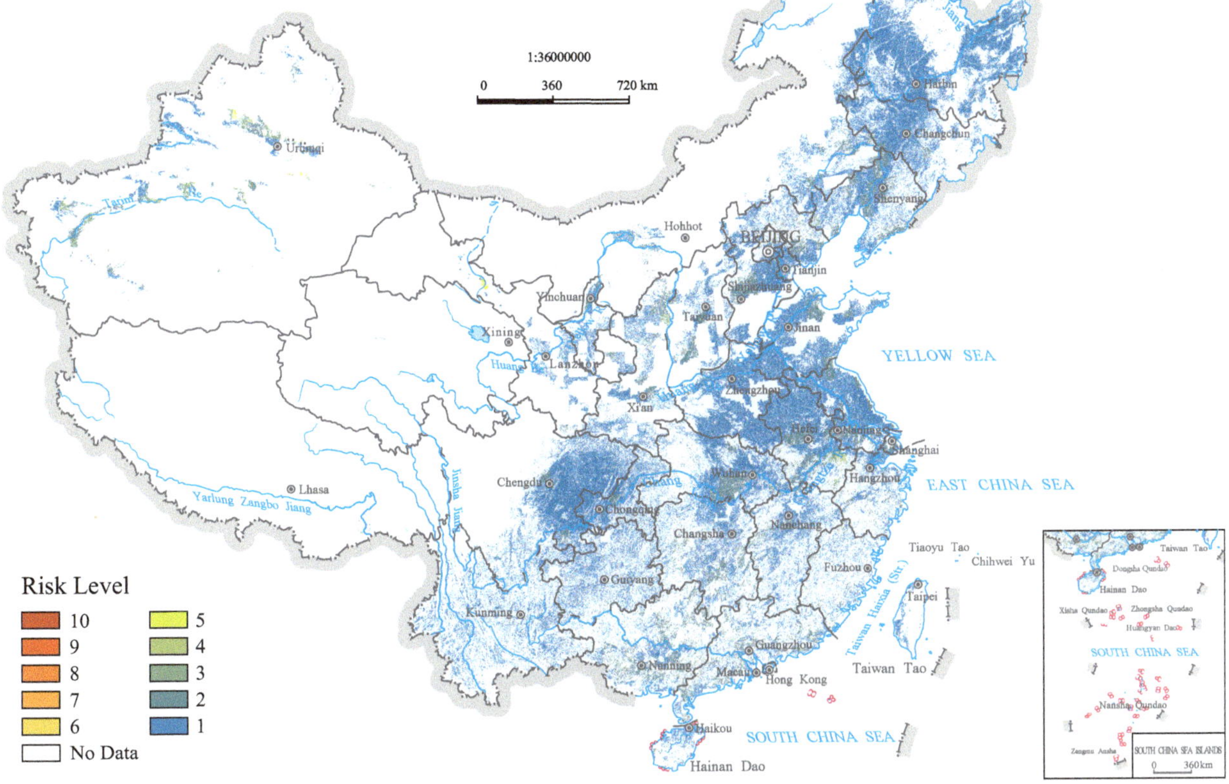

Risk Levels of Soybean Yield over 2011-2040 for RCP4.5

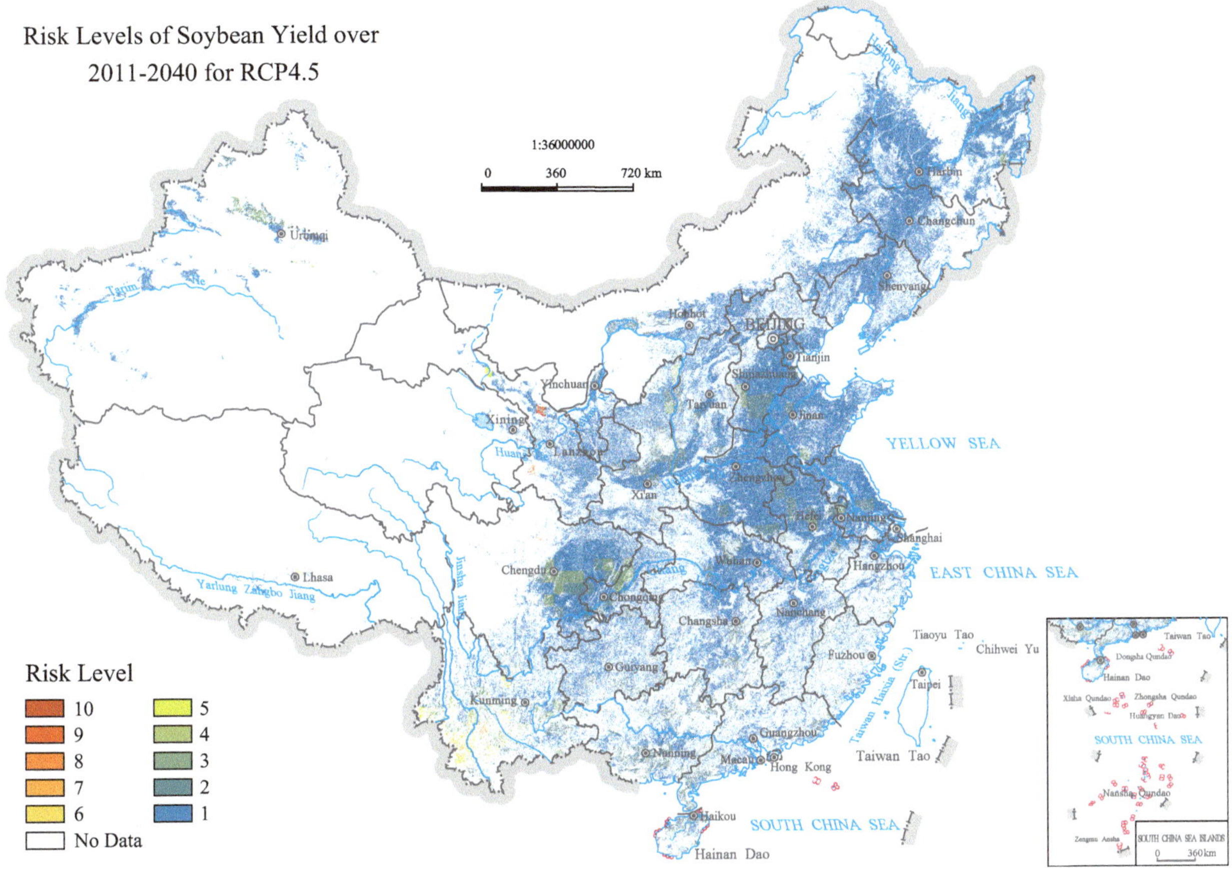

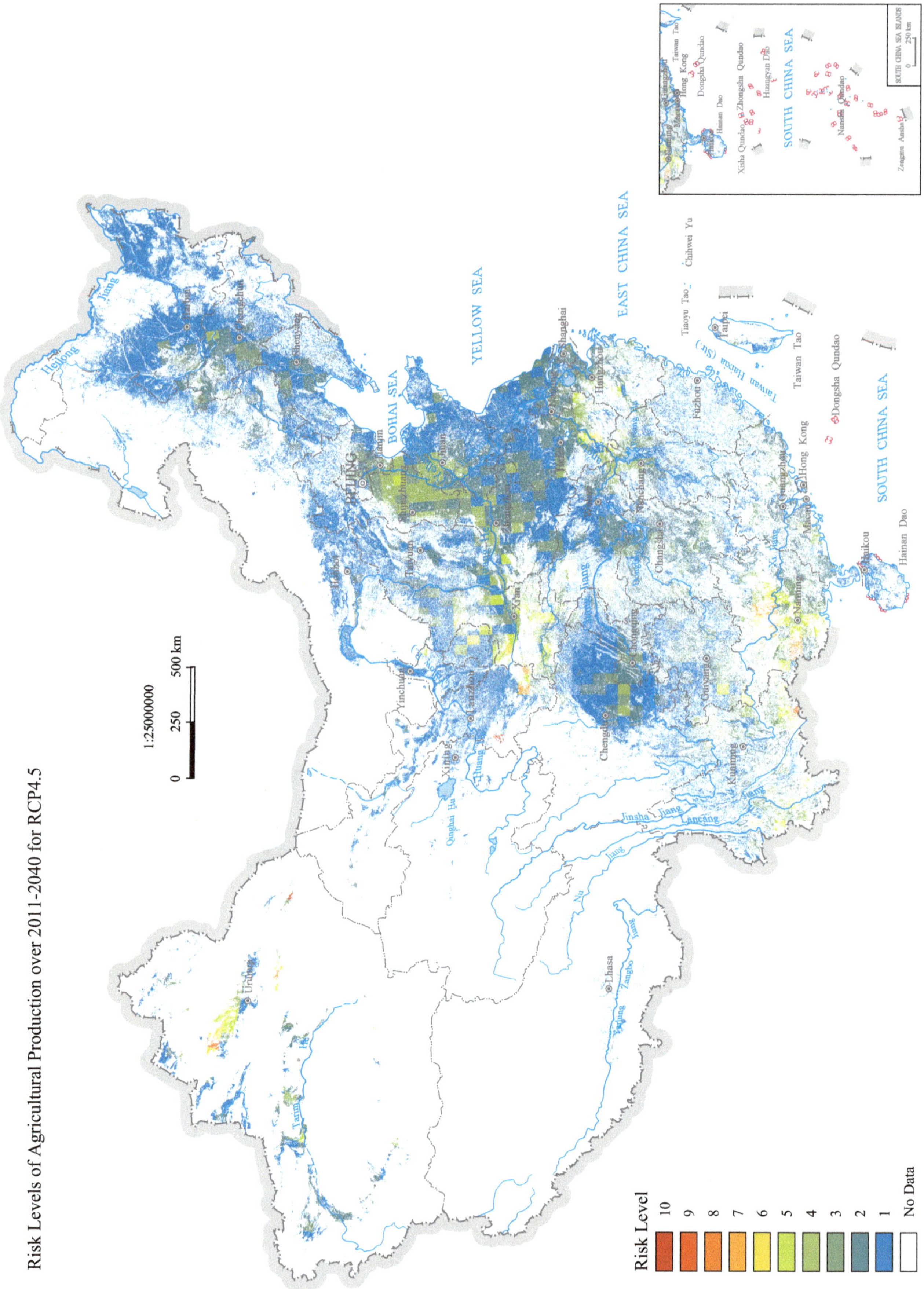
Risk Levels of Agricultural Production over 2011-2040 for RCP4.5
1:25000000
0 250 500 km
Risk Level
10
9
8
7
6
5
4
3
2
1
No Data
BEIJING
BOHAI SEA
YELLOW SEA
EAST CHINA SEA
SOUTH CHINA SEA
Shanghai
Hangzhou
Fuzhou
Taipei
Taiwan Tao
Diaoyu Tao
Chiwei Yu
Taiwan Haixia (Str.)
Dongsha Qundao
Hong Kong
Macao
Haikou
Hainan Dao
Nanning
Guangzhou
Changsha
Nanchang
Guiyang
Kunming
Chengdu
Chongqing
Xi'an
Lanzhou
Xining
Yinchuan
Taiyuan
Shijiazhuang
Tianjin
Jinan
Shenyang
Changchun
Harbin
Heilong Jiang
Qinghai Hu
Jinsha Jiang
Lancang Jiang
Nu Jiang
Yarlung Zangbo Jiang
Lhasa
Urumqi
Tarim He
Hainan Dao
Xisha Qundao
Zhongsha Qundao
Huangyan Dao
Nansha Qundao
Zengmu Ansha
SOUTH CHINA SEA ISLANDS

Risk Levels of Maize Yield over 2041-2070 for RCP4.5

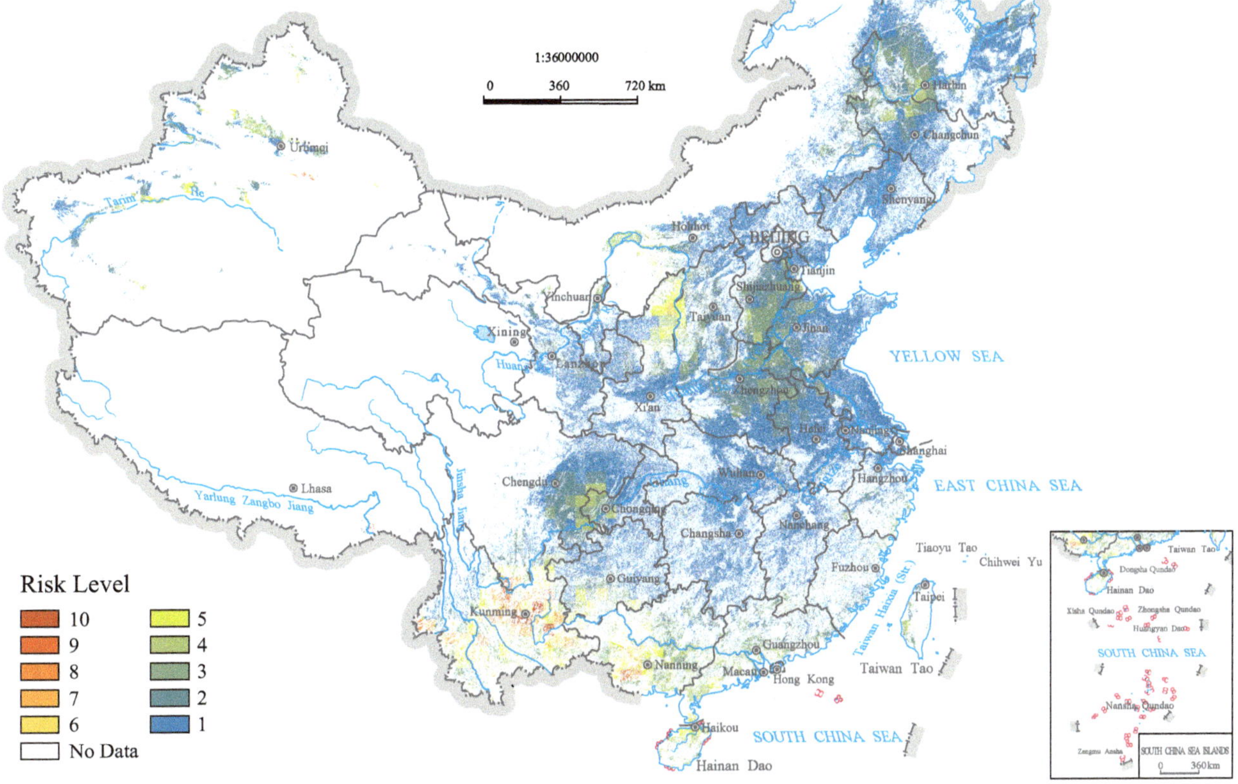

Risk Levels of Wheat Yield over 2041-2070 for RCP4.5

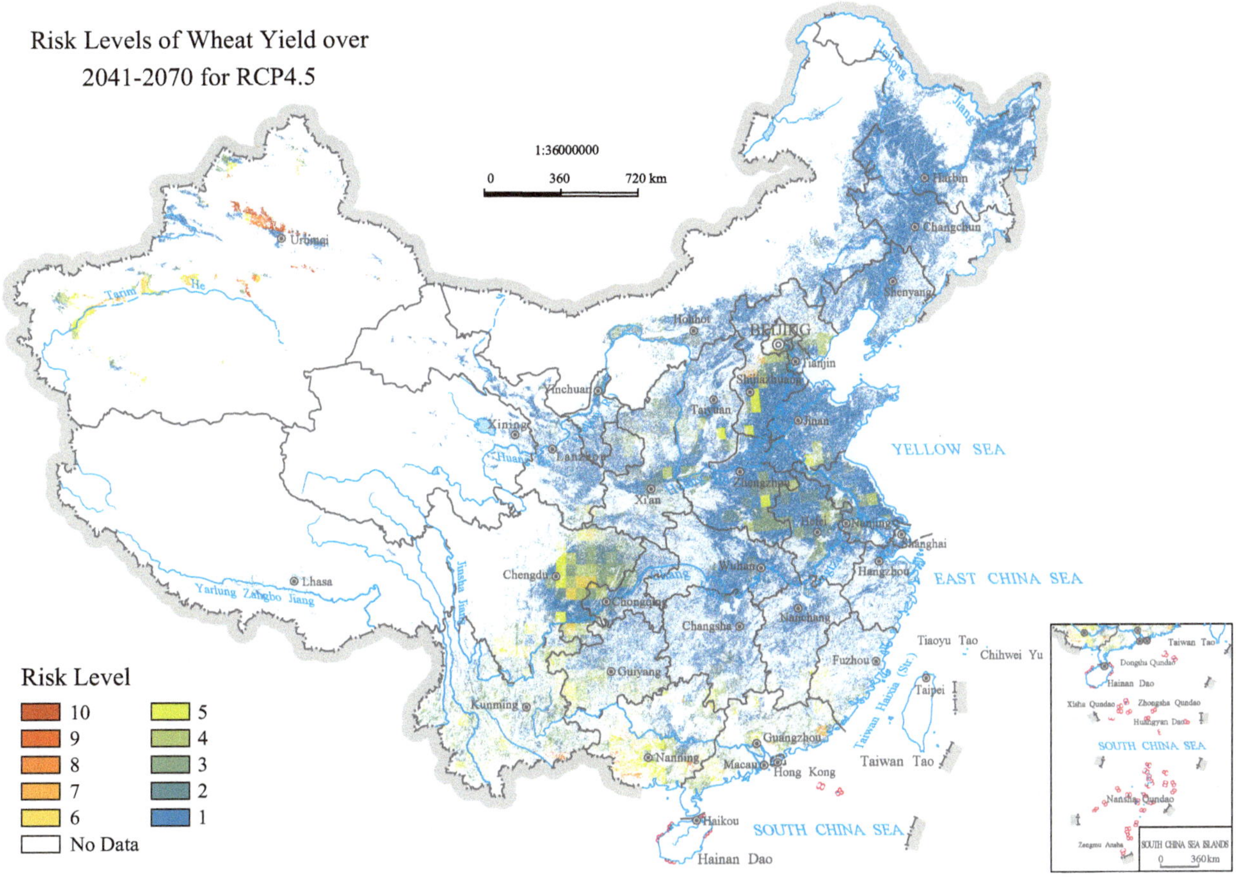

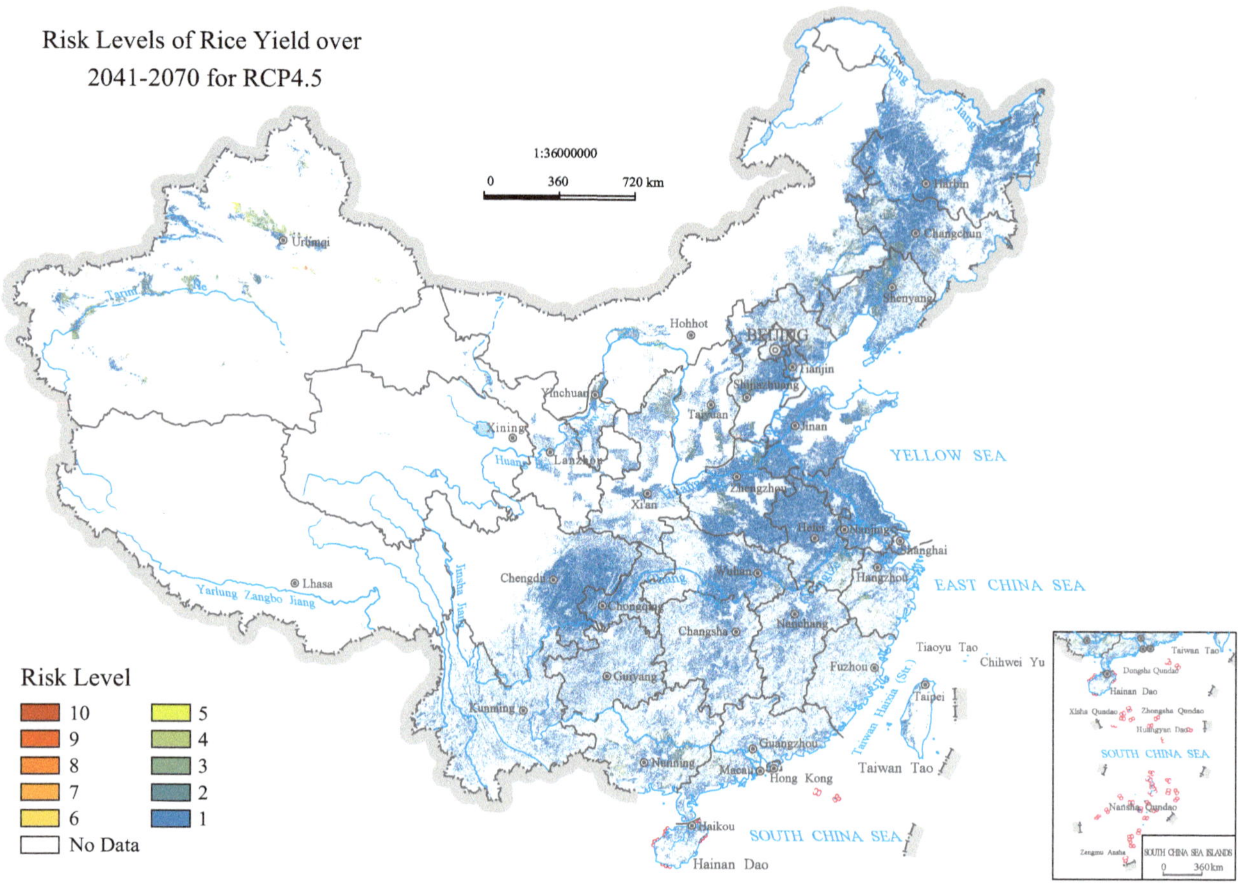
Risk Levels of Rice Yield over
2041-2070 for RCP4.5
1:36000000
0 360 720 km
Risk Level
10
9
8
7
6
5
4
3
2
1
No Data
Urümqi
Hohhot
BEIJING
Tianjin
Shijiazhuang
Taiyuan
Yinchuan
Xining
Lanzhou
Jinan
Zhengzhou
Xi'an
Hefei
Nanjing
Shanghai
Wuhan
Hangzhou
Chengdu
Chongqing
Changsha
Nanchang
Lhasa
Yarlung Zangbo Jiang
Guiyang
Kunming
Fuzhou
Taipei
Guangzhou
Nanning
Macau
Hong Kong
Haikou
Hainan Dao
Harbin
Changchun
Shenyang
YELLOW SEA
EAST CHINA SEA
SOUTH CHINA SEA
Taiwan Tao
Tiaoyu Tao
Chihwei Yu
SOUTH CHINA SEA ISLANDS
0 360km

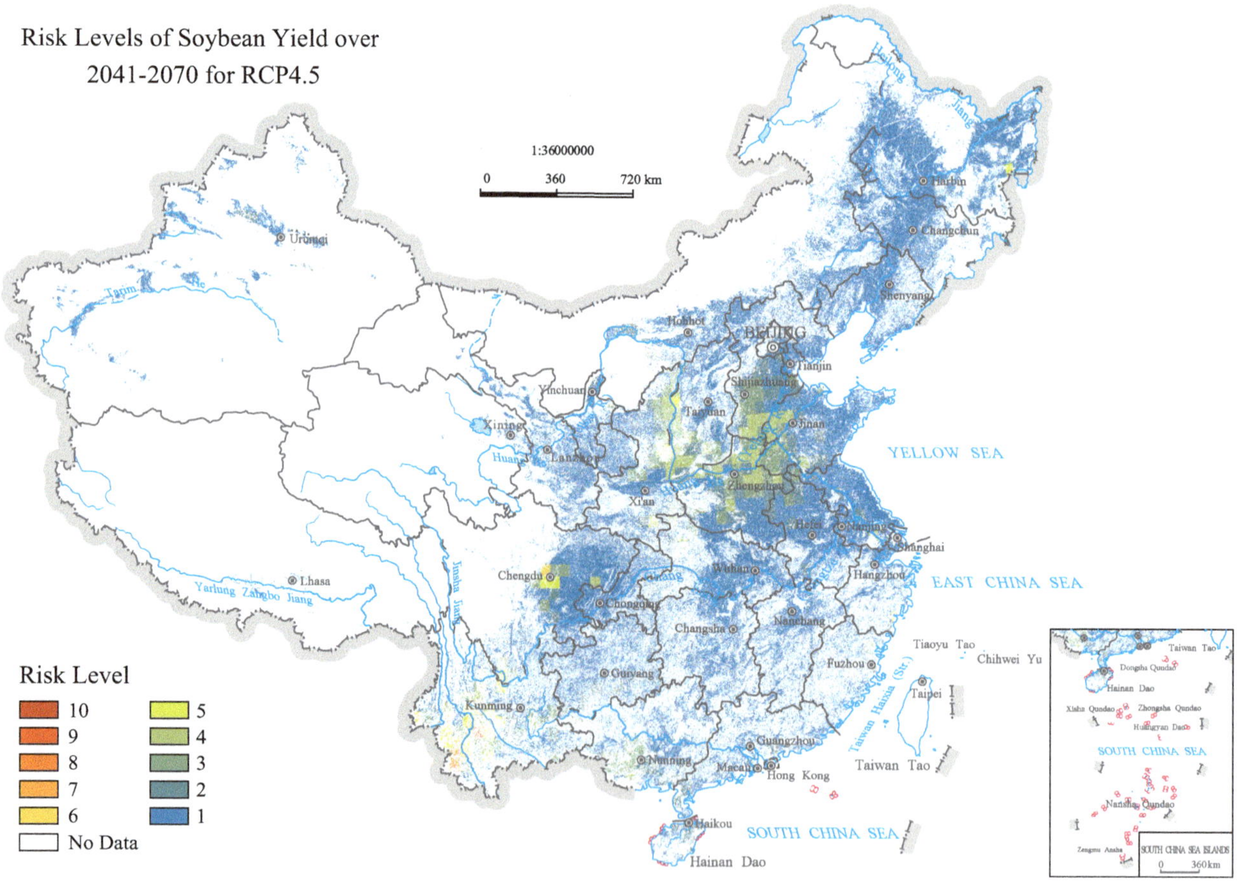
Risk Levels of Soybean Yield over
2041-2070 for RCP4.5
1:36000000
0 360 720 km
Risk Level
10
9
8
7
6
5
4
3
2
1
No Data
Urümqi
Hohhot
BEIJING
Tianjin
Shijiazhuang
Taiyuan
Yinchuan
Xining
Lanzhou
Jinan
Zhengzhou
Xi'an
Hefei
Nanjing
Shanghai
Wuhan
Hangzhou
Chengdu
Chongqing
Changsha
Nanchang
Lhasa
Yarlung Zangbo Jiang
Guiyang
Kunming
Fuzhou
Taipei
Guangzhou
Nanning
Macau
Hong Kong
Haikou
Hainan Dao
Harbin
Changchun
Shenyang
YELLOW SEA
EAST CHINA SEA
SOUTH CHINA SEA
Taiwan Tao
Tiaoyu Tao
Chihwei Yu
SOUTH CHINA SEA ISLANDS
0 360km

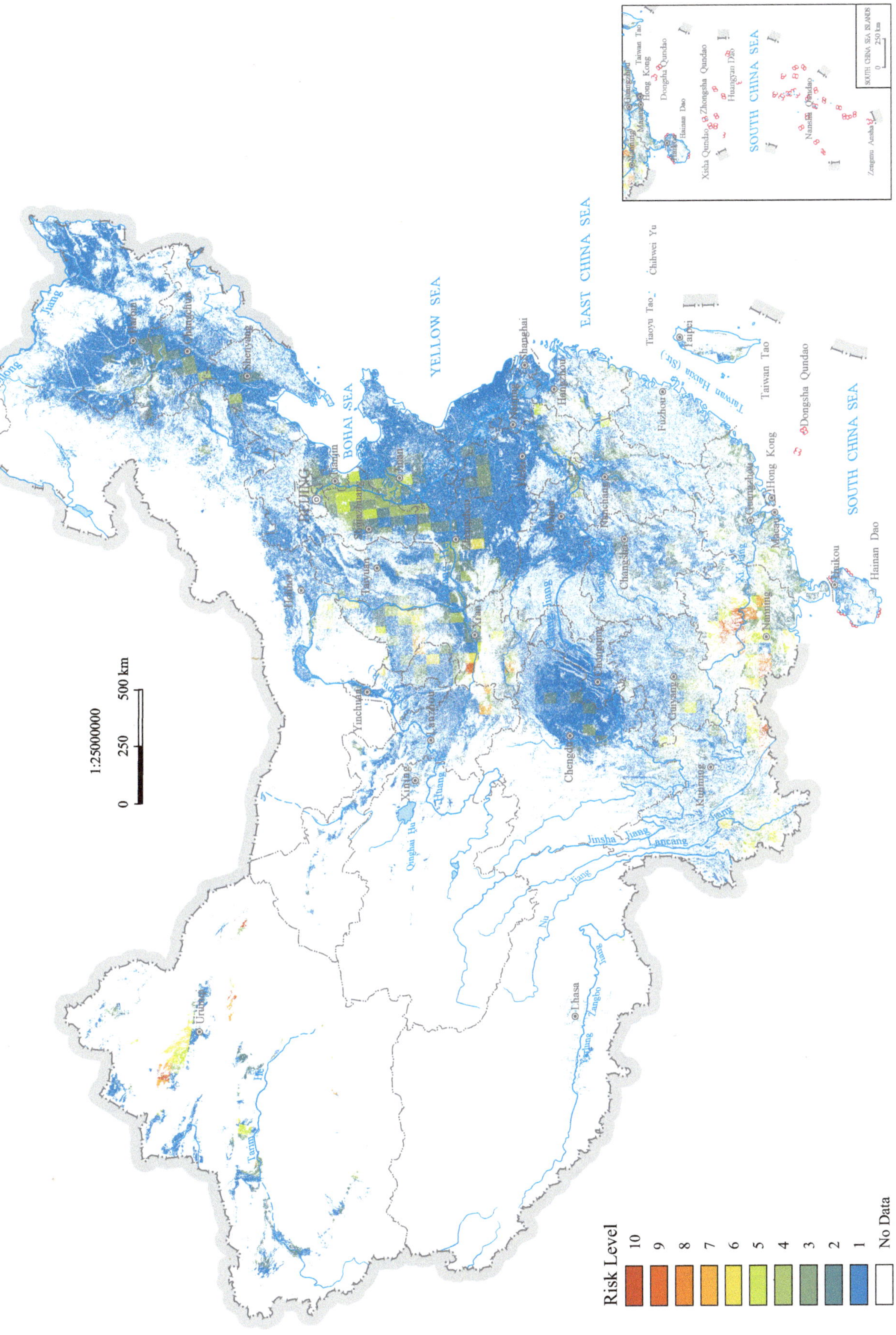

Risk Levels of Agricultural Production over 2041-2070 for RCP4.5

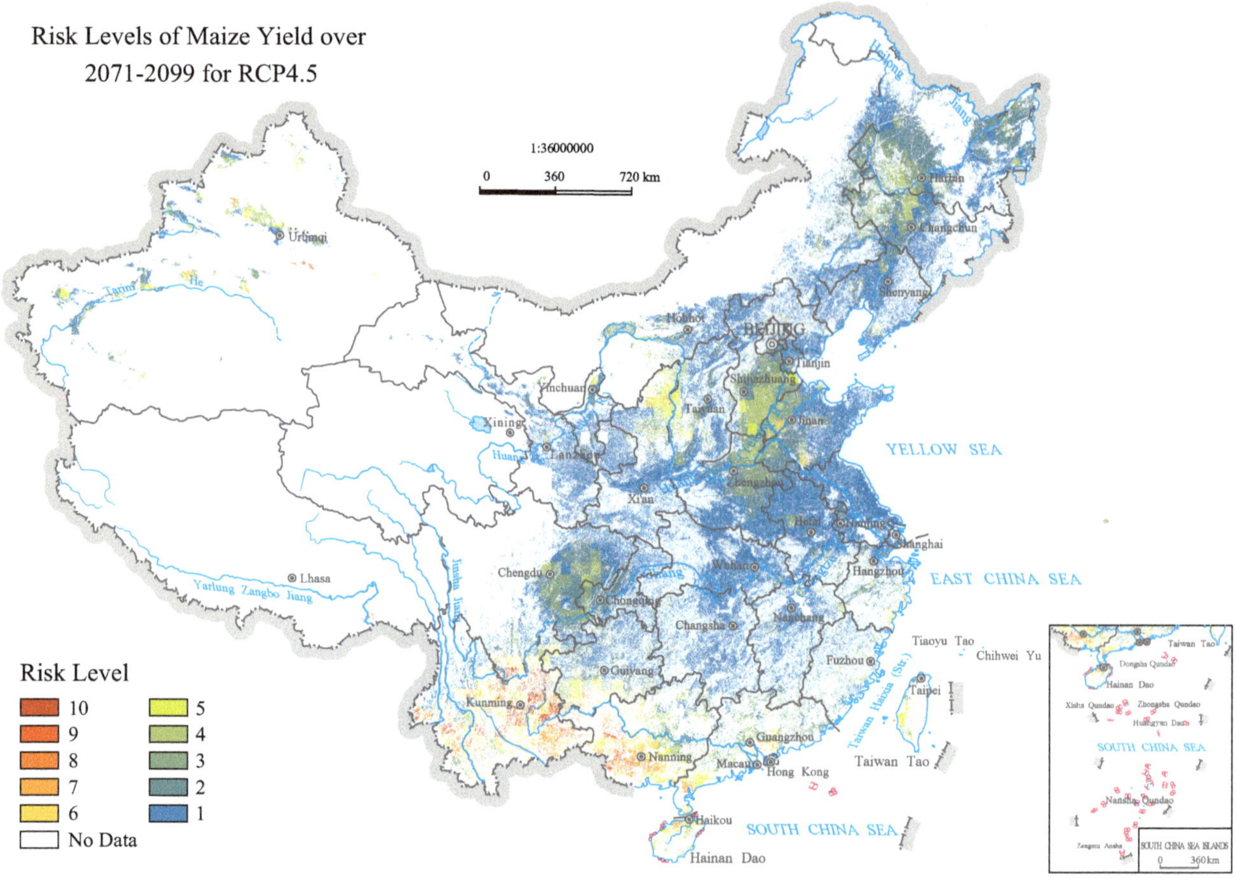
Risk Levels of Maize Yield over
2071-2099 for RCP4.5
1:36000000
0 360 720 km
Risk Level
10
9
8
7
6
5
4
3
2
1
No Data
YELLOW SEA
EAST CHINA SEA
SOUTH CHINA SEA
Beijing
Lhasa
Chengdu
Kunming
Taiwan Tao
Hainan Dao

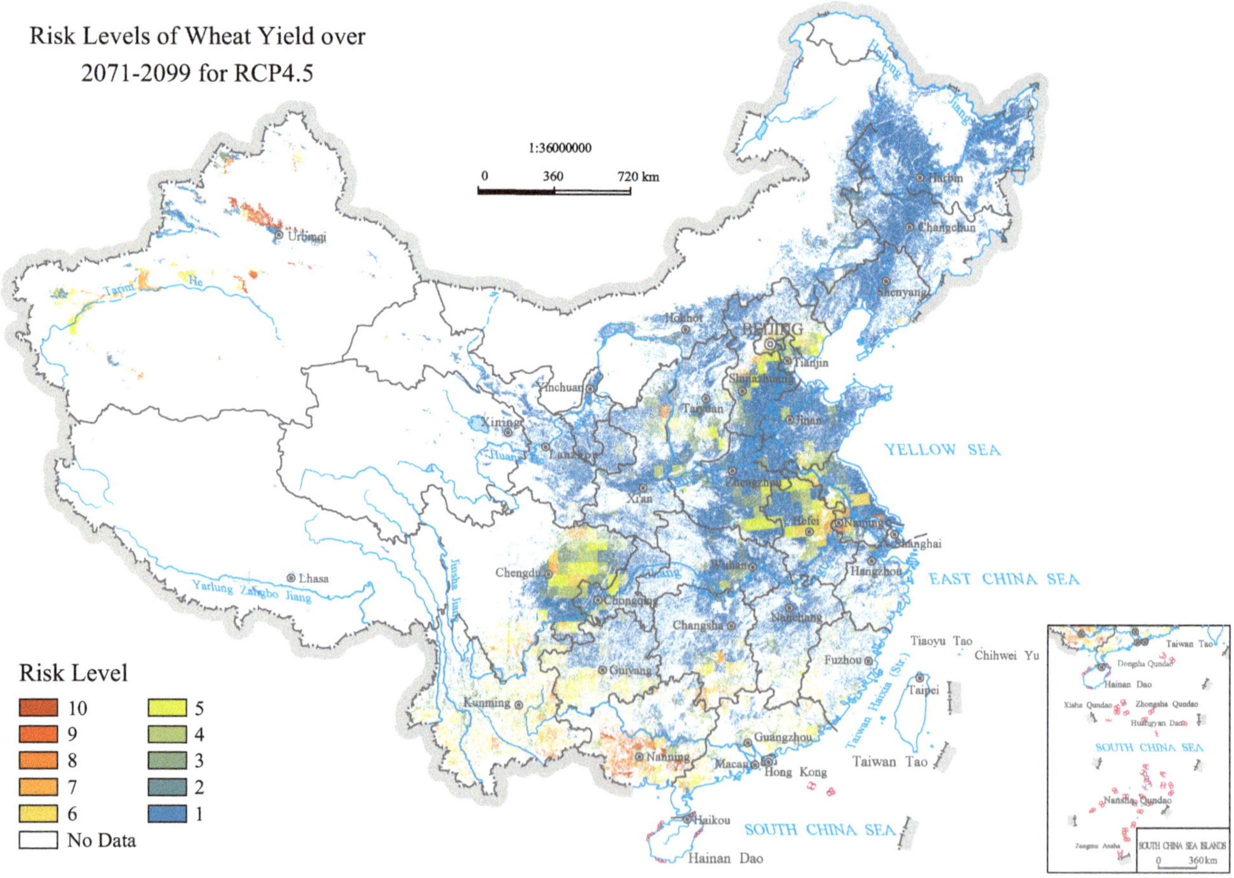
Risk Levels of Wheat Yield over
2071-2099 for RCP4.5
1:36000000
0 360 720 km
Risk Level
10
9
8
7
6
5
4
3
2
1
No Data
YELLOW SEA
EAST CHINA SEA
SOUTH CHINA SEA
Beijing
Lhasa
Chengdu
Kunming
Taiwan Tao
Hainan Dao

Risk Levels of Rice Yield over 2071-2099 for RCP4.5

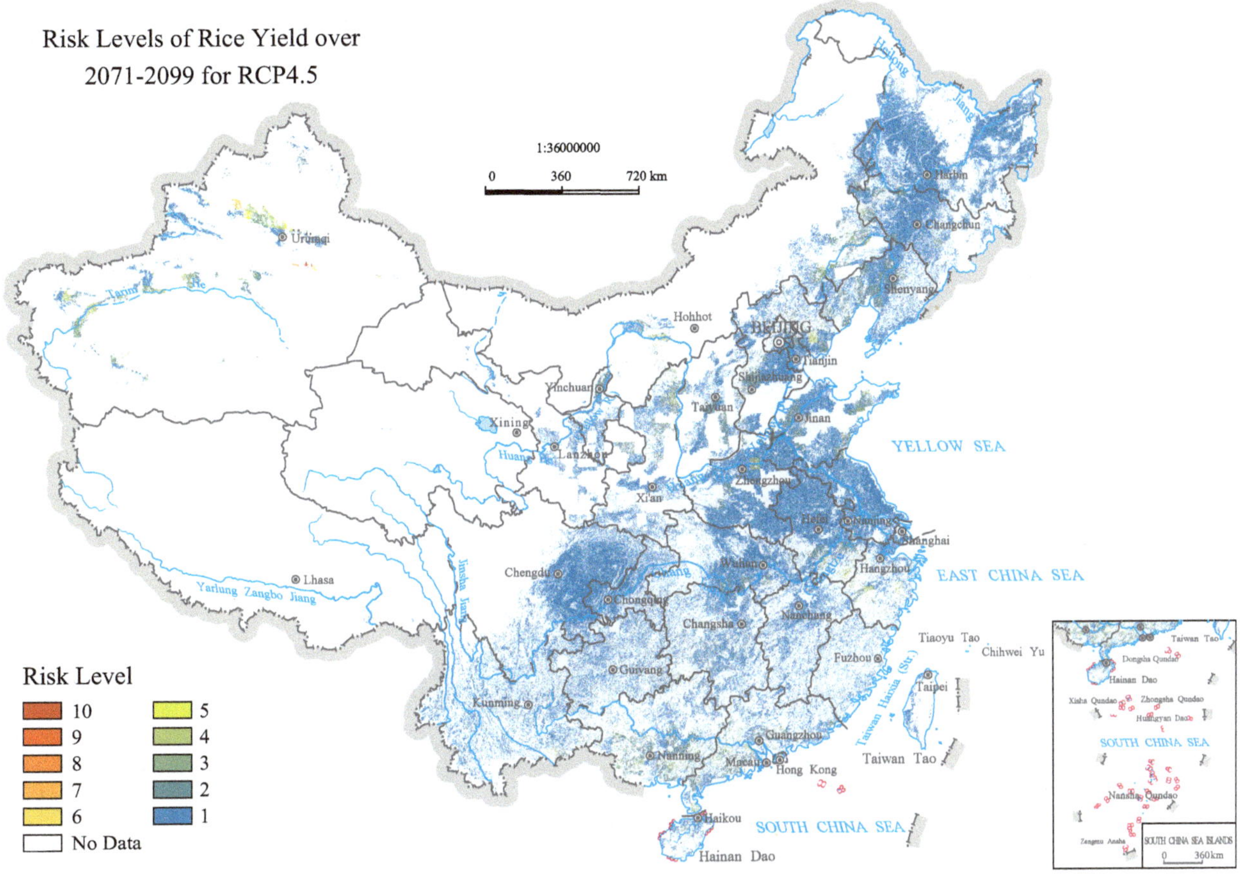

Risk Levels of Soybean Yield over 2071-2099 for RCP4.5

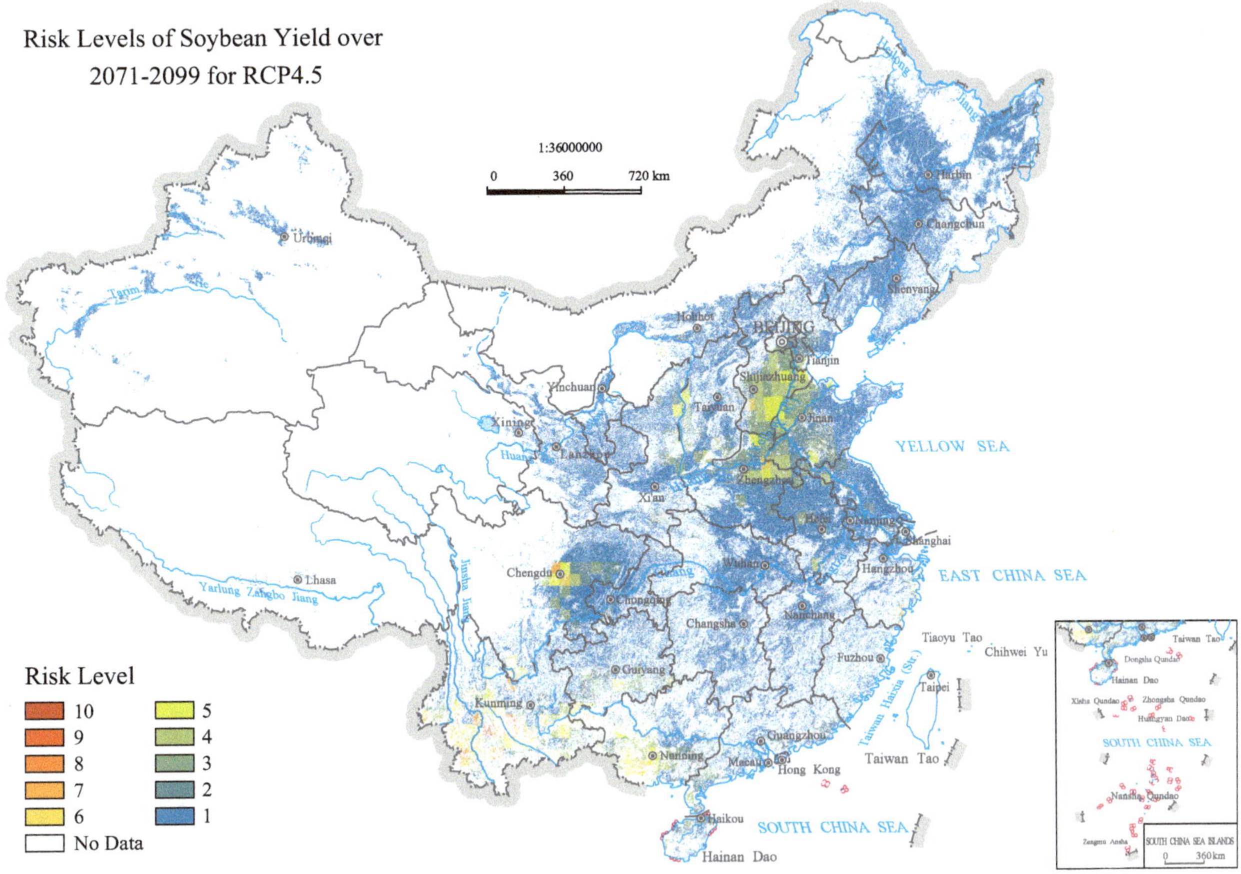

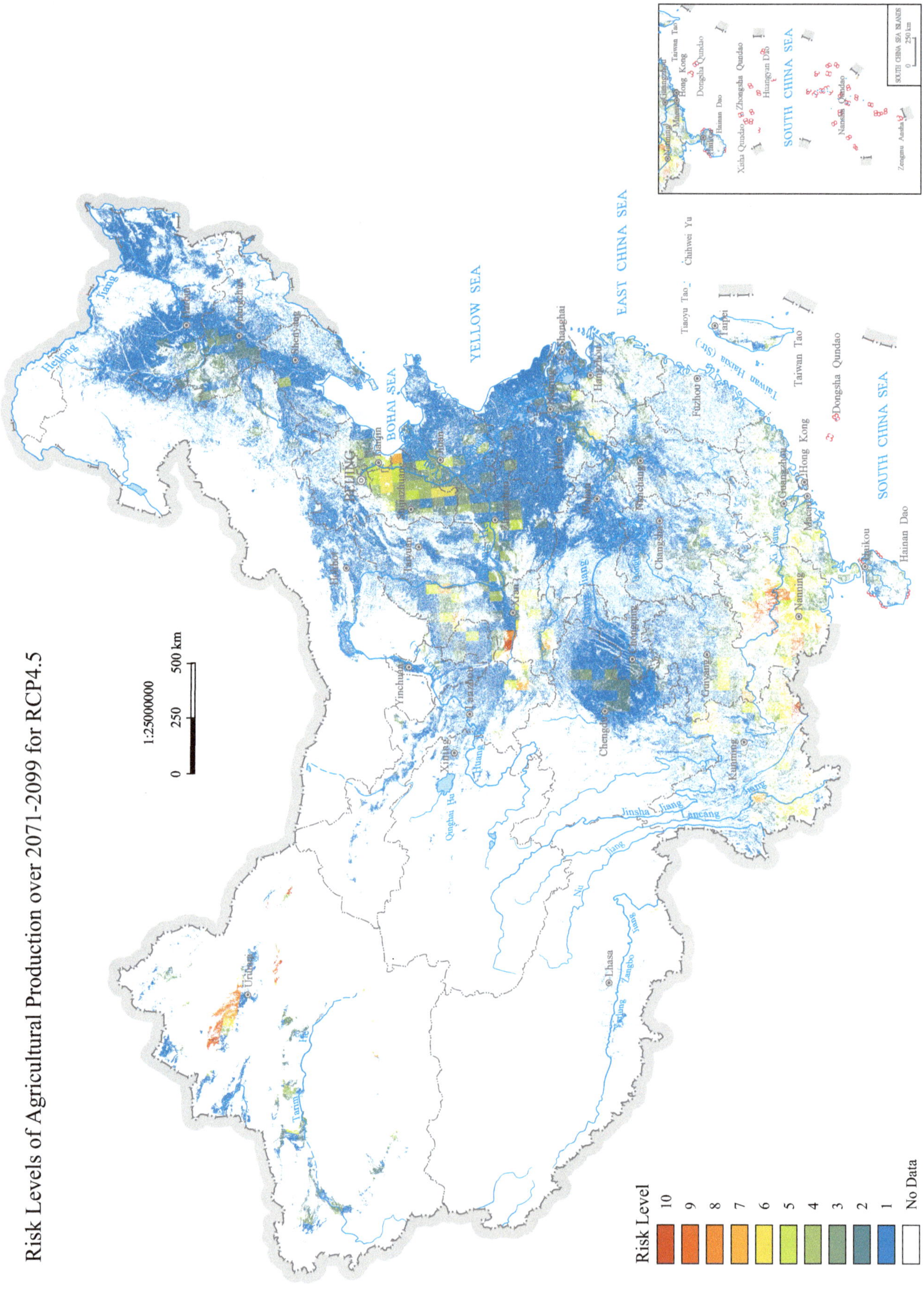

Risk Levels of Agricultural Production over 2071-2099 for RCP4.5

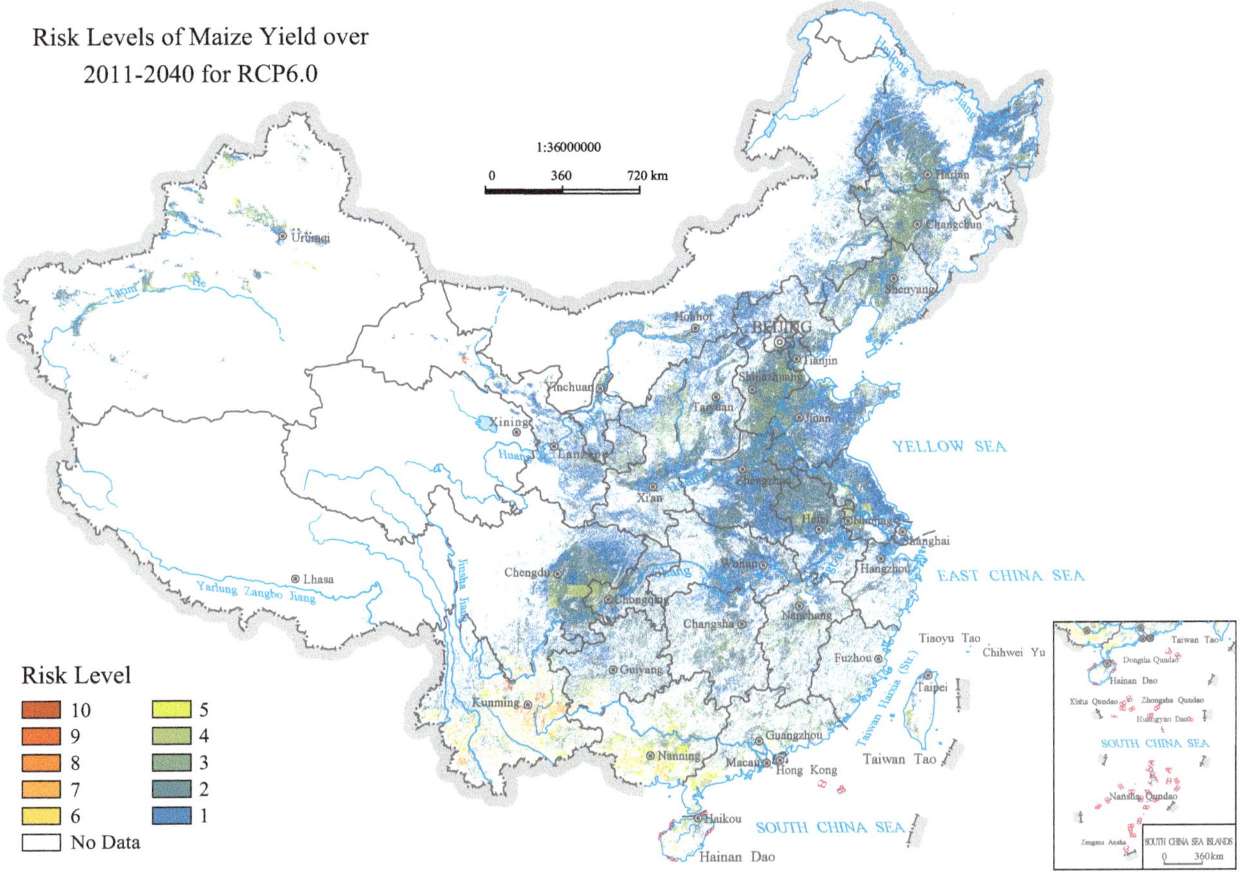
Risk Levels of Maize Yield over
2011-2040 for RCP6.0
1:36000000
0 360 720 km
Risk Level
10
9
8
7
6
5
4
3
2
1
No Data
Ürümqi
Hohhot
BEIJING
Tianjin
Shijiazhuang
Taiyuan
Jinan
Yinchuan
Xining
Lanzhou
Xi'an
Zhengzhou
Hefei
Nanjing
Shanghai
Hangzhou
Wuhan
Chengdu
Chongqing
Changsha
Nanchang
Fuzhou
Taipei
Guiyang
Kunming
Nanning
Guangzhou
Macau
Hong Kong
Haikou
Hainan Dao
Lhasa
Harbin
Changchun
Shenyang
YELLOW SEA
EAST CHINA SEA
SOUTH CHINA SEA
Taiwan Tao
Tiaoyu Tao
Chihwei Yu
Yarlung Zangbo Jiang
Jinsha Jiang
SOUTH CHINA SEA ISLANDS

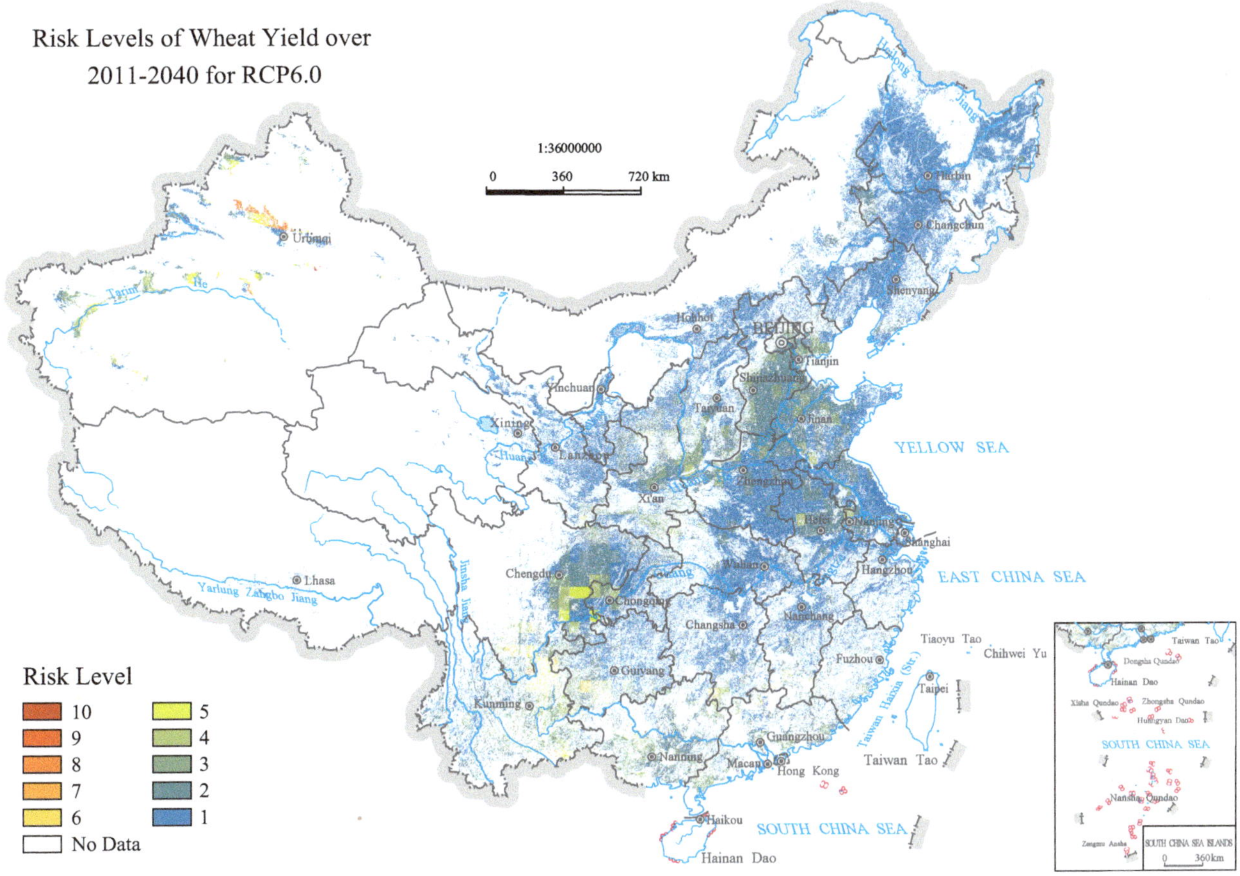
Risk Levels of Wheat Yield over
2011-2040 for RCP6.0
1:36000000
0 360 720 km
Risk Level
10
9
8
7
6
5
4
3
2
1
No Data
Ürümqi
Hohhot
BEIJING
Tianjin
Shijiazhuang
Taiyuan
Jinan
Yinchuan
Xining
Lanzhou
Xi'an
Zhengzhou
Hefei
Nanjing
Shanghai
Hangzhou
Wuhan
Chengdu
Chongqing
Changsha
Nanchang
Fuzhou
Taipei
Guiyang
Kunming
Nanning
Guangzhou
Macau
Hong Kong
Haikou
Hainan Dao
Lhasa
Harbin
Changchun
Shenyang
YELLOW SEA
EAST CHINA SEA
SOUTH CHINA SEA
Taiwan Tao
Tiaoyu Tao
Chihwei Yu
Yarlung Zangbo Jiang
Jinsha Jiang
SOUTH CHINA SEA ISLANDS

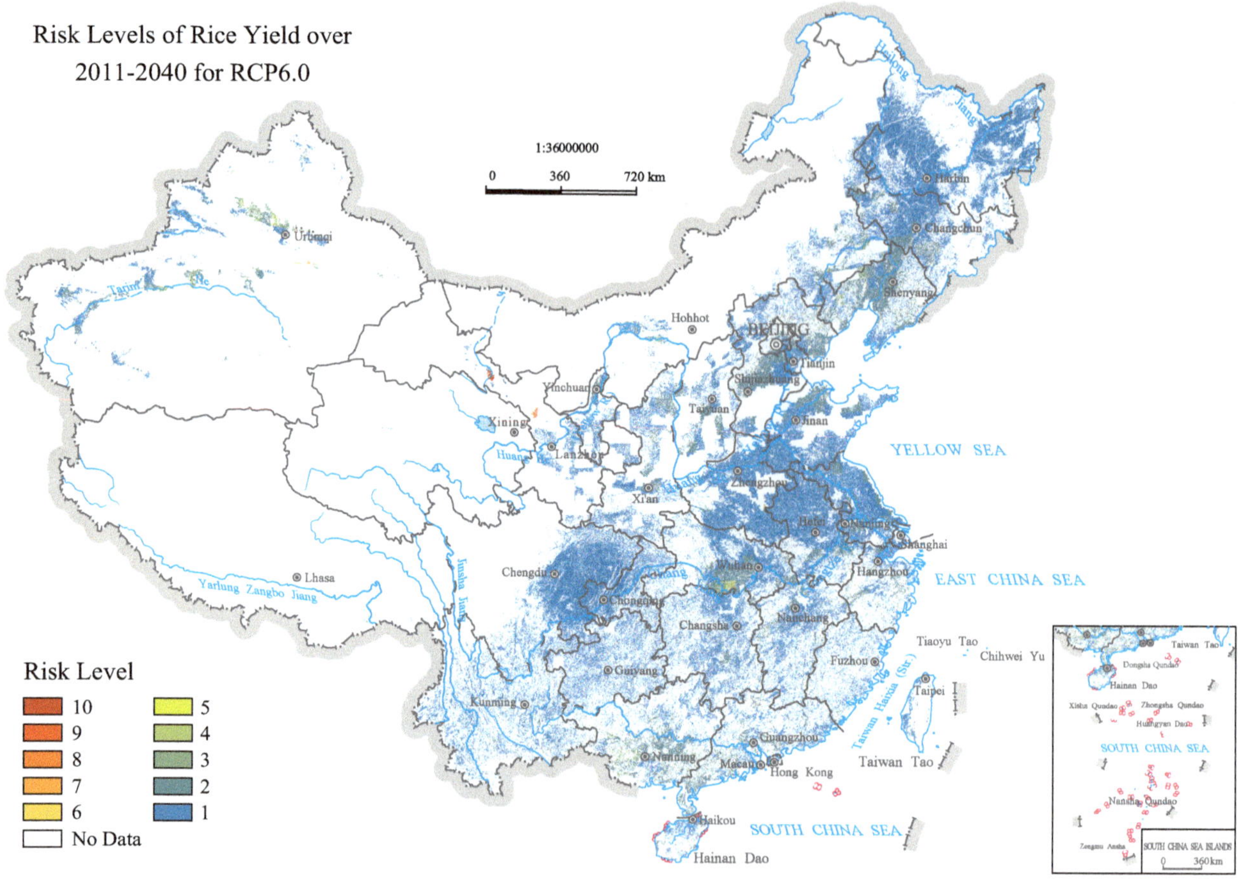
Risk Levels of Rice Yield over
2011-2040 for RCP6.0
1:36000000
0 360 720 km
Risk Level
10
9
8
7
6
5
4
3
2
1
No Data
YELLOW SEA
EAST CHINA SEA
SOUTH CHINA SEA
Tiaoyu Tao
Chihwei Yu
Taiwan Tao
Hainan Dao
BEIJING
Harbin
Changchun
Shenyang
Hohhot
Tianjin
Shijiazhuang
Taiyuan
Jinan
Yinchuan
Xining
Lanzhou
Xi'an
Zhengzhou
Hefei
Nanjing
Shanghai
Hangzhou
Wuhan
Chengdu
Chongqing
Changsha
Nanchang
Fuzhou
Taipei
Guiyang
Kunming
Nanning
Guangzhou
Macau
Hong Kong
Haikou
Lhasa
Urümqi
Yarlung Zangbo Jiang
SOUTH CHINA SEA ISLANDS

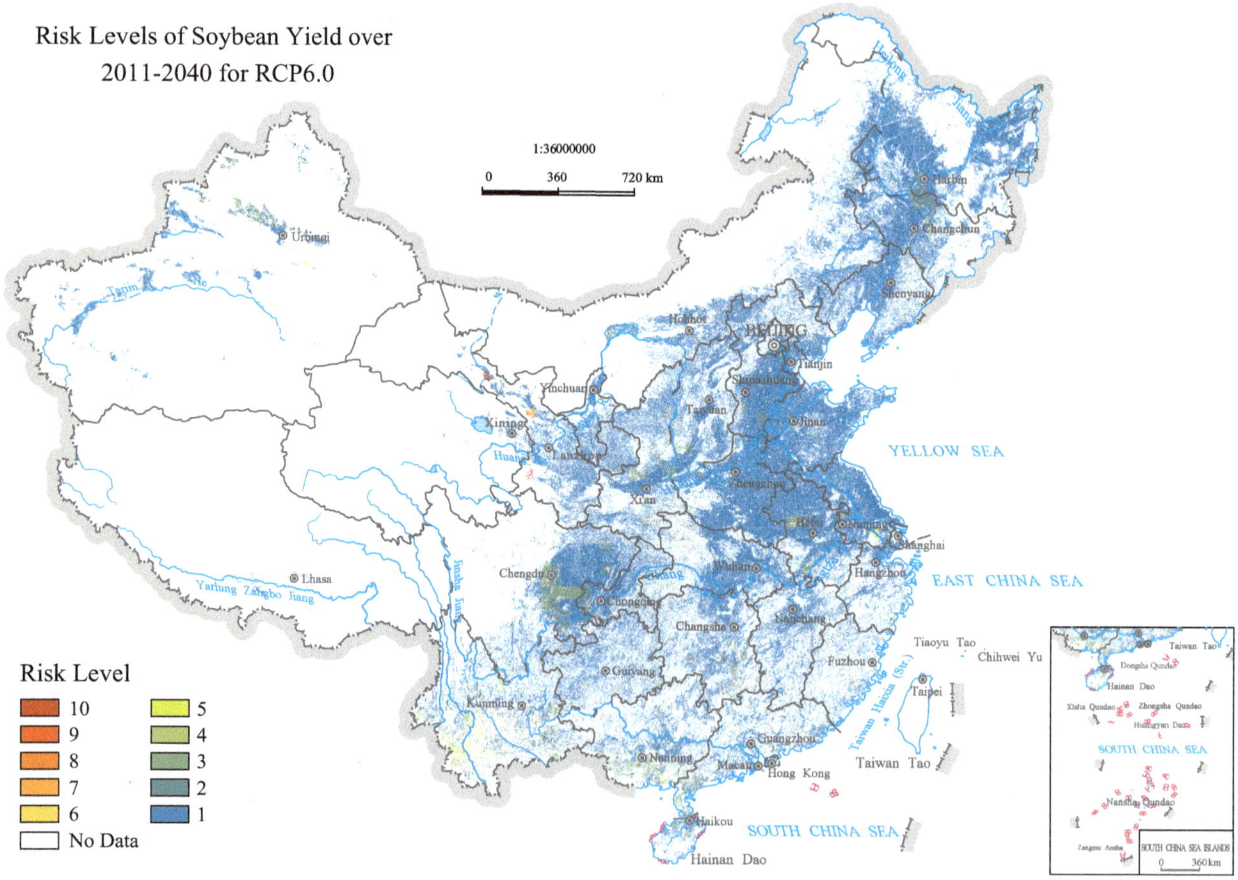
Risk Levels of Soybean Yield over
2011-2040 for RCP6.0
1:36000000
0 360 720 km
Risk Level
10
9
8
7
6
5
4
3
2
1
No Data
YELLOW SEA
EAST CHINA SEA
SOUTH CHINA SEA
Tiaoyu Tao
Chihwei Yu
Taiwan Tao
Hainan Dao
BEIJING
Harbin
Changchun
Shenyang
Hohhot
Tianjin
Shijiazhuang
Taiyuan
Jinan
Yinchuan
Xining
Lanzhou
Xi'an
Zhengzhou
Hefei
Nanjing
Shanghai
Hangzhou
Wuhan
Chengdu
Chongqing
Changsha
Nanchang
Fuzhou
Taipei
Guiyang
Kunming
Nanning
Guangzhou
Macau
Hong Kong
Haikou
Lhasa
Urümqi
Yarlung Zangbo Jiang
SOUTH CHINA SEA ISLANDS

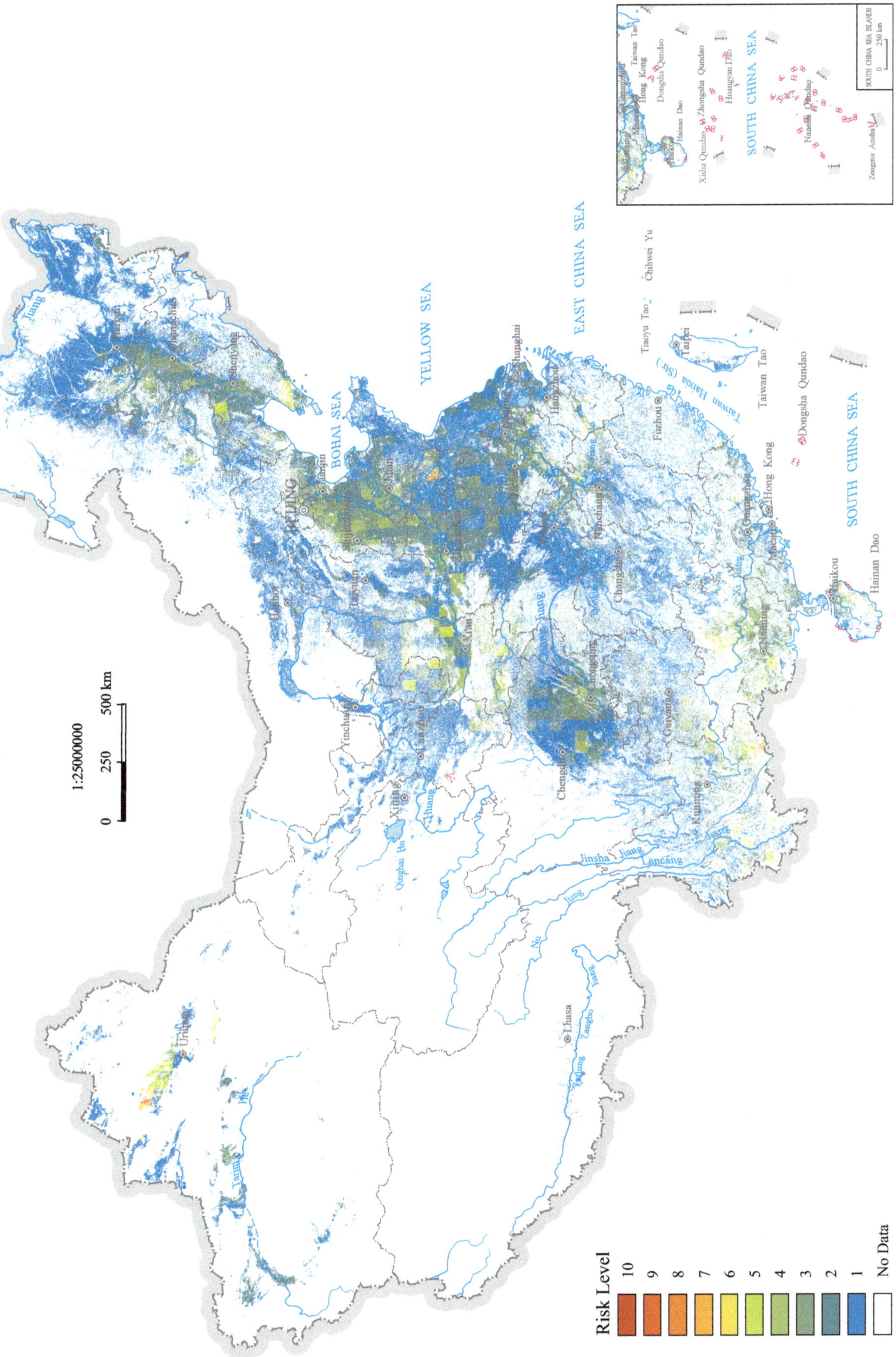

Risk Levels of Agricultural Production over 2011-2040 for RCP6.0

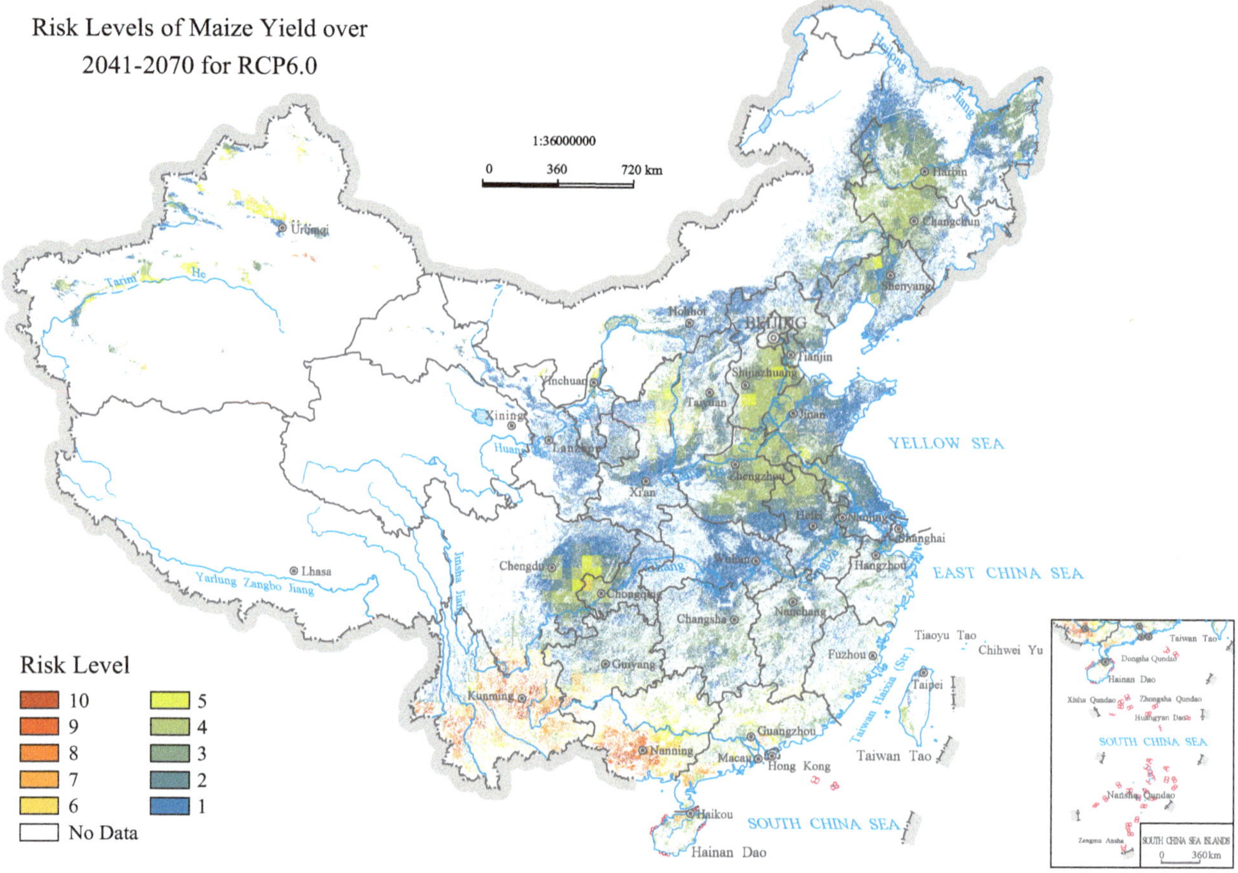
Risk Levels of Maize Yield over 2041-2070 for RCP6.0
1:36000000
0 360 720 km
Risk Level
10
9
8
7
6
5
4
3
2
1
No Data
Urümqi
Tarim He
Harbin
Changchun
Shenyang
Hohhot
BEIJING
Tianjin
Shijiazhuang
Taiyuan
Yinchuan
Xining
Lanzhou
Jinan
Zhengzhou
Xi'an
Hefei
Nanjing
Shanghai
Hangzhou
Wuhan
Chengdu
Chongqing
Lhasa
Yarlung Zangbo Jiang
Jinsha Jiang
Changsha
Nanchang
Fuzhou
Guiyang
Kunming
Nanning
Guangzhou
Macau
Hong Kong
Haikou
Hainan Dao
Taipei
Taiwan Tao
Tiaoyu Tao
Chihwei Yu
YELLOW SEA
EAST CHINA SEA
SOUTH CHINA SEA

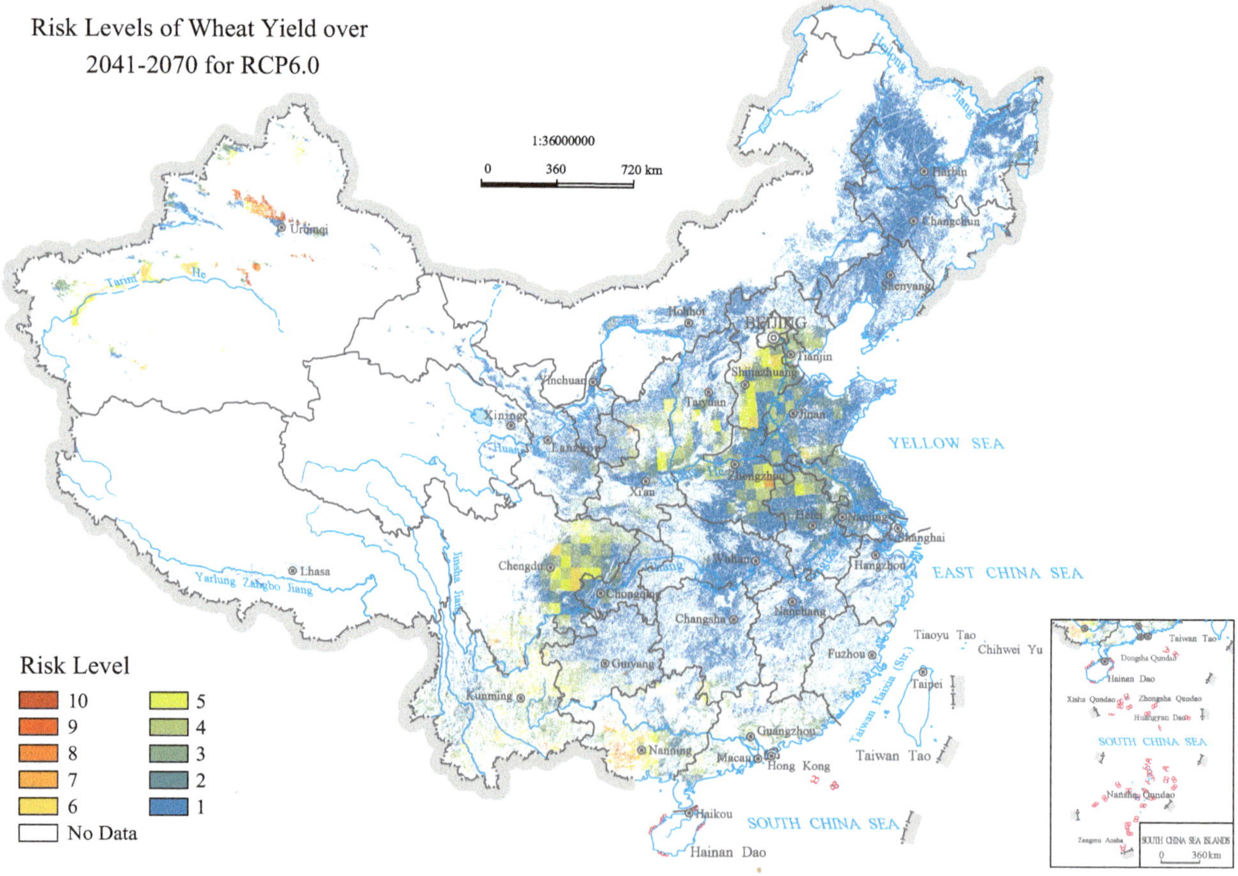
Risk Levels of Wheat Yield over 2041-2070 for RCP6.0
1:36000000
0 360 720 km
Risk Level
10
9
8
7
6
5
4
3
2
1
No Data
Urümqi
Tarim He
Harbin
Changchun
Shenyang
Hohhot
BEIJING
Tianjin
Shijiazhuang
Taiyuan
Yinchuan
Xining
Lanzhou
Jinan
Zhengzhou
Xi'an
Hefei
Nanjing
Shanghai
Hangzhou
Wuhan
Chengdu
Chongqing
Lhasa
Yarlung Zangbo Jiang
Jinsha Jiang
Changsha
Nanchang
Fuzhou
Guiyang
Kunming
Nanning
Guangzhou
Macau
Hong Kong
Haikou
Hainan Dao
Taipei
Taiwan Tao
Tiaoyu Tao
Chihwei Yu
YELLOW SEA
EAST CHINA SEA
SOUTH CHINA SEA

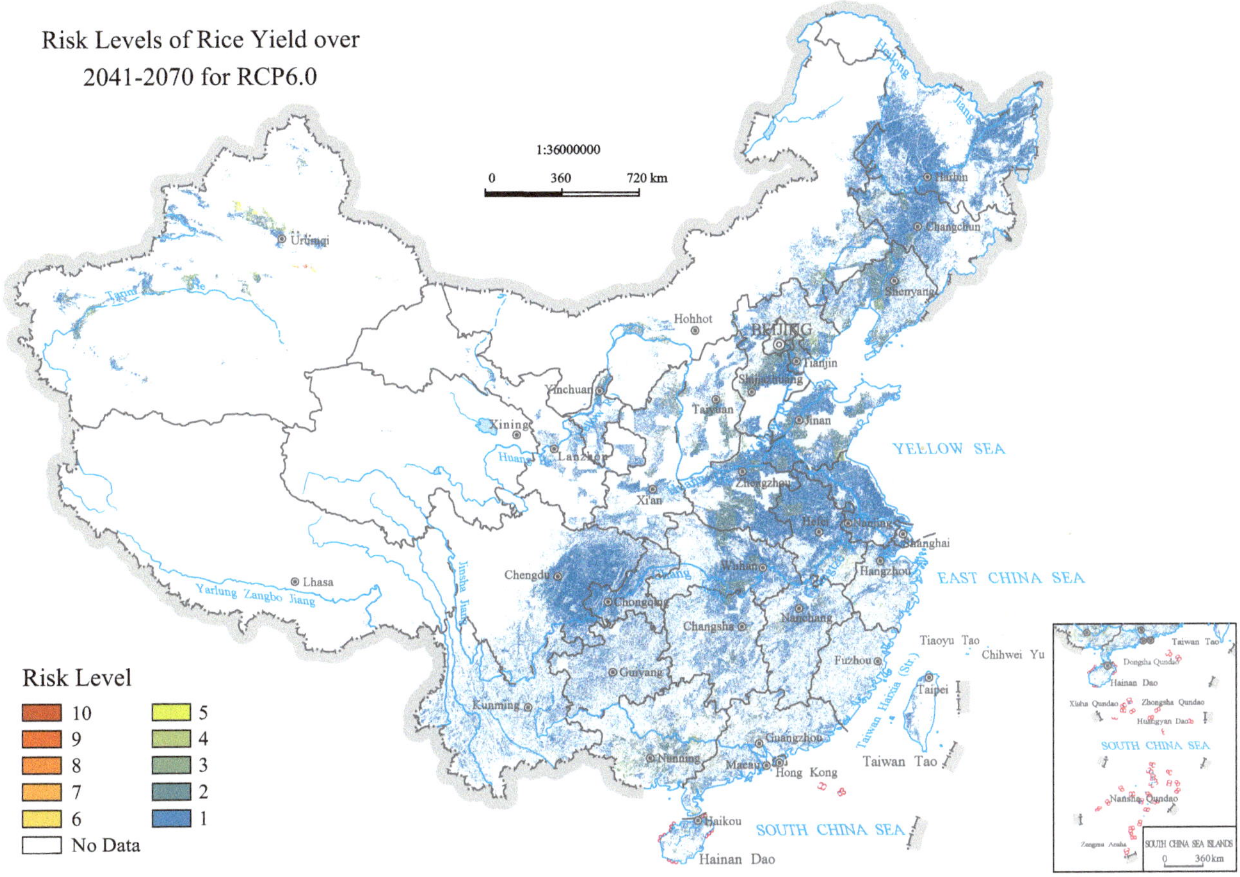
Risk Levels of Rice Yield over
2041-2070 for RCP6.0
1:36000000
0 360 720 km
Risk Level
10
9
8
7
6
5
4
3
2
1
No Data
Harbin
Changchun
Shenyang
Hohhot
BEIJING
Tianjin
Shijiazhuang
Taiyuan
Jinan
Yinchuan
Xining
Lanzhou
Xi'an
Zhengzhou
Hefei
Nanjing
Shanghai
Hangzhou
Wuhan
Chengdu
Chongqing
Changsha
Nanchang
Fuzhou
Guiyang
Kunming
Nanning
Guangzhou
Macau
Hong Kong
Haikou
Taipei
Lhasa
Urümqi
Yarlung Zangbo Jiang
YELLOW SEA
EAST CHINA SEA
SOUTH CHINA SEA
Tiaoyu Tao
Chihwei Yu
Taiwan Tao
Hainan Dao

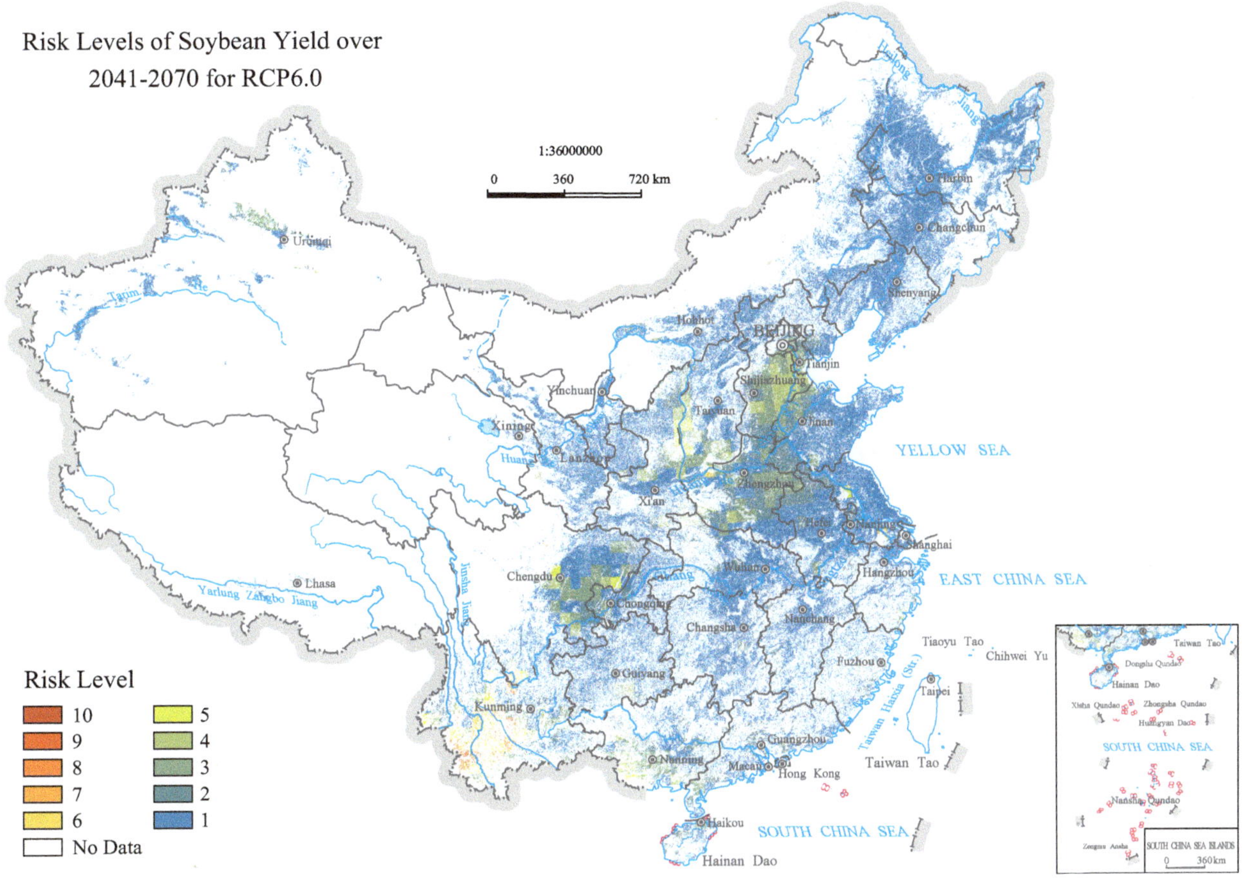
Risk Levels of Soybean Yield over
2041-2070 for RCP6.0
1:36000000
0 360 720 km
Risk Level
10
9
8
7
6
5
4
3
2
1
No Data
Harbin
Changchun
Shenyang
Hohhot
BEIJING
Tianjin
Shijiazhuang
Taiyuan
Jinan
Yinchuan
Xining
Lanzhou
Xi'an
Zhengzhou
Hefei
Nanjing
Shanghai
Hangzhou
Wuhan
Chengdu
Chongqing
Changsha
Nanchang
Fuzhou
Guiyang
Kunming
Nanning
Guangzhou
Macau
Hong Kong
Haikou
Taipei
Lhasa
Urümqi
Yarlung Zangbo Jiang
YELLOW SEA
EAST CHINA SEA
SOUTH CHINA SEA
Tiaoyu Tao
Chihwei Yu
Taiwan Tao
Hainan Dao

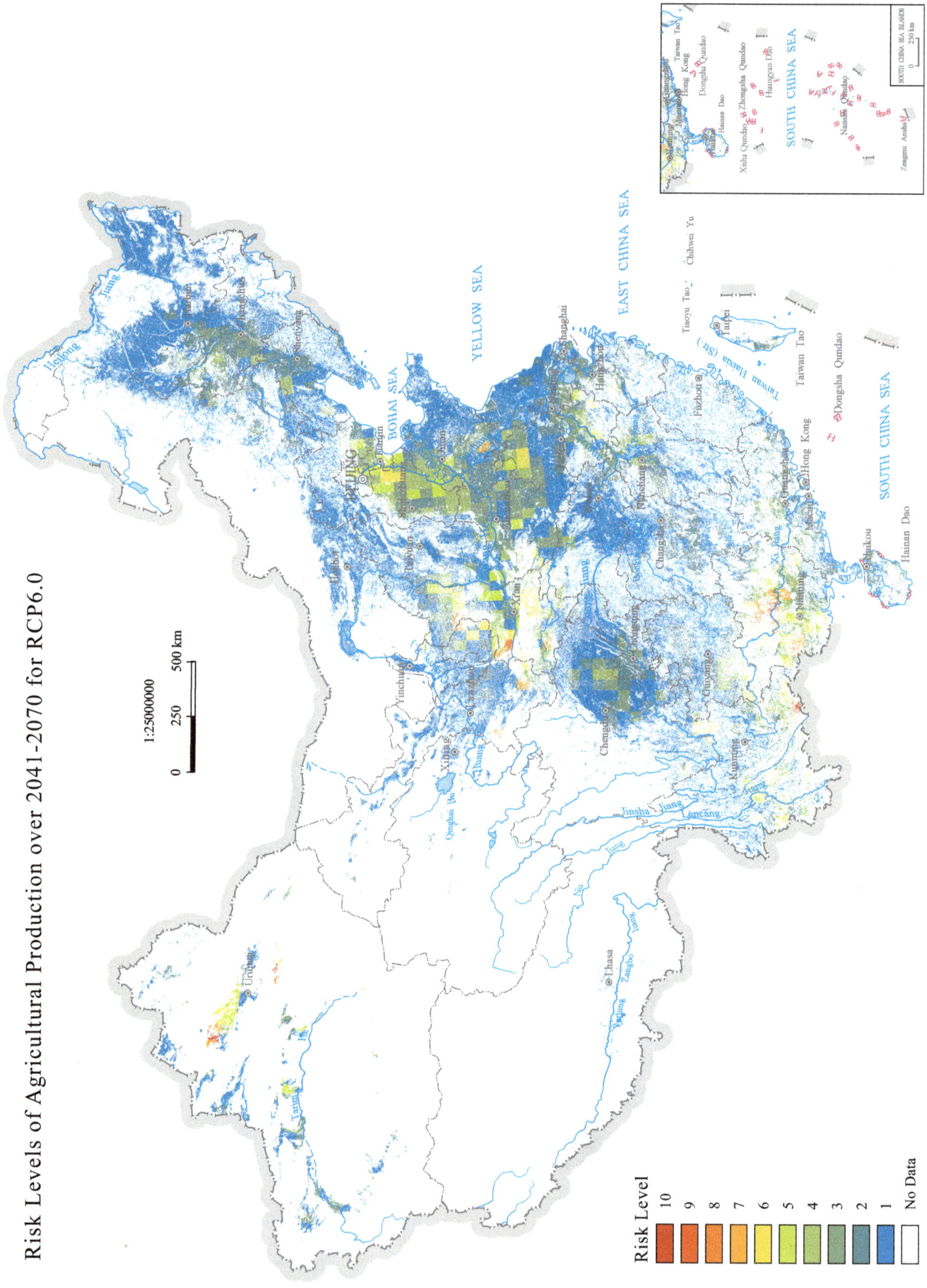
Risk Levels of Agricultural Production over 2041-2070 for RCP6.0
1:25000000
0 250 500 km
Risk Level
10
9
8
7
6
5
4
3
2
1
No Data
BEIJING
BOHAI SEA
YELLOW SEA
EAST CHINA SEA
SOUTH CHINA SEA
Heilong Jiang
Harbin
Changchun
Shenyang
Tianjin
Jinan
Shanghai
Hangzhou
Nanjing
Hefei
Shijiazhuang
Taiyuan
Hohhot
Zhengzhou
Xi'an
Wuhan
Nanchang
Changsha
Fuzhou
Taipei
Taiwan Tao
Taiwan Haixia (Str.)
Chihwei Yu
Dongsha Qundao
Guangzhou
Hong Kong
Macao
Haikou
Hainan Dao
Nanning
Guiyang
Chongqing
Chengdu
Kunming
Lanzhou
Xining
Yinchuan
Qinghai Hu
Jinsha Jiang
Lancang Jiang
Nu Jiang
Lhasa
Yarlung Zangbo Jiang
Urumqi
Tarim
SOUTH CHINA SEA ISLANDS

Risk Levels of Maize Yield over 2071-2099 for RCP6.0

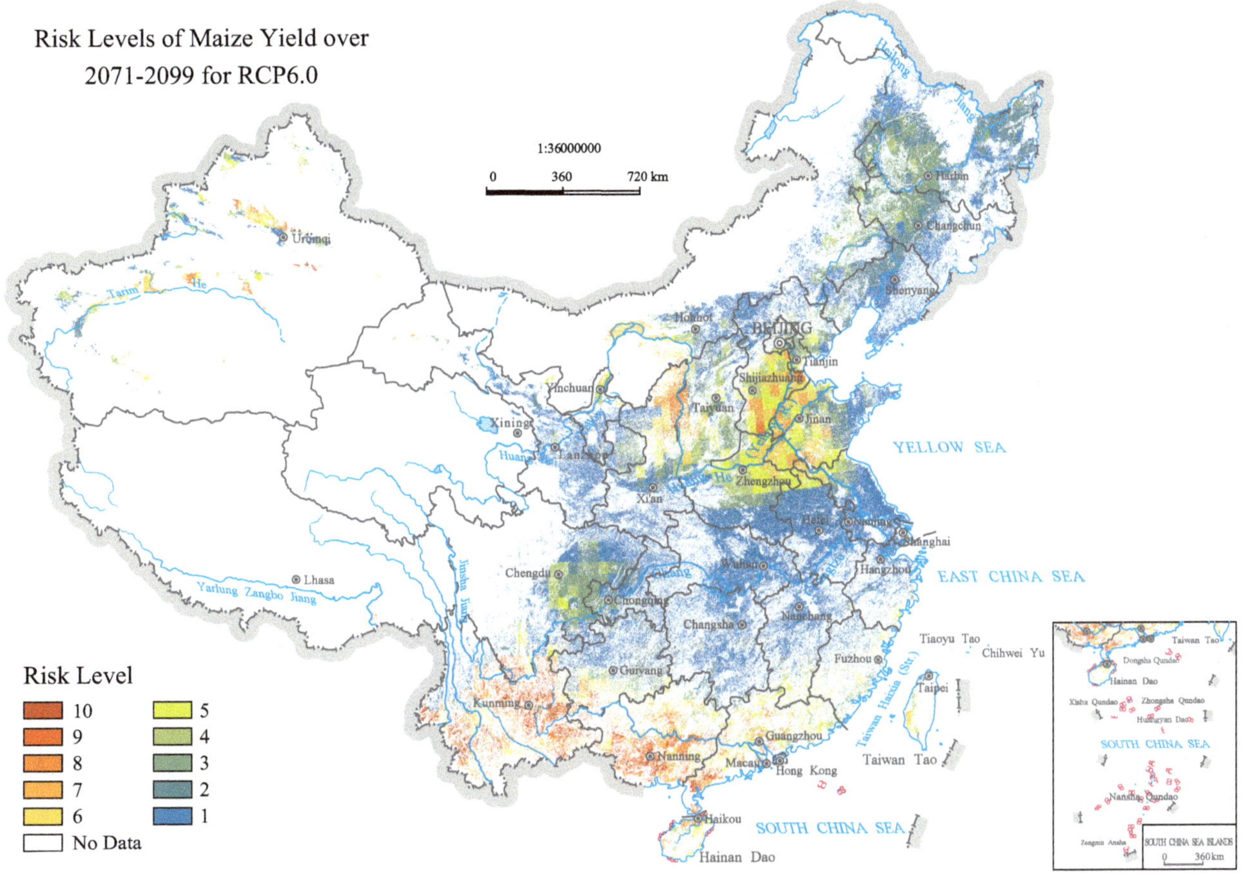

Risk Levels of Wheat Yield over 2071-2099 for RCP6.0

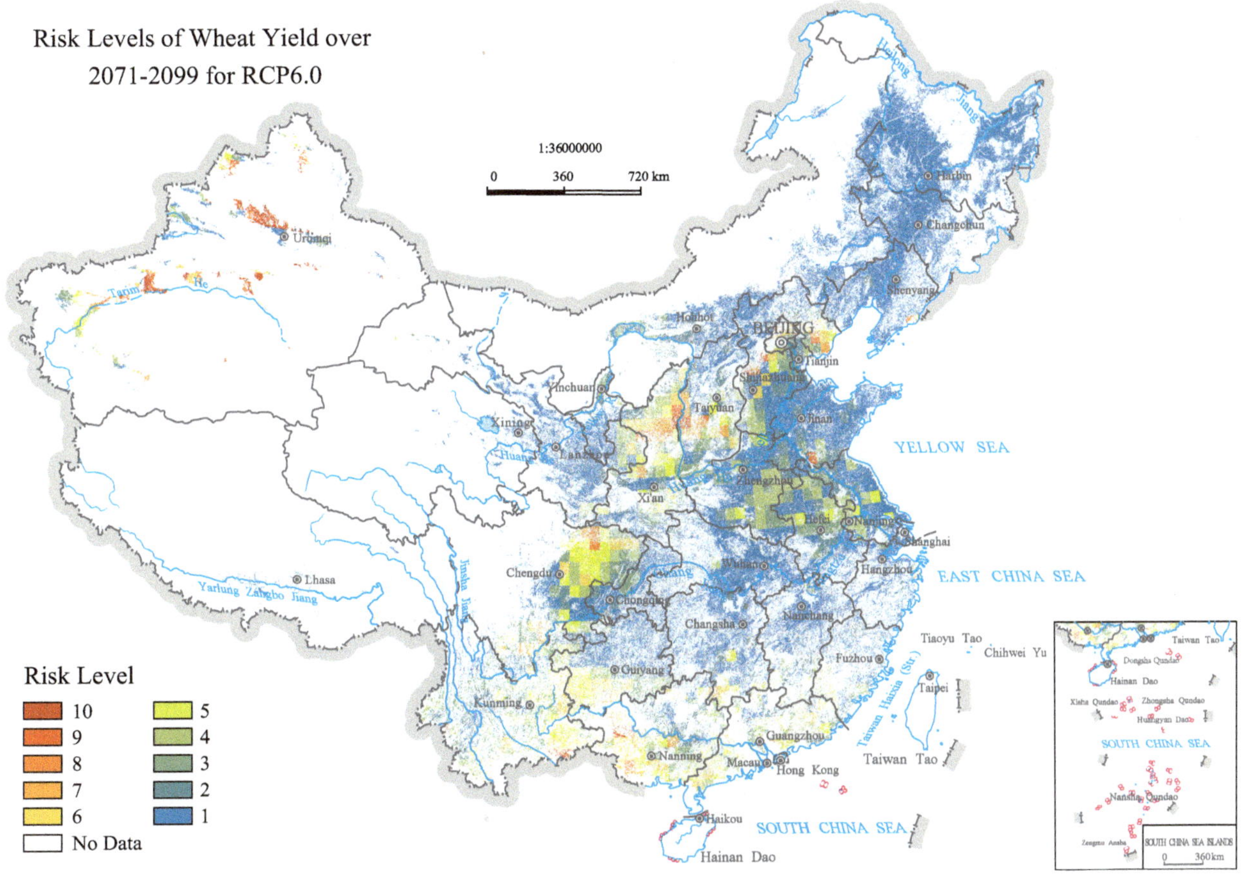

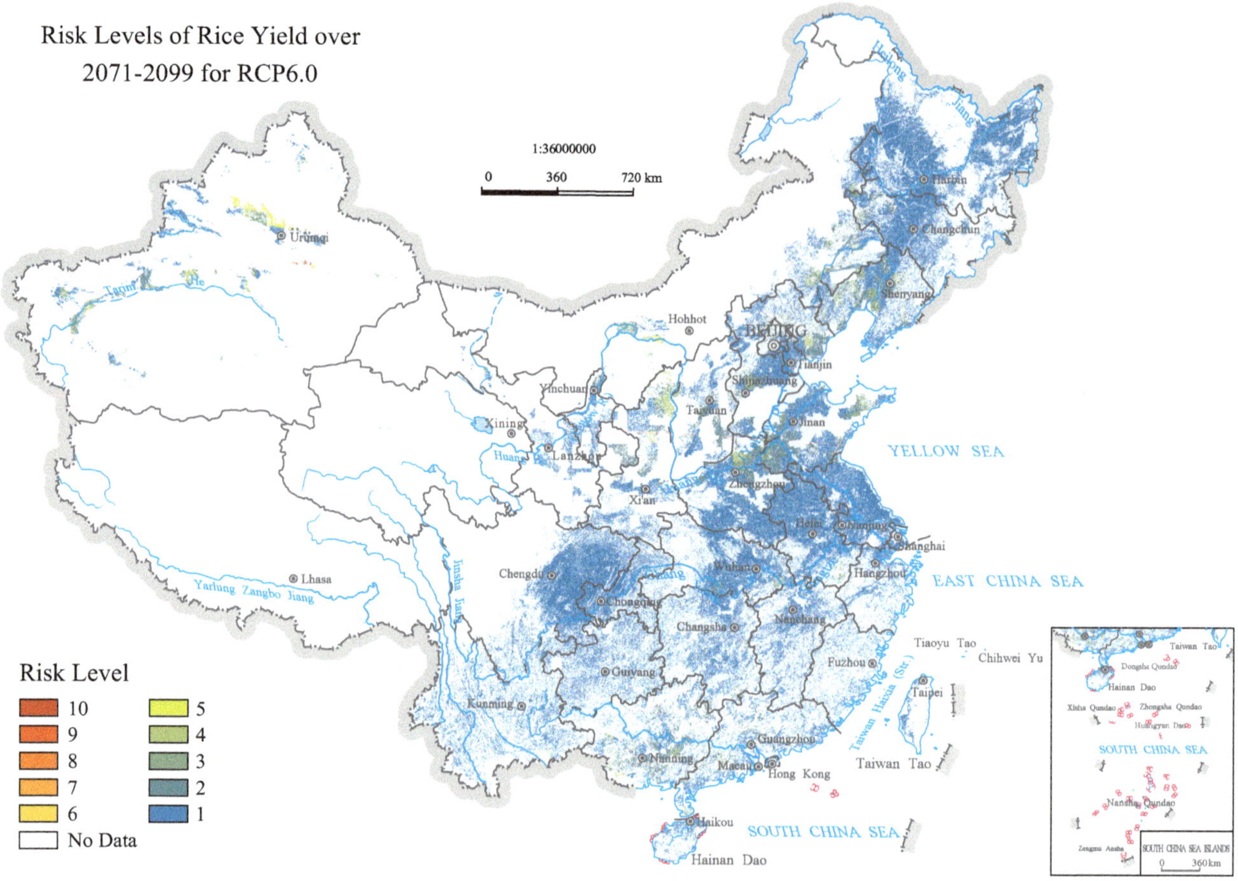
Risk Levels of Rice Yield over
2071-2099 for RCP6.0
1:36000000
0 360 720 km
Risk Level
10
9
8
7
6
5
4
3
2
1
No Data
YELLOW SEA
EAST CHINA SEA
SOUTH CHINA SEA
Urümqi
Hohhot
BEIJING
Tianjin
Shijiazhuang
Taiyuan
Yinchuan
Xining
Lanzhou
Jinan
Zhengzhou
Xi'an
Hefei
Nanjing
Shanghai
Hangzhou
Wuhan
Chengdu
Chongqing
Lhasa
Changsha
Nanchang
Fuzhou
Taipei
Guiyang
Kunming
Nanning
Guangzhou
Macao
Hong Kong
Haikou
Hainan Dao
Harbin
Changchun
Shenyang
Tiaoyu Tao
Chihwei Yu
Taiwan Tao
Yarlung Zangbo Jiang
Jinsha Jiang
SOUTH CHINA SEA ISLANDS
0 360km

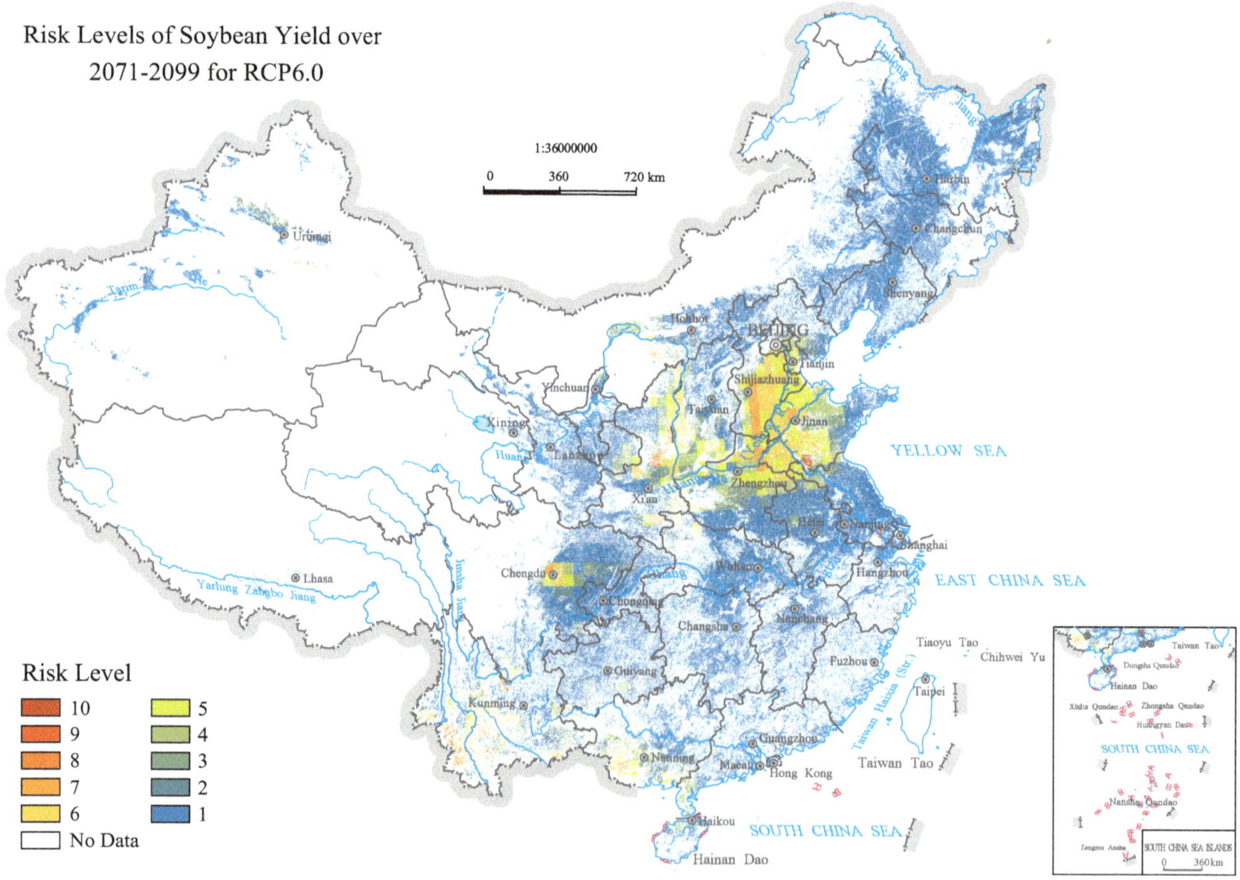
Risk Levels of Soybean Yield over
2071-2099 for RCP6.0
1:36000000
0 360 720 km
Risk Level
10
9
8
7
6
5
4
3
2
1
No Data
YELLOW SEA
EAST CHINA SEA
SOUTH CHINA SEA
Urümqi
Hohhot
BEIJING
Tianjin
Shijiazhuang
Taiyuan
Yinchuan
Xining
Lanzhou
Jinan
Zhengzhou
Xi'an
Hefei
Nanjing
Shanghai
Hangzhou
Wuhan
Chengdu
Chongqing
Lhasa
Changsha
Nanchang
Fuzhou
Taipei
Guiyang
Kunming
Nanning
Guangzhou
Macao
Hong Kong
Haikou
Hainan Dao
Harbin
Changchun
Shenyang
Tiaoyu Tao
Chihwei Yu
Taiwan Tao
Yarlung Zangbo Jiang
Jinsha Jiang
SOUTH CHINA SEA ISLANDS
0 360km

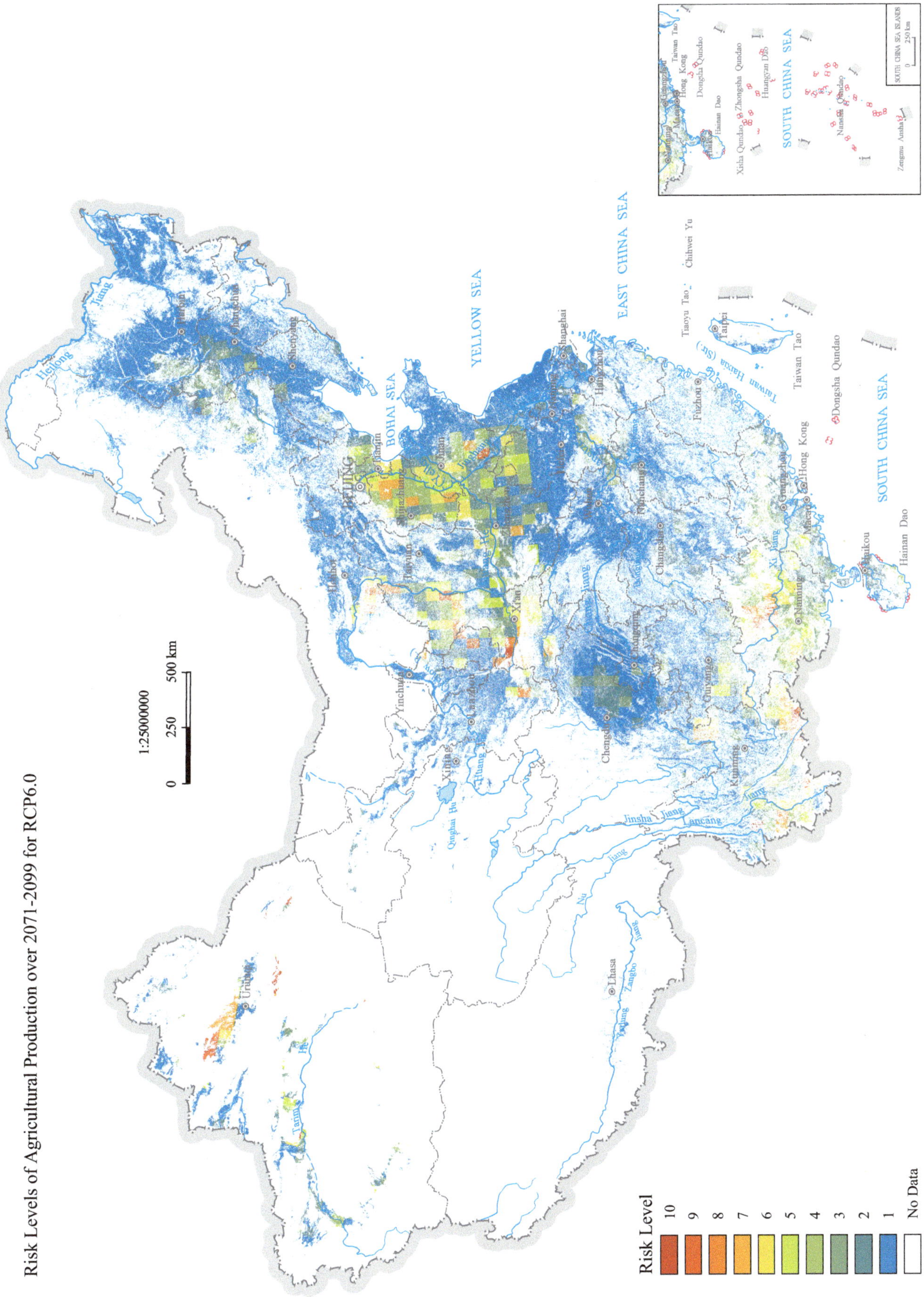

Risk Levels of Agricultural Production over 2071-2099 for RCP6.0

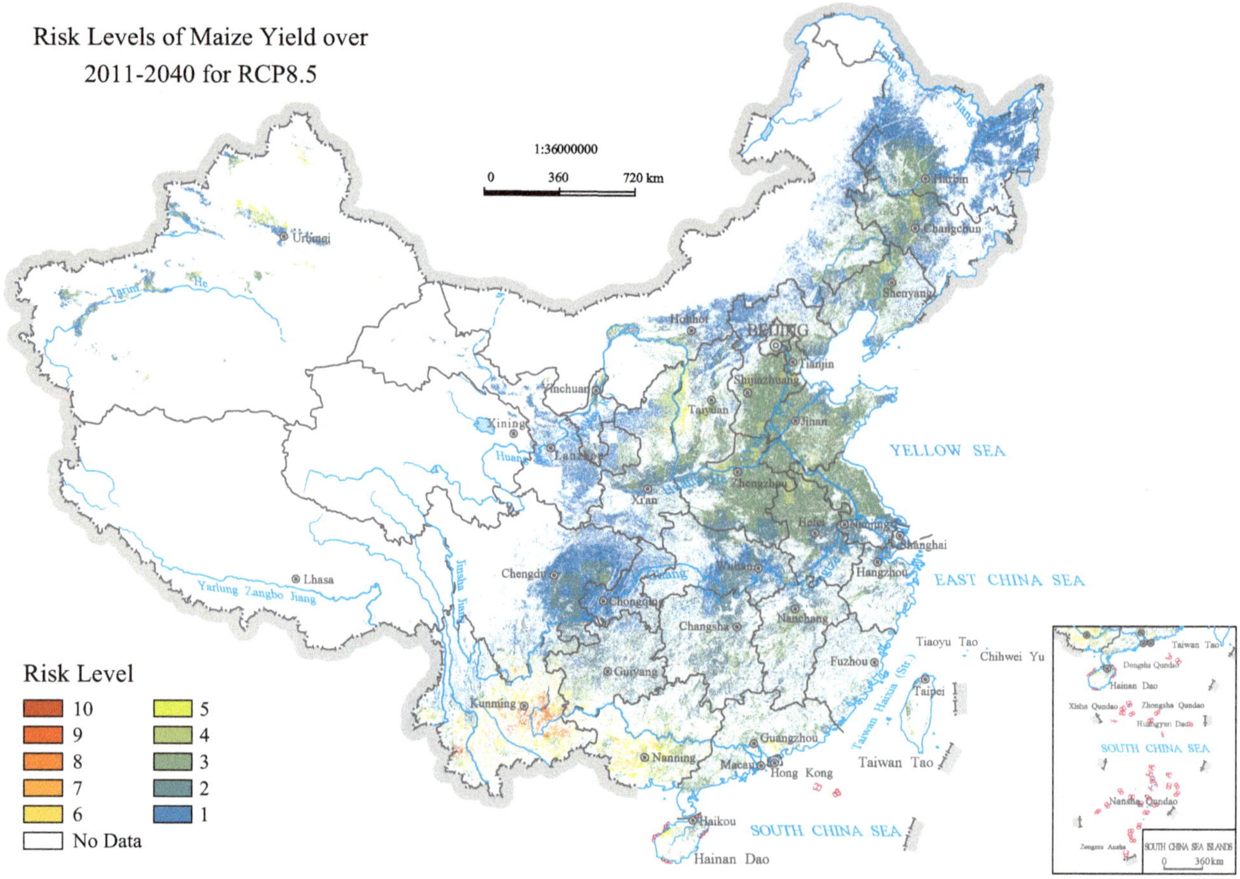
Risk Levels of Maize Yield over
2011-2040 for RCP8.5
1:36000000
0 360 720 km
Risk Level
10
9
8
7
6
5
4
3
2
1
No Data
YELLOW SEA
EAST CHINA SEA
SOUTH CHINA SEA
Hainan Dao
Taiwan Tao
BEIJING
Lhasa
Yarlung Zangbo Jiang
SOUTH CHINA SEA ISLANDS

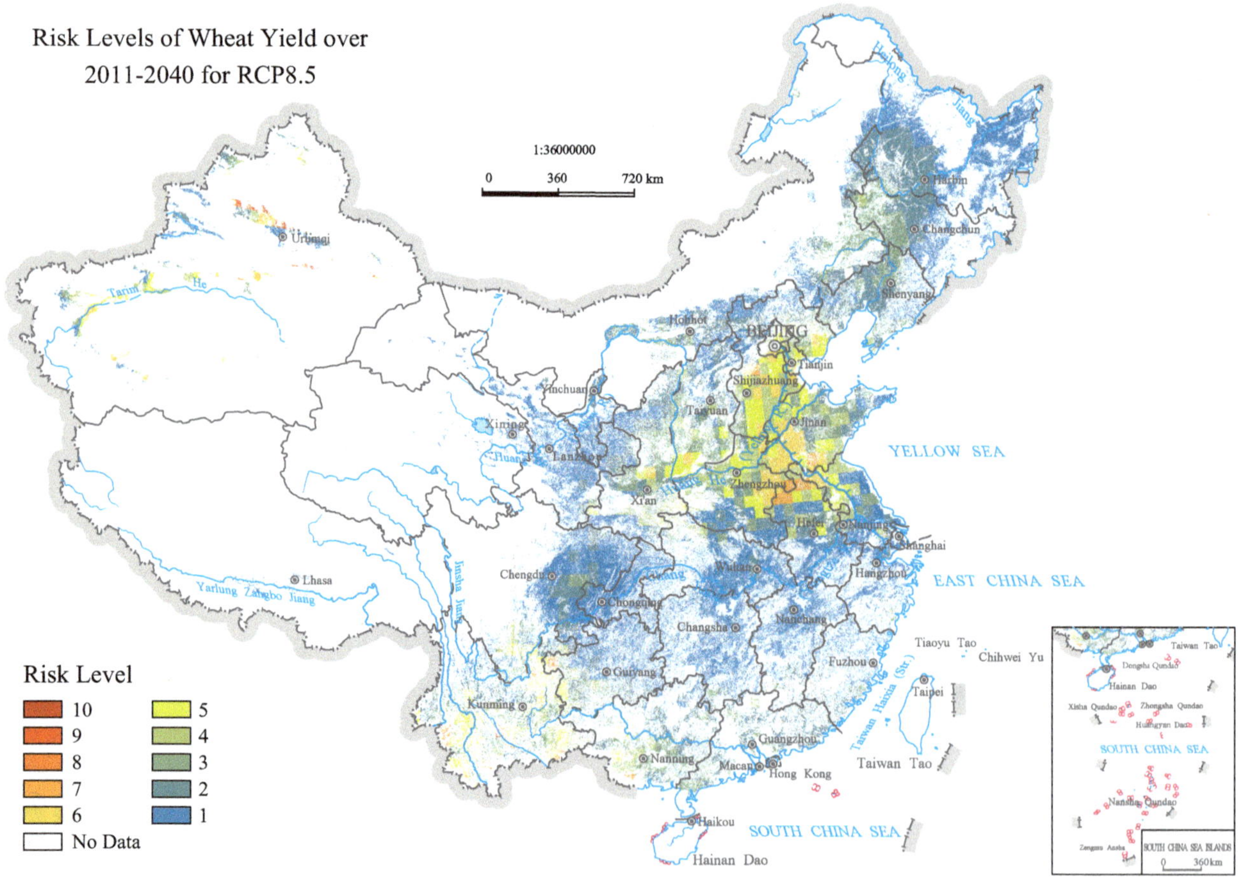
Risk Levels of Wheat Yield over
2011-2040 for RCP8.5
1:36000000
0 360 720 km
Risk Level
10
9
8
7
6
5
4
3
2
1
No Data
YELLOW SEA
EAST CHINA SEA
SOUTH CHINA SEA
Hainan Dao
Taiwan Tao
BEIJING
Lhasa
SOUTH CHINA SEA ISLANDS

Risk Levels of Rice Yield over 2011-2040 for RCP8.5

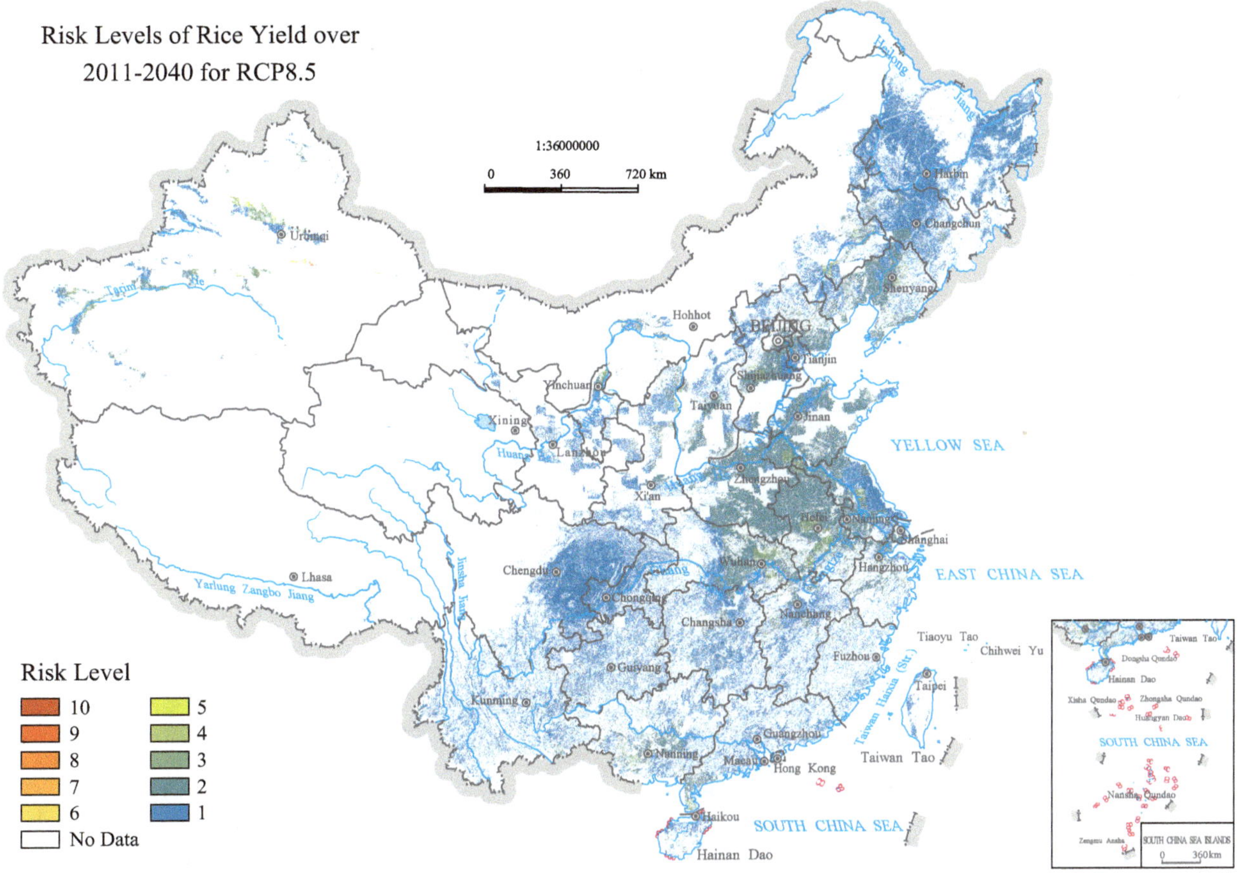

Risk Levels of Soybean Yield over 2011-2040 for RCP8.5

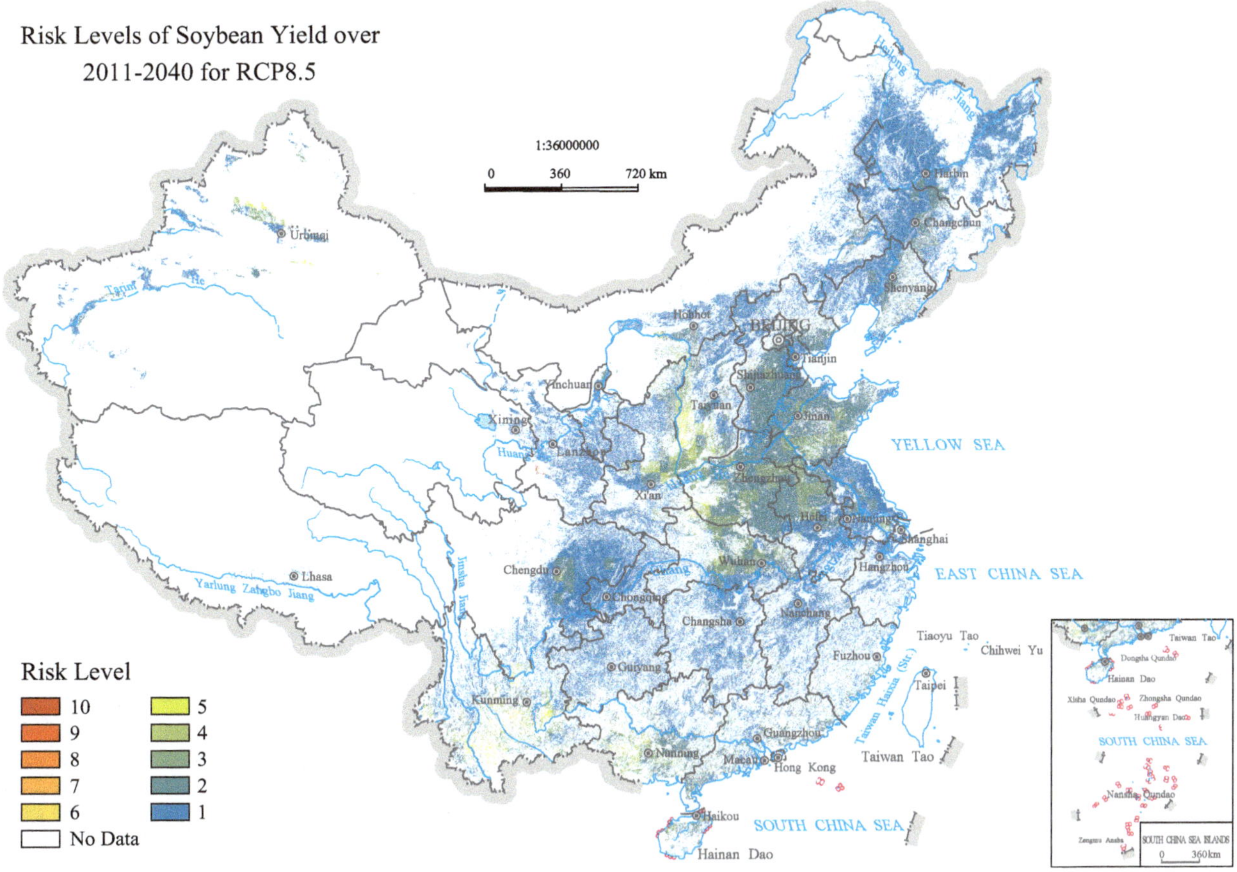

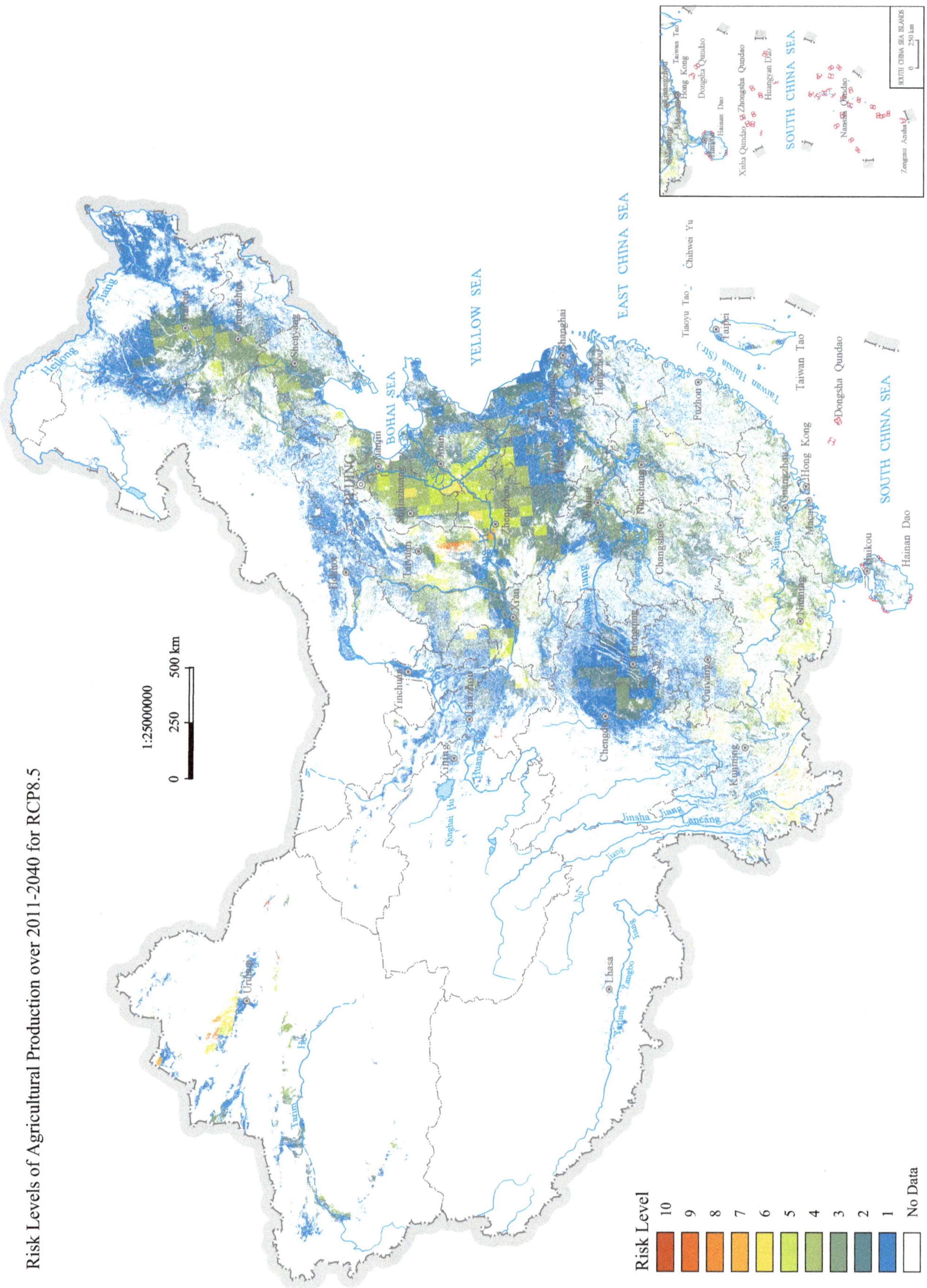
Risk Levels of Agricultural Production over 2011-2040 for RCP8.5
1:25000000
0 250 500 km
Risk Level
10
9
8
7
6
5
4
3
2
1
No Data
YELLOW SEA
EAST CHINA SEA
BOHAI SEA
SOUTH CHINA SEA
BEIJING
Shanghai
Lhasa
Chengdu
Kunming
Guiyang
Changsha
Nanchang
Fuzhou
Hong Kong
Taibei
Taiwan Tao
Hainan Dao
Haikou
Nanning
Xining
Yinchuan
Lanzhou
Shenyang
Heilong Jiang
Jinsha Jiang
Lancang Jiang
Qinghai Hu
Dongsha Qundao
Chihwei Yu
Tiaoyu Tao

Risk Levels of Maize Yield over 2041-2070 for RCP8.5

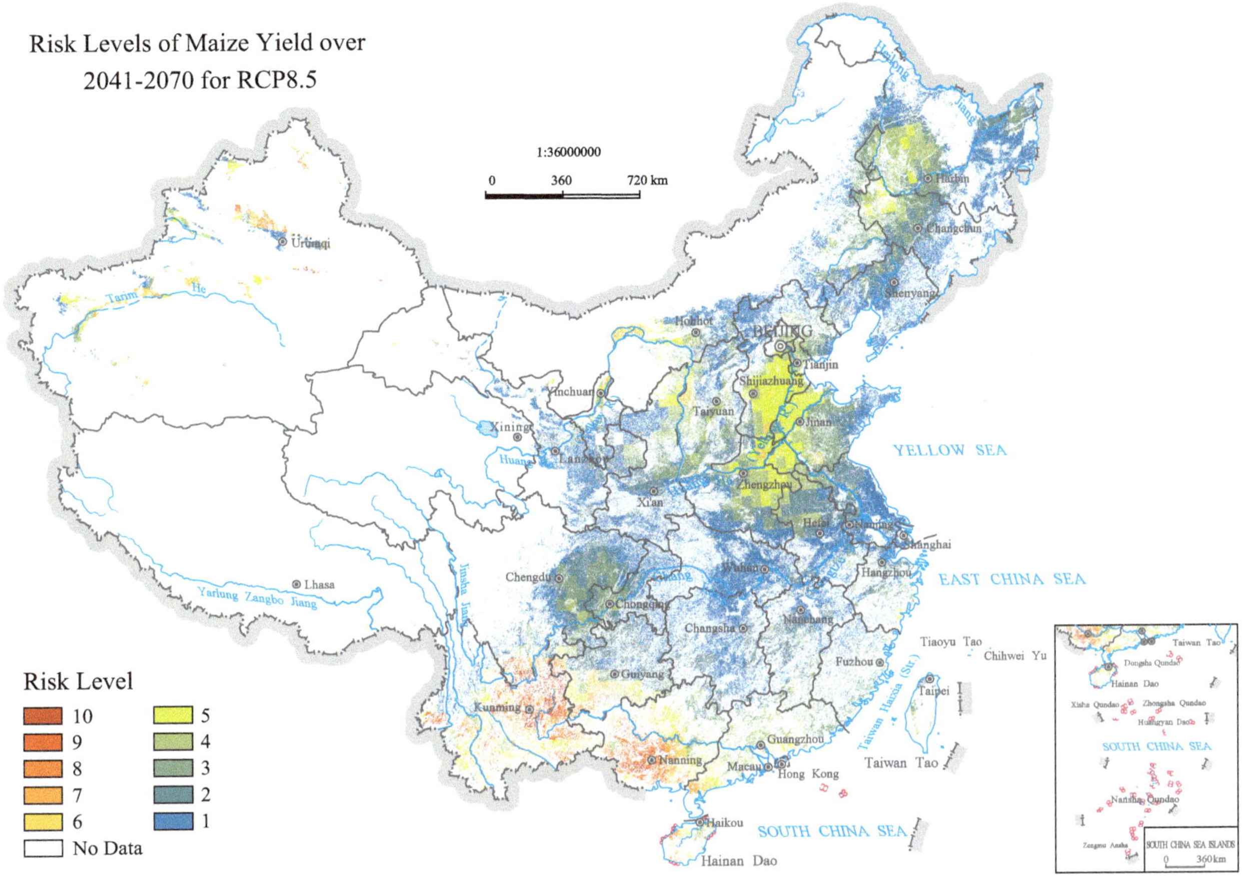

Risk Levels of Wheat Yield over 2041-2070 for RCP8.5

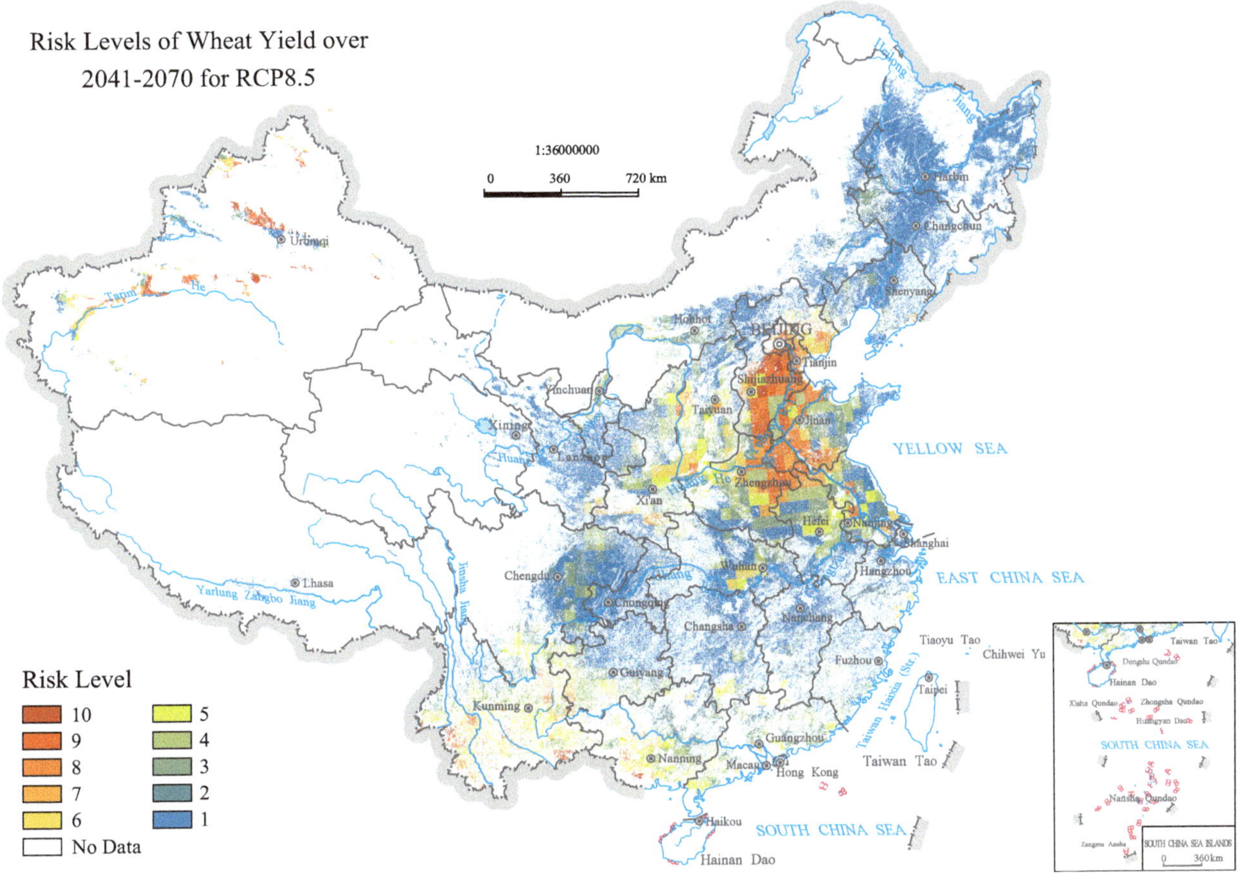

Risk Levels of Rice Yield over 2041-2070 for RCP8.5

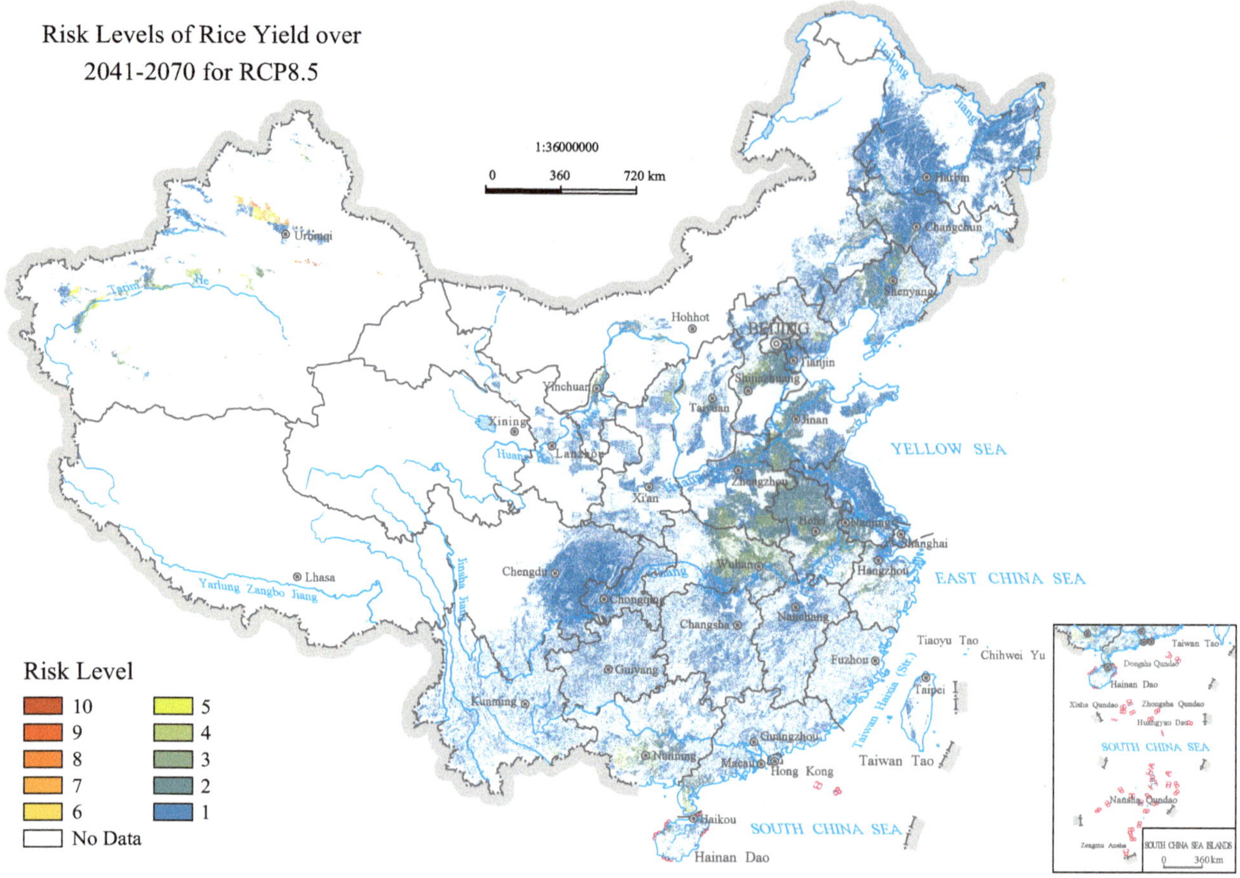

Risk Levels of Soybean Yield over 2041-2070 for RCP8.5

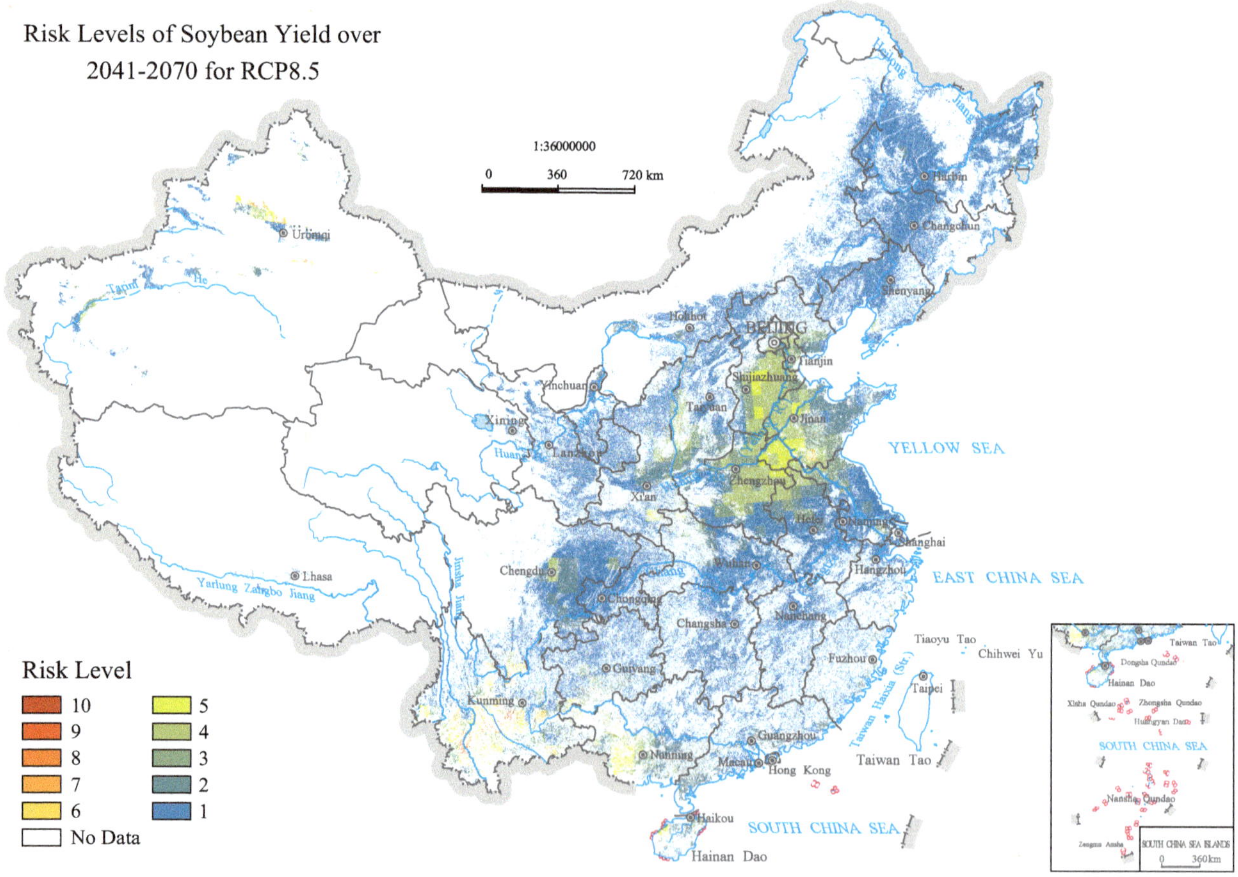

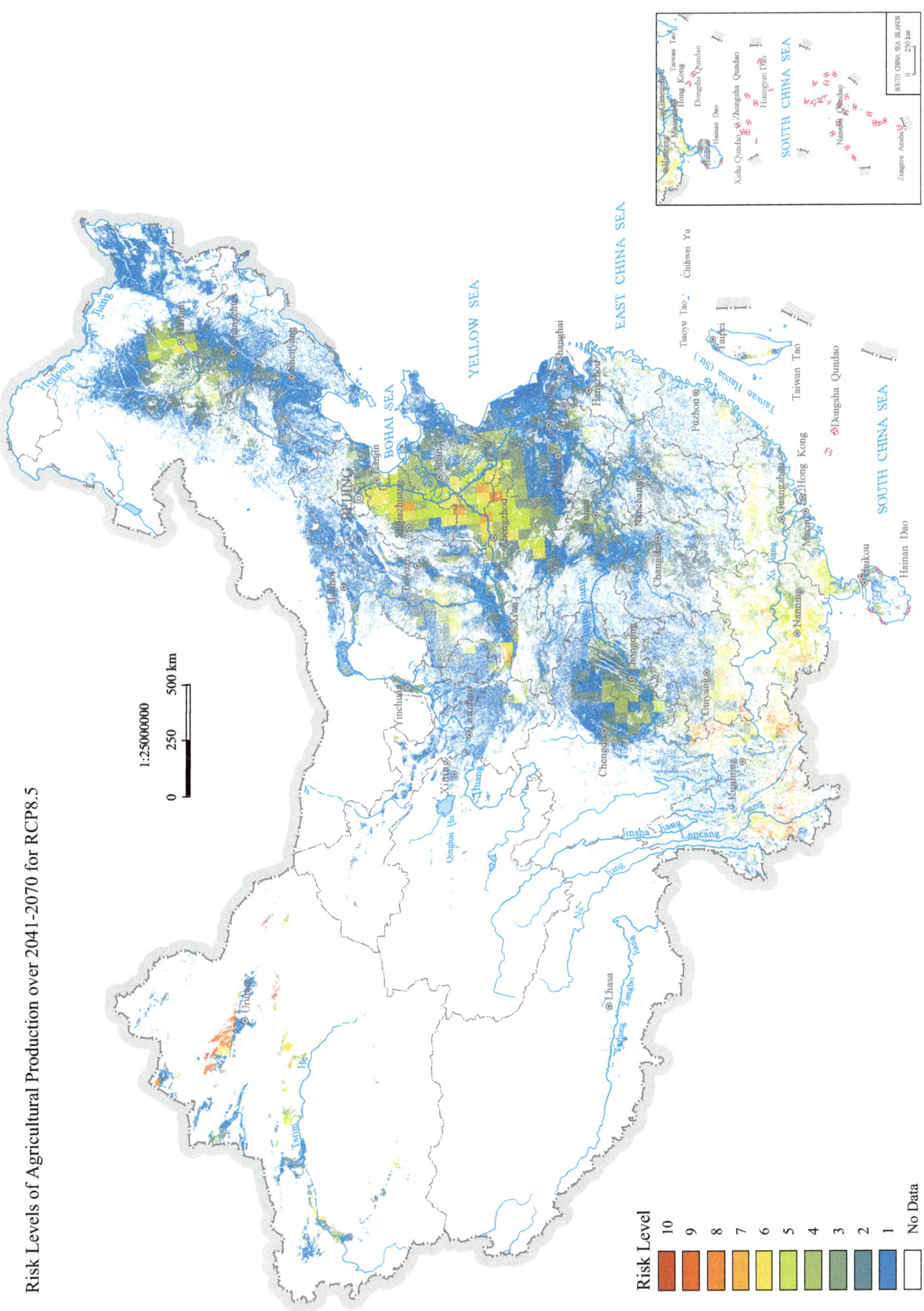
Risk Levels of Agricultural Production over 2041-2070 for RCP8.5
1:25000000
0 250 500 km
Risk Level
10
9
8
7
6
5
4
3
2
1
No Data
BEIJING
Tianjin
Shijiazhuang
Jinan
Zhengzhou
Taiyuan
Xi'an
Yinchuan
Lanzhou
Xining
Qinghai Hu
Chengdu
Chongqing
Guiyang
Kunming
Nanning
Guangzhou
Hong Kong
Macau
Haikou
Hainan Dao
Changsha
Nanchang
Wuhan
Hefei
Nanjing
Shanghai
Hangzhou
Fuzhou
Taibei
Taiwan Tao
Diaoyu Dao
Chiwei Yu
Harbin
Changchun
Shenyang
Heilong Jiang
Urumqi
Lhasa
Yarlung Zangbo Jiang
Tarim He
Jinsha Jiang
Lancang Jiang
Nu Jiang
YELLOW SEA
BOHAI SEA
EAST CHINA SEA
SOUTH CHINA SEA
Dongsha Qundao
Taiwan Haixia (Str.)
Xi Jiang
Huang He
Hong Kong
Dongsha Qundao
Hainan Dao
Xisha Qundao
Zhongsha Qundao
Huangyan Dao
Nansha Qundao
Zengmu Ansha
SOUTH CHINA SEA ISLANDS
0 250 km

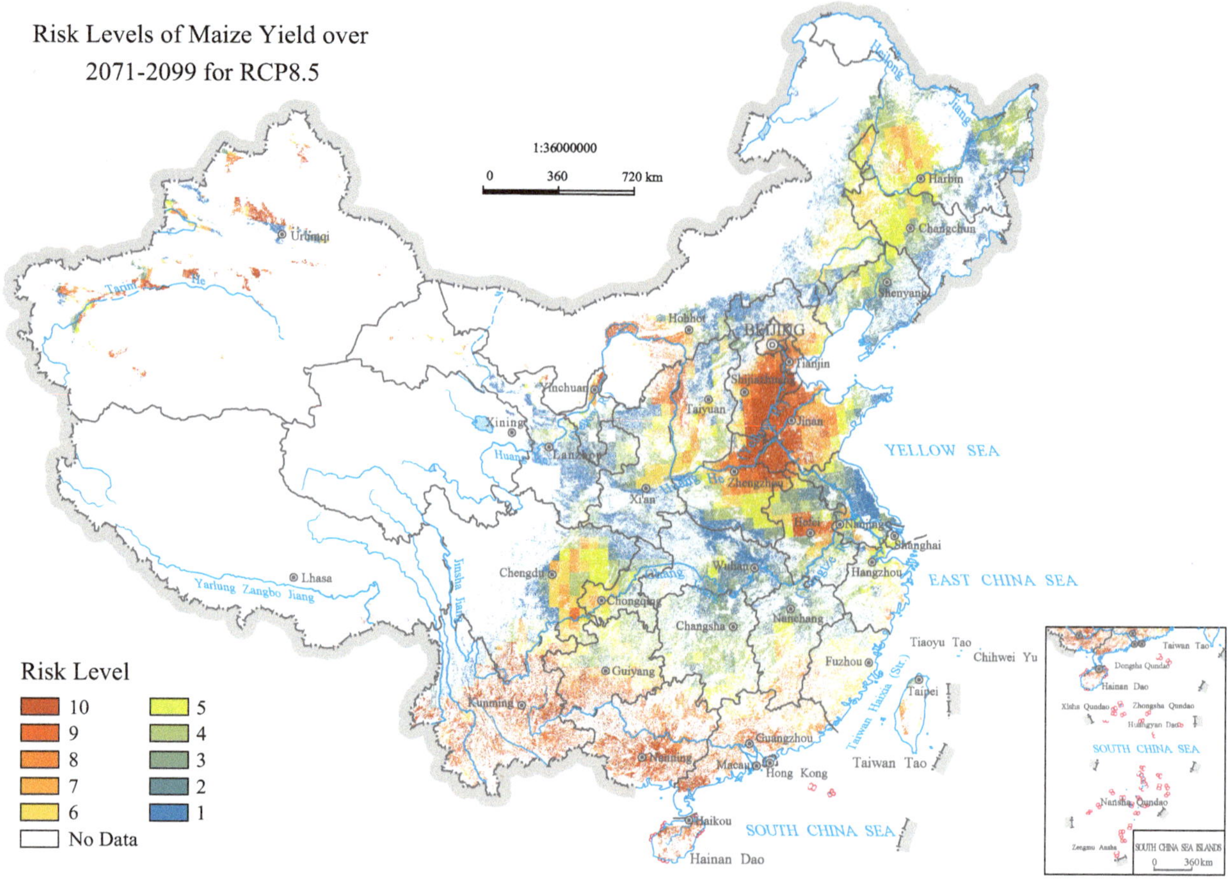
Risk Levels of Maize Yield over
2071-2099 for RCP8.5
1:36000000
0 360 720 km
Urümqi
Hohhot
BEIJING
Tianjin
Shijiazhuang
Taiyuan
Jinan
Yinchuan
Xining
Lanzhou
Zhengzhou
Xi'an
Hefei
Nanjing
Shanghai
Hangzhou
Wuhan
Chengdu
Chongqing
Changsha
Nanchang
Lhasa
Guiyang
Kunming
Nanning
Guangzhou
Fuzhou
Taipei
Macau
Hong Kong
Haikou
Harbin
Changchun
Shenyang
YELLOW SEA
EAST CHINA SEA
SOUTH CHINA SEA
Hainan Dao
Taiwan Tao
Tiaoyu Tao
Chihwei Yu
Risk Level
10
9
8
7
6
5
4
3
2
1
No Data

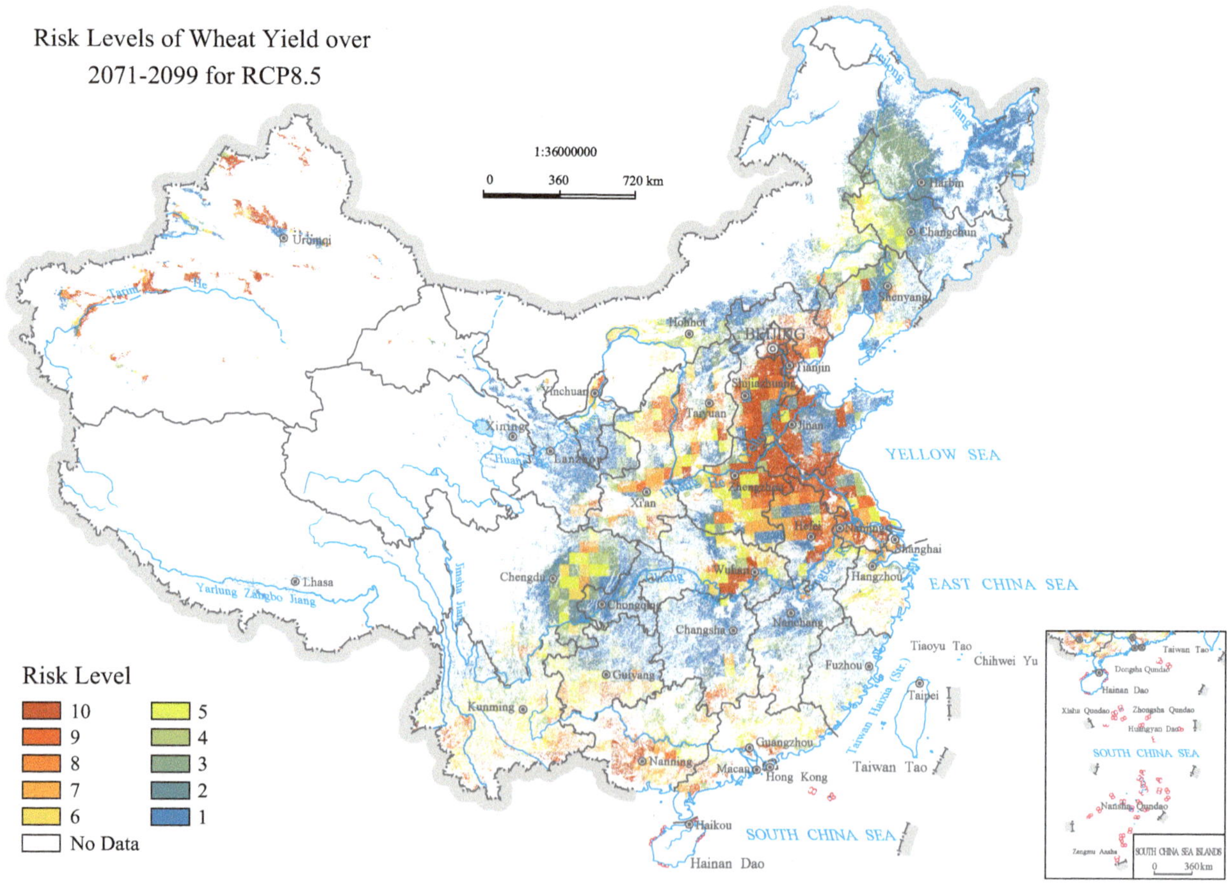
Risk Levels of Wheat Yield over
2071-2099 for RCP8.5
1:36000000
0 360 720 km
Urümqi
Hohhot
BEIJING
Tianjin
Taiyuan
Jinan
Yinchuan
Xining
Lanzhou
Xi'an
Hefei
Shanghai
Hangzhou
Wuhan
Chengdu
Chongqing
Changsha
Nanchang
Lhasa
Guiyang
Kunming
Nanning
Guangzhou
Fuzhou
Taipei
Macau
Hong Kong
Haikou
Harbin
Changchun
Shenyang
YELLOW SEA
EAST CHINA SEA
SOUTH CHINA SEA
Hainan Dao
Taiwan Tao
Tiaoyu Tao
Chihwei Yu
Risk Level
10
9
8
7
6
5
4
3
2
1
No Data

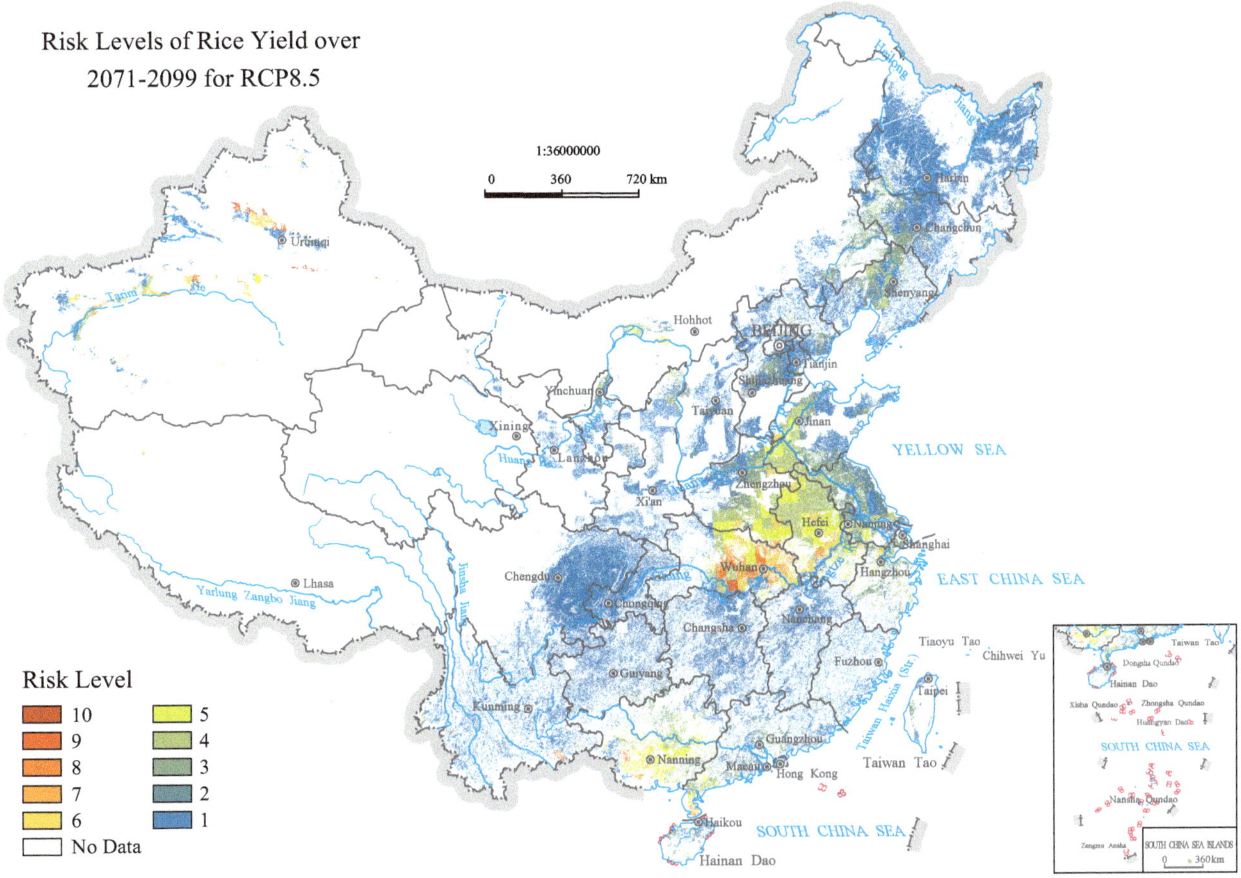
Risk Levels of Rice Yield over
2071-2099 for RCP8.5
1:36000000
0 360 720 km
Risk Level
10
9
8
7
6
5
4
3
2
1
No Data
Harbin
Changchun
Shenyang
Hohhot
BEIJING
Tianjin
Shijiazhuang
Taiyuan
Jinan
Yinchuan
Xining
Lanzhou
Xi'an
Zhengzhou
Hefei
Nanjing
Shanghai
Hangzhou
Wuhan
Chengdu
Chongqing
Changsha
Nanchang
Lhasa
Yarlung Zangbo Jiang
Guiyang
Kunming
Fuzhou
Taipei
Nanning
Guangzhou
Macau
Hong Kong
Haikou
Hainan Dao
Taiwan Tao
Tiaoyu Tao
Chihwei Yu
Urümqi
YELLOW SEA
EAST CHINA SEA
SOUTH CHINA SEA

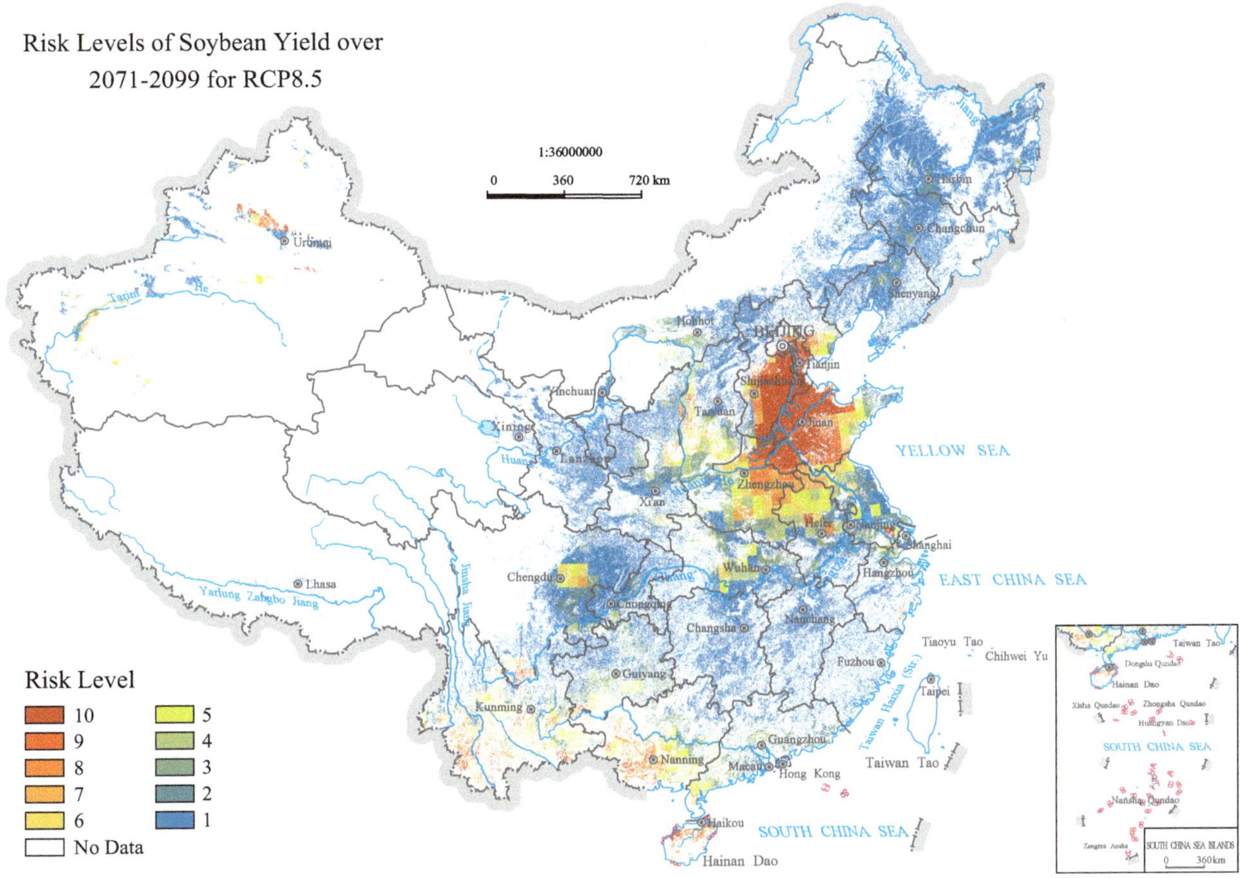
Risk Levels of Soybean Yield over
2071-2099 for RCP8.5
1:36000000
0 360 720 km
Risk Level
10
9
8
7
6
5
4
3
2
1
No Data
Harbin
Changchun
Shenyang
Hohhot
BEIJING
Tianjin
Shijiazhuang
Taiyuan
Jinan
Yinchuan
Xining
Lanzhou
Xi'an
Zhengzhou
Hefei
Nanjing
Shanghai
Hangzhou
Wuhan
Chengdu
Chongqing
Changsha
Nanchang
Lhasa
Yarlung Zangbo Jiang
Guiyang
Kunming
Fuzhou
Taipei
Nanning
Guangzhou
Macau
Hong Kong
Haikou
Hainan Dao
Taiwan Tao
Tiaoyu Tao
Chihwei Yu
Urümqi
YELLOW SEA
EAST CHINA SEA
SOUTH CHINA SEA

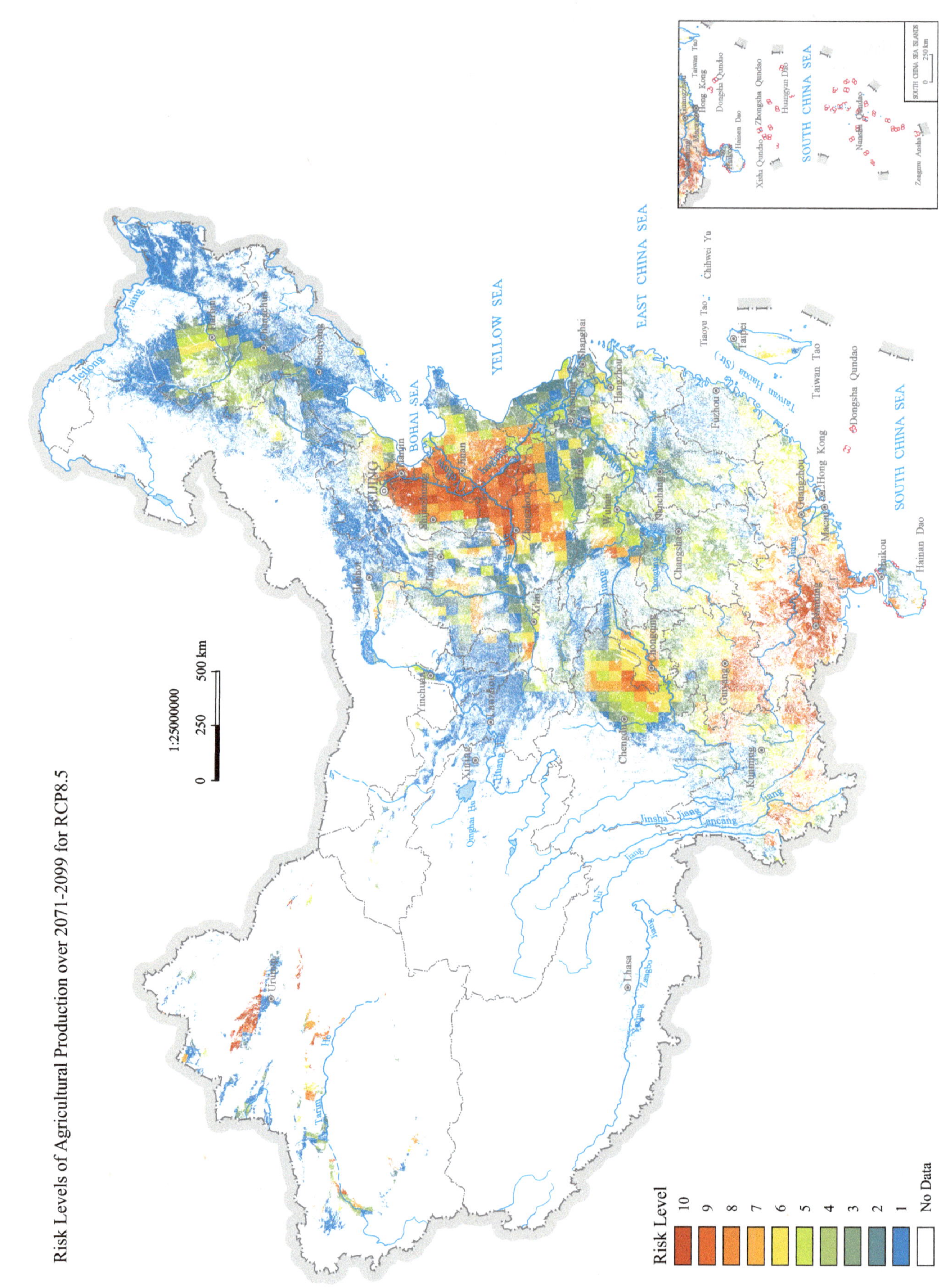
Risk Levels of Agricultural Production over 2071-2099 for RCP8.5
1:25000000
0 250 500 km
Risk Level
10
9
8
7
6
5
4
3
2
1
No Data
BEIJING
YELLOW SEA
BOHAI SEA
EAST CHINA SEA
SOUTH CHINA SEA
Heilong Jiang
Harbin
Changchun
Shenyang
Tianjin
Jinan
Shijiazhuang
Taiyuan
Hohhot
Yinchuan
Lanzhou
Xining
Xi'an
Zhengzhou
Hefei
Nanjing
Shanghai
Hangzhou
Wuhan
Nanchang
Fuzhou
Taipei
Changsha
Guangzhou
Hong Kong
Macau
Haikou
Hainan Dao
Nanning
Guiyang
Kunming
Chongqing
Chengdu
Lhasa
Urumqi
Qinghai Hu
Jinsha Jiang
Lancang
Nu Jiang
Yarlung Zangbo Jiang
Tarim He
Taiwan Tao
Dongsha Qundao
Diaoyu Dao
Chihwei Yu
Taiwan Haixia (Str.)
Xi Jiang
Xisha Qundao
Zhongsha Qundao
Huangyan Dao
Nansha Qundao
Zengmu Ansha
SOUTH CHINA SEA ISLANDS
0 250 km

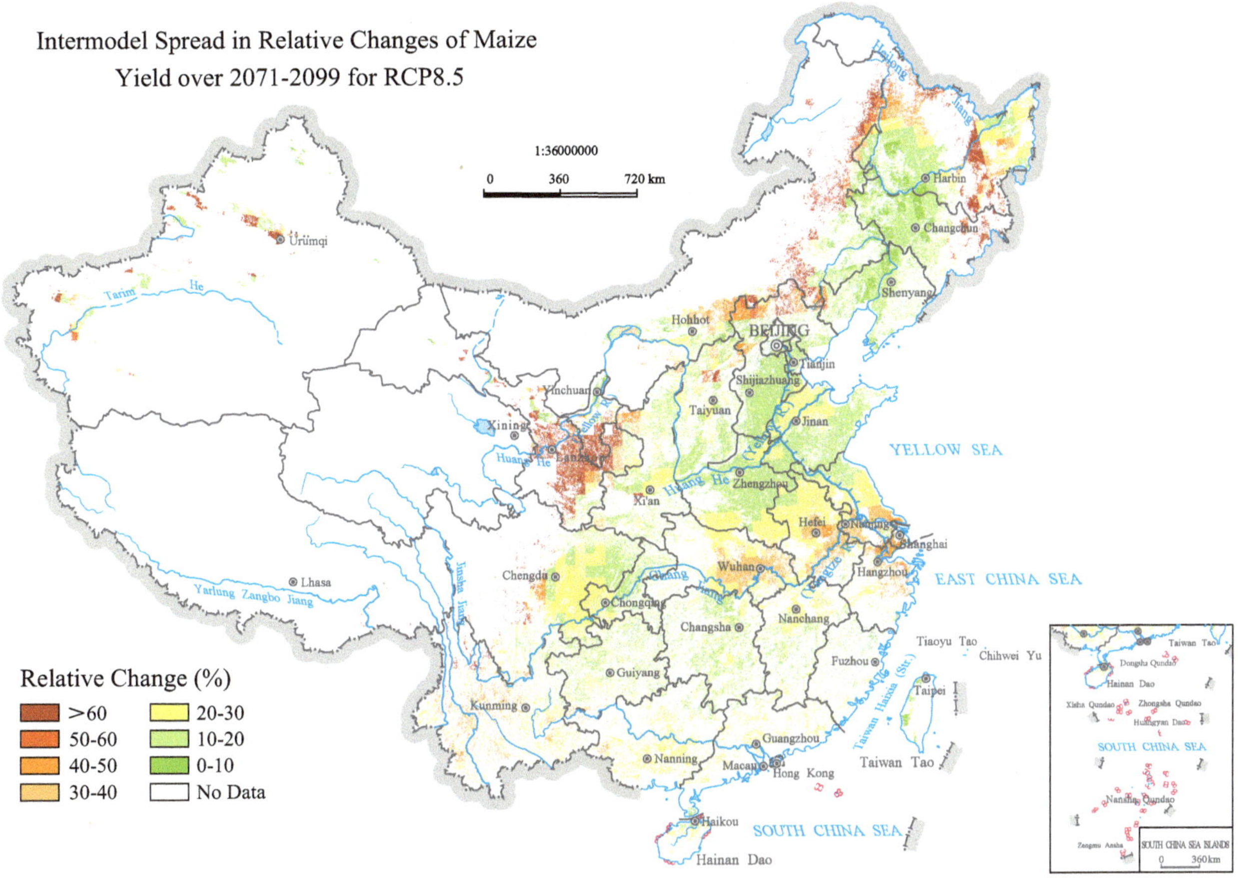
Intermodel Spread in Relative Changes of Maize
Yield over 2071-2099 for RCP8.5
1:36000000
0 360 720 km
Relative Change (%)
>60
50-60
40-50
30-40
20-30
10-20
0-10
No Data
Ürümqi
Harbin
Changchun
Shenyang
Hohhot
BEIJING
Tianjin
Shijiazhuang
Taiyuan
Jinan
Xining
Lanzhou
Xi'an
Zhengzhou
Hefei
Nanjing
Shanghai
Hangzhou
Wuhan
Chengdu
Chongqing
Lhasa
Changsha
Nanchang
Fuzhou
Taipei
Guiyang
Kunming
Nanning
Guangzhou
Macau
Hong Kong
Haikou
Hainan Dao
YELLOW SEA
EAST CHINA SEA
SOUTH CHINA SEA
Tiaoyu Tao
Chihwei Yu
Taiwan Tao
Yarlung Zangbo Jiang
Tarim He
Huang He
SOUTH CHINA SEA ISLANDS
0 360km

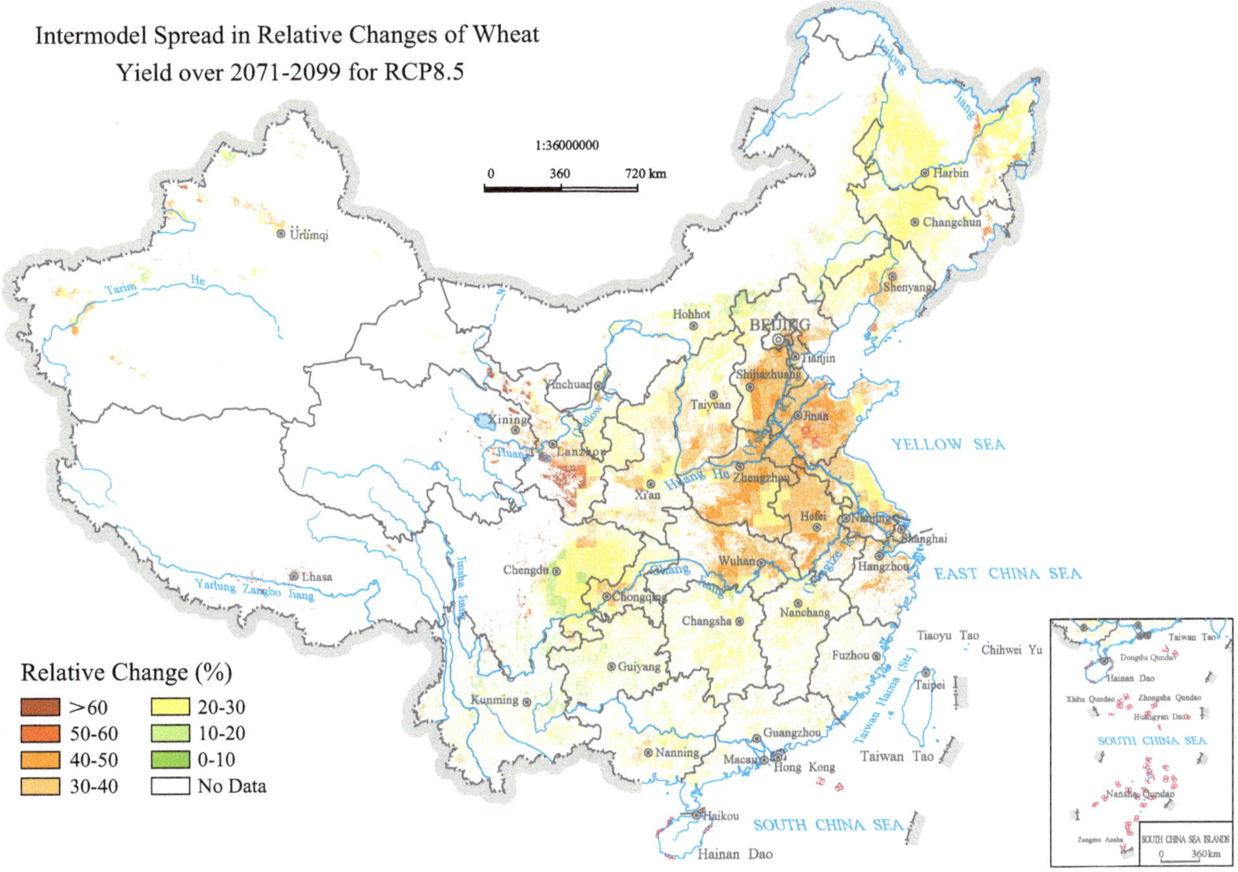
Intermodel Spread in Relative Changes of Wheat
Yield over 2071-2099 for RCP8.5
1:36000000
0 360 720 km
Relative Change (%)
>60
50-60
40-50
30-40
20-30
10-20
0-10
No Data
Ürümqi
Harbin
Changchun
Shenyang
Hohhot
BEIJING
Tianjin
Shijiazhuang
Taiyuan
Jinan
Xining
Lanzhou
Xi'an
Zhengzhou
Hefei
Nanjing
Shanghai
Hangzhou
Wuhan
Chengdu
Chongqing
Lhasa
Changsha
Nanchang
Fuzhou
Taipei
Guiyang
Kunming
Nanning
Guangzhou
Macau
Hong Kong
Haikou
Hainan Dao
YELLOW SEA
EAST CHINA SEA
SOUTH CHINA SEA
Tiaoyu Tao
Chihwei Yu
Taiwan Tao
Yarlung Zangbo Jiang
Tarim He
Huang He
SOUTH CHINA SEA ISLANDS
0 360km

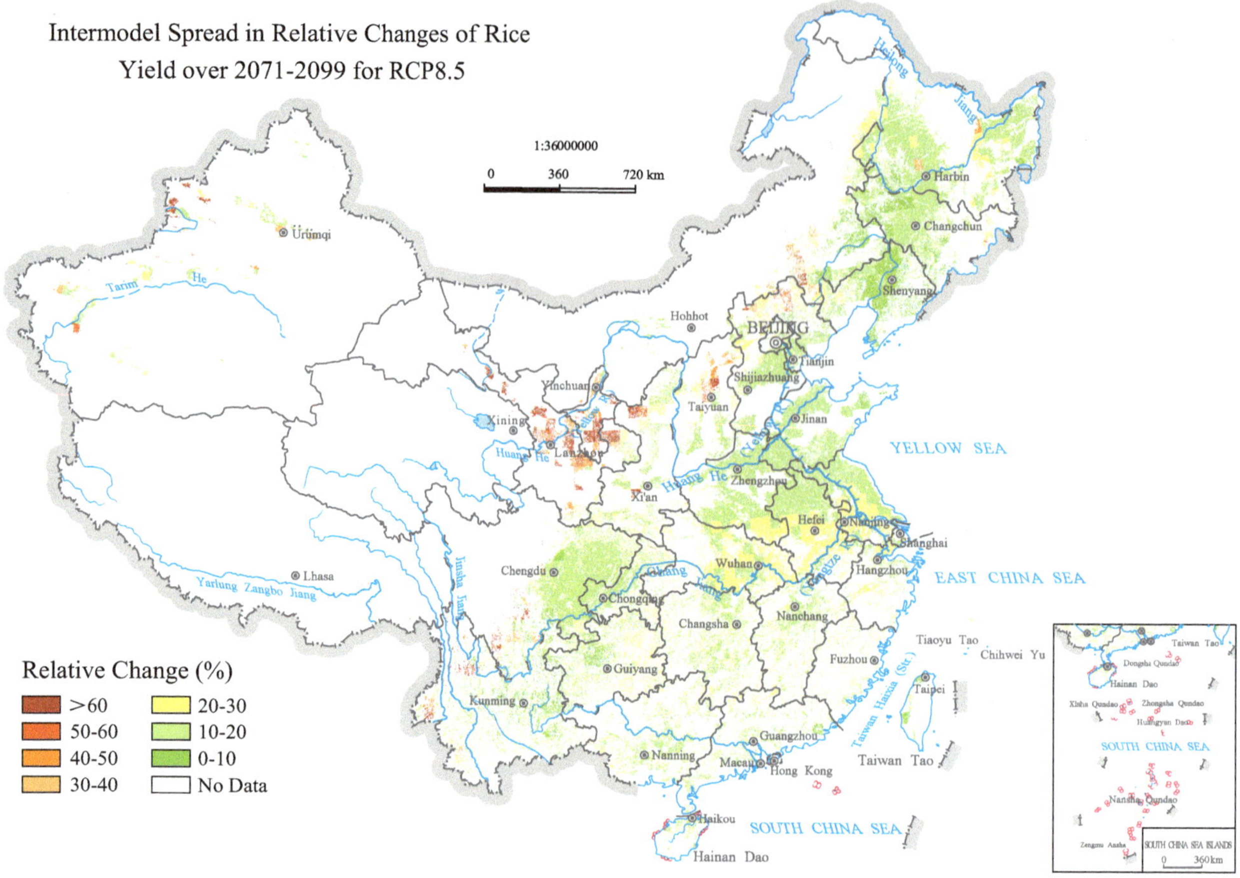
Intermodel Spread in Relative Changes of Rice
Yield over 2071-2099 for RCP8.5
1:36000000
0 360 720 km
Relative Change (%)
>60
50-60
40-50
30-40
20-30
10-20
0-10
No Data
YELLOW SEA
EAST CHINA SEA
SOUTH CHINA SEA
Hainan Dao
Taiwan Tao
SOUTH CHINA SEA ISLANDS

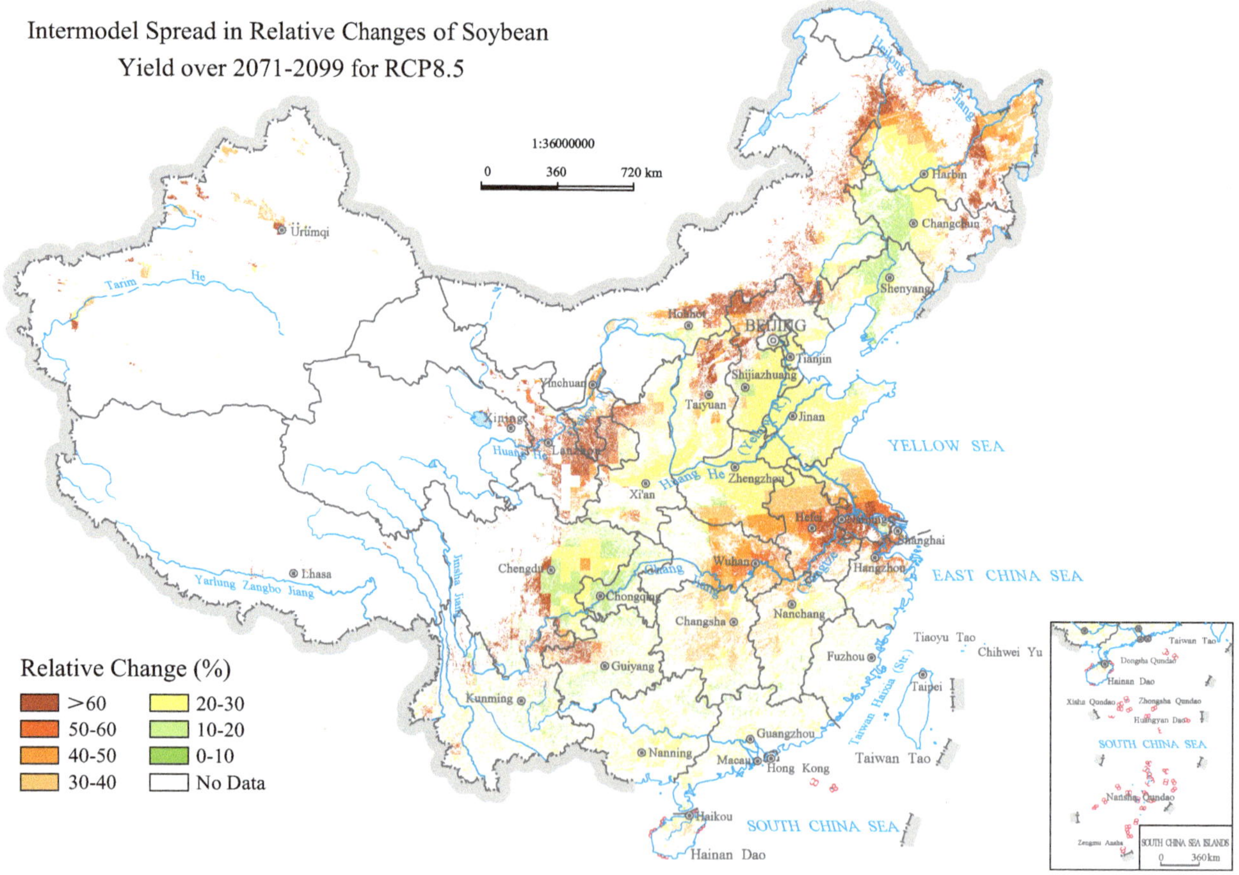
Intermodel Spread in Relative Changes of Soybean
Yield over 2071-2099 for RCP8.5
1:36000000
0 360 720 km
Relative Change (%)
>60
50-60
40-50
30-40
20-30
10-20
0-10
No Data
YELLOW SEA
EAST CHINA SEA
SOUTH CHINA SEA
Hainan Dao
Taiwan Tao
SOUTH CHINA SEA ISLANDS

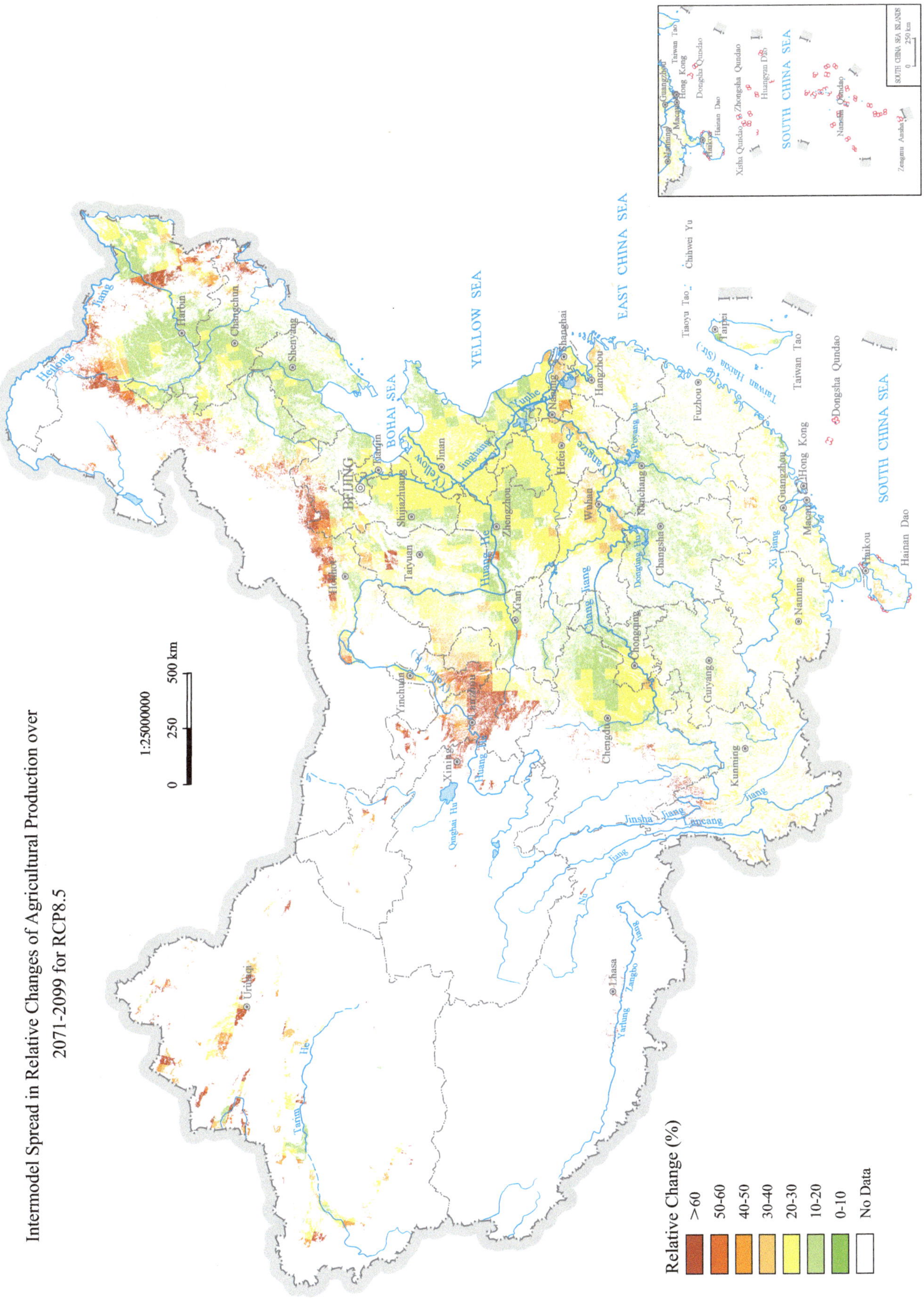

Intermodel Spread in Relative Changes of Agricultural Production over 2071-2099 for RCP8.5

References

Asseng, S., Ewert, F., Rosenzweig, C., Jones, J. W., Hatfield, J. L., Ruane, A. C., et al. (2013). Uncertainty in simulating wheat yields under climate change. *Nature Climate Change, 3,* 827–832.

Challinor, A. J., Simelton, E. S., Fraser, E. D., Hemming, D., & Collins, M. (2010). Increased crop failure due to climate change: Assessing adaptation options using models and socio-economic data for wheat in China. *Environmental Research Letters, 5,* 1–8.

Hempel, S., Frieler, K., Warszawski, L., Schewe, J., & Piontek, F. (2013). A trend-preserving bias correction—the ISI-MIP approach. *Earth System Dynamics, 4,* 219–236.

Ju, H., Lin, E. D., Wheeler, T., Andrew, C., & Jiang, S. (2013). Climate change modeling and its roles to Chinese crops yield. *Journal of Integrative Agriculture, 12,* 892–902.

Portmann, F. T., Siebert, S., & Döll, P. (2010). MIRCA2000—Global monthly irrigated and rain-fed crop areas around the year 2000: A new high-resolution data set for agricultural and hydro-logical modeling. *Global Biogeochemical Cycles, 24,* 1–24.

Taylor, K., Stouffer, R., & Meehl, G. (2012). An overview of CMIP5 and the experiment design. *Bulletin of the American Meteorological Society, 93,* 485–498.

Warszawski, L., Frieler, K., Huber, V., Piontek, F., Serdeczny, O., & Schewe, J. (2014). The inter-sectoral impact model intercomparison projection (ISI-MIP): Project framework. *Proceedings of the National Academy of Sciences of the United States of America, 111,* 3228–3232.

Yin, Y., Tang, Q., & Liu, X. (2015). A multi-model analysis of change in potential yield of major crops in China under climate change. *Earth System Dynamics, 6,* 45–59.

5 Ecological Risks

Yuanyuan Yin, Qiuhong Tang, Xingcai Liu and Huijuan Cui

5.1 Background

Climate change has strong influence on terrestrial ecosystems, and the influence is almost certain to grow in the near future (IPCC 2014). Continuing climate change and climate extremes may cause significant impacts on the terrestrial ecosystems, such as decrease in regional carbon stocks and reduction in vegetation leaf cover (Donohue et al. 2013). China has a variety of ecosystems ranging from alpine tundra to evergreen tropics and from desert to forest. The impact of climate change on terrestrial ecosystems in China has received considerable attention.

Many impact assessments focused on one or a few aspects such as vegetation structure and distribution (Tao and Zhang 2010; Wang 2014), or carbon cycle [e.g., carbon flux and stock, net primary productivity (NPP)] (Tao and Zhang 2010; Piao et al. 2009; Ni 2011). However, it is difficult to identify the main factors influencing ecosystem shifts and to know the degree of climate change tolerated prior to the shift of complex ecological systems. In order to quantify the comprehensive risks of ecosystem alterations under climate change, Heyder et al. (2011) proposed an aggregated metric, Γ, of joint changes in macroscopic ecosystem features. This metric is based on a specific subset of macroscopic variables (e.g., carbon fluxes, carbon stocks, and water fluxes) that characterize the ecosystem state. Essentially, this metric Γ uses aggregated changes in the biogeochemical ecosystem state as a proxy for the risk of ecosystem shifts. Ecosystem shift in this study refers to the state that the ecosystem has been pushed beyond the point of recovering (Scheffer and Carpenter 2003). Based on the specific subsets of variables, this metric could also be used to identify the main factors influencing ecosystem shifts. To identify the high-risk areas and the main contributing factors of the predicted ecosystem shifts in China, a multi-model analysis of ecosystem shift risk using the metric Γ was presented over the naturally vegetated land of China in this chapter (Yin et al. 2016).

Y. Yin · Q. Tang (✉) · X. Liu
Key Laboratory of Water Cycle and Related Land Surface Processes, Institute of Geographic Sciences and Natural Resources Research, Chinese Academy of Sciences, Beijing, 100101, China
e-mail: tangqh@igsnrr.ac.cn

H. Cui
Key Laboratory of Land Surface Pattern and Simulation, Institute of Geographic Sciences and Natural Resources Research, Chinese Academy of Sciences, Beijing, 100101, China

5.2 Method

In this chapter, the ecosystem change metric Γ was calculated following Heyder et al. (2011) (Eq. 5.1). The metric calculates the difference between an ecosystem state under climate change and the current state. Ecosystem states are characterized as vectors in a multi-dimensional state space by the variables (Table 5.1) simulated by four global gridded vegetation models (GGVMs) in ISI-MIP (Table 5.2), with each dimension representing a specific fluxes change, stock, or process variable. The simulation outputs of the GGVMs were provided from 1971 to 2099. Three GGVMs (JeDi, JULES, and LPJmL) presented simulation results under four representative concentration pathways (RCPs) namely, RCP2.6, RCP4.5, RCP6.0, and RCP8.5 of five general circulation models (GCMs) from the Fifth Coupled Model Inter-comparison Project (CMIP5) while VISIT model offered simulation results exclusively for RCP2.6 and RCP8.5 of three GCMs. The model outputs of LPJmL and VISIT were provided on a 0.5° × 0.5° grid, and the model outputs of JeDi and JULES were provided on a 1.25° × 1.85° grid. We assigned the value of a 0.5° × 0.5° grid or 1.25° × 1.85° grid, to which the central point of a 1 km grid belong, to the 1 km grid (Taylor et al. 2012). All the GGVMs took into account of the CO_2 fertilization effects. As the land surface was assumed to be covered by natural vegetation only during these model simulations, the ecosystem shifts suggested by the GGVMs were only driven through climate change rather than land-use change. Therefore, we focused on the risk of only natural vegetation by excluding sandy desert, swamp, and cultivated areas.

Q. Tang and Q. Ge (eds.), *Atlas of Environmental Risks Facing China Under Climate Change*,
IHDP/Future Earth-Integrated Risk Governance Project Series, DOI 10.1007/978-981-10-4199-0_5

Table 5.1 Variable subsets used for analysis

Subset	Variables
Carbon fluxes	NPP; fire carbon
Carbon stocks	Carbon contained in vegetation and soil
Water fluxes	Transpiration; evaporation; runoff
All	Carbon fluxes; carbon stocks; water fluxes; soil water content (SWC)

Table 5.2 Overview of the GCMs and GGVMs

	Name	Institute
GCMs	HadGEM2-ES	Met Office Hadley Centre
	IPSL-CM5A-LR	Institute Pierre-Simon Laplace
	MIROC-ESM-CHEM	Japan Agency for Marine-Earth Science and Technology, Atmosphere and Ocean Research Institute (The University of Tokyo), and National Institute for Environmental Studies
	GFDL-ESM2M	Geophysical Fluid Dynamics Laboratory
	NorESM1-M	Norwegian Climate Centre
GGVMs	JeDi	Max-Planck-Institut für Biogeochemie (Germany)
	JULES	Centre for Ecology and Hydrology (UK); Met Office Hadley Centre (UK); University of Exeter (UK)
	LPJmL	PIK (Germany)
	VISIT	National Institute for Environmental Studies (Japan)

$$\Gamma = \{\Delta V + cS(c, \sigma_c) + gS(g, \sigma_g) + bS(b, \sigma_b)\}/4 \quad (5.1)$$

where ΔV is change in vegetation structure; c is relative, local change of ecosystem; g is absolute, global change of ecosystem; b is changes in the relative magnitude of key biogeochemical exchange fluxes; and $S(x, \sigma_x)$ is change in inter-annual variability (a normalized sigmoid function of the ratio to standard deviation in the reference period, and computed for c, g and b). The range of value for the above dimensions is 0–1. Since the vegetation composition data are unavailable, ΔV is not included in the calculation. The remaining three components (c, b, and g) were scaled up accordingly in Eq. (5.1) and the factor in the denominator becomes 3 instead of 4.

The period of 1981–2010 was used as the reference period to estimate the current state of ecosystem. The reference period was also used in calculation of c, g, and $S(x, \sigma_x)$. The g was calculated based on data in China. The relative change of NPP and the ecosystem change metric Γ were computed between future periods (2011–2040, 2041–2070, and 2071–2099, respectively) and the reference period. The metric Γ was first calculated for each of the variable subset 'carbon fluxes', 'carbon stocks', or 'water fluxes' individually and then was calculated for the variable subset 'all' (Table 1). There were five GCMs and four GGVMs, making a maximum number of 20 model pairs under each RCP. However, this study used 18 model pairs for RCP2.6 and RCP8.5, and 15 model pairs for RCP4.5 and RCP6.0 due to unavailability of some model simulations in ISI-MIP. The Multimodel-ensemble Medians (MMs) of relative change of NPP or Γ were calculated for the periods of 2011–2040, 2041–2070, and 2071–2099, respectively. The MMs of Γ were divided into ten levels (Table 5.3). The standard deviation (SD) of the relative change or Γ from all the available GCM-GGVM pairs was used to quantify the model uncertainty.

5.3 Results

5.3.1 Relative Change of NPP and Risk Ecological Risk Under Climate Change

The relative change of NPP is less than zero in certain regions of transition zone between cropping area and nomadic area and northwest China. The largest increase of NPP is in Tibetan Plateau and in most regions of northwest China. The relative change is predicted to be greater under higher emission scenario.

Depending on the variable subset considered, the spatial patterns of the metric vary greatly. It suggests that the contributing factors to the ecosystem shifts are dissimilar for different regions. For a specified variable subset considered, the spatial pattern of the metric under different RCPs is similar. However, the metric is generally small under the

Table 5.3 Ecological risk levels

Risk level	1	2	3	4	5	6	7	8	9	10
MM of Γ	≤0.1	0.1–0.15	0.15–0.2	0.2–0.25	0.25–0.3	0.3–0.35	0.35–0.4	0.4–0.45	0.45–0.5	>0.5

lowest emission scenario and becomes greater under higher emission scenario. The metric Γ for carbon fluxes under RCP8.5 is considerable in the Tibetan Plateau and temperate humid/subhumid regions. The metric Γ for carbon stocks under RCP8.5 is greater than 0.3 in the Tibetan Plateau and the eastern China regions. Based on simulations of the GGVMs, the metric Γ for water fluxes is small in most parts of China under all emission scenarios. The only exception is the Tibetan Plateau region, where the median of the metric Γ for water fluxes is greater than 0.25, indicating a considerable influence of water fluxes change on ecosystem shift, under RCP8.5 scenario.

For all variable subsets, the ecosystem change metric Γ is small in most parts of China excluding for an area in the Tibetan Plateau region under the lowest emission scenario RCP2.6. Under RCP8.5 scenario, the Tibetan Plateau region, part of the temperate humid/subhumid, and cold temperate humid regions would have severe risks of ecosystem shifts. The high value of Γ over the Tibetan Plateau region results from the combined changes of carbon fluxes, carbon stocks, and water fluxes. The temperate humid/subhumid and cold temperate humid regions would have moderate risks of ecosystem shifts. The risk of ecosystem shifts is generally low in the northwest arid region and tropical and subtropical humid regions.

5.3.2 Model Spread and Uncertainty

The standard deviation of relative changes in NPP and Γ (for carbon fluxes, carbon stocks, water fluxes, and all variables) calculated from all available GCM-GGVM pairs over 2071–2099 for RCP8.5 are illustrated in five maps. The model spread of the relative change of NPP is high in the west China, and low in the east China. The model spread of the Γ for carbon fluxes is high in transition zone between cropping area and nomadic area and certain regions in the Tibetan Plateau. The model spread of the Γ for carbon stocks is large than 0.2 in most regions of east China as well as in the Tibetan Plateau region, while it is smaller than 0.2 only in some regions in northwest China. The model spread of the integrated metric Γ is less than 0.1 under all RCPs in the northwest arid region and southern part of the tropical and subtropical humid region, where the risk of ecosystem shift is generally low. The model spread is generally high in those areas, where the estimated risk of ecosystem shift is high. The largest model spread is related to the Tibetan Plateau region and northeast China. The standard deviation of the Γ estimates is about 0.25 under RCP8.5 scenario.

5.4 Maps

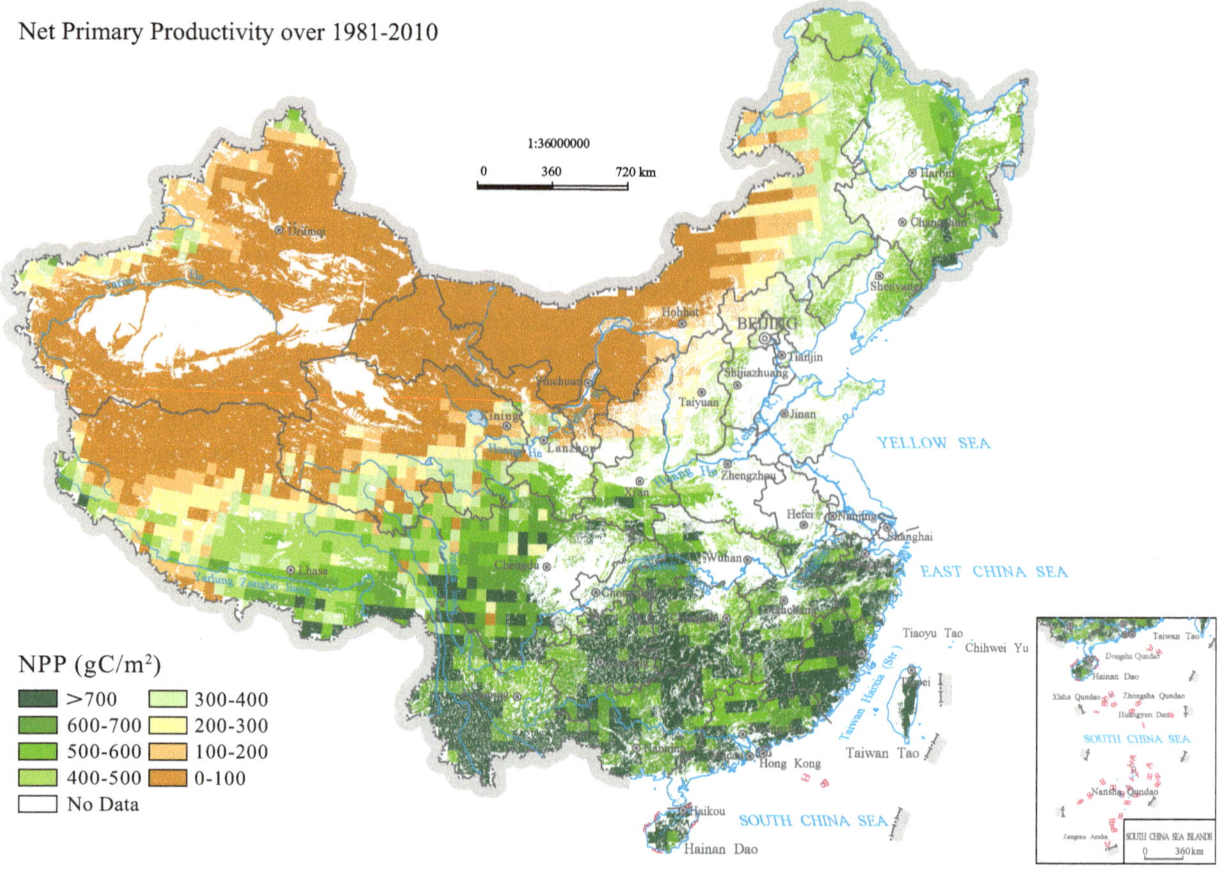
Net Primary Productivity over 1981-2010
1:36000000
0 360 720 km
NPP (gC/m²)
>700
600-700
500-600
400-500
No Data
300-400
200-300
100-200
0-100
YELLOW SEA
EAST CHINA SEA
SOUTH CHINA SEA
Hainan Dao
Taiwan Tao
Hong Kong
Tiaoyu Tao
Chihwei Yu
BEIJING
Tianjin
Shijiazhuang
Taiyuan
Jinan
Zhengzhou
Hefei
Nanjing
Shanghai
Wuhan
Chengdu
Lanzhou
Xining
Yinchuan
Hohhot
Harbin
Changchun
Shenyang
Lhasa
Urumqi
Haikou
SOUTH CHINA SEA ISLANDS
0 360km

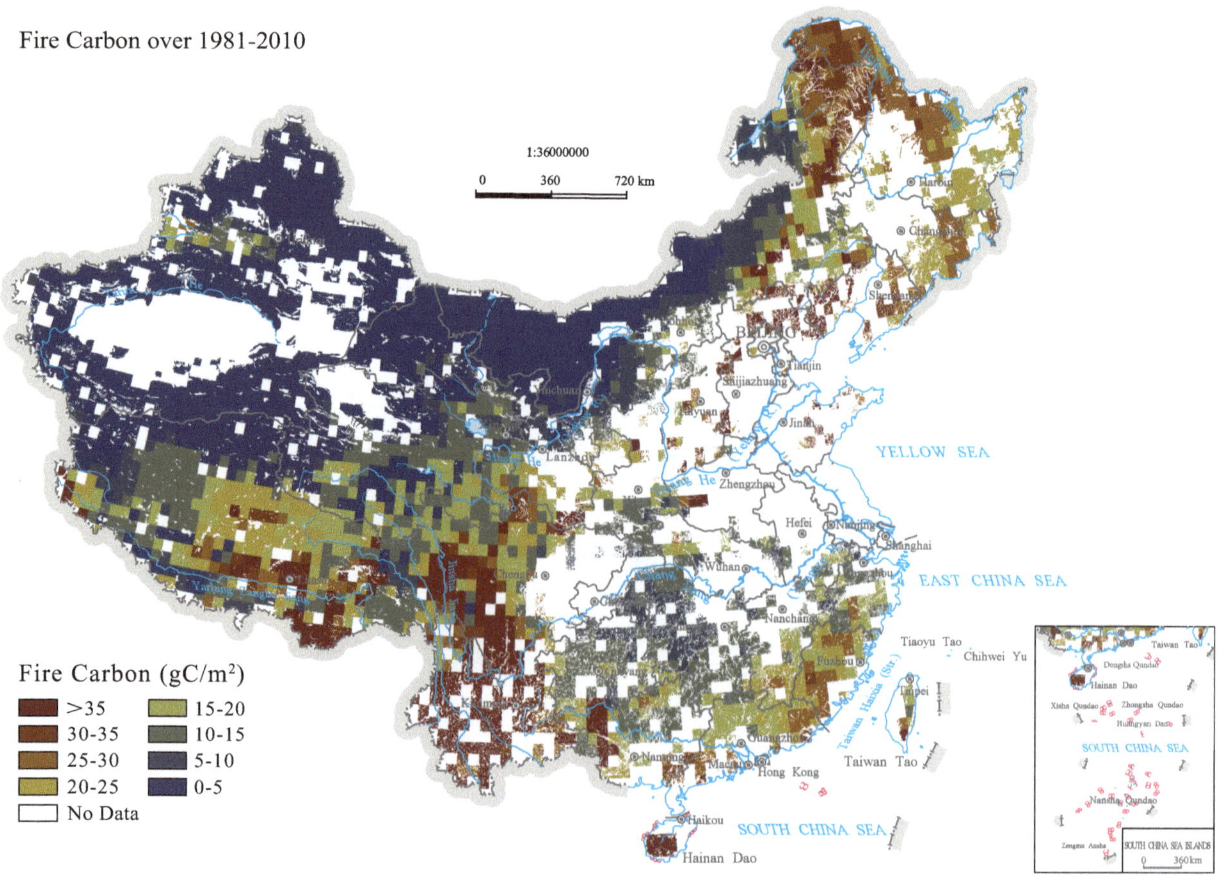
Fire Carbon over 1981-2010
1:36000000
0 360 720 km
Fire Carbon (gC/m²)
>35
30-35
25-30
20-25
No Data
15-20
10-15
5-10
0-5
YELLOW SEA
EAST CHINA SEA
SOUTH CHINA SEA
Hainan Dao
Taiwan Tao
Hong Kong
Tiaoyu Tao
Chihwei Yu
Tianjin
Shijiazhuang
Zhengzhou
Hefei
Shanghai
Wuhan
Lanzhou
Haikou
SOUTH CHINA SEA ISLANDS
0 360km

Carbon Contained in Vegetation over 1981-2010

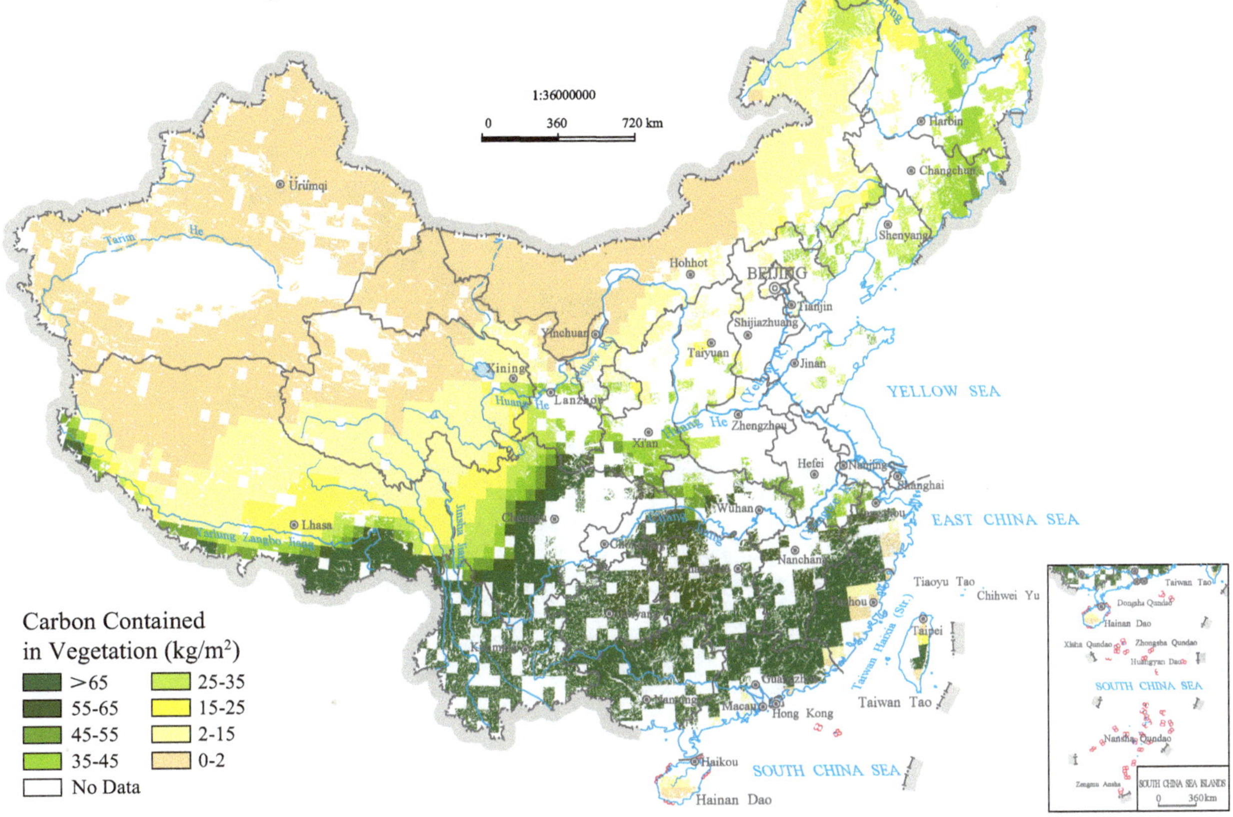

Carbon Contained in Soil over 1981-2010

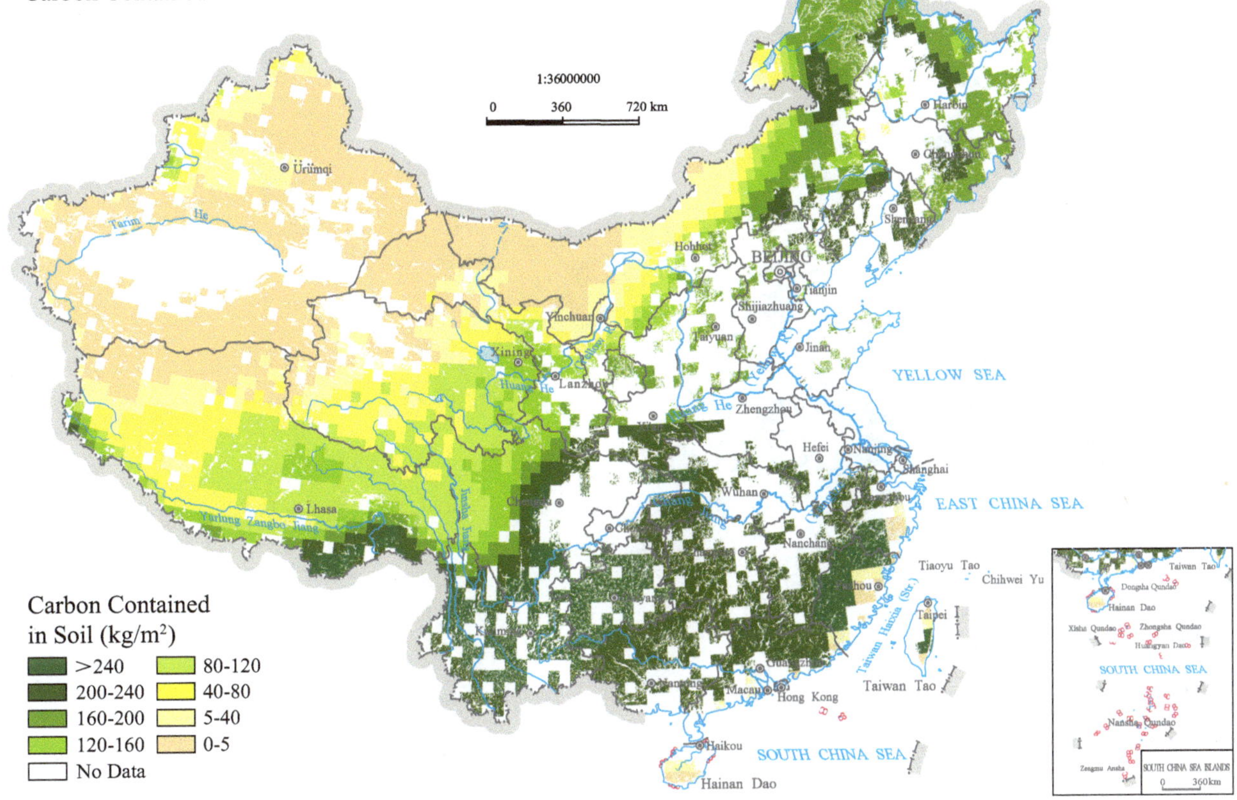

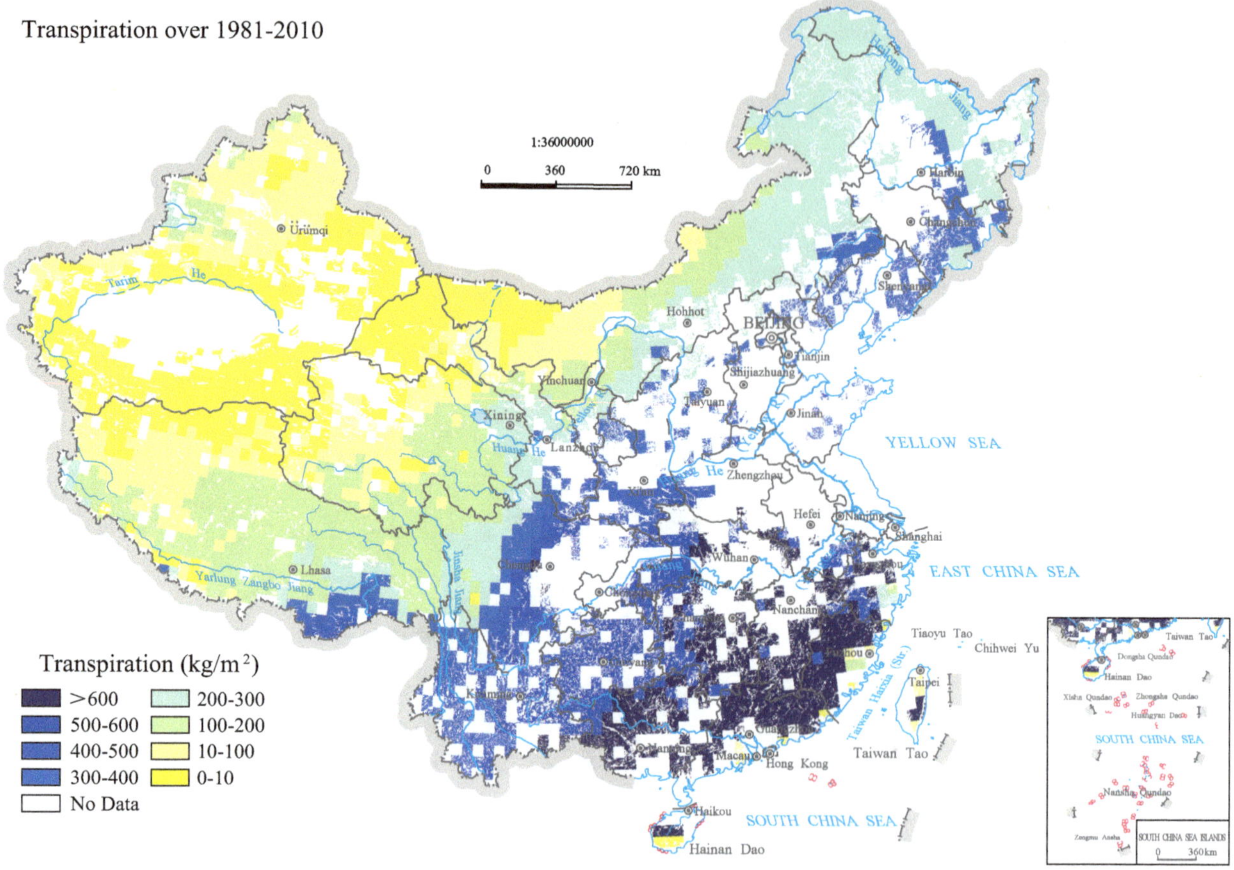
Transpiration over 1981-2010
1:36000000
0 360 720 km
Transpiration (kg/m²)
>600
500-600
400-500
300-400
No Data
200-300
100-200
10-100
0-10
YELLOW SEA
EAST CHINA SEA
SOUTH CHINA SEA
Hainan Dao
Taiwan Tao
Tiaoyu Tao
Chihwei Yu
Hong Kong
Haikou
Lhasa
Ürümqi
Hohhot
BEIJING
Tianjin
Shijiazhuang
Xining
Shanghai
Zhengzhou
Hefei
Taipei
Macau
Tarim He
Yarlung Zangbo Jiang
SOUTH CHINA SEA ISLANDS

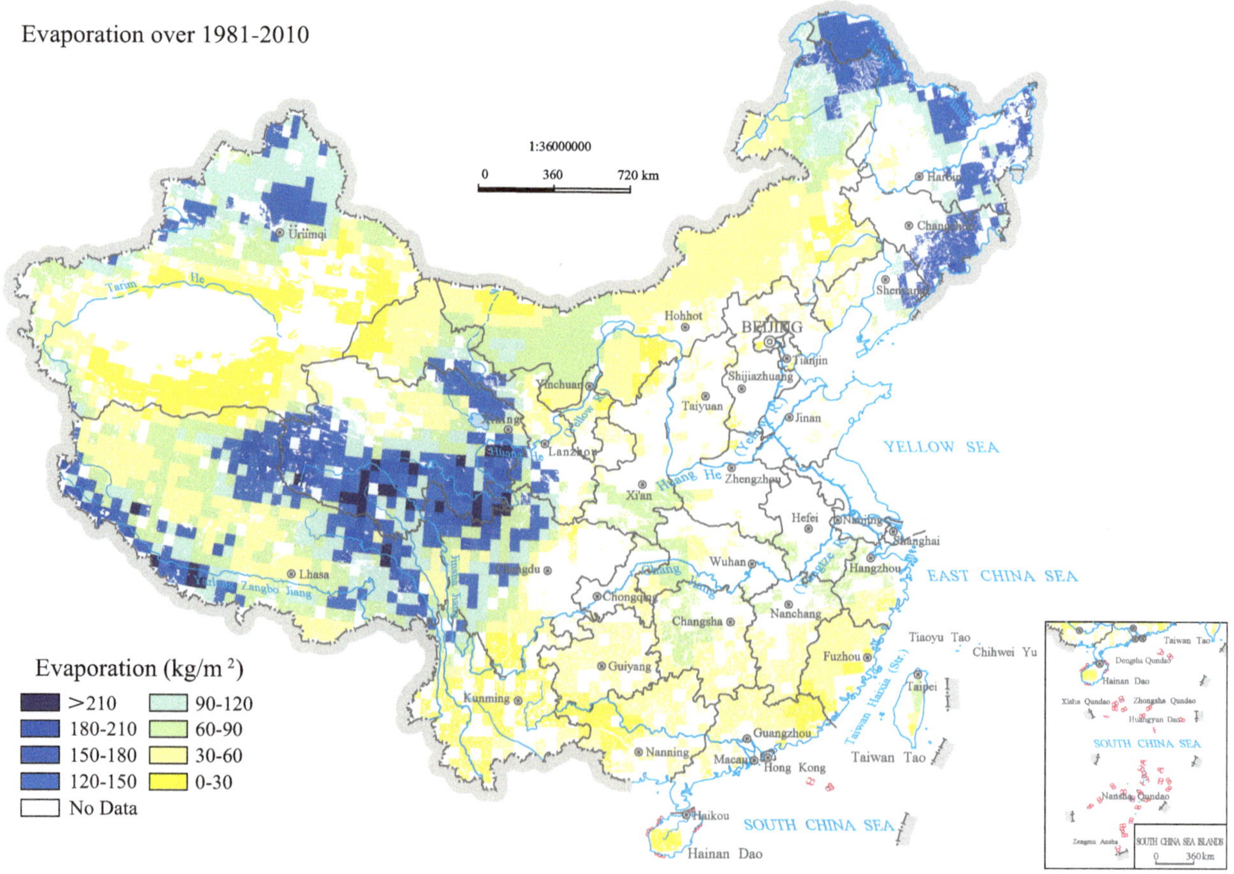
Evaporation over 1981-2010
1:36000000
0 360 720 km
Evaporation (kg/m²)
>210
180-210
150-180
120-150
No Data
90-120
60-90
30-60
0-30
YELLOW SEA
EAST CHINA SEA
SOUTH CHINA SEA
Hainan Dao
Taiwan Tao
Tiaoyu Tao
Chihwei Yu
Hong Kong
Haikou
Lhasa
Ürümqi
Hohhot
BEIJING
Tianjin
Shijiazhuang
Taiyuan
Jinan
Zhengzhou
Hefei
Shanghai
Hangzhou
Wuhan
Nanchang
Changsha
Chongqing
Guiyang
Kunming
Nanning
Guangzhou
Fuzhou
Taipei
Macau
Lanzhou
Xi'an
Harbin
Shenyang
Yinchuan
Tarim He
SOUTH CHINA SEA ISLANDS

Runoff over 1981-2010

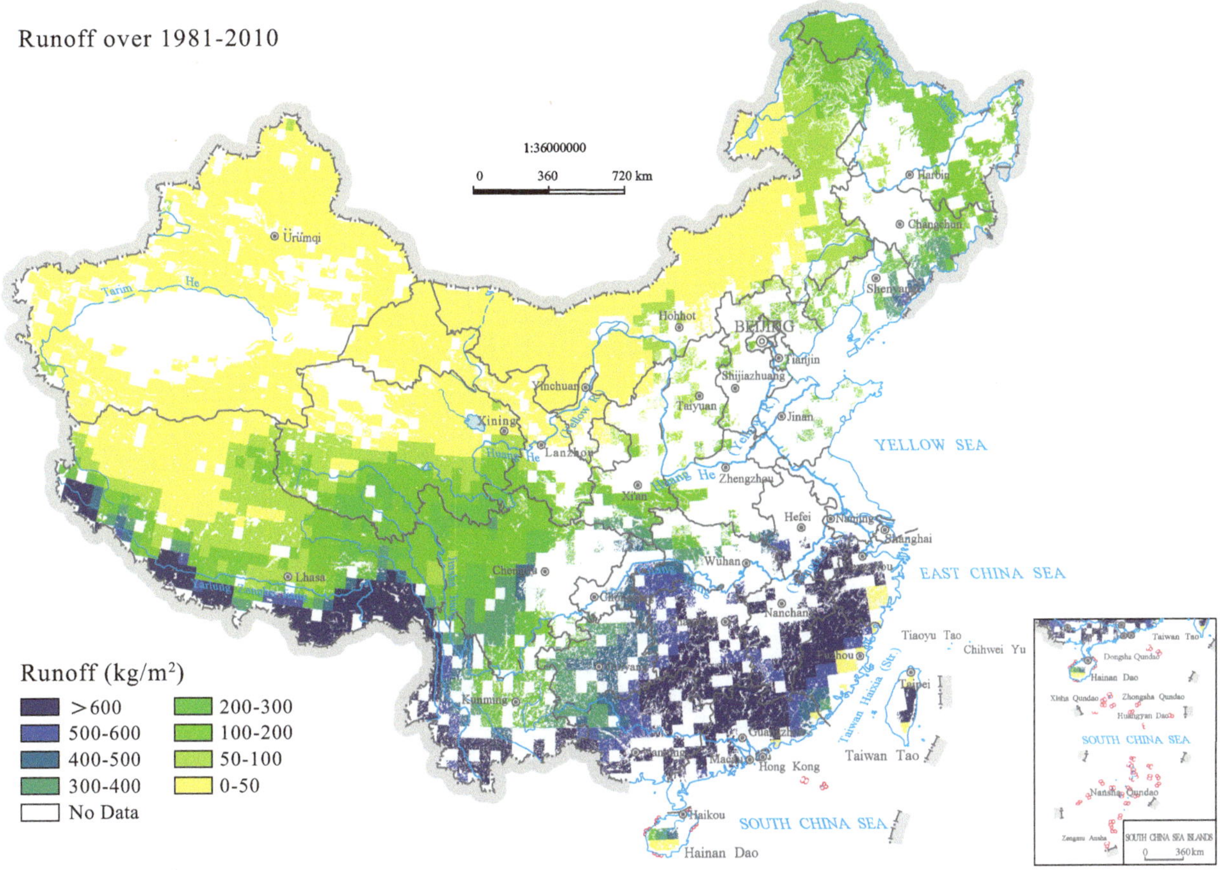

Soil Water Content over 1981-2010

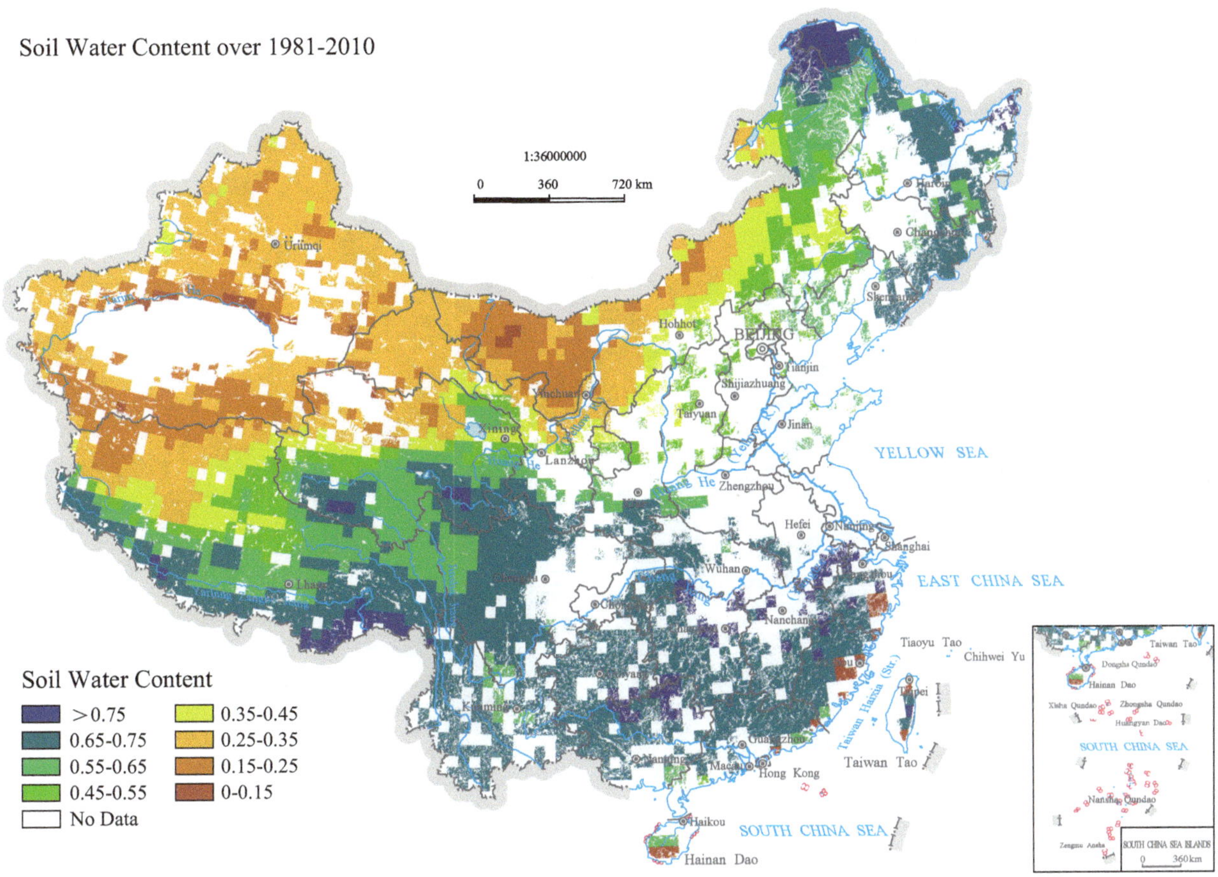

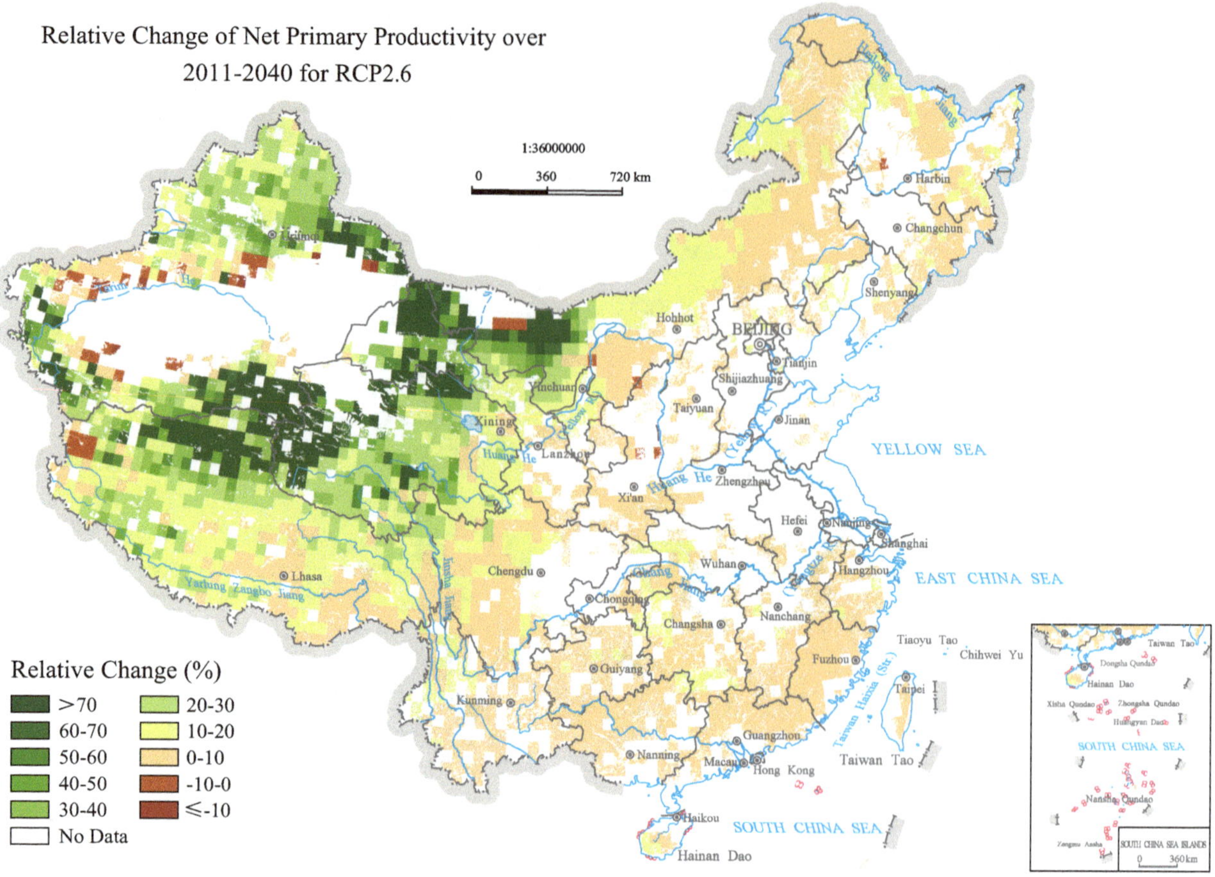
Relative Change of Net Primary Productivity over
2011-2040 for RCP2.6
1:36000000
0 360 720 km
Relative Change (%)
>70
60-70
50-60
40-50
30-40
20-30
10-20
0-10
-10-0
≤-10
No Data
Harbin
Changchun
Shenyang
Hohhot
BEIJING
Tianjin
Shijiazhuang
Taiyuan
Yinchuan
Xining
Lanzhou
Jinan
Zhengzhou
Xi'an
Hefei
Nanjing
Shanghai
Wuhan
Hangzhou
Chengdu
Chongqing
Lhasa
Changsha
Nanchang
Fuzhou
Guiyang
Kunming
Nanning
Guangzhou
Macau
Hong Kong
Haikou
Taipei
YELLOW SEA
EAST CHINA SEA
SOUTH CHINA SEA
Tiaoyu Tao
Chihwei Yu
Taiwan Tao
Hainan Dao

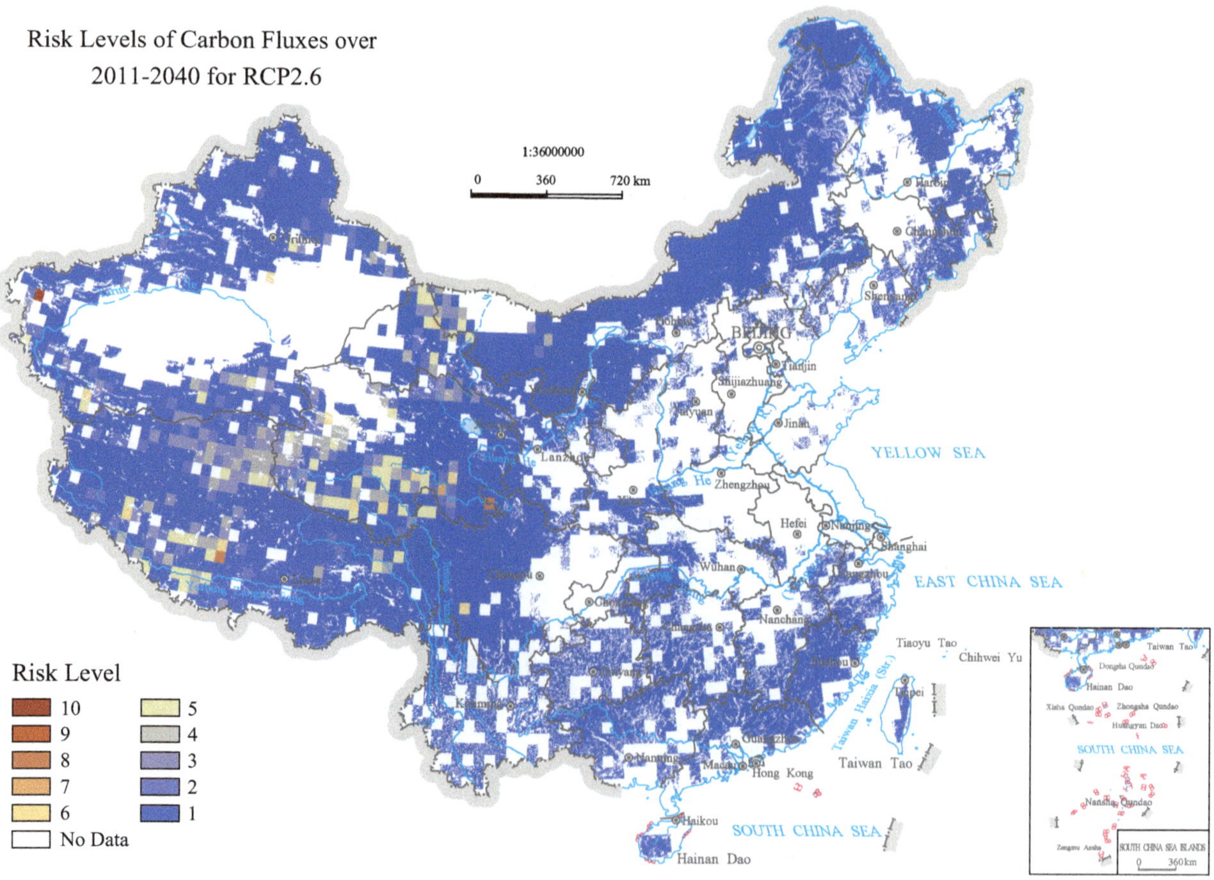
Risk Levels of Carbon Fluxes over
2011-2040 for RCP2.6
1:36000000
0 360 720 km
Risk Level
10
9
8
7
6
5
4
3
2
1
No Data
BEIJING
Tianjin
Shijiazhuang
Jinan
Zhengzhou
Hefei
Shanghai
Wuhan
Hong Kong
Haikou
YELLOW SEA
EAST CHINA SEA
SOUTH CHINA SEA
Tiaoyu Tao
Chihwei Yu
Taiwan Tao
Hainan Dao

Risk Levels of Carbon Stocks over 2011-2040 for RCP2.6

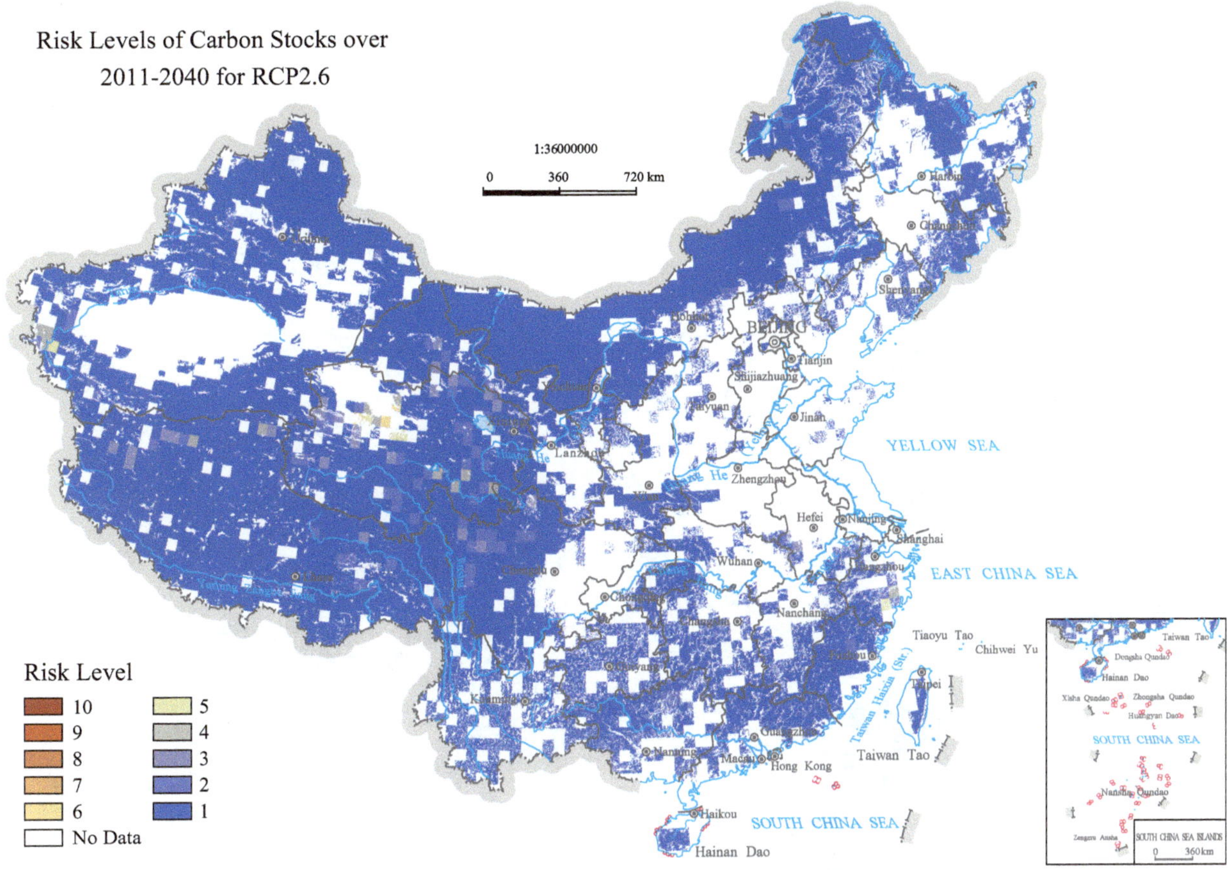

Risk Levels of Water Fluxes over 2011-2040 for RCP2.6

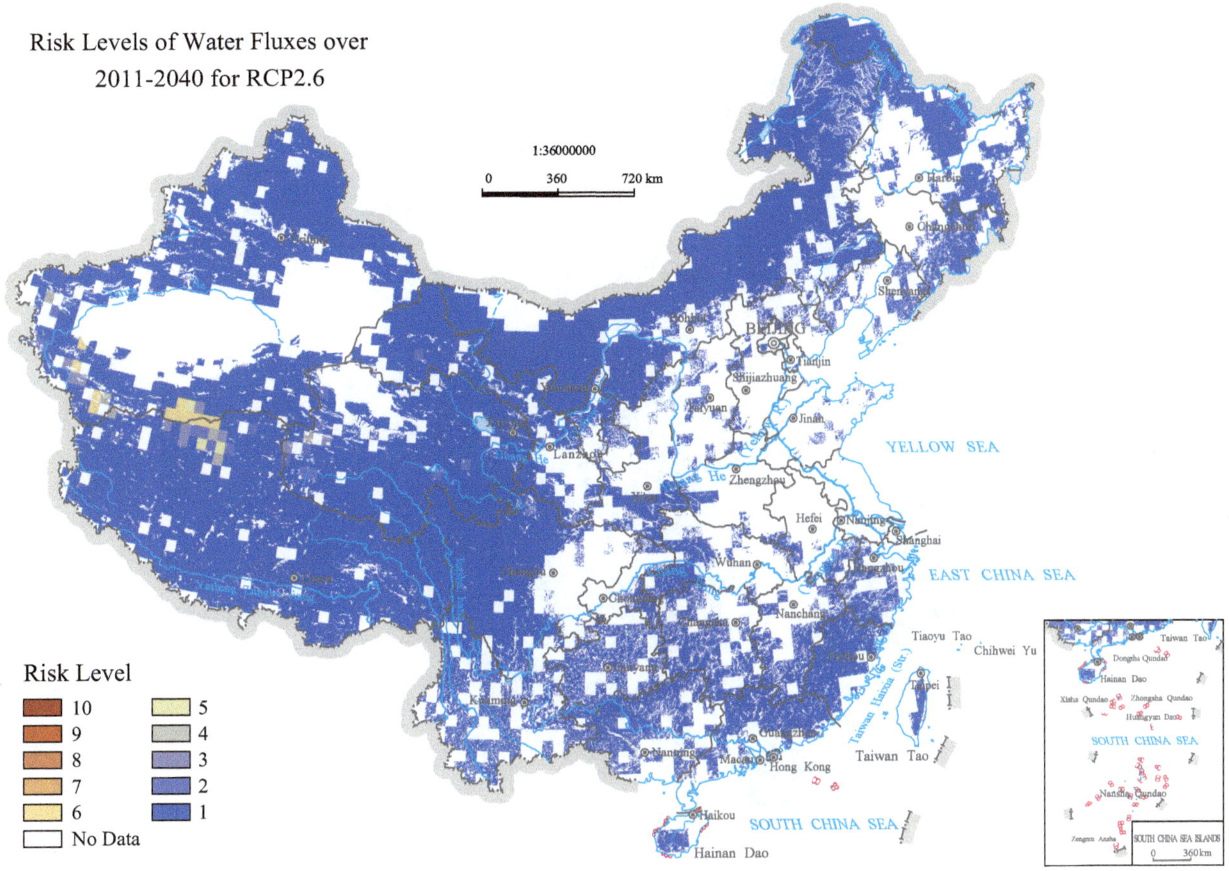

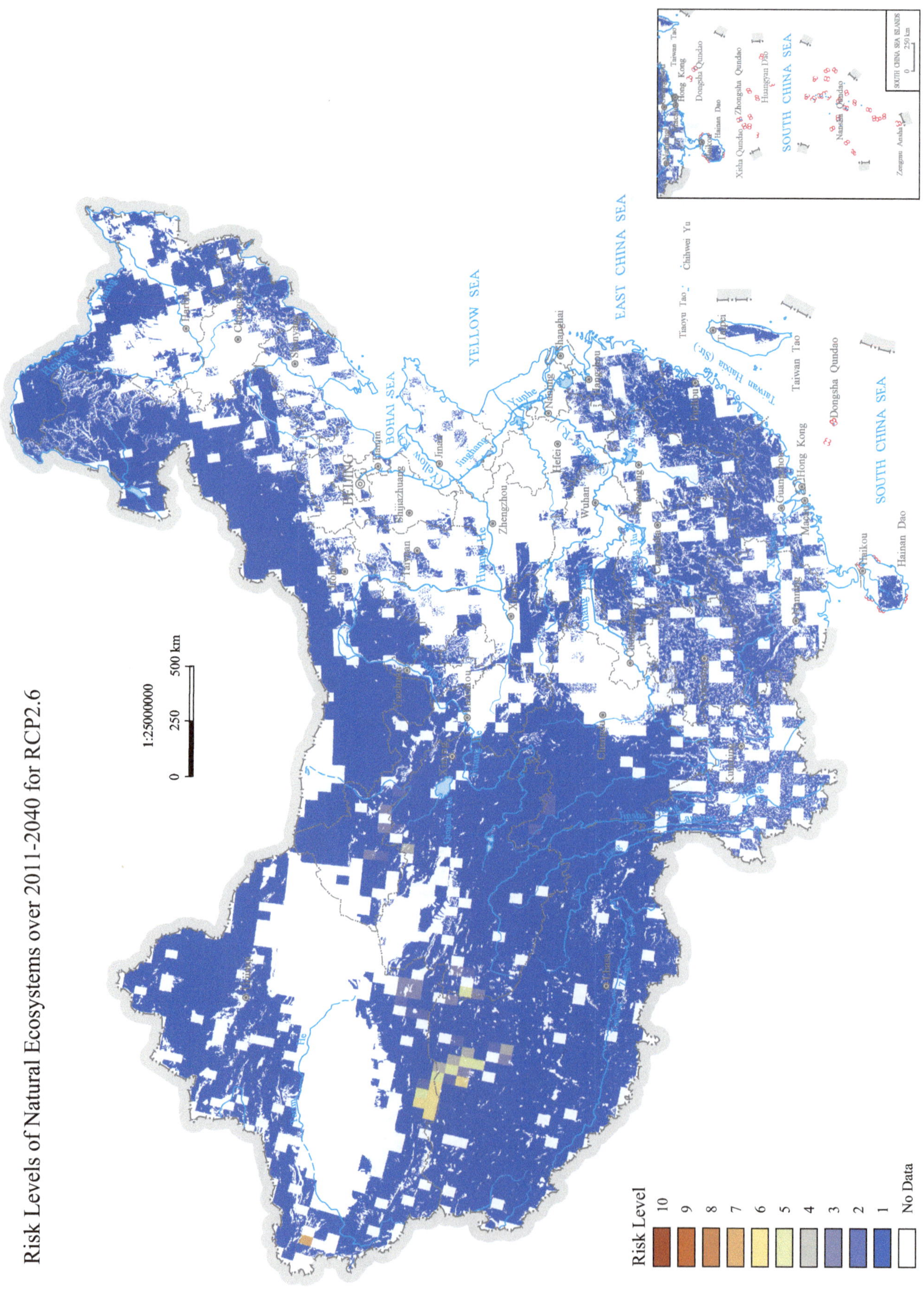
Risk Levels of Natural Ecosystems over 2011-2040 for RCP2.6
1:25000000
0 250 500 km
Risk Level
10
9
8
7
6
5
4
3
2
1
No Data
BOHAI SEA
YELLOW SEA
EAST CHINA SEA
SOUTH CHINA SEA
BEIJING
Shanghai
Taiwan Tao
Hainan Dao

Relative Change of Net Primary Productivity over 2041-2070 for RCP2.6

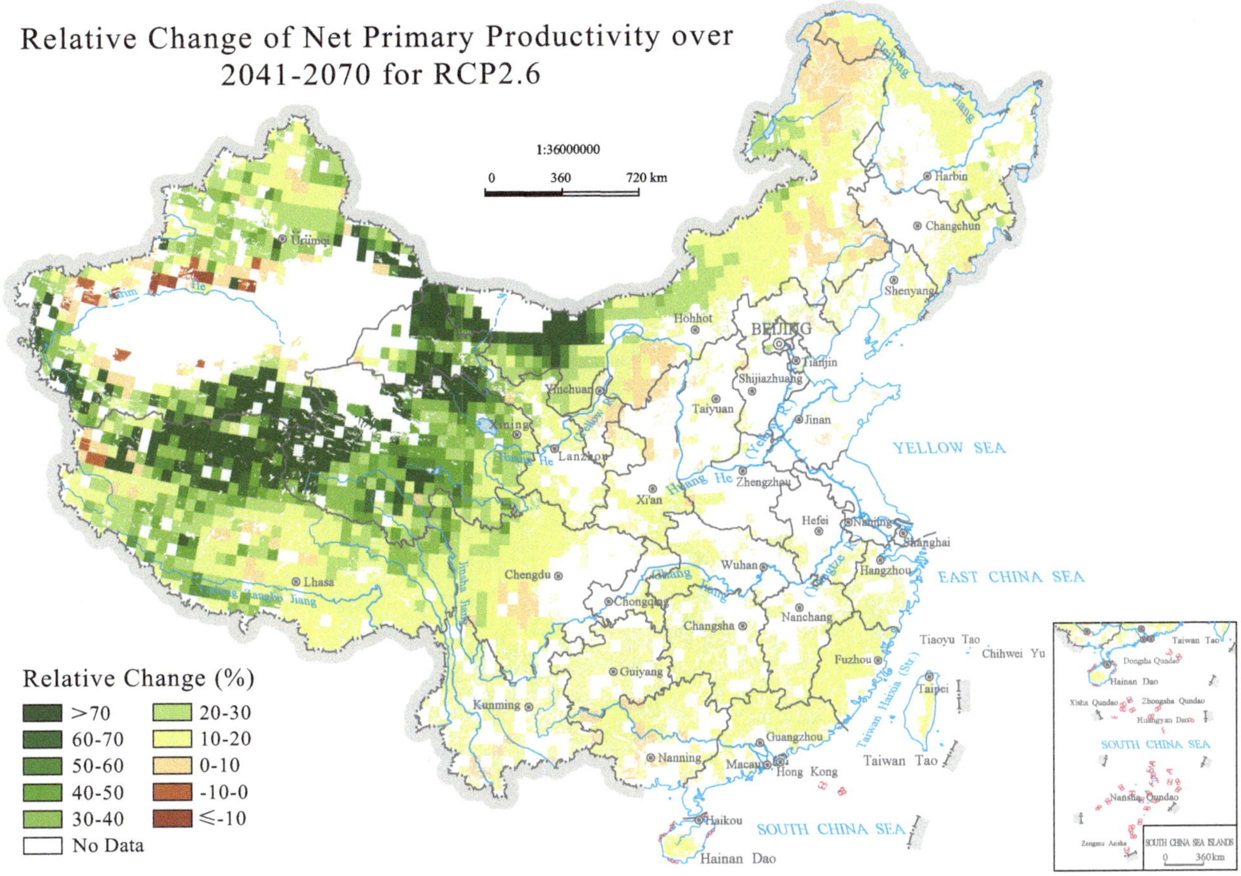

Risk Levels of Carbon Fluxes over 2041-2070 for RCP2.6

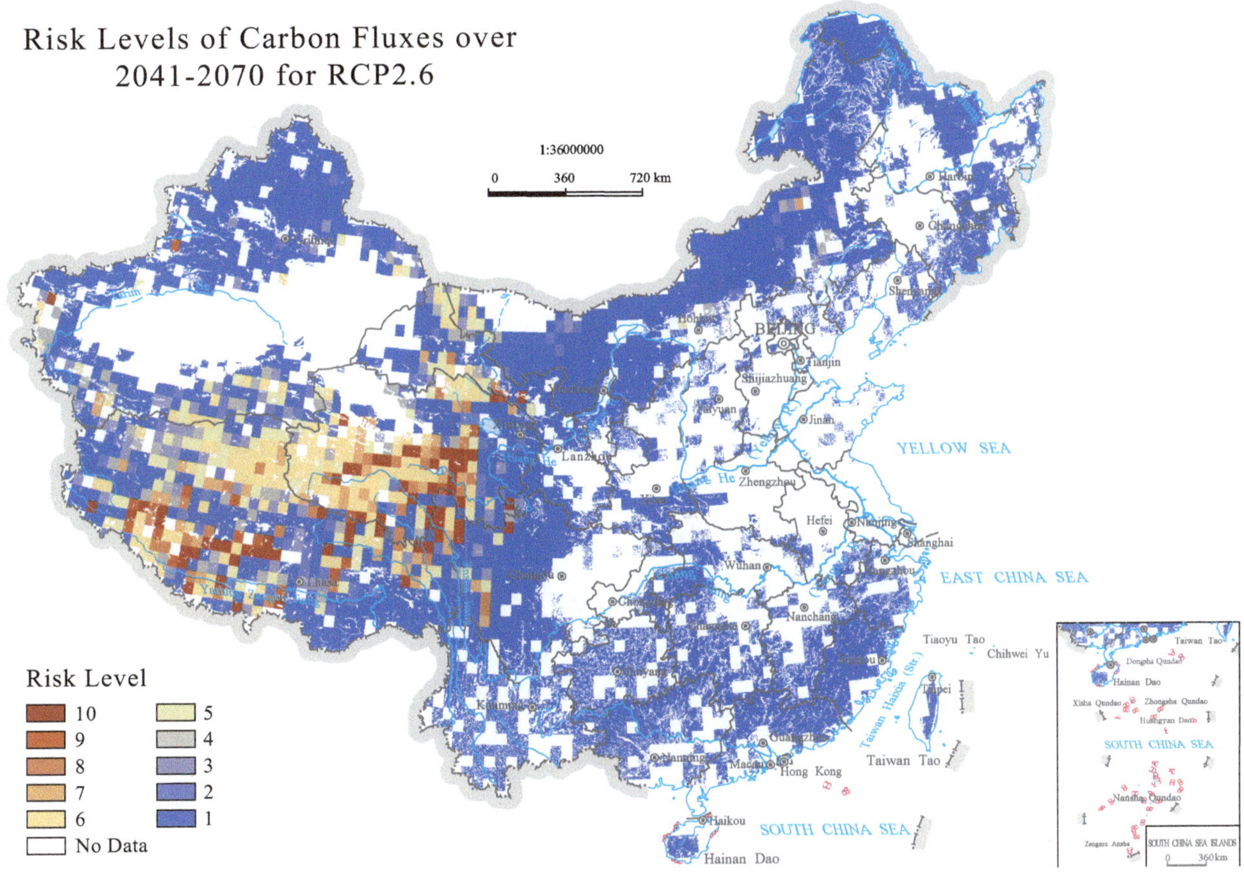

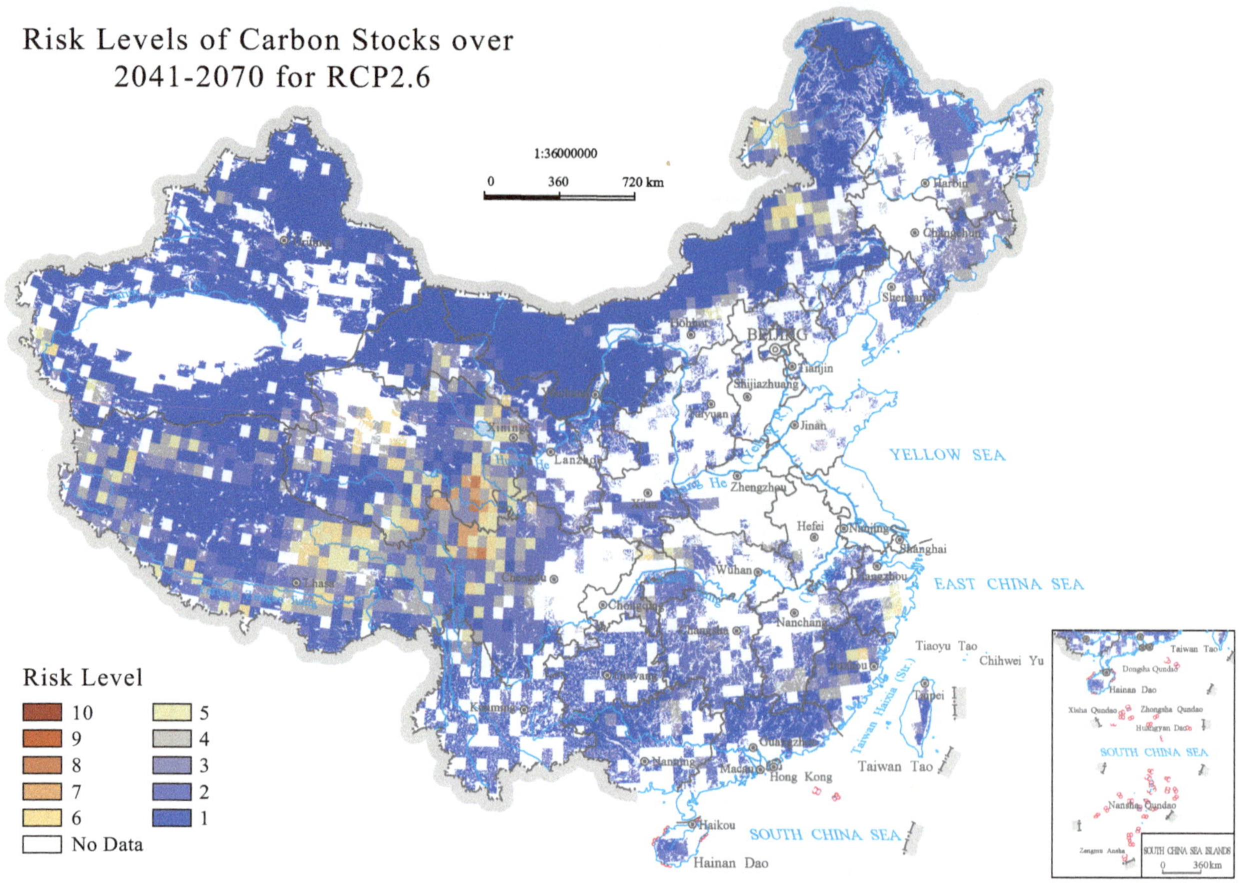
Risk Levels of Carbon Stocks over
2041-2070 for RCP2.6
1:36000000
0 360 720 km
Risk Level
10
9
8
7
6
5
4
3
2
1
No Data
BEIJING
Tianjin
Shijiazhuang
Jinan
Zhengzhou
Hefei
Nanjing
Shanghai
Wuhan
Nanchang
Lanzhou
Harbin
Hong Kong
Macao
Haikou
Hainan Dao
Taiwan Tao
Tiaoyu Tao
Chihwei Yu
YELLOW SEA
EAST CHINA SEA
SOUTH CHINA SEA

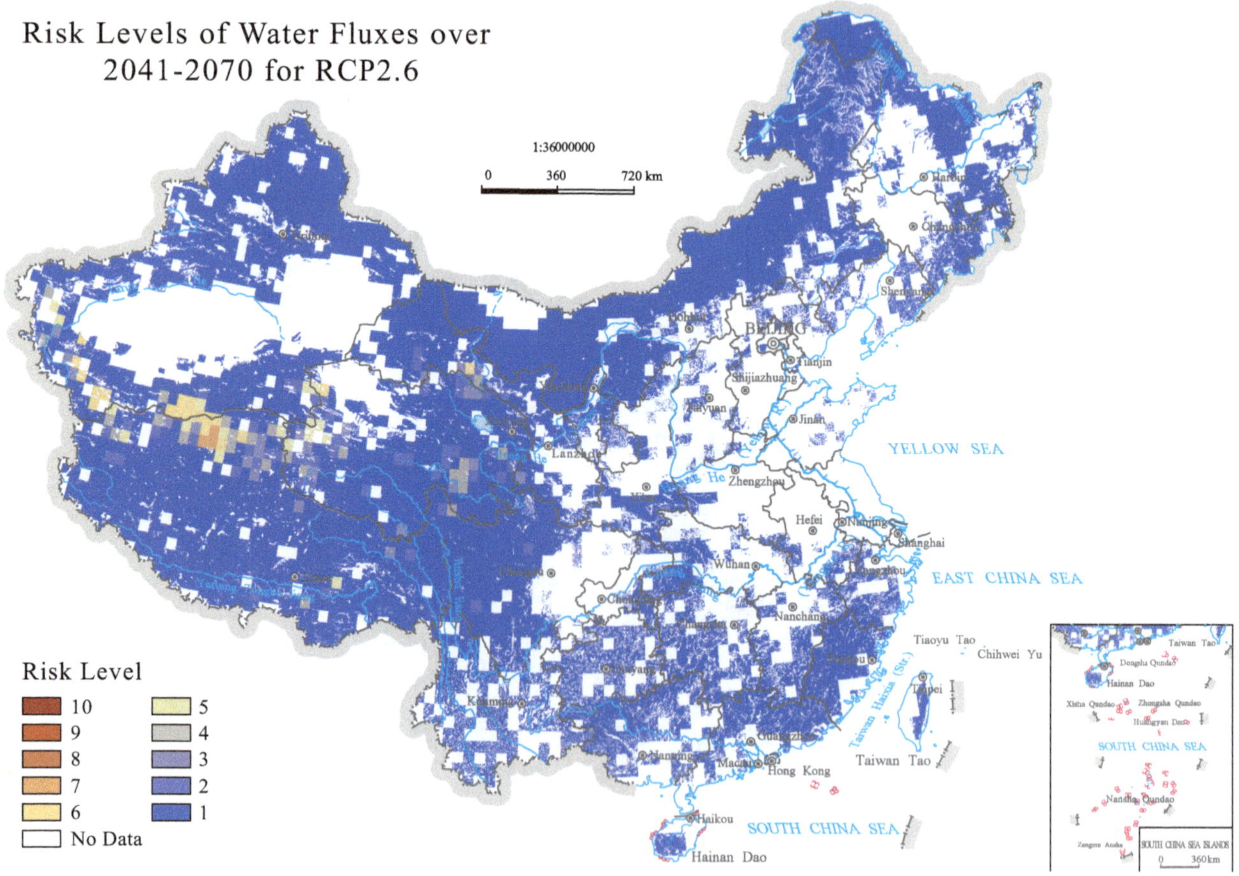
Risk Levels of Water Fluxes over
2041-2070 for RCP2.6
1:36000000
0 360 720 km
Risk Level
10
9
8
7
6
5
4
3
2
1
No Data
Tianjin
Shijiazhuang
Zhengzhou
Hefei
Shanghai
Wuhan
Hong Kong
Haikou
Hainan Dao
Taiwan Tao
Tiaoyu Tao
Chihwei Yu
YELLOW SEA
EAST CHINA SEA
SOUTH CHINA SEA

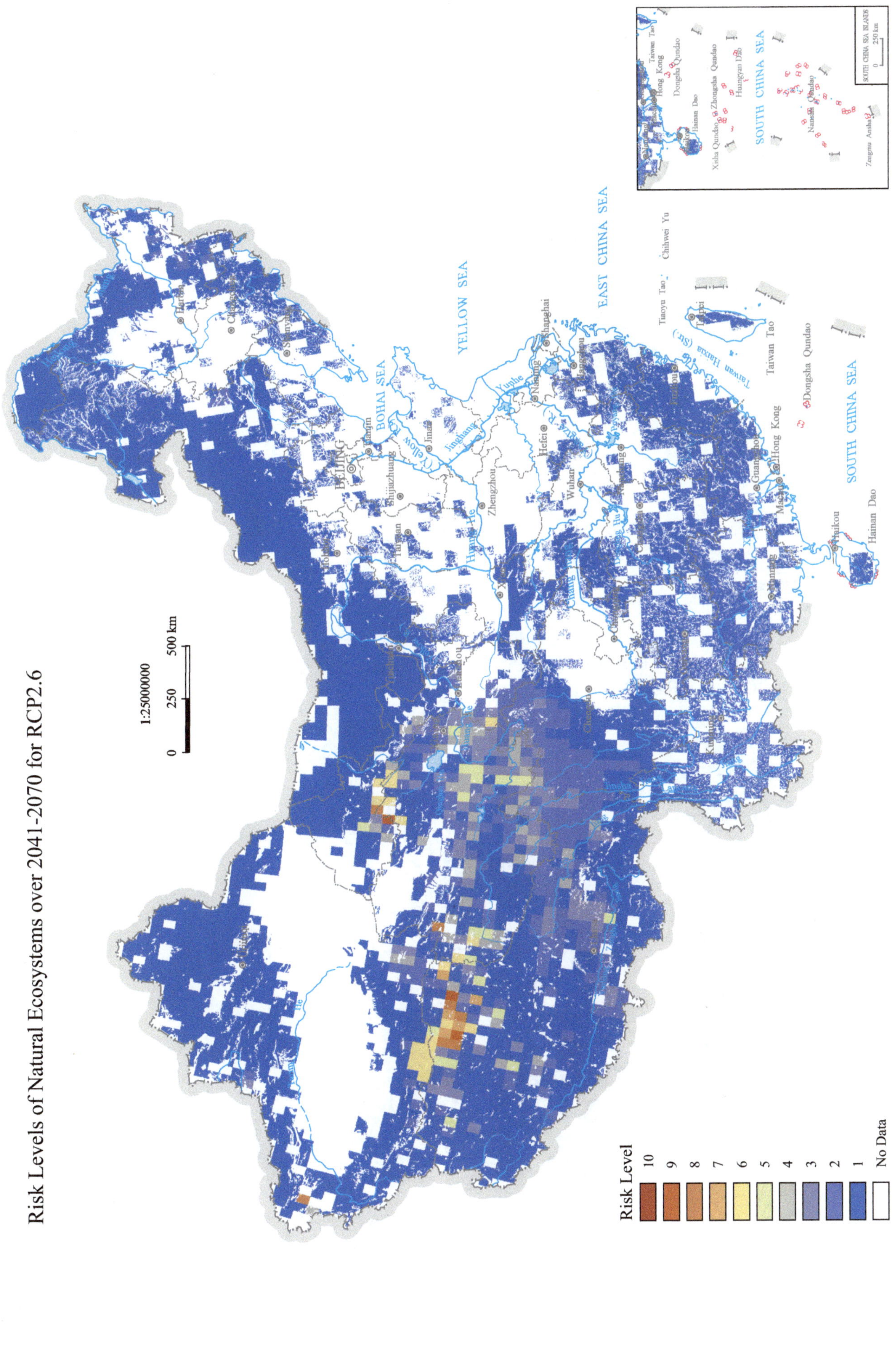

Risk Levels of Natural Ecosystems over 2041-2070 for RCP2.6

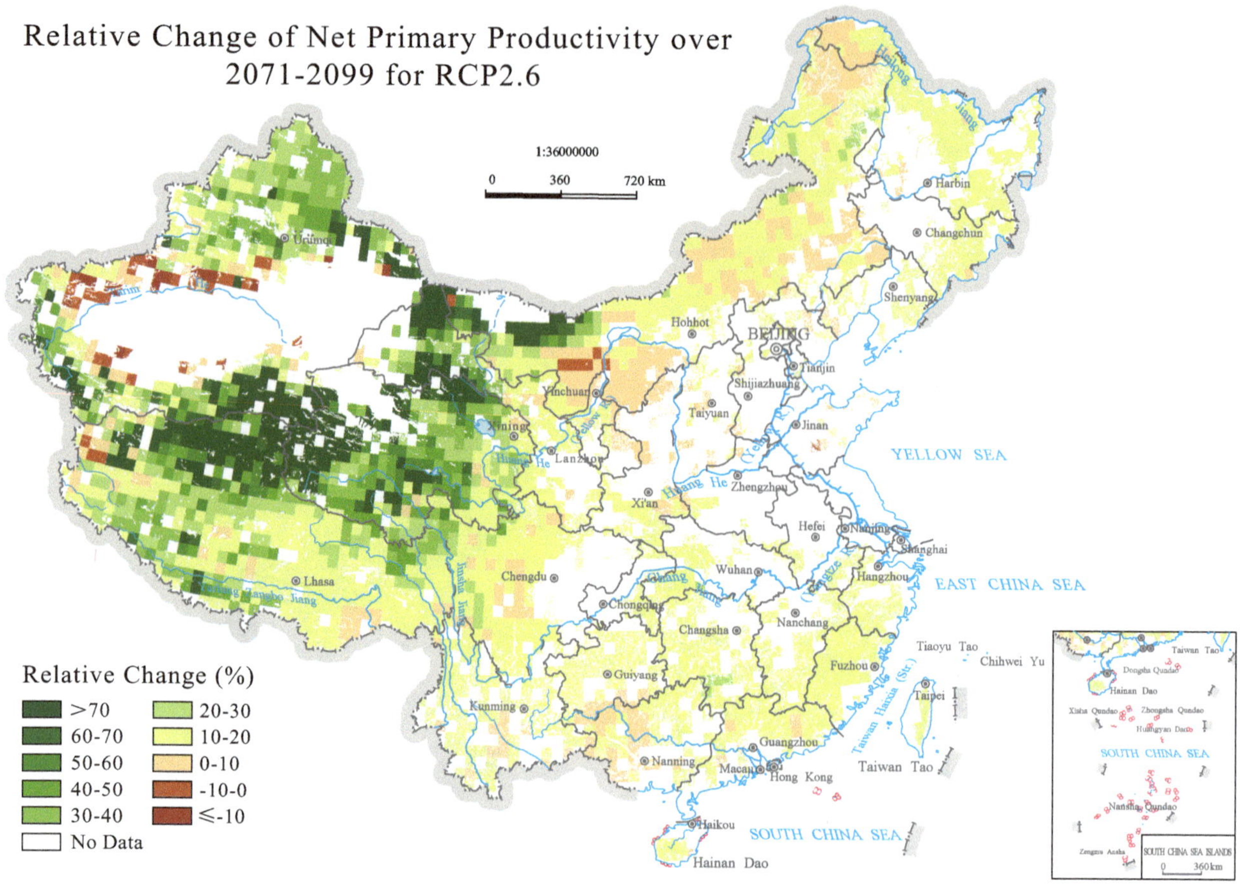
Relative Change of Net Primary Productivity over 2071-2099 for RCP2.6
1:36000000
0 360 720 km
Relative Change (%)
>70
60-70
50-60
40-50
30-40
20-30
10-20
0-10
-10-0
≤-10
No Data
YELLOW SEA
EAST CHINA SEA
SOUTH CHINA SEA
Harbin
Changchun
Shenyang
Hohhot
BEIJING
Tianjin
Shijiazhuang
Taiyuan
Jinan
Yinchuan
Xining
Lanzhou
Zhengzhou
Xi'an
Hefei
Nanjing
Shanghai
Wuhan
Hangzhou
Chengdu
Chongqing
Changsha
Nanchang
Lhasa
Guiyang
Kunming
Fuzhou
Taipei
Nanning
Guangzhou
Macau
Hong Kong
Haikou
Hainan Dao
Taiwan Tao
Tiaoyu Tao
Chihwei Yu

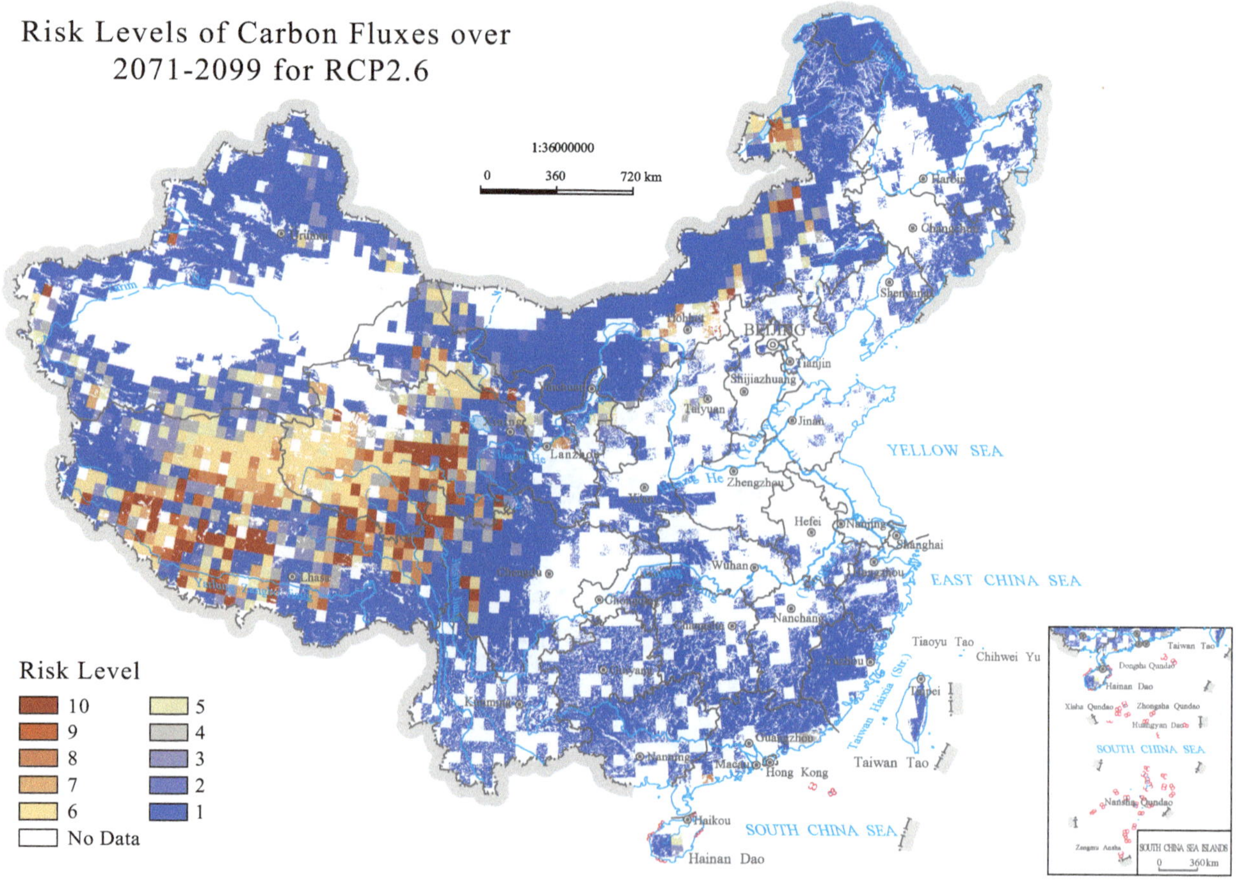
Risk Levels of Carbon Fluxes over 2071-2099 for RCP2.6
1:36000000
0 360 720 km
Risk Level
10
9
8
7
6
5
4
3
2
1
No Data
YELLOW SEA
EAST CHINA SEA
SOUTH CHINA SEA
BEIJING
Tianjin
Shijiazhuang
Taiyuan
Jinan
Zhengzhou
Hefei
Nanjing
Shanghai
Wuhan
Hong Kong
Haikou
Hainan Dao
Taiwan Tao
Tiaoyu Tao
Chihwei Yu

Risk Levels of Carbon Stocks over 2071-2099 for RCP2.6

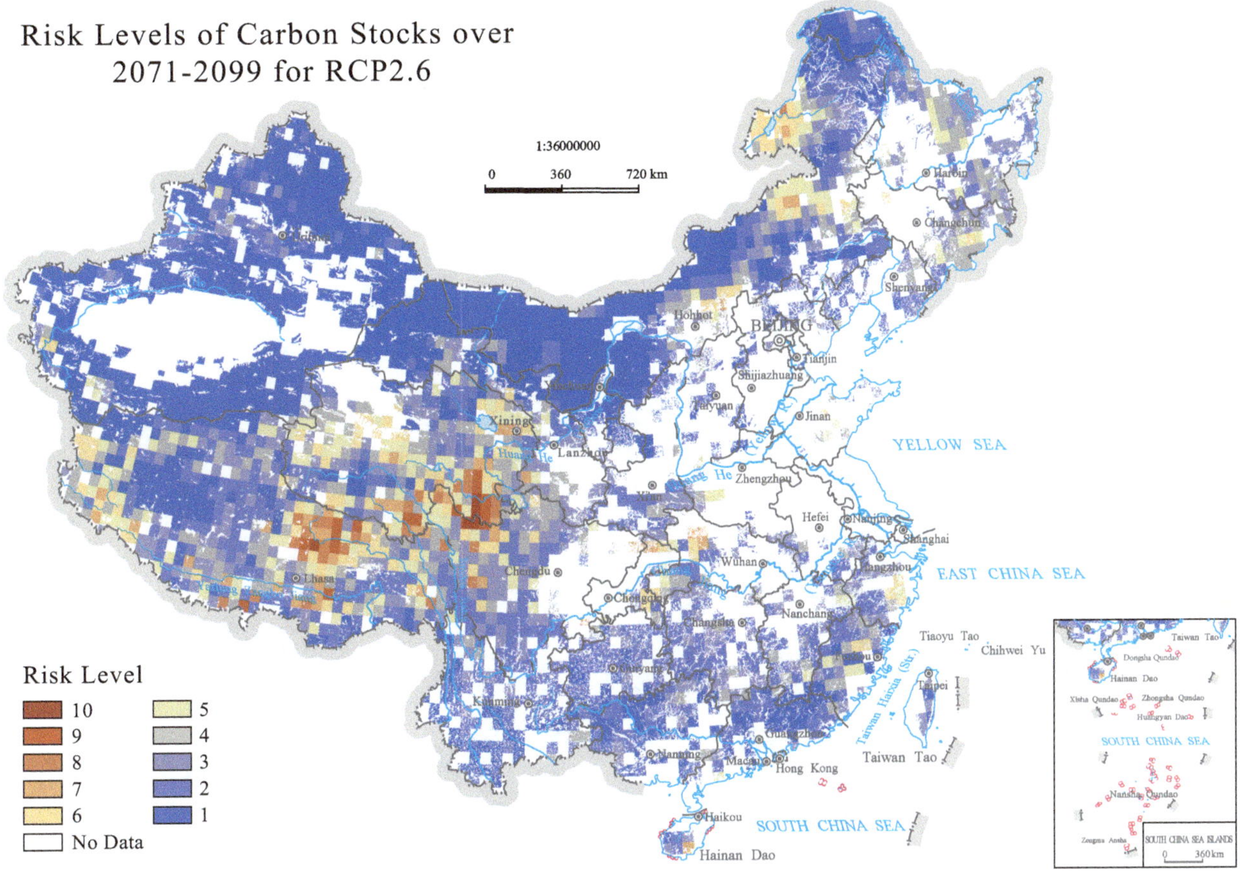

Risk Levels of Water Fluxes over 2071-2099 for RCP2.6

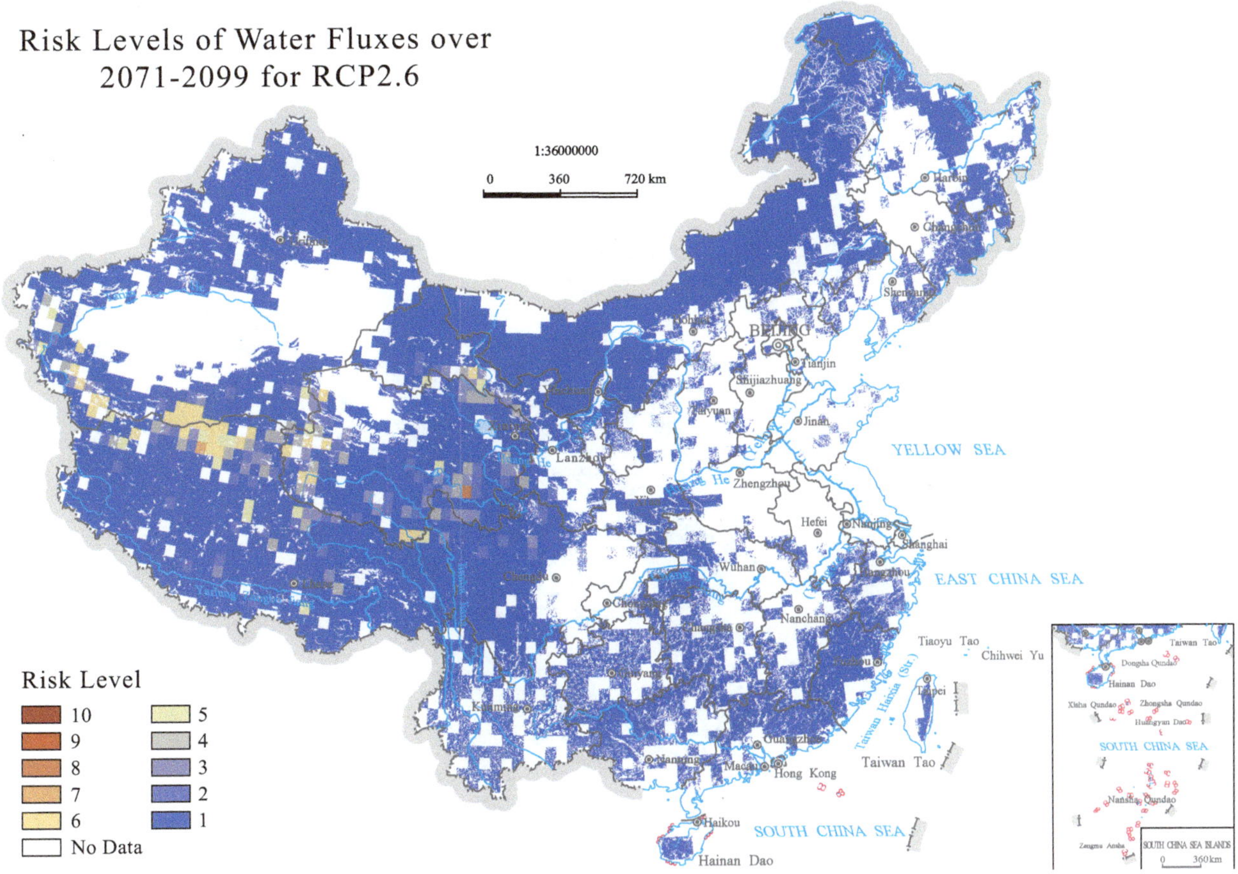

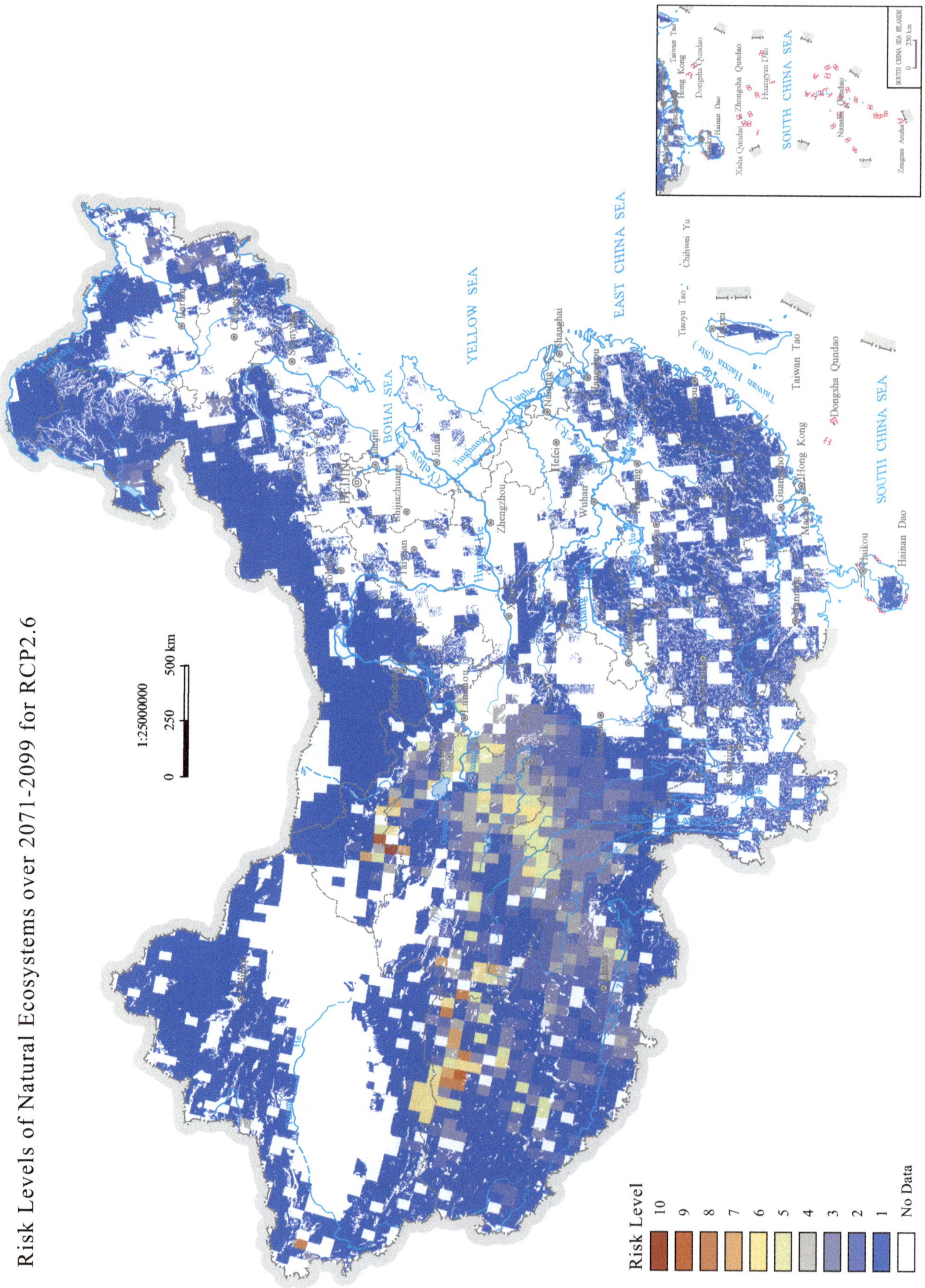
Risk Levels of Natural Ecosystems over 2071-2099 for RCP2.6
1:25000000
0 250 500 km
Risk Level
10
9
8
7
6
5
4
3
2
1
No Data
BEIJING
YELLOW SEA
BOHAI SEA
EAST CHINA SEA
SOUTH CHINA SEA
Taiwan Tao
Dongsha Qundao
Hainan Dao
Shanghai
Zhengzhou
Hong Kong

Relative Change of Net Primary Productivity over 2011-2040 for RCP4.5

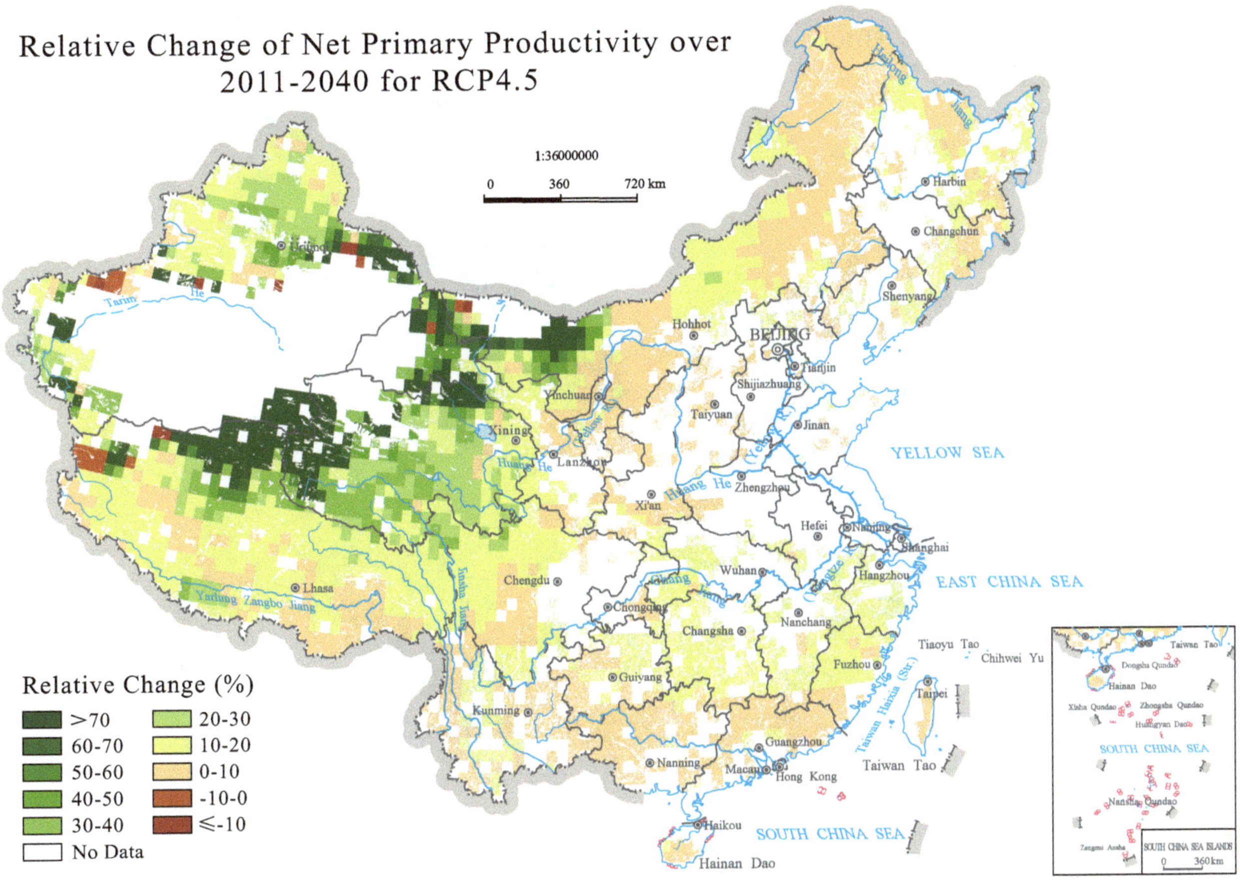

Risk Levels of Carbon Fluxes over 2011-2040 for RCP4.5

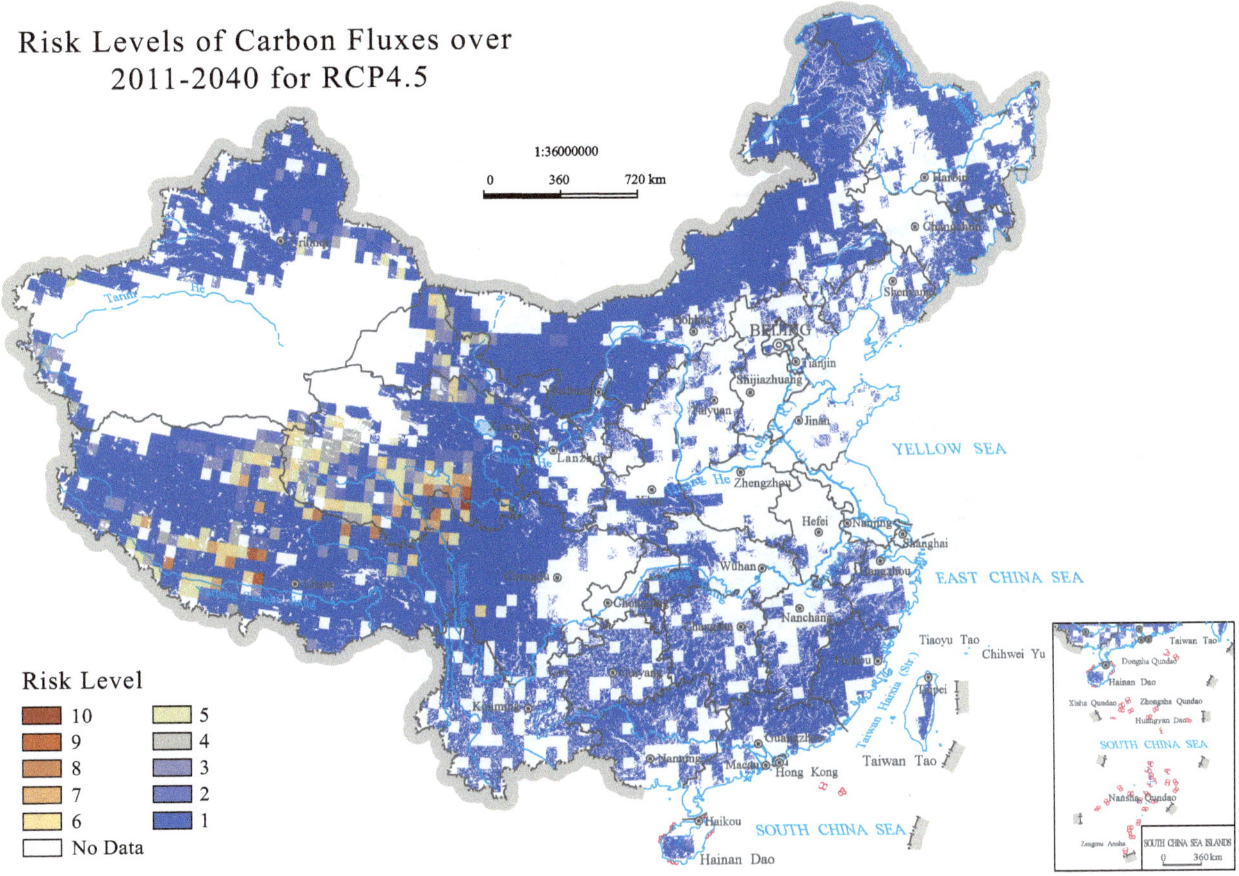

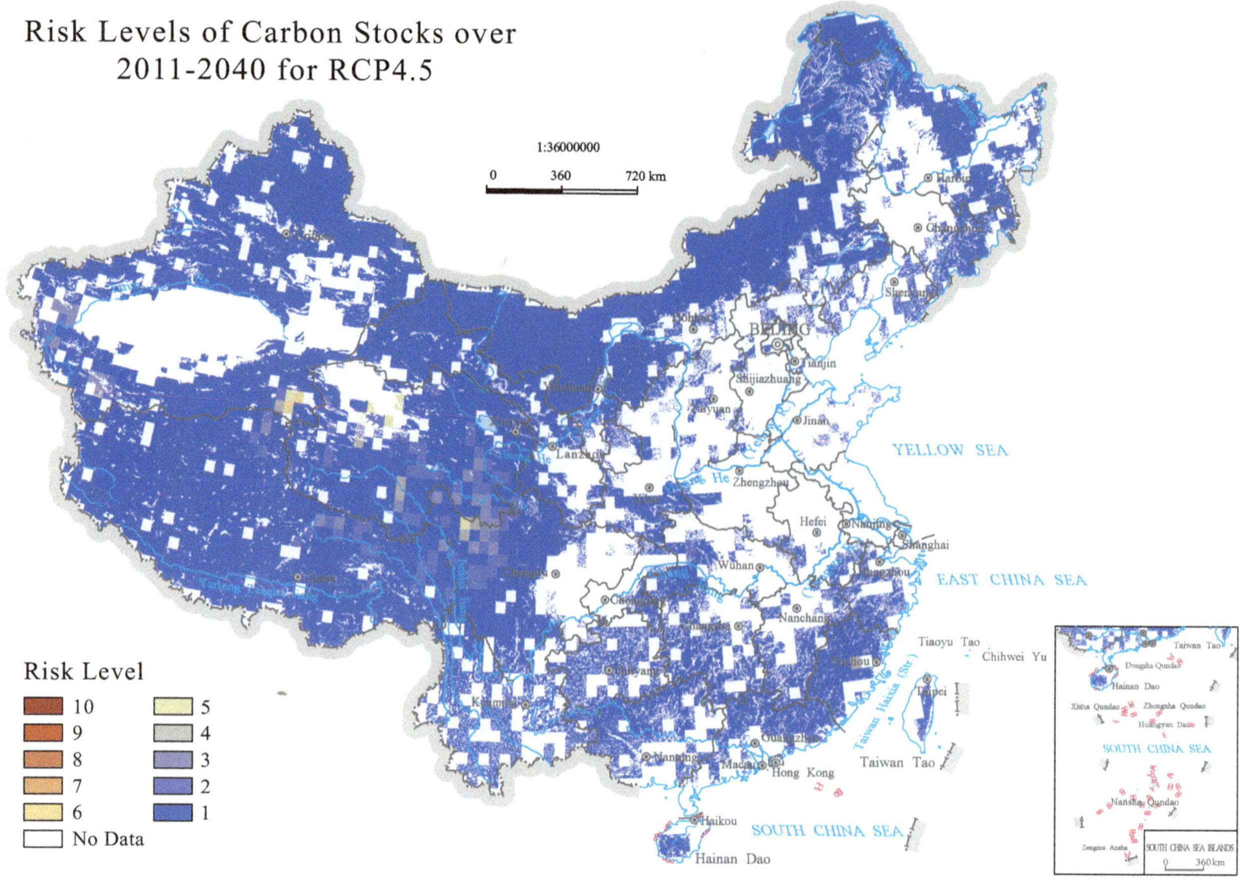
Risk Levels of Carbon Stocks over 2011-2040 for RCP4.5
1:36000000
0 360 720 km
YELLOW SEA
EAST CHINA SEA
SOUTH CHINA SEA
Tiaoyu Tao
Chihwei Yu
Taiwan Tao
Hainan Dao
Hong Kong
Haikou
Shanghai
Zhengzhou
Hefei
Wuhan
Nanchang
Lanzhou
Jinan
Tianjin
Shijiazhuang
Taiyuan
BEIJING
Risk Level
10
9
8
7
6
5
4
3
2
1
No Data
SOUTH CHINA SEA
SOUTH CHINA SEA ISLANDS

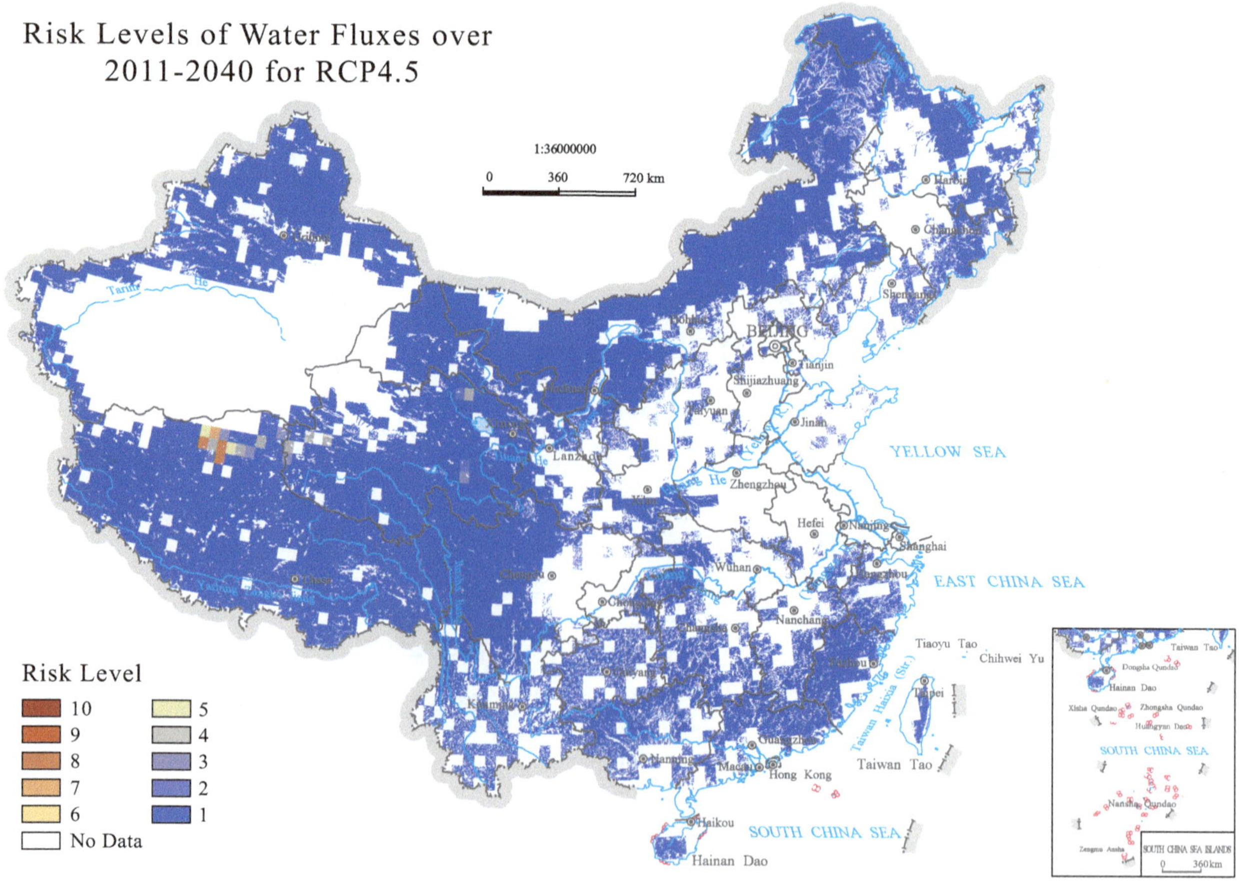
Risk Levels of Water Fluxes over 2011-2040 for RCP4.5
1:36000000
0 360 720 km
YELLOW SEA
EAST CHINA SEA
SOUTH CHINA SEA
Tiaoyu Tao
Chihwei Yu
Taiwan Tao
Hainan Dao
Hong Kong
Haikou
Shanghai
Zhengzhou
Hefei
Wuhan
Nanchang
Lanzhou
Jinan
Tianjin
Shijiazhuang
Taiyuan
BEIJING
Risk Level
10
9
8
7
6
5
4
3
2
1
No Data
SOUTH CHINA SEA
SOUTH CHINA SEA ISLANDS

Risk Levels of Natural Ecosystems over 2011-2040 for RCP4.5

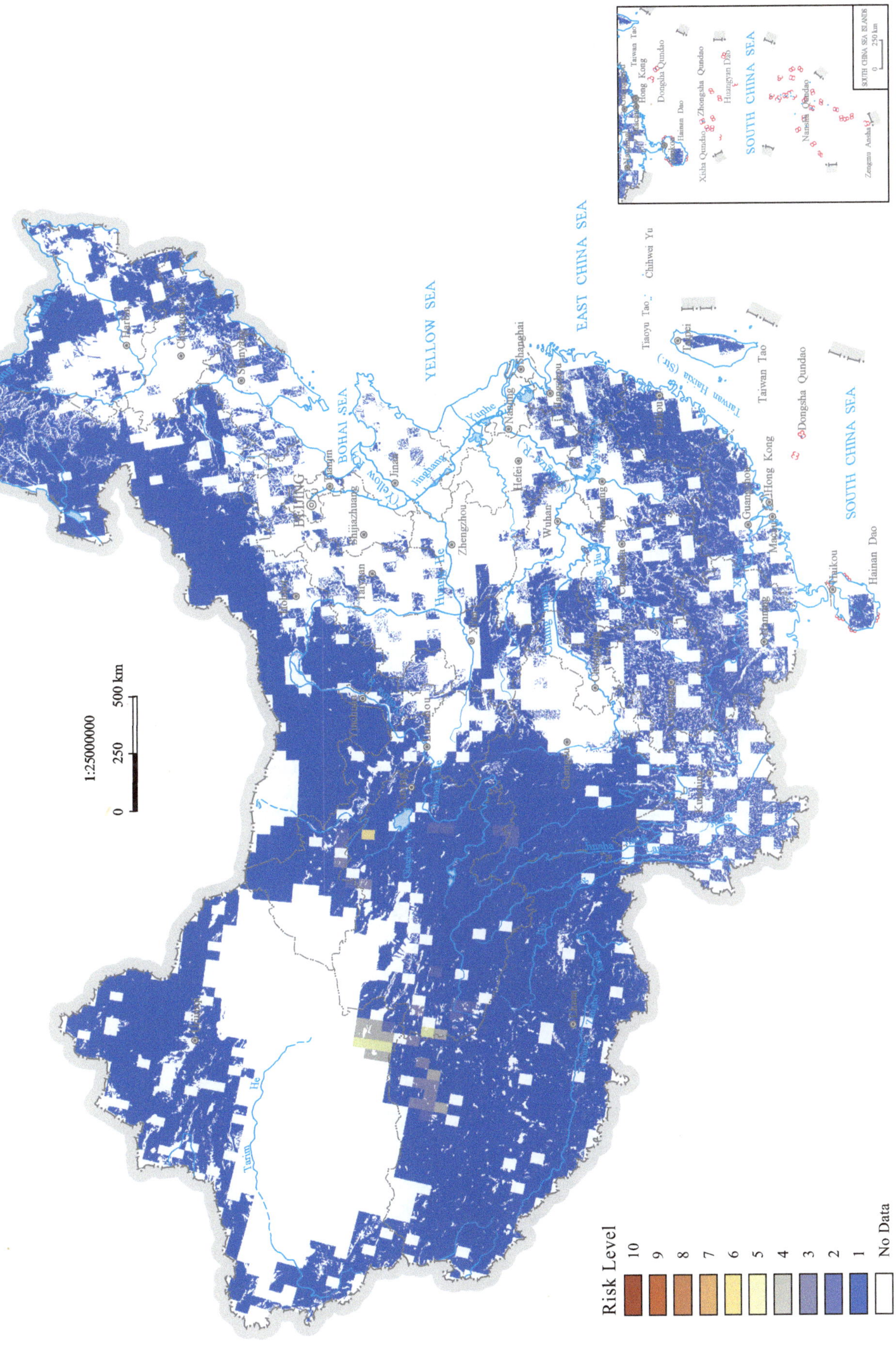

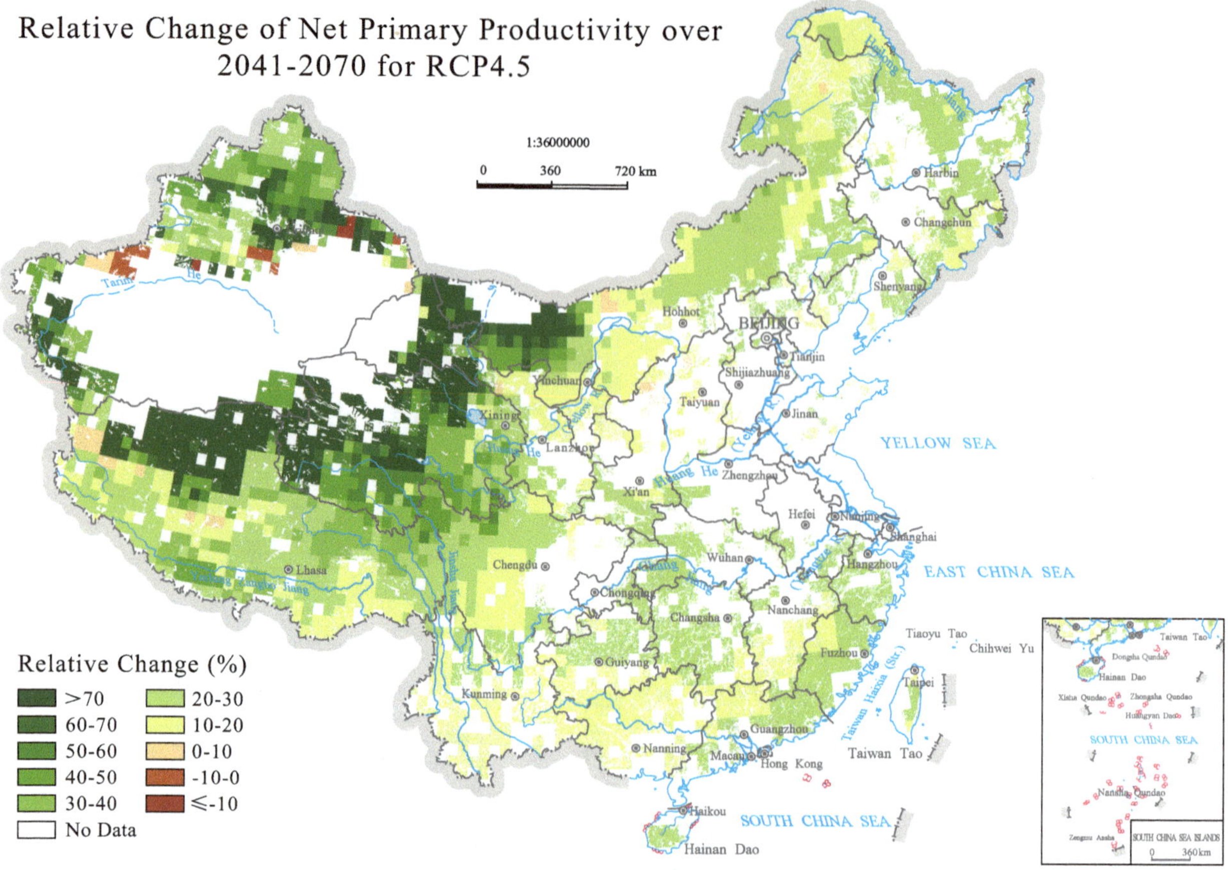
Relative Change of Net Primary Productivity over 2041-2070 for RCP4.5
1:36000000
0 360 720 km
Relative Change (%)
>70
60-70
50-60
40-50
30-40
20-30
10-20
0-10
-10-0
≤-10
No Data
YELLOW SEA
EAST CHINA SEA
SOUTH CHINA SEA
Hainan Dao
Taiwan Tao
Tiaoyu Tao
Chihwei Yu
Beijing
Harbin
Changchun
Shenyang
Hohhot
Tianjin
Shijiazhuang
Taiyuan
Jinan
Yinchuan
Xining
Lanzhou
Zhengzhou
Xi'an
Hefei
Nanjing
Shanghai
Wuhan
Hangzhou
Chengdu
Chongqing
Lhasa
Changsha
Nanchang
Fuzhou
Taipei
Guiyang
Kunming
Nanning
Guangzhou
Macau
Hong Kong
Haikou

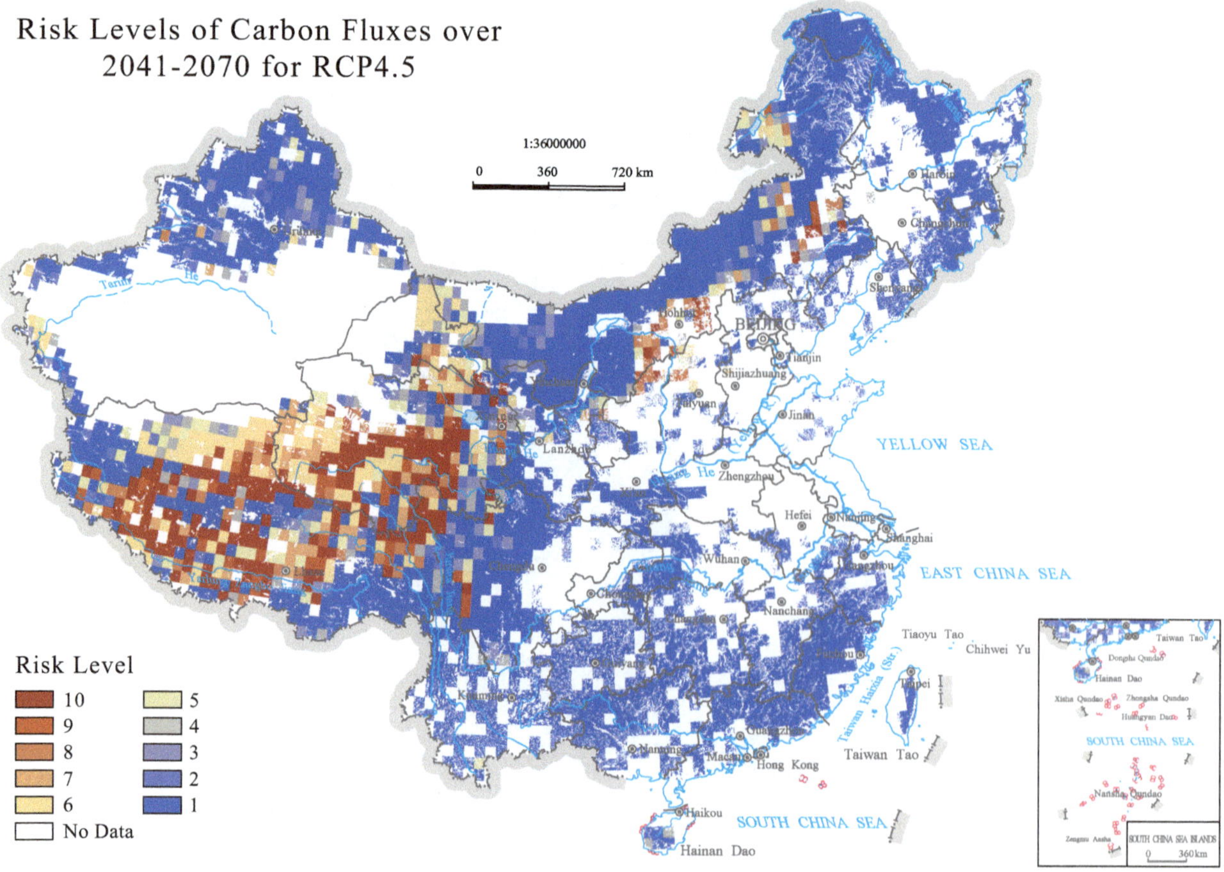
Risk Levels of Carbon Fluxes over 2041-2070 for RCP4.5
1:36000000
0 360 720 km
Risk Level
10
9
8
7
6
5
4
3
2
1
No Data
YELLOW SEA
EAST CHINA SEA
SOUTH CHINA SEA
Hainan Dao
Taiwan Tao
Tiaoyu Tao
Chihwei Yu
Beijing
Tianjin
Shijiazhuang
Jinan
Zhengzhou
Hefei
Nanjing
Shanghai
Wuhan
Nanchang
Taipei
Macau
Hong Kong
Haikou

Risk Levels of Carbon Stocks over 2041-2070 for RCP4.5

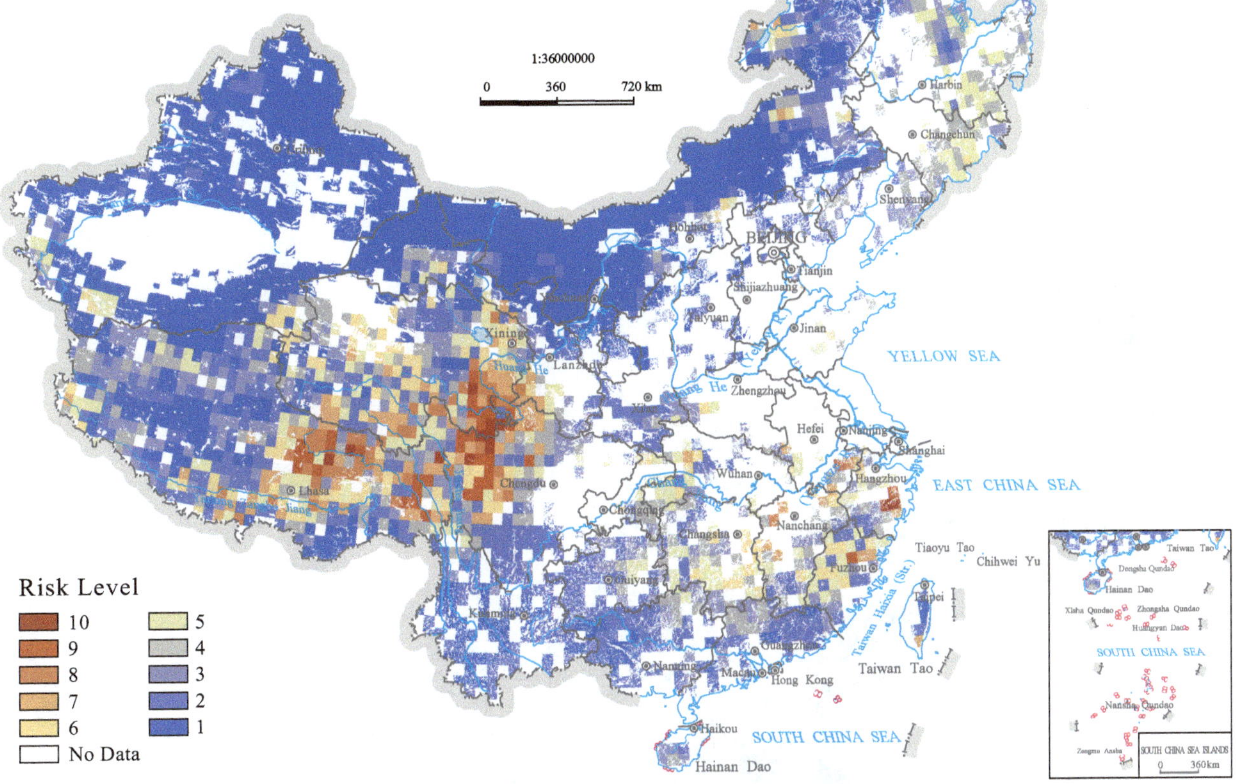

Risk Levels of Water Fluxes over 2041-2070 for RCP4.5

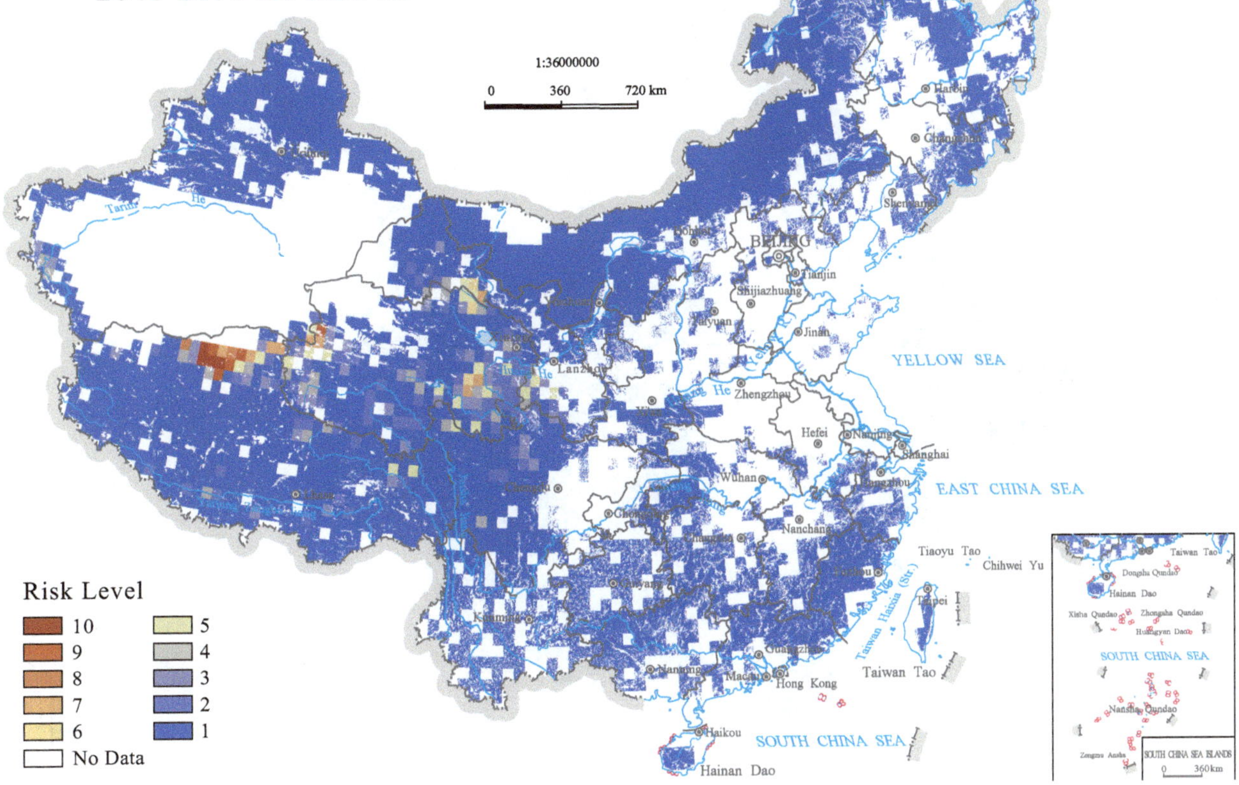

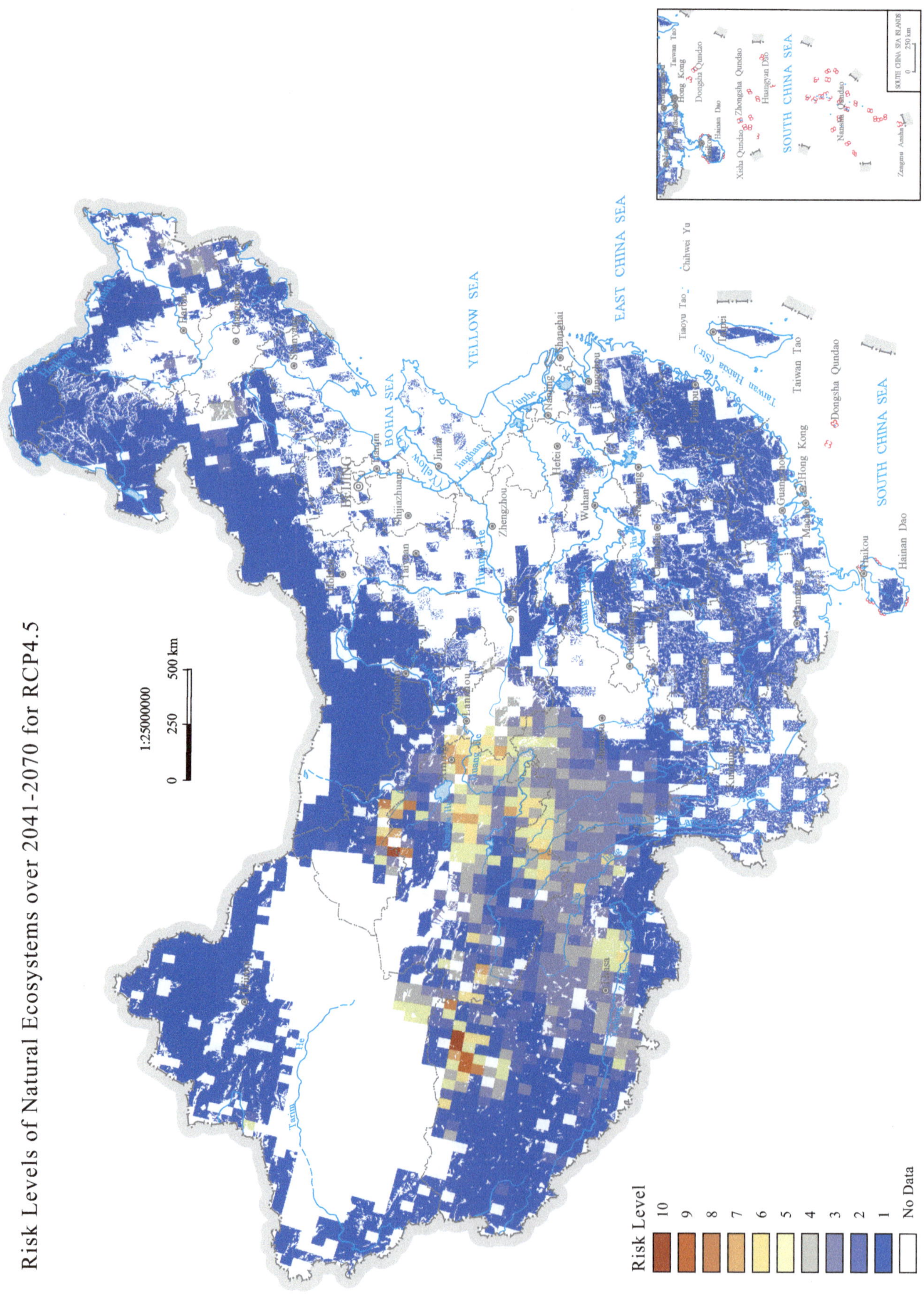
Risk Levels of Natural Ecosystems over 2041-2070 for RCP4.5
1:25000000
0 250 500 km
Risk Level
10
9
8
7
6
5
4
3
2
1
No Data
BEIJING
Shijiazhuang
Taiyuan
Zhengzhou
Hefei
Wuhan
Nanjing
Shanghai
Lhasa
Hong Kong
Macao
Haikou
Hainan Dao
Taiwan Tao
Dongsha Qundao
Diaoyu Dao
Chiwei Yu
BOHAI SEA
YELLOW SEA
EAST CHINA SEA
SOUTH CHINA SEA
Taiwan Haixia
Xisha Qundao
Zhongsha Qundao
Huangyan Dao
Nansha Qundao
SOUTH CHINA SEA ISLANDS

Relative Change of Net Primary Productivity over 2071-2099 for RCP4.5

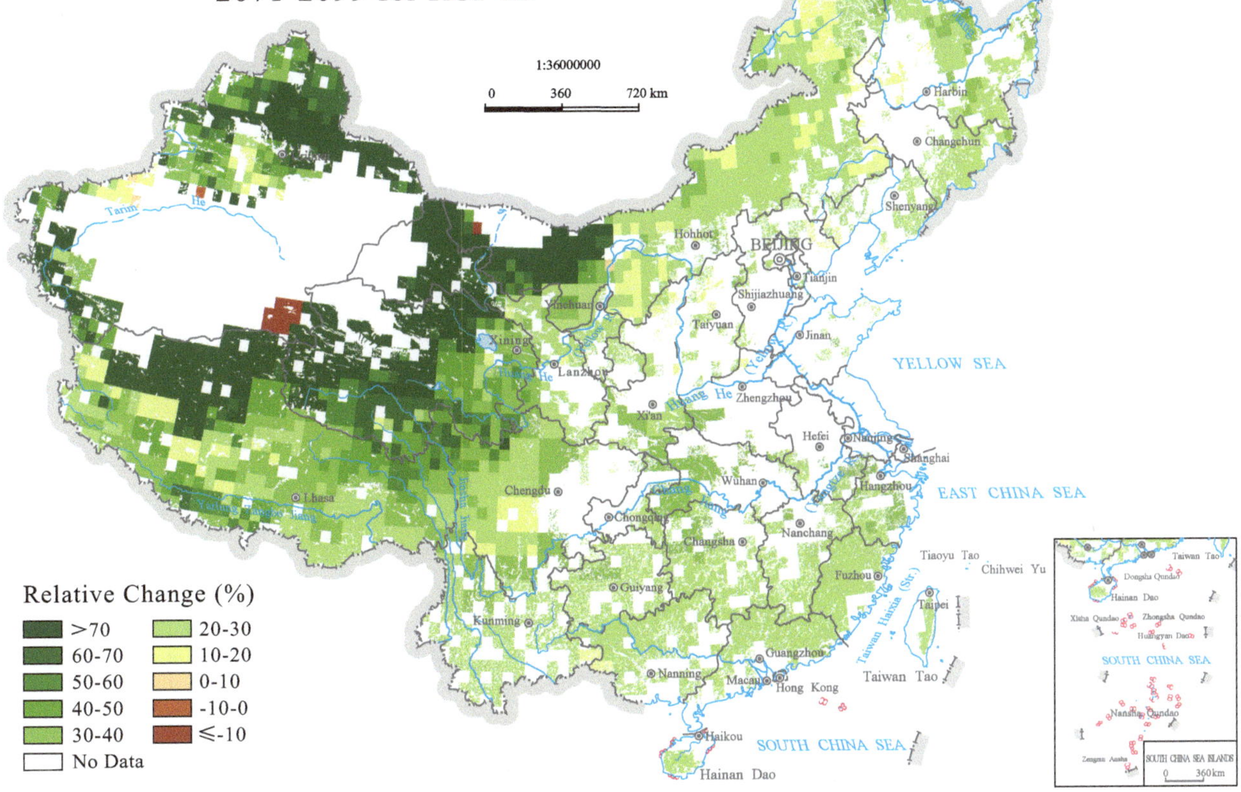

Risk Levels of Carbon Fluxes over 2071-2099 for RCP4.5

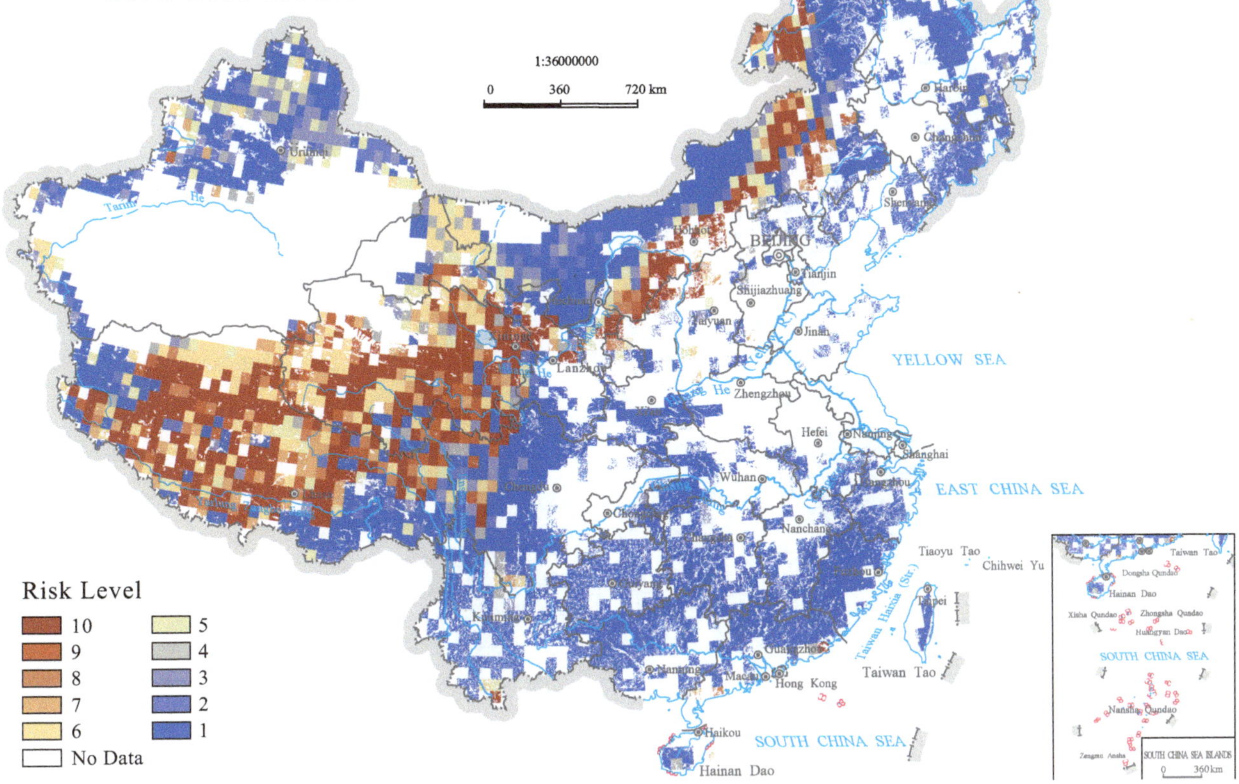

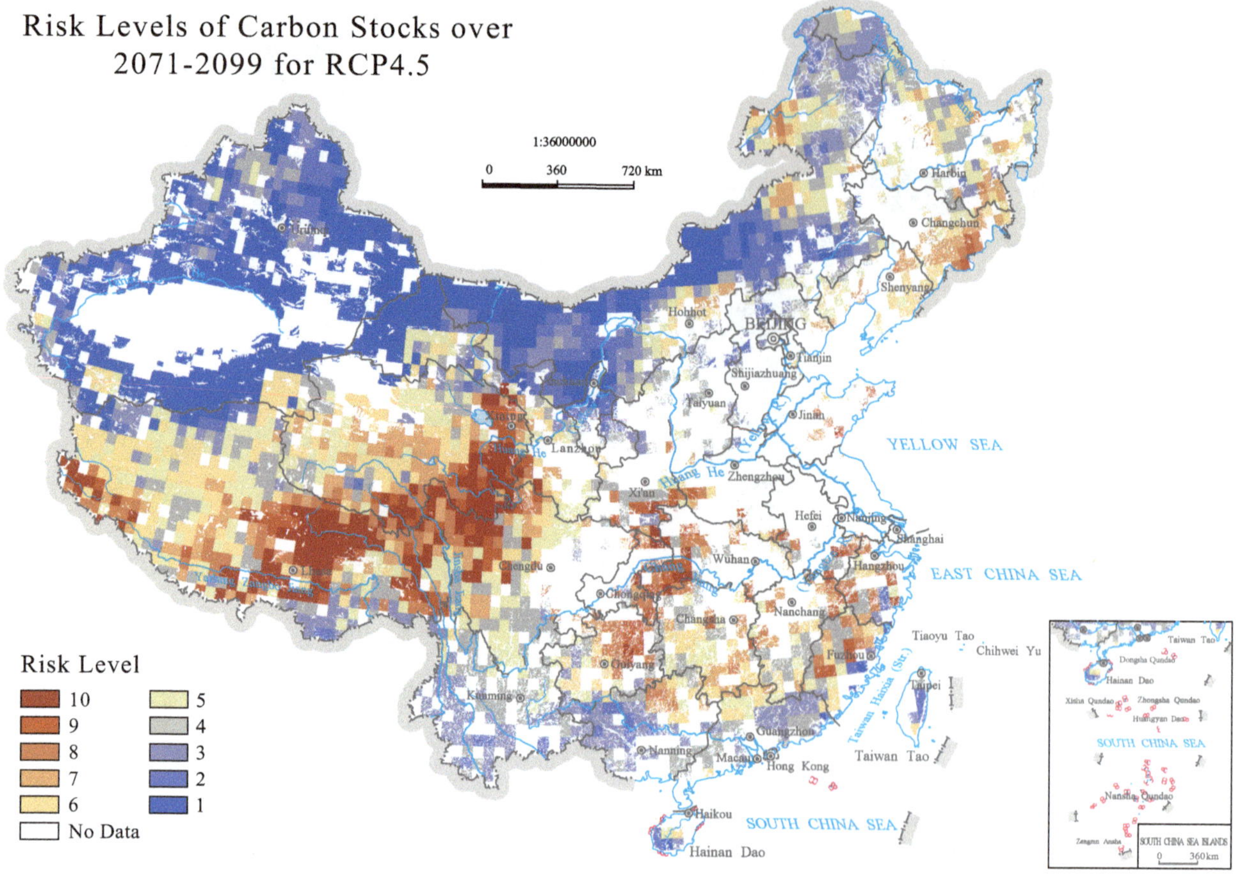
Risk Levels of Carbon Stocks over
2071-2099 for RCP4.5
1:36000000
0 360 720 km
Risk Level
10
9
8
7
6
5
4
3
2
1
No Data
YELLOW SEA
EAST CHINA SEA
SOUTH CHINA SEA
Hainan Dao
Taiwan Tao
Tiaoyu Tao
Chihwei Yu
Hong Kong
Macao
BEIJING
Tianjin
Shijiazhuang
Taiyuan
Jinan
Hohhot
Shenyang
Changchun
Harbin
Lanzhou
Xi'an
Zhengzhou
Hefei
Nanjing
Shanghai
Hangzhou
Wuhan
Nanchang
Changsha
Chengdu
Guiyang
Kunming
Nanning
Guangzhou
Fuzhou
Taipei
Haikou
SOUTH CHINA SEA

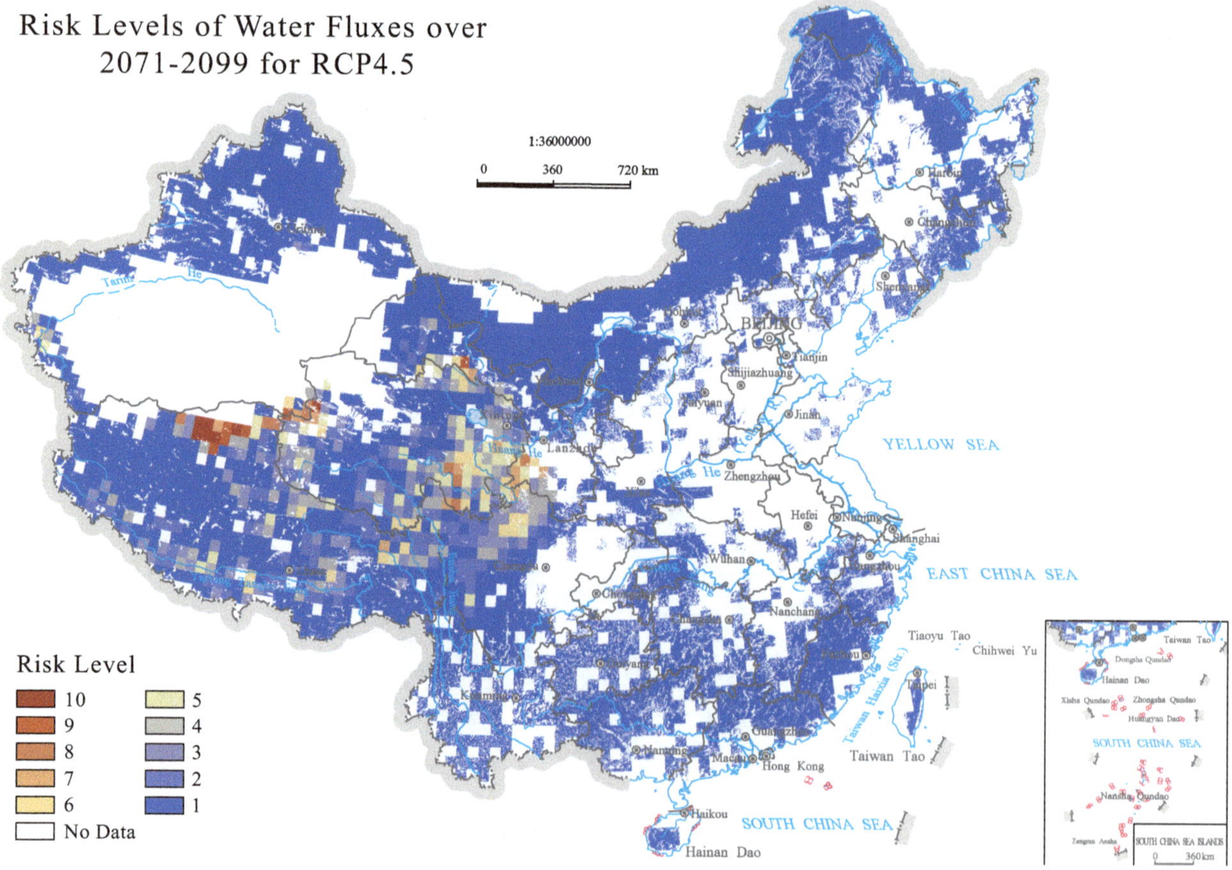
Risk Levels of Water Fluxes over
2071-2099 for RCP4.5
1:36000000
0 360 720 km
Risk Level
10
9
8
7
6
5
4
3
2
1
No Data
YELLOW SEA
EAST CHINA SEA
SOUTH CHINA SEA
Hainan Dao
Taiwan Tao
Tiaoyu Tao
Chihwei Yu
Hong Kong
Macao
BEIJING
Tianjin
Shijiazhuang
Taiyuan
Jinan
Zhengzhou
Hefei
Nanjing
Shanghai
Wuhan
Nanchang
Lanzhou
Haikou
SOUTH CHINA SEA

Risk Levels of Natural Ecosystems over 2071-2099 for RCP4.5

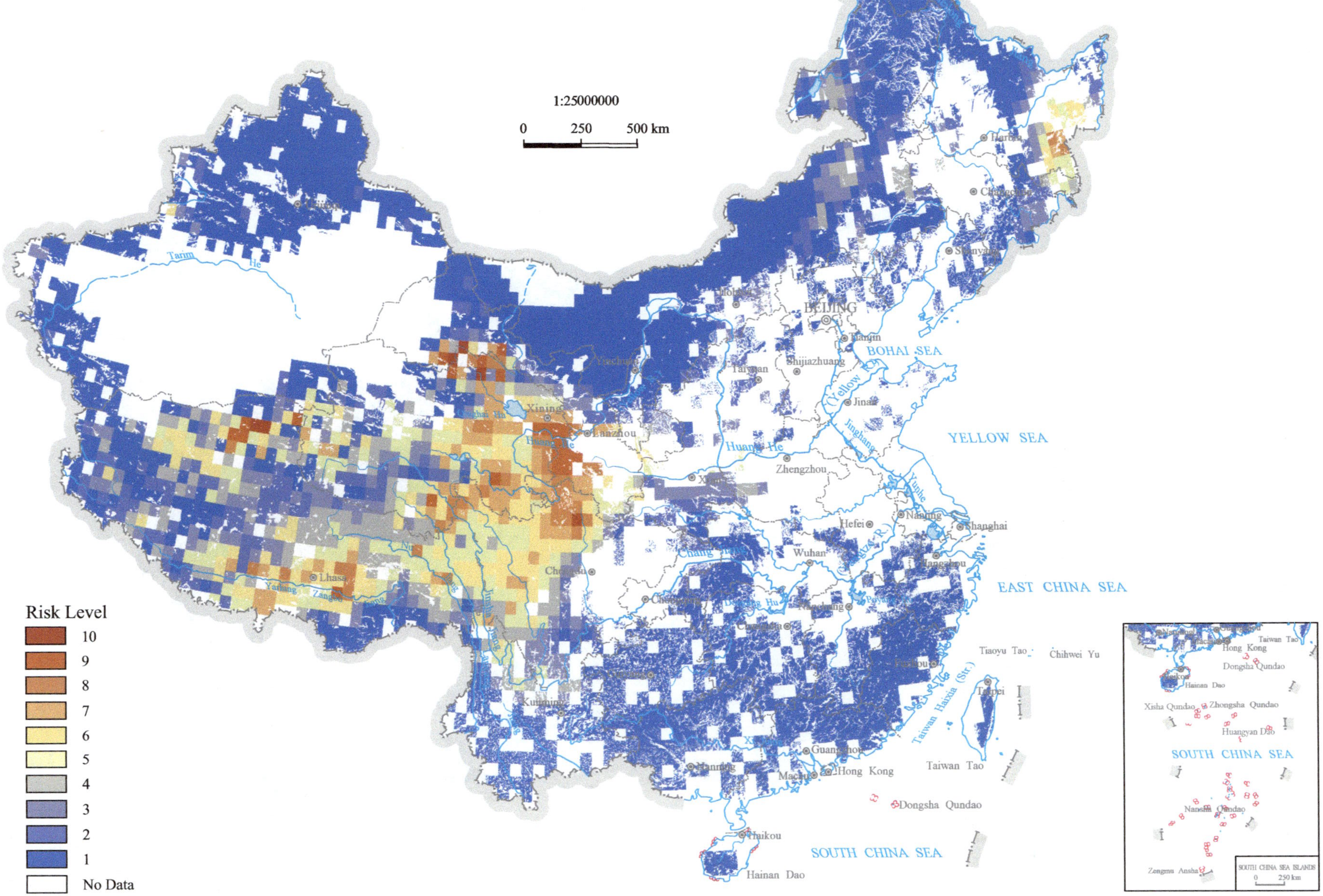

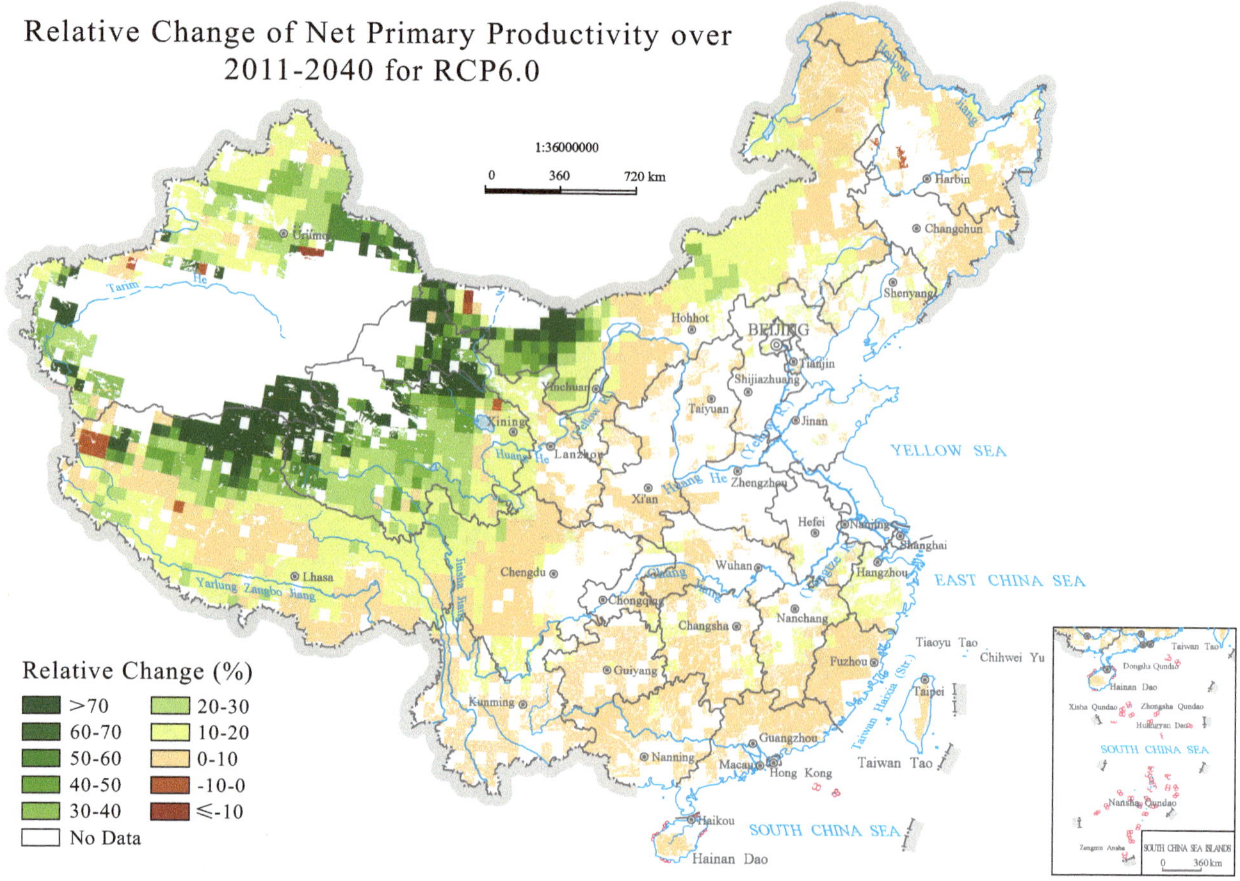
Relative Change of Net Primary Productivity over
2011-2040 for RCP6.0
1:36000000
0 360 720 km
Relative Change (%)
>70
60-70
50-60
40-50
30-40
No Data
20-30
10-20
0-10
-10-0
≤-10
Harbin
Changchun
Shenyang
Hohhot
BEIJING
Tianjin
Shijiazhuang
Taiyuan
Jinan
Yinchuan
Xining
Lanzhou
Xi'an
Zhengzhou
Hefei
Nanjing
Shanghai
Hangzhou
Wuhan
Chengdu
Chongqing
Changsha
Nanchang
Lhasa
Guiyang
Kunming
Fuzhou
Taipei
Nanning
Guangzhou
Macau
Hong Kong
Haikou
Hainan Dao
YELLOW SEA
EAST CHINA SEA
SOUTH CHINA SEA
Tiaoyu Tao
Chihwei Yu
Taiwan Tao
Tarim He
Yarlung Zangbo Jiang
Huang He
Heilong Jiang
Jinsha Jiang
Taiwan Haixia (Str.)

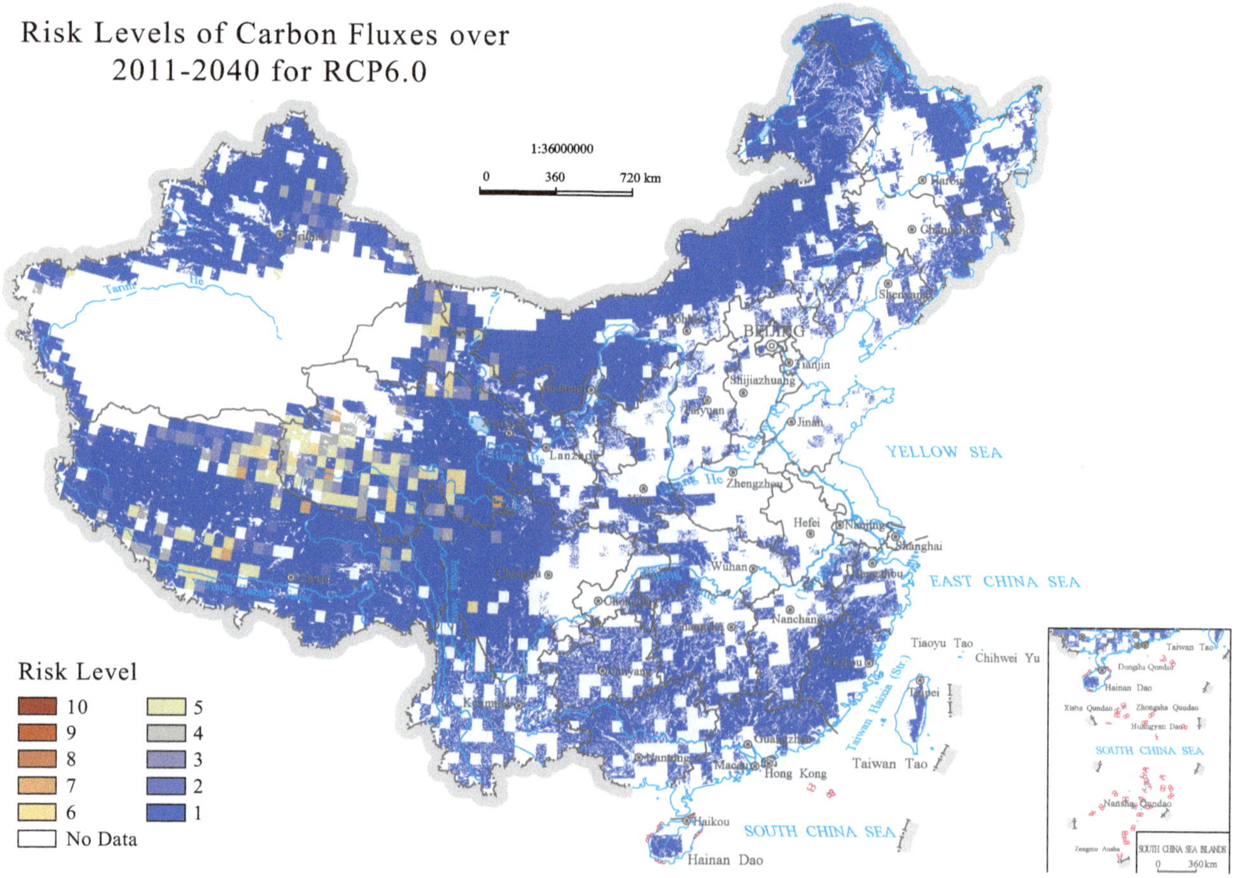
Risk Levels of Carbon Fluxes over
2011-2040 for RCP6.0
1:36000000
0 360 720 km
Risk Level
10
9
8
7
6
No Data
5
4
3
2
1
BEIJING
Tianjin
Shijiazhuang
Jinan
Zhengzhou
Hefei
Nanjing
Shanghai
Wuhan
Nanchang
Taipei
Macau
Hong Kong
Haikou
Hainan Dao
YELLOW SEA
EAST CHINA SEA
SOUTH CHINA SEA
Tiaoyu Tao
Chihwei Yu
Taiwan Tao

Risk Levels of Carbon Stocks over 2011-2040 for RCP6.0

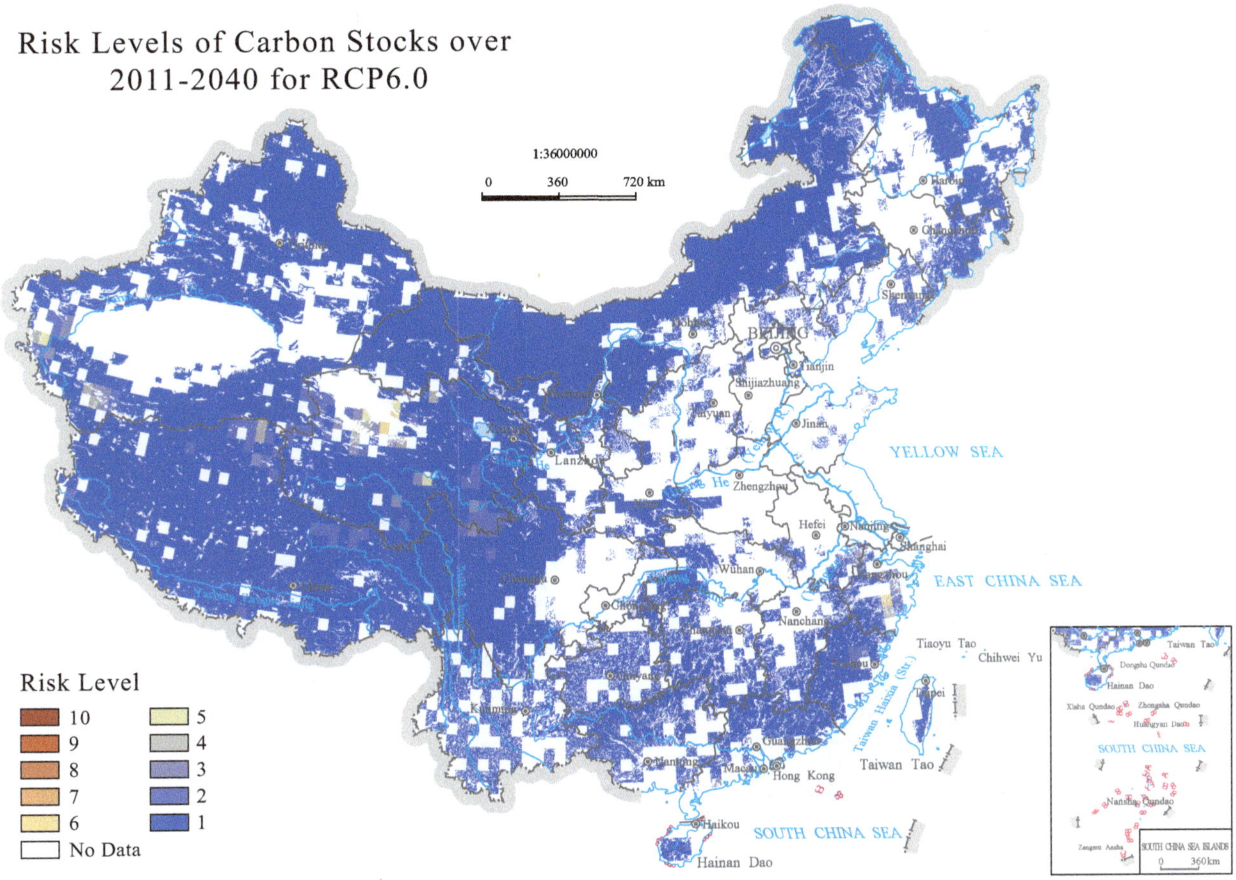

Risk Levels of Water Fluxes over 2011-2040 for RCP6.0

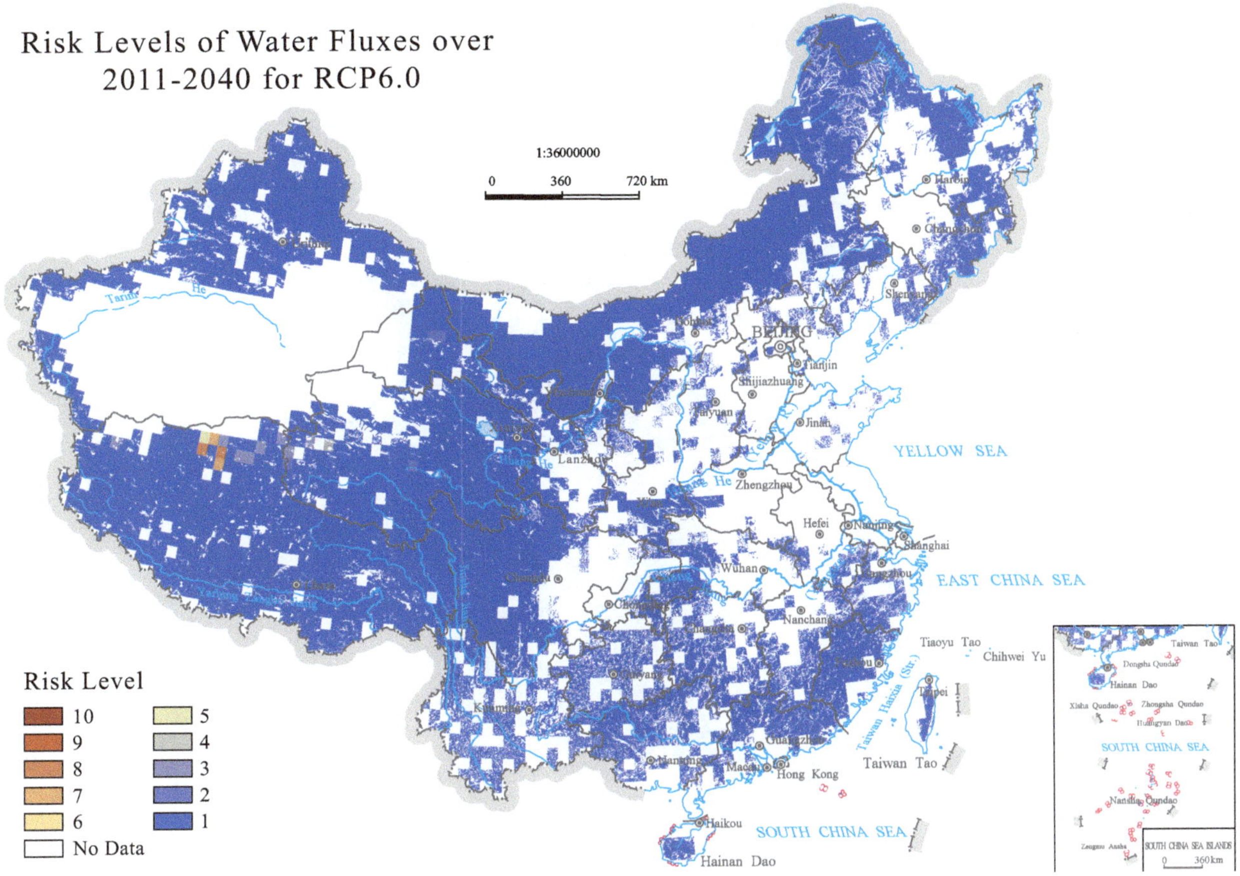

Risk Levels of Natural Ecosystems over 2011-2040 for RCP6.0

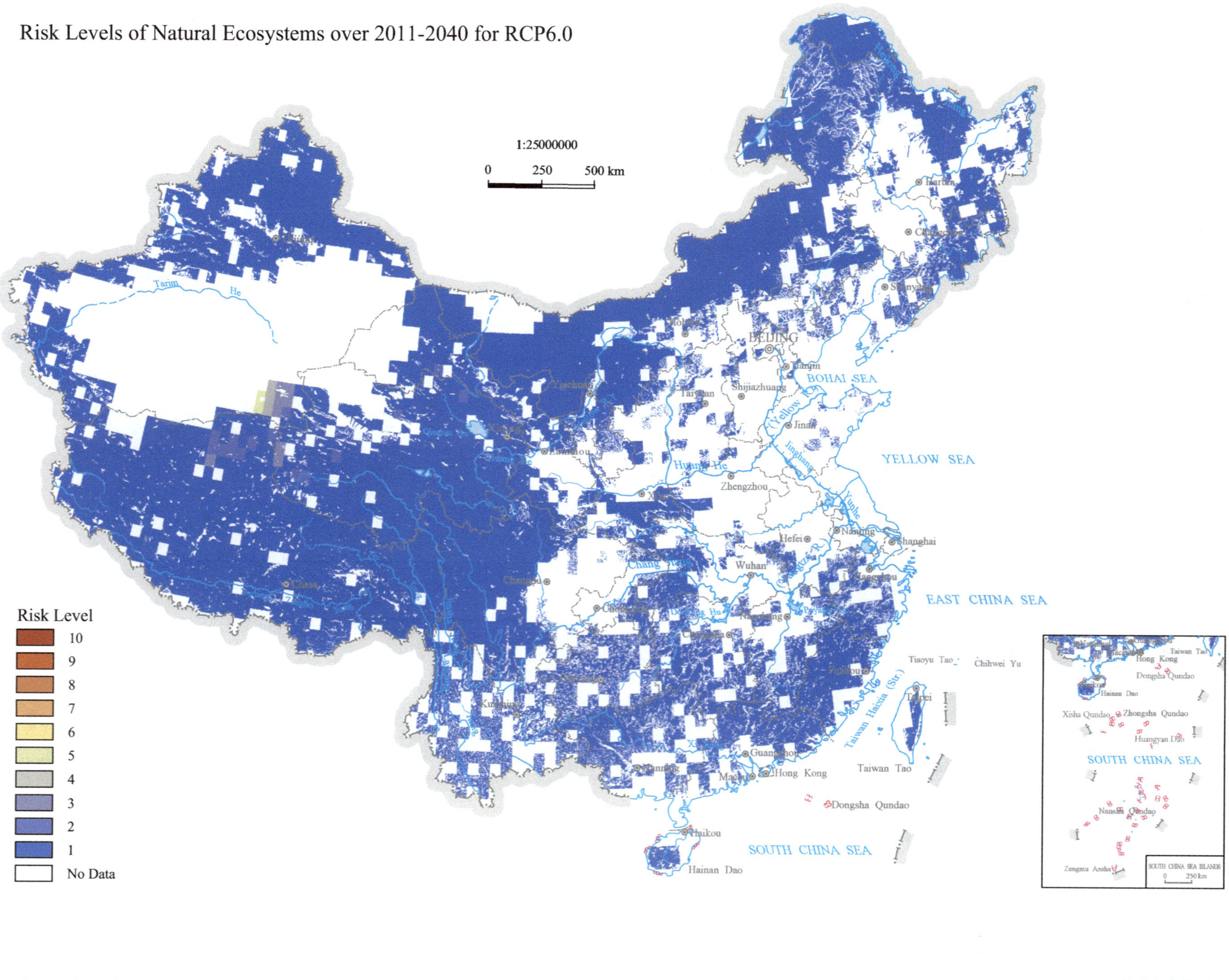

Relative Change of Net Primary Productivity over 2041-2070 for RCP6.0

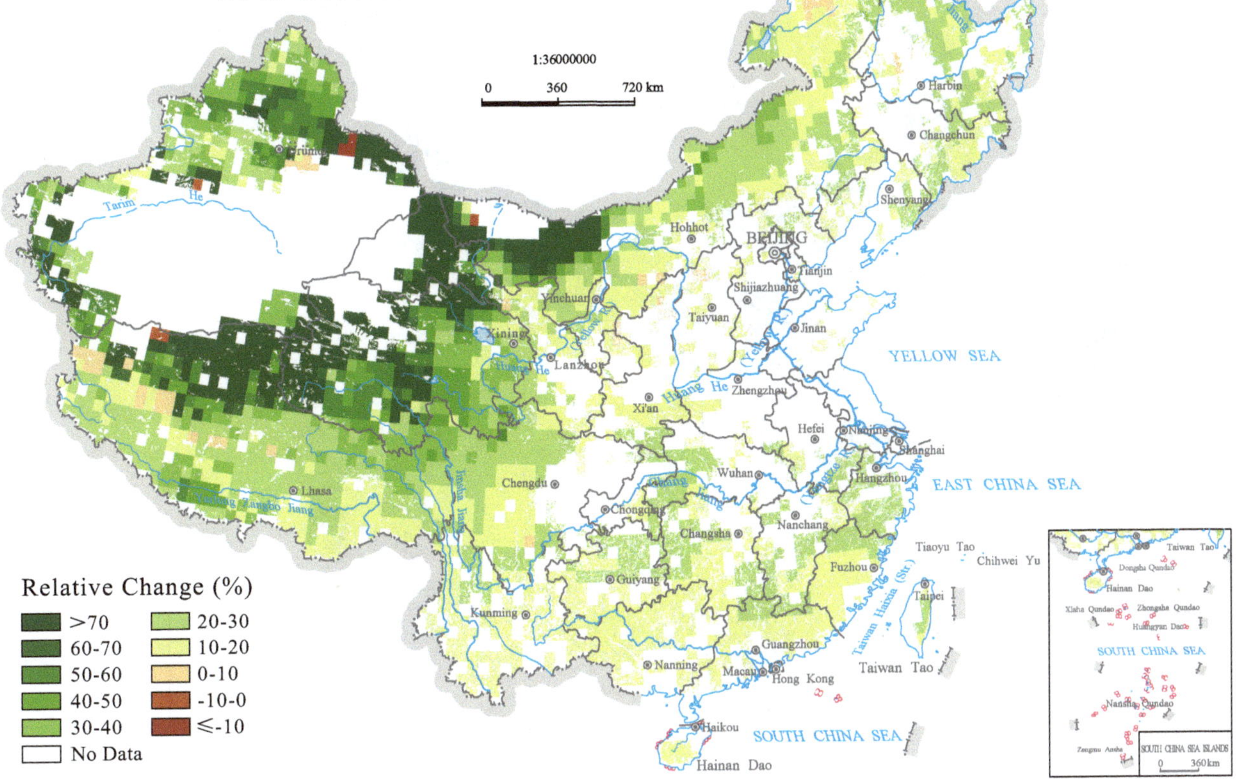

Risk Levels of Carbon Fluxes over 2041-2070 for RCP6.0

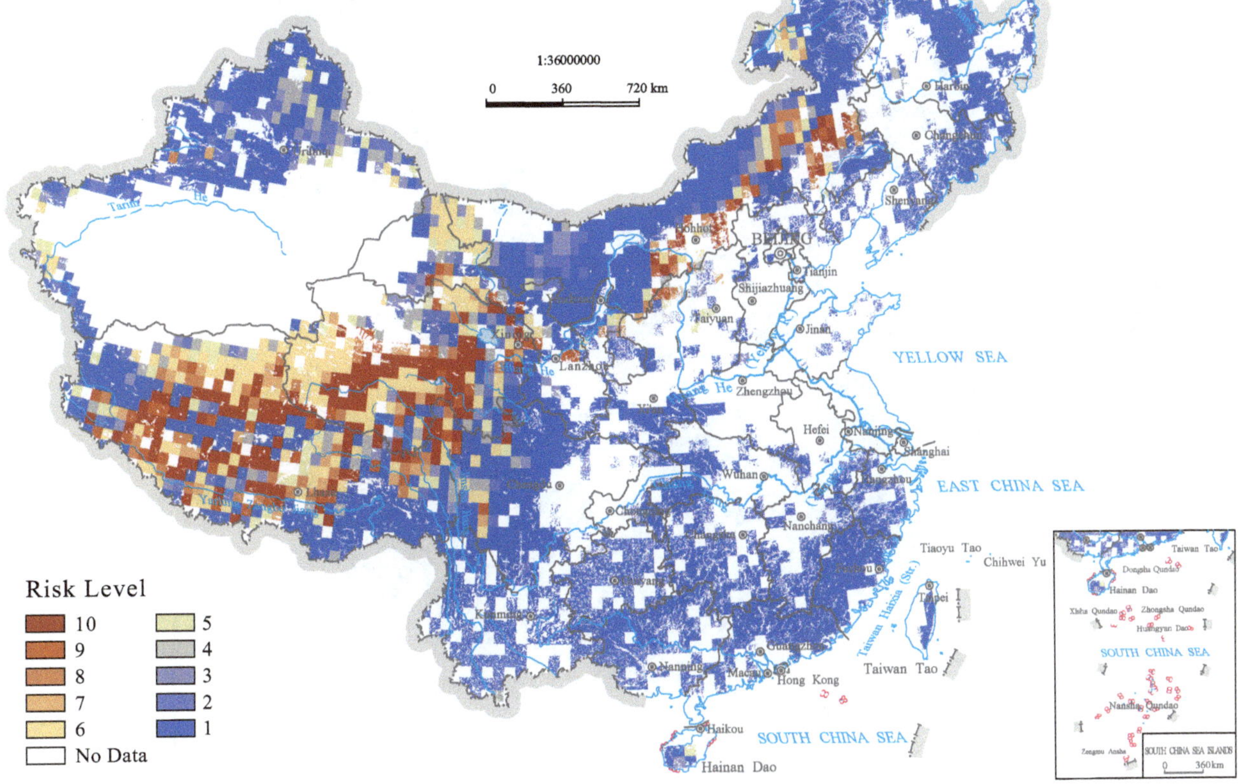

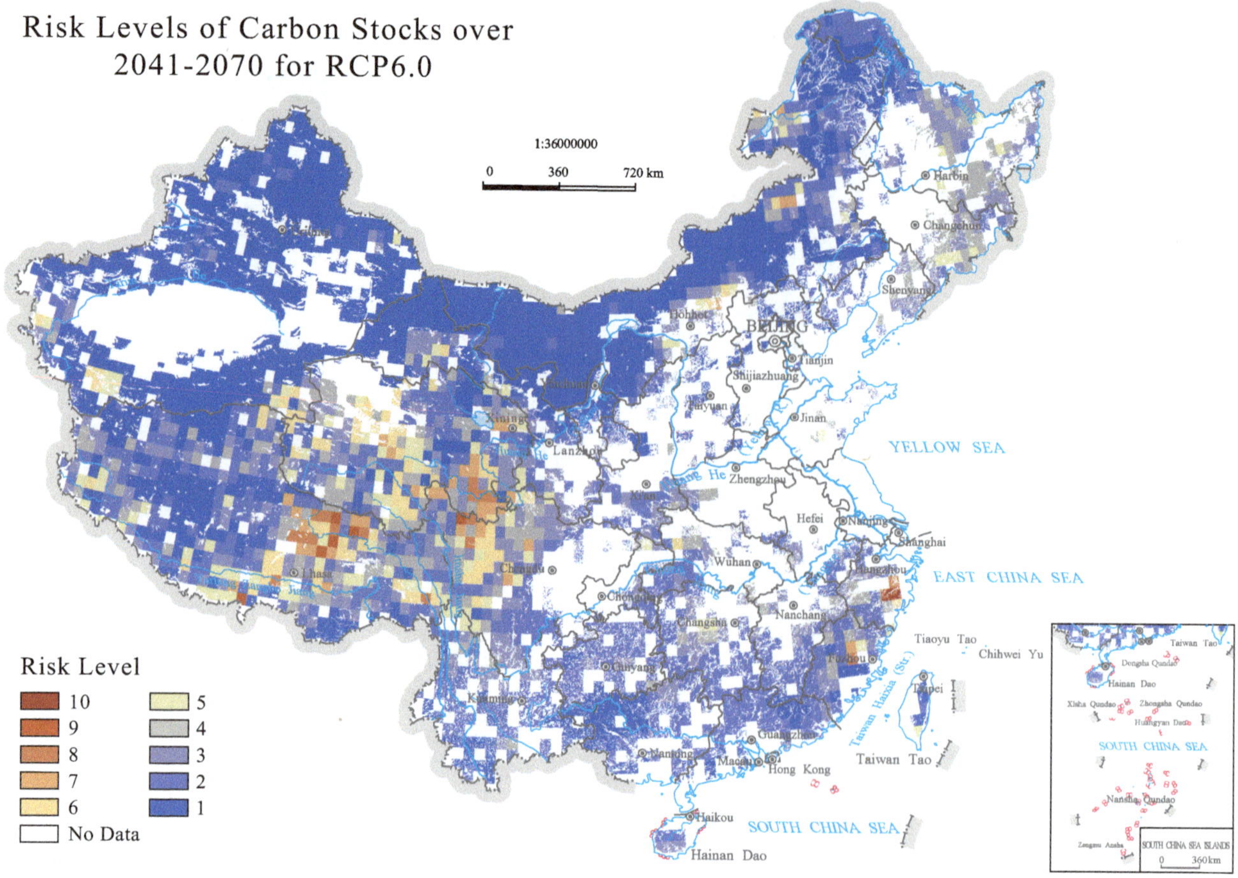
Risk Levels of Carbon Stocks over
2041-2070 for RCP6.0
1:36000000
0 360 720 km
Risk Level
10
9
8
7
6
5
4
3
2
1
No Data
Harbin
Changchun
Shenyang
BEIJING
Tianjin
Shijiazhuang
Taiyuan
Jinan
Lanzhou
Zhengzhou
Hefei
Nanjing
Shanghai
Wuhan
Hangzhou
Chengdu
Changsha
Nanchang
Kunming
Guangzhou
Macao
Hong Kong
Haikou
Hainan Dao
Taipei
Taiwan Tao
Tiaoyu Tao
Chihwei Yu
YELLOW SEA
EAST CHINA SEA
SOUTH CHINA SEA
SOUTH CHINA SEA ISLANDS

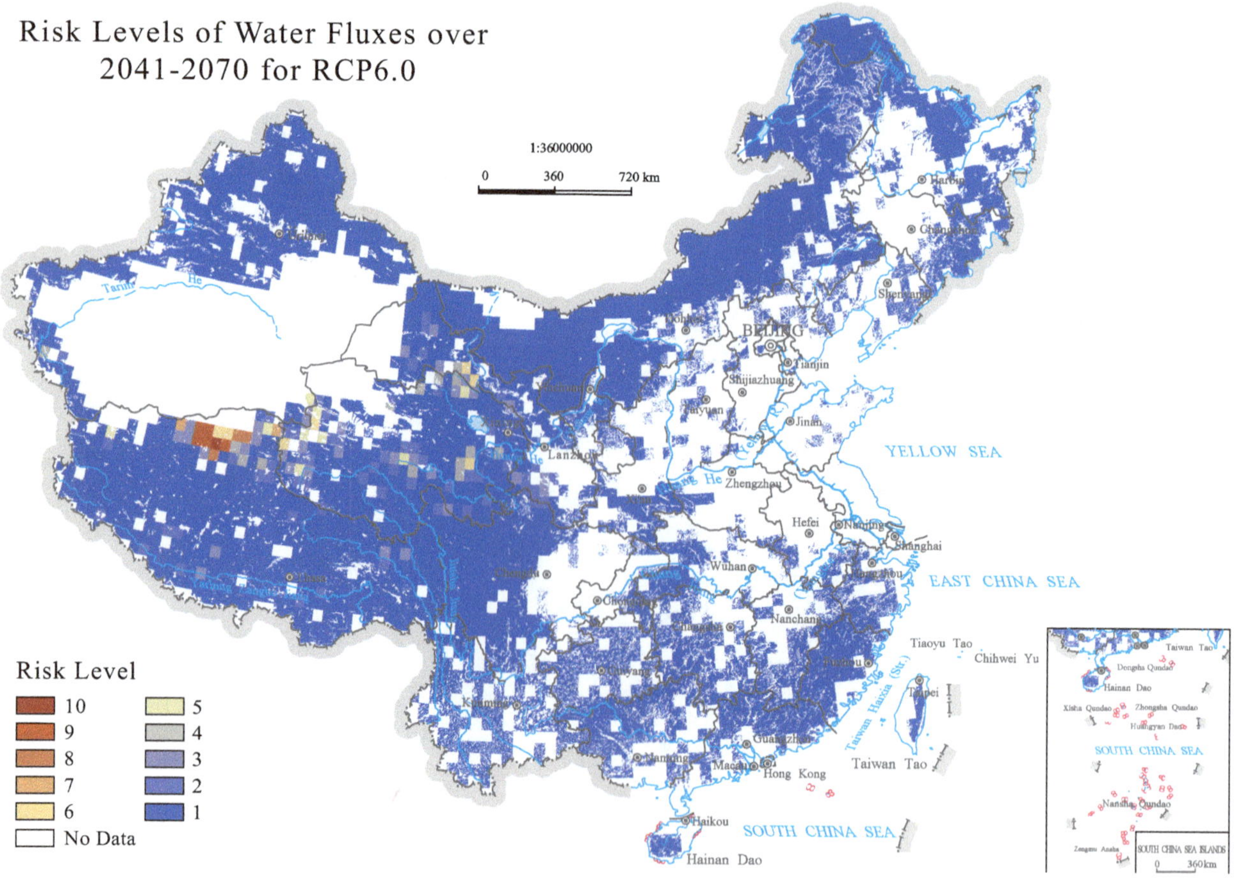
Risk Levels of Water Fluxes over
2041-2070 for RCP6.0
1:36000000
0 360 720 km
Risk Level
10
9
8
7
6
5
4
3
2
1
No Data
BEIJING
Tianjin
Shijiazhuang
Taiyuan
Jinan
Lanzhou
Zhengzhou
Hefei
Nanjing
Shanghai
Wuhan
Nanchang
Guangzhou
Macao
Hong Kong
Haikou
Hainan Dao
Taipei
Taiwan Tao
Tiaoyu Tao
Chihwei Yu
YELLOW SEA
EAST CHINA SEA
SOUTH CHINA SEA
SOUTH CHINA SEA ISLANDS

Risk Levels of Natural Ecosystems over 2041-2070 for RCP6.0

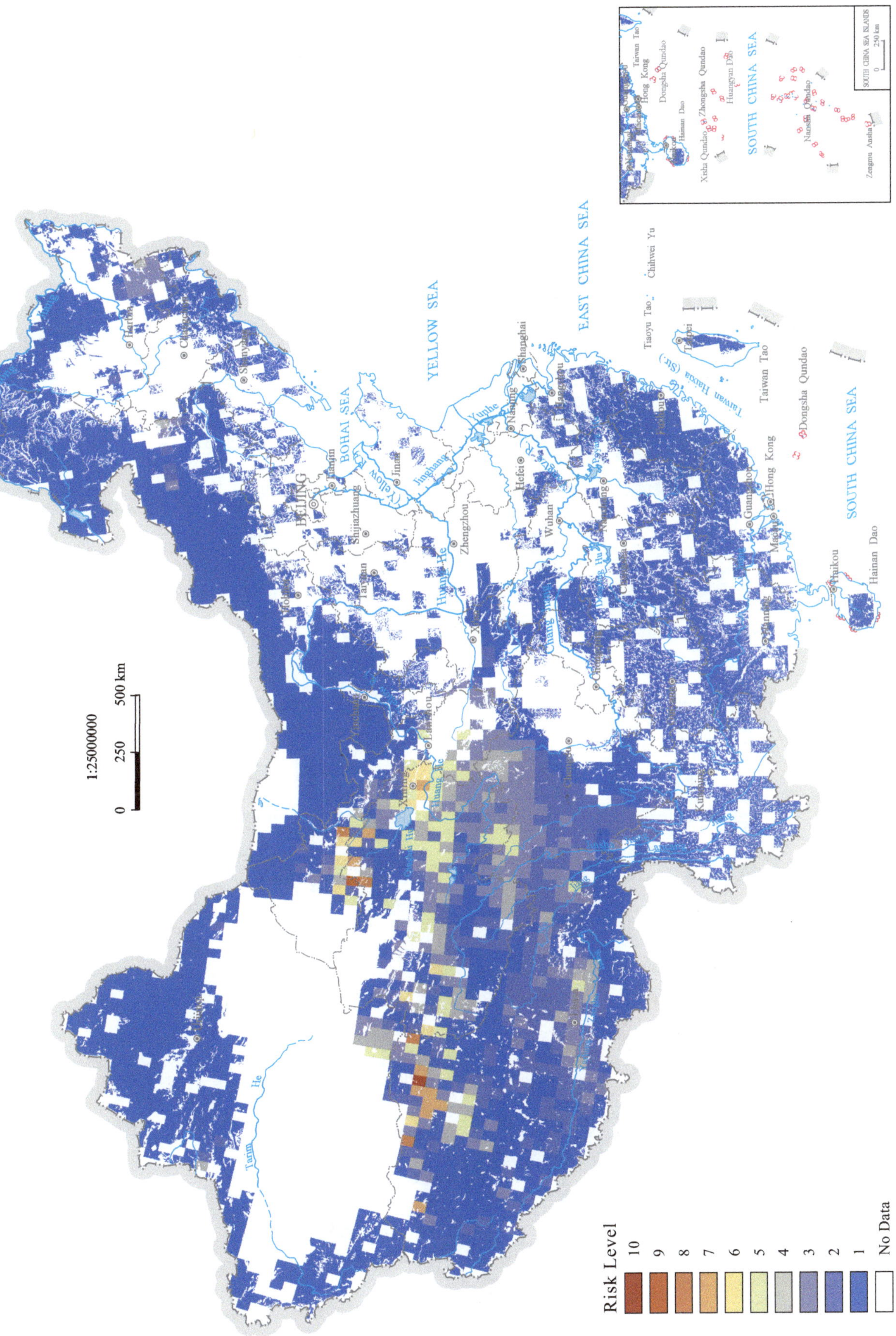

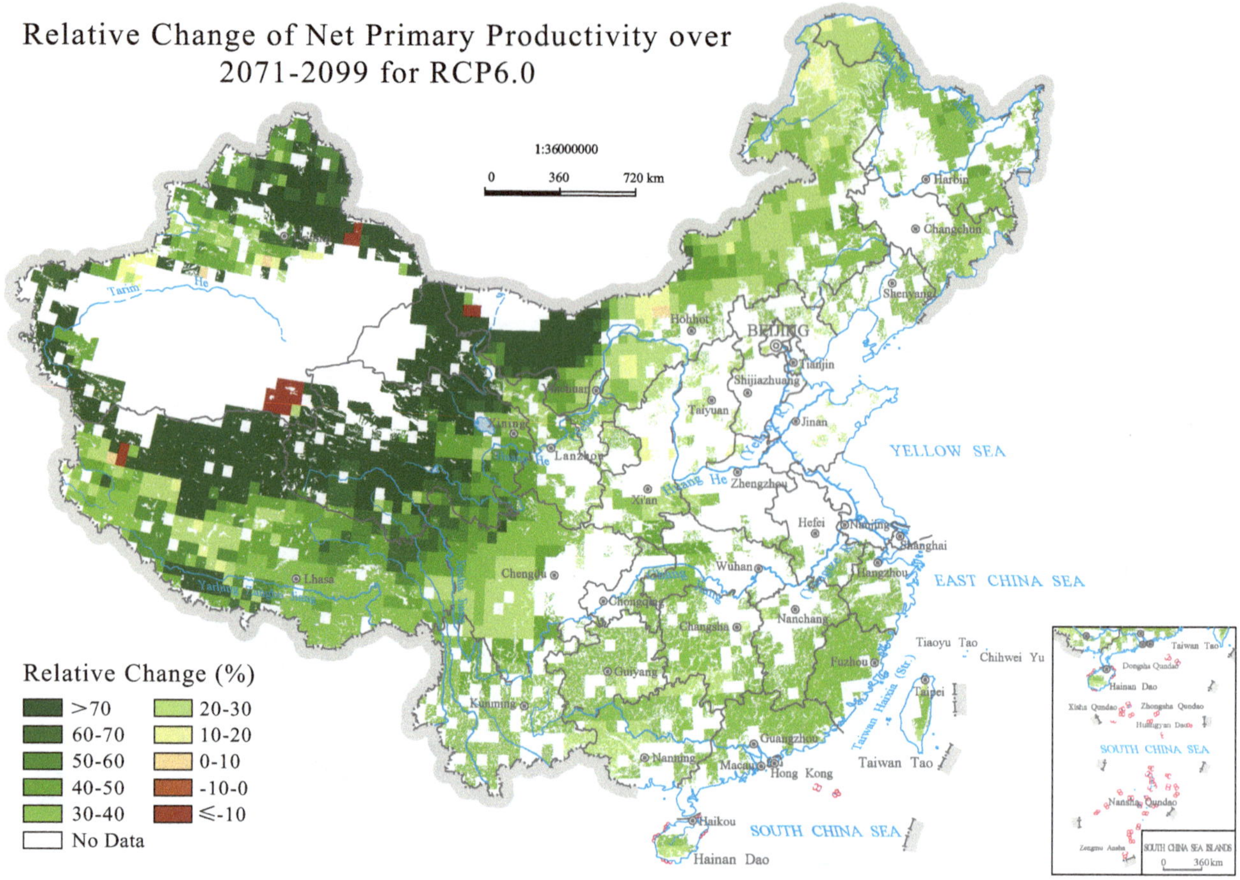
Relative Change of Net Primary Productivity over 2071-2099 for RCP6.0
1:36000000
0 360 720 km
Relative Change (%)
>70
60-70
50-60
40-50
30-40
20-30
10-20
0-10
-10-0
≤-10
No Data
Harbin
Changchun
Shenyang
Hohhot
BEIJING
Tianjin
Shijiazhuang
Taiyuan
Jinan
Xining
Lanzhou
Zhengzhou
Xi'an
Hefei
Nanjing
Shanghai
Wuhan
Hangzhou
Chengdu
Lhasa
Nanchang
Changsha
Guiyang
Fuzhou
Kunming
Taipei
Guangzhou
Nanning
Macao
Hong Kong
Haikou
Hainan Dao
Tarim He
YELLOW SEA
EAST CHINA SEA
SOUTH CHINA SEA
Tiaoyu Tao
Chihwei Yu
Taiwan Tao

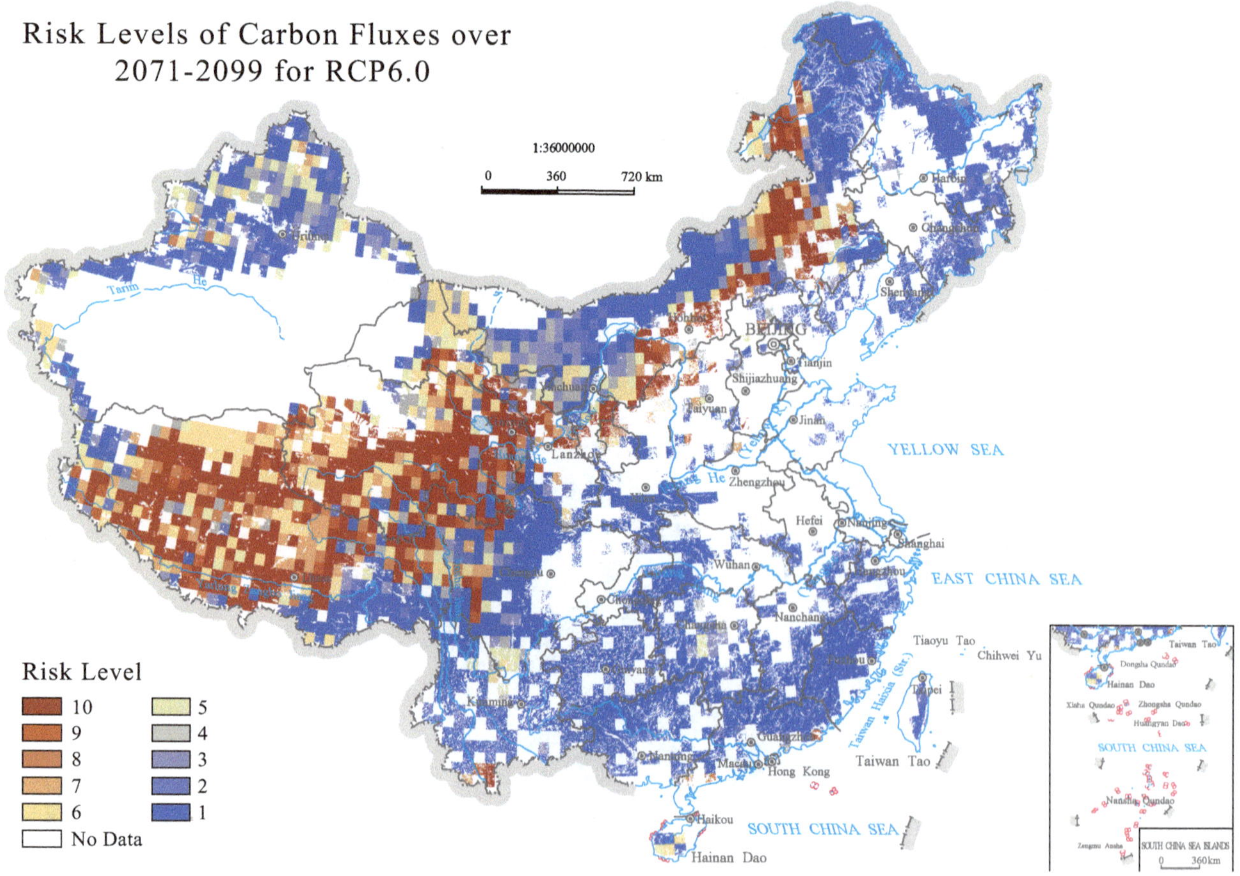
Risk Levels of Carbon Fluxes over 2071-2099 for RCP6.0
1:36000000
0 360 720 km
Risk Level
10
9
8
7
6
5
4
3
2
1
No Data
Harbin
Shijiazhuang
Zhengzhou
Hefei
Nanjing
Shanghai
Wuhan
Nanchang
Taipei
Macao
Hong Kong
Haikou
Hainan Dao
Tarim He
YELLOW SEA
EAST CHINA SEA
SOUTH CHINA SEA
Tiaoyu Tao
Chihwei Yu
Taiwan Tao

Risk Levels of Carbon Stocks over 2071-2099 for RCP6.0

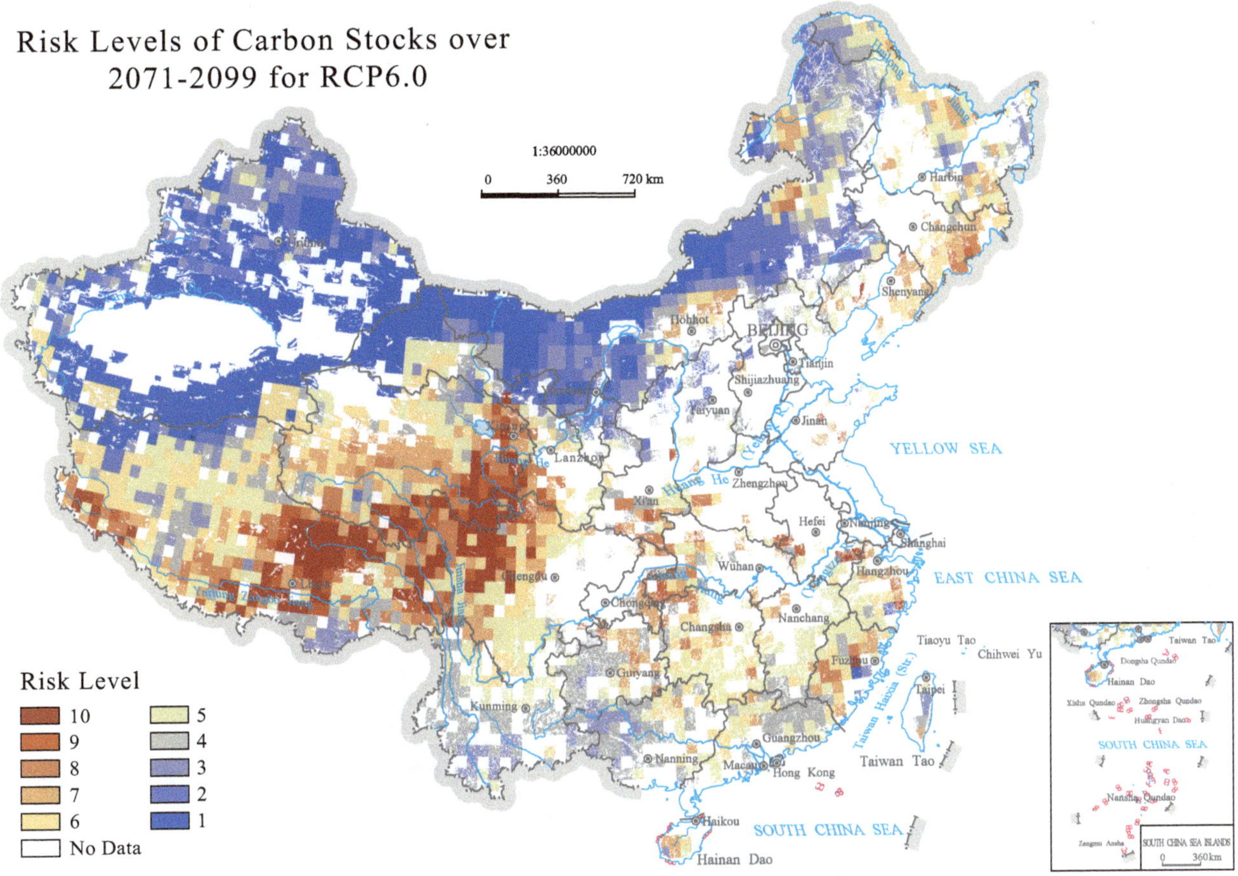

Risk Levels of Water Fluxes over 2071-2099 for RCP6.0

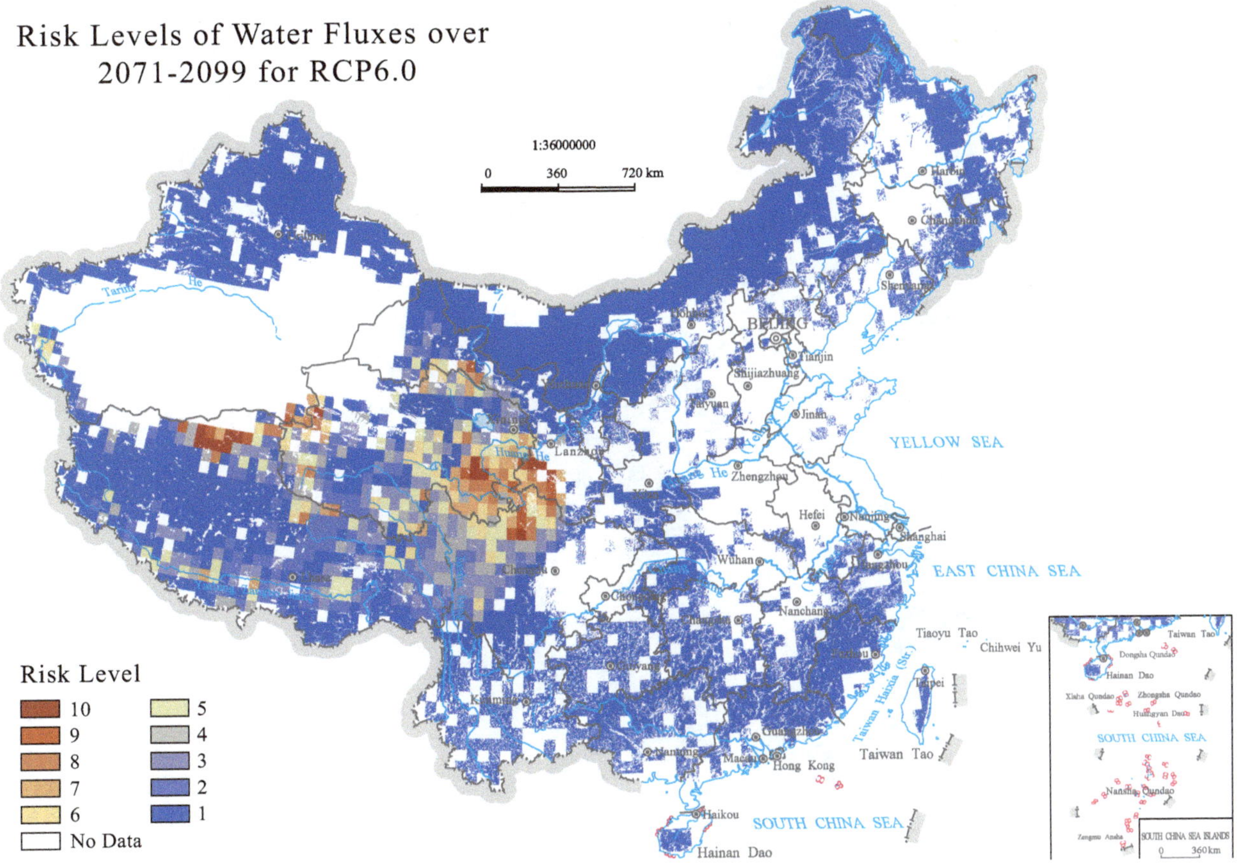

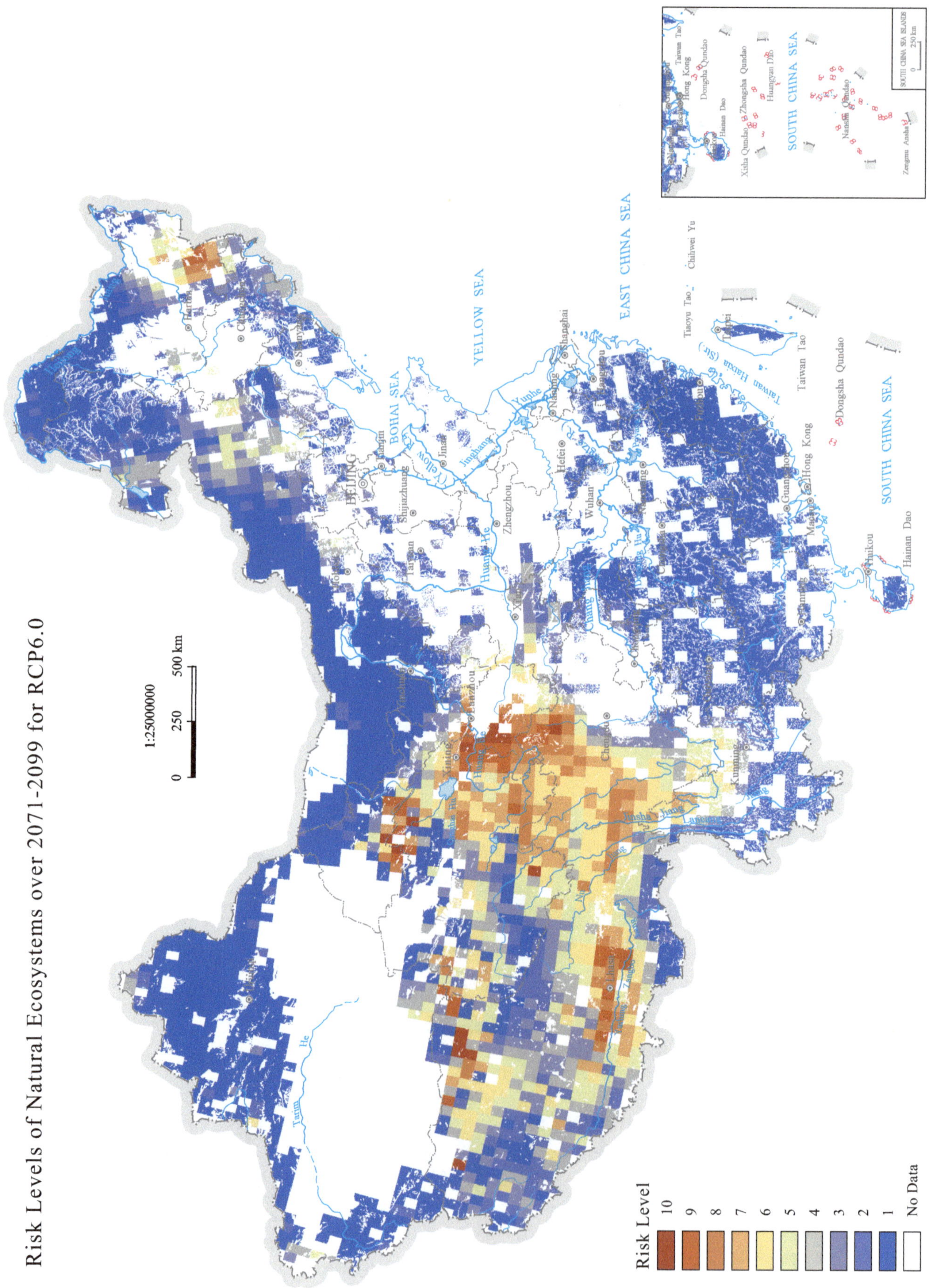
Risk Levels of Natural Ecosystems over 2071-2099 for RCP6.0
1:25000000
0 250 500 km
Risk Level
10
9
8
7
6
5
4
3
2
1
No Data
BOHAI SEA
YELLOW SEA
EAST CHINA SEA
SOUTH CHINA SEA

Relative Change of Net Primary Productivity over 2011-2040 for RCP8.5

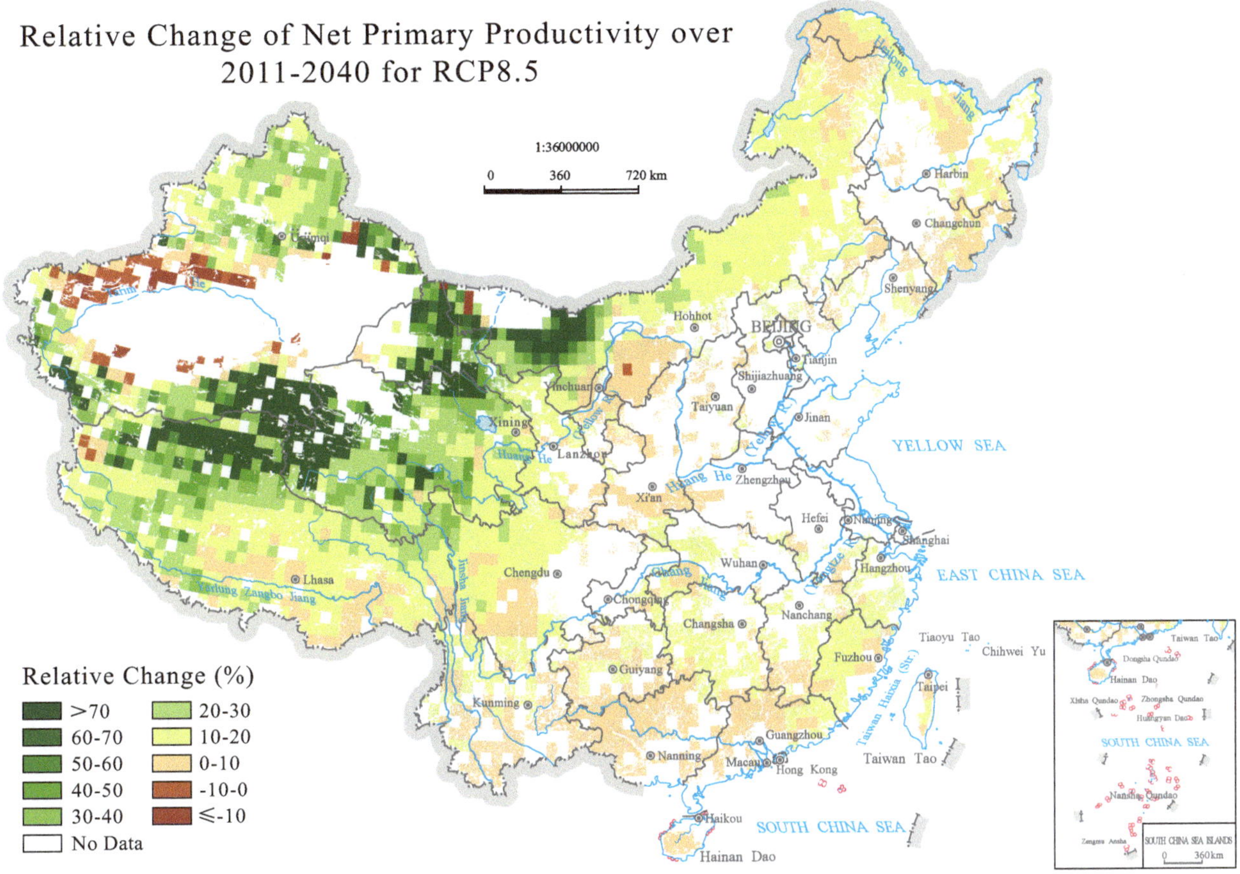

Risk Levels of Carbon Fluxes over 2011-2040 for RCP8.5

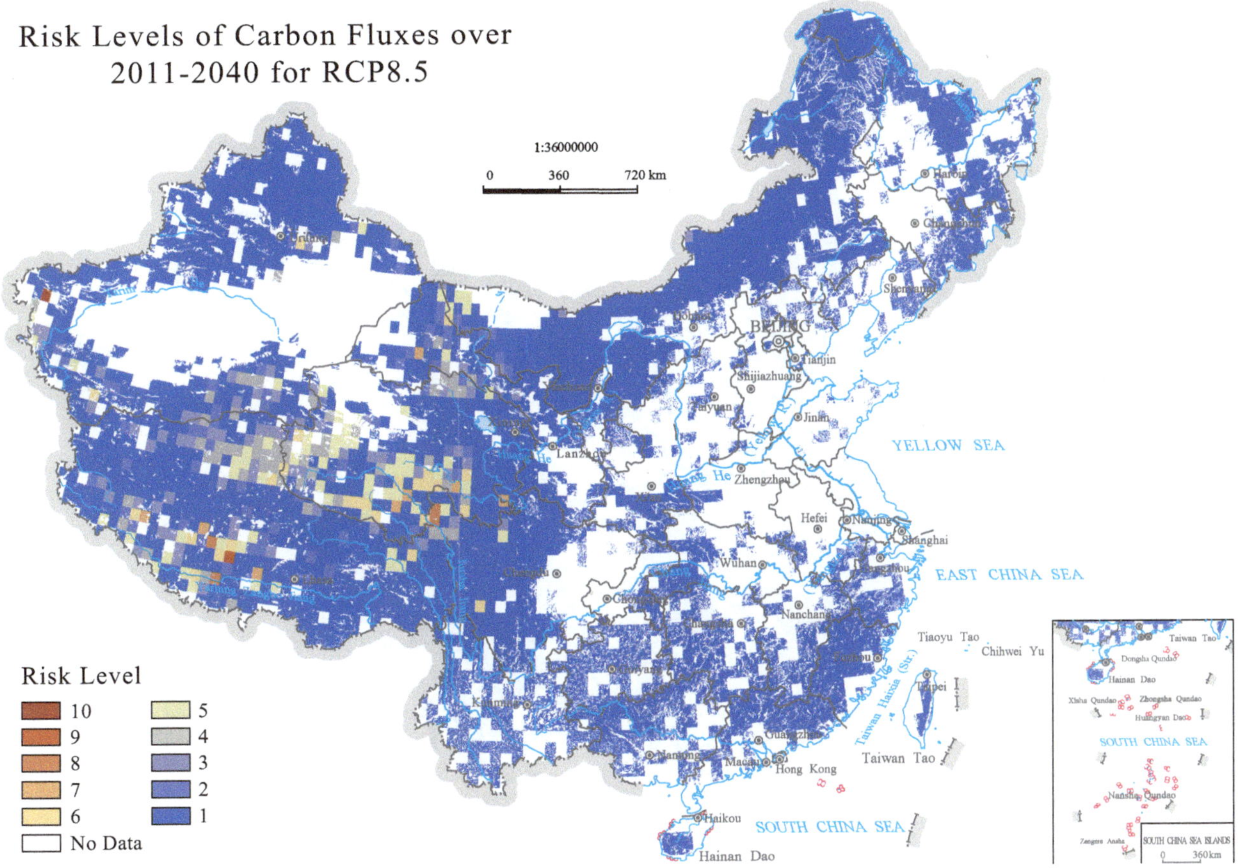

Risk Levels of Carbon Stocks over 2011-2040 for RCP8.5

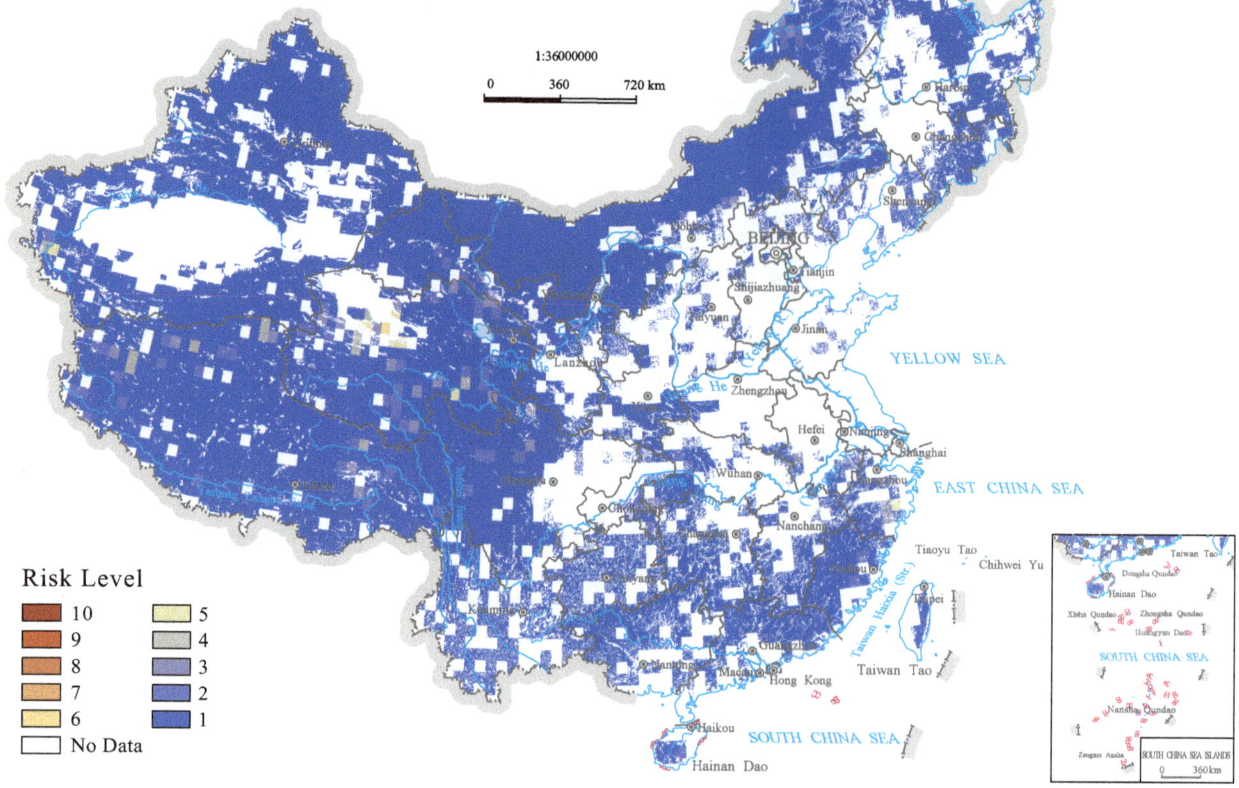

Risk Levels of Water Fluxes over 2011-2040 for RCP8.5

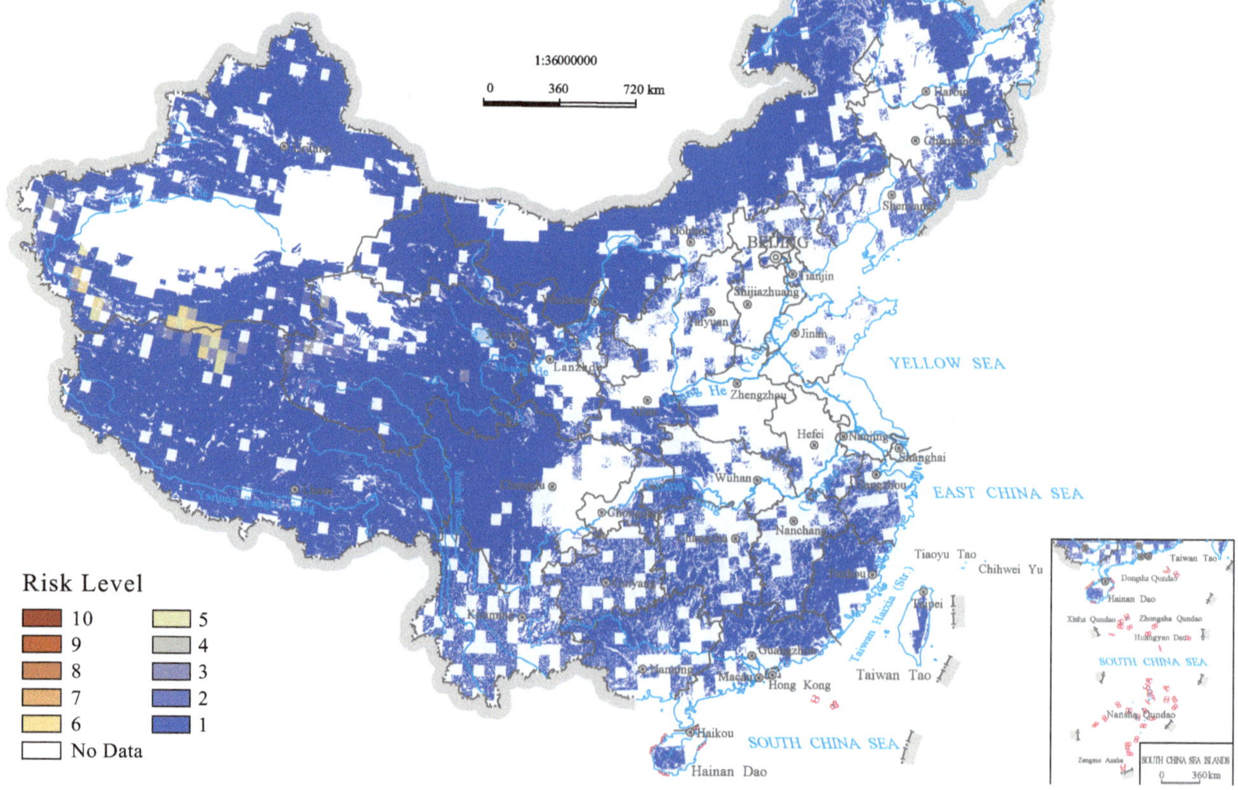

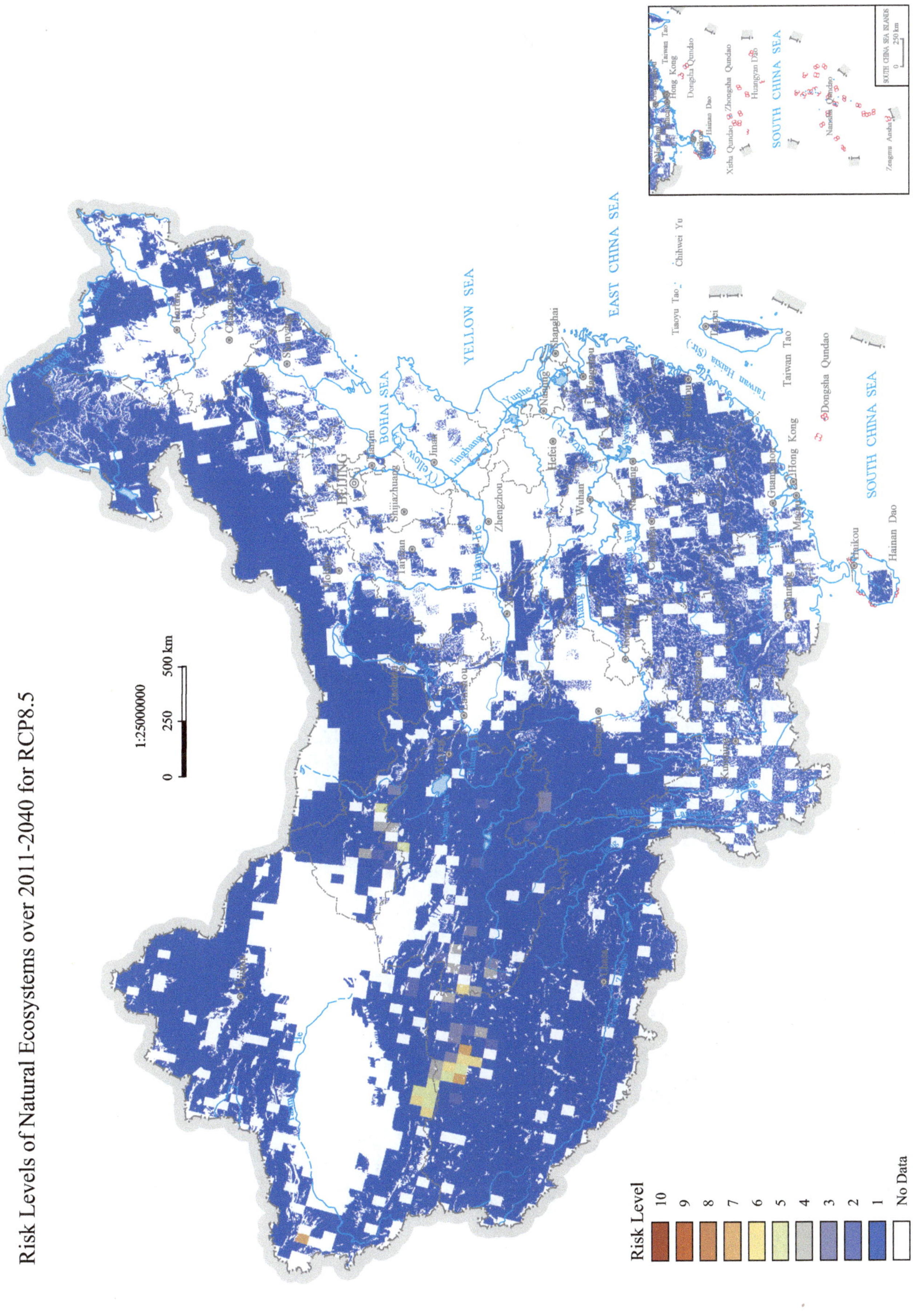
Risk Levels of Natural Ecosystems over 2011-2040 for RCP8.5
1:25000000
0 250 500 km
Risk Level
10
9
8
7
6
5
4
3
2
1
No Data
YELLOW SEA
EAST CHINA SEA
BOHAI SEA
SOUTH CHINA SEA
BEIJING
Shanghai
Zhengzhou
Wuhan
Hefei
Nanjing
Shijiazhuang
Hong Kong
Haikou
Hainan Dao
Taiwan Tao
Dongsha Qundao
Taiwan Haixia (Str.)
Chihwei Yu
Tiaoyu Tao
SOUTH CHINA SEA ISLANDS

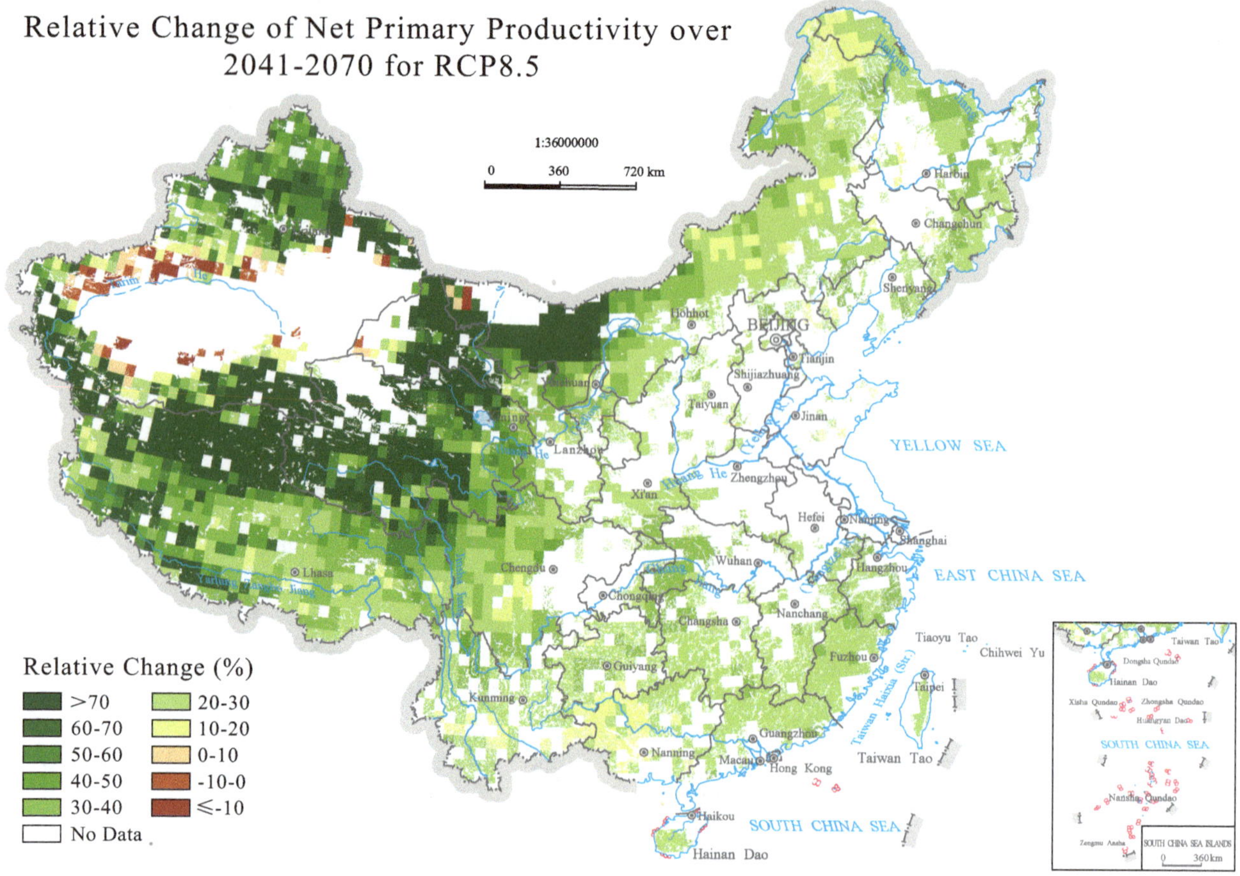
Relative Change of Net Primary Productivity over
2041-2070 for RCP8.5
1:36000000
0 360 720 km
Relative Change (%)
>70
60-70
50-60
40-50
30-40
No Data
20-30
10-20
0-10
-10-0
≤-10
Harbin
Changchun
Shenyang
Hohhot
BEIJING
Tianjin
Shijiazhuang
Taiyuan
Jinan
YELLOW SEA
Lanzhou
Zhengzhou
Xi'an
Hefei
Nanjing
Shanghai
Chengdu
Wuhan
Hangzhou
EAST CHINA SEA
Lhasa
Chongqing
Changsha
Nanchang
Tiaoyu Tao
Chihwei Yu
Fuzhou
Guiyang
Taipei
Kunming
Guangzhou
Nanning
Macao
Hong Kong
Taiwan Tao
Haikou
SOUTH CHINA SEA
Hainan Dao
Taiwan Tao
Hainan Dao
SOUTH CHINA SEA
Nansha Qundao
SOUTH CHINA SEA ISLANDS
0 360km

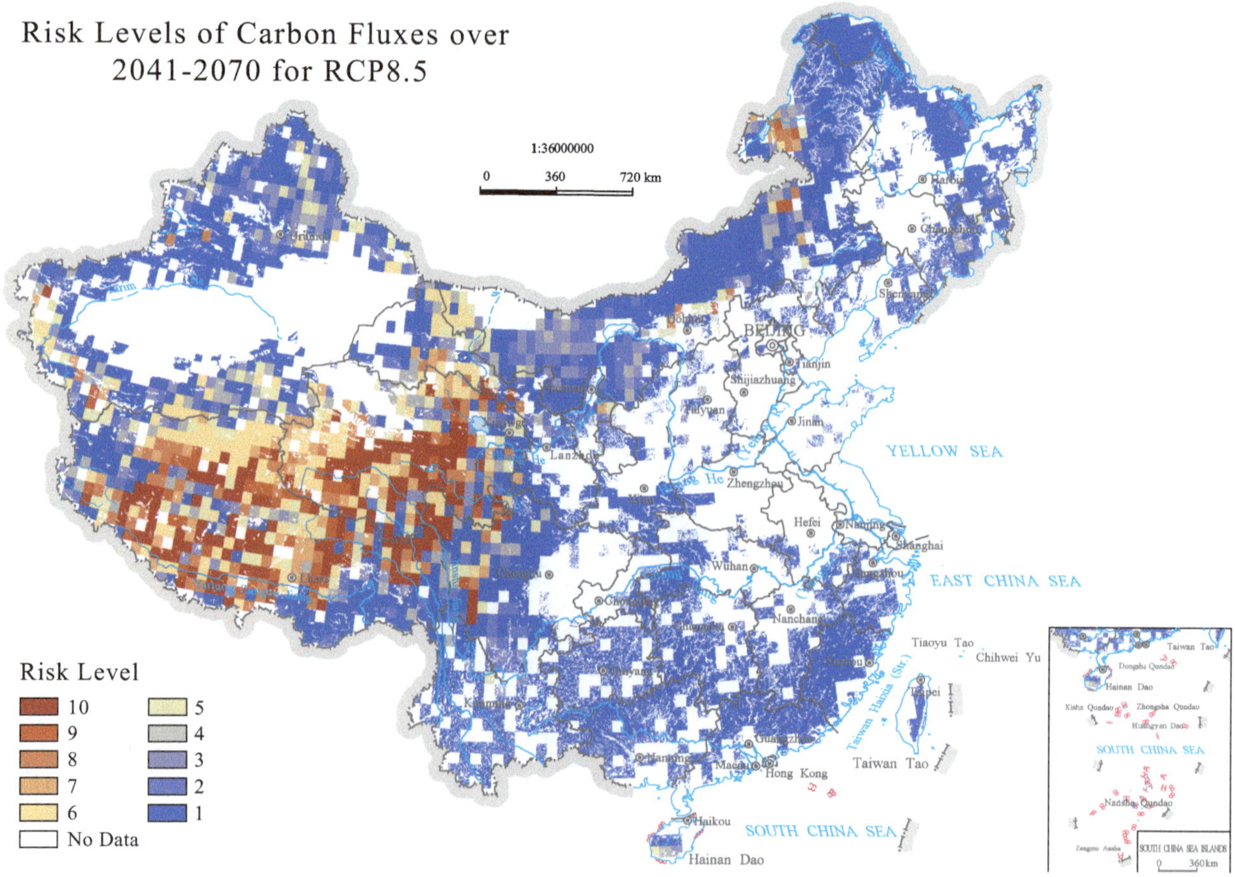
Risk Levels of Carbon Fluxes over
2041-2070 for RCP8.5
1:36000000
0 360 720 km
Risk Level
10
9
8
7
6
No Data
5
4
3
2
1
BEIJING
Tianjin
Shijiazhuang
Taiyuan
Jinan
YELLOW SEA
Lanzhou
Zhengzhou
Hefei
Nanjing
Shanghai
Wuhan
EAST CHINA SEA
Nanchang
Tiaoyu Tao
Chihwei Yu
Taipei
Macao
Hong Kong
Taiwan Tao
Haikou
SOUTH CHINA SEA
Hainan Dao
Taiwan Tao
Hainan Dao
SOUTH CHINA SEA
Nansha Qundao
SOUTH CHINA SEA ISLANDS
0 360km

Risk Levels of Carbon Stocks over 2041-2070 for RCP8.5

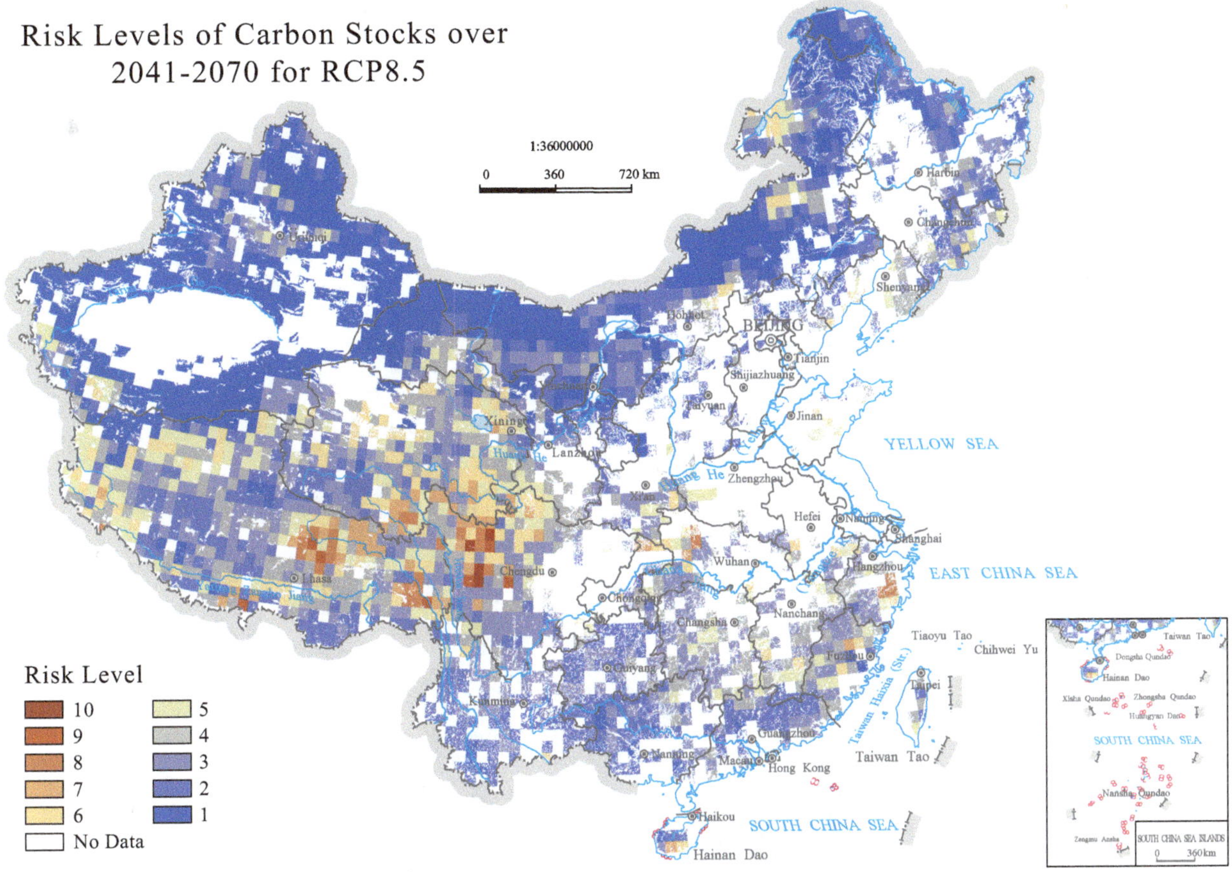

Risk Levels of Water Fluxes over 2041-2070 for RCP8.5

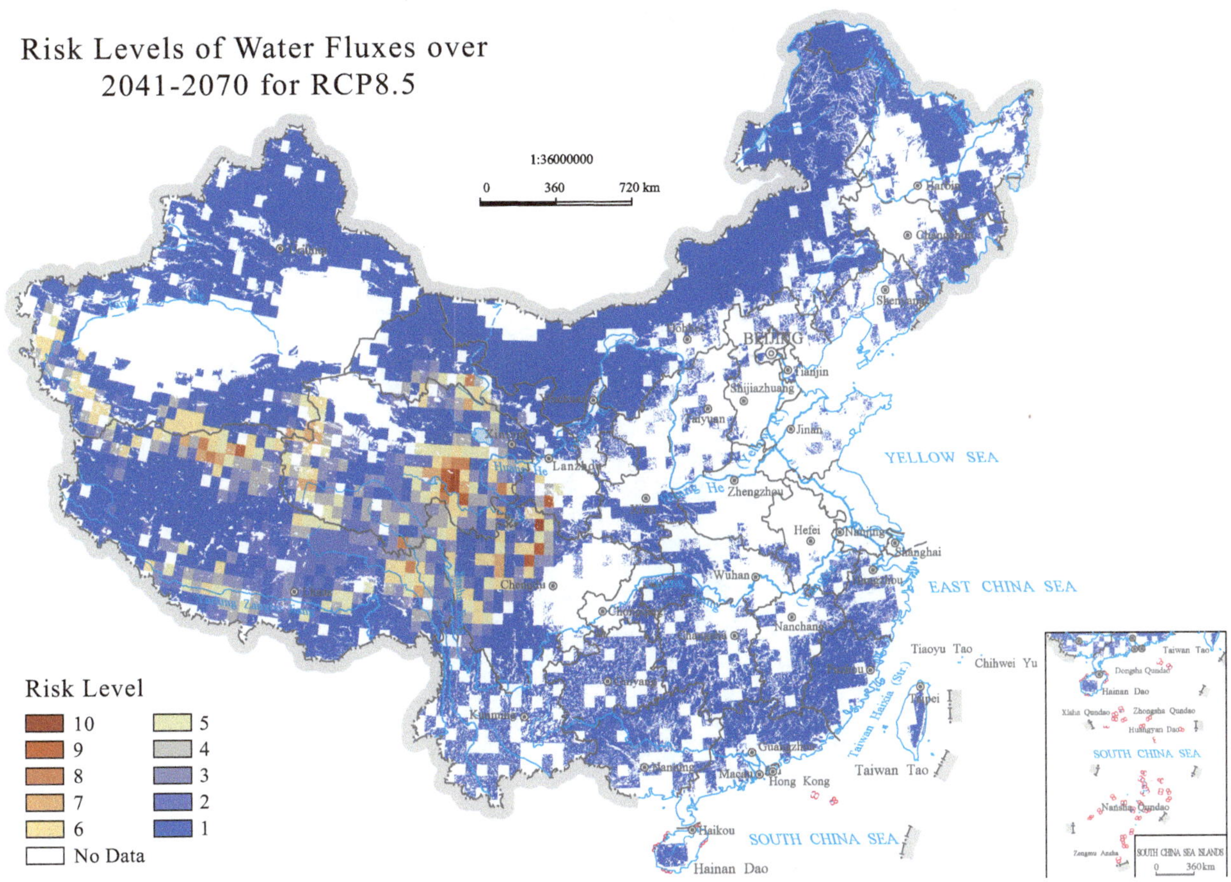

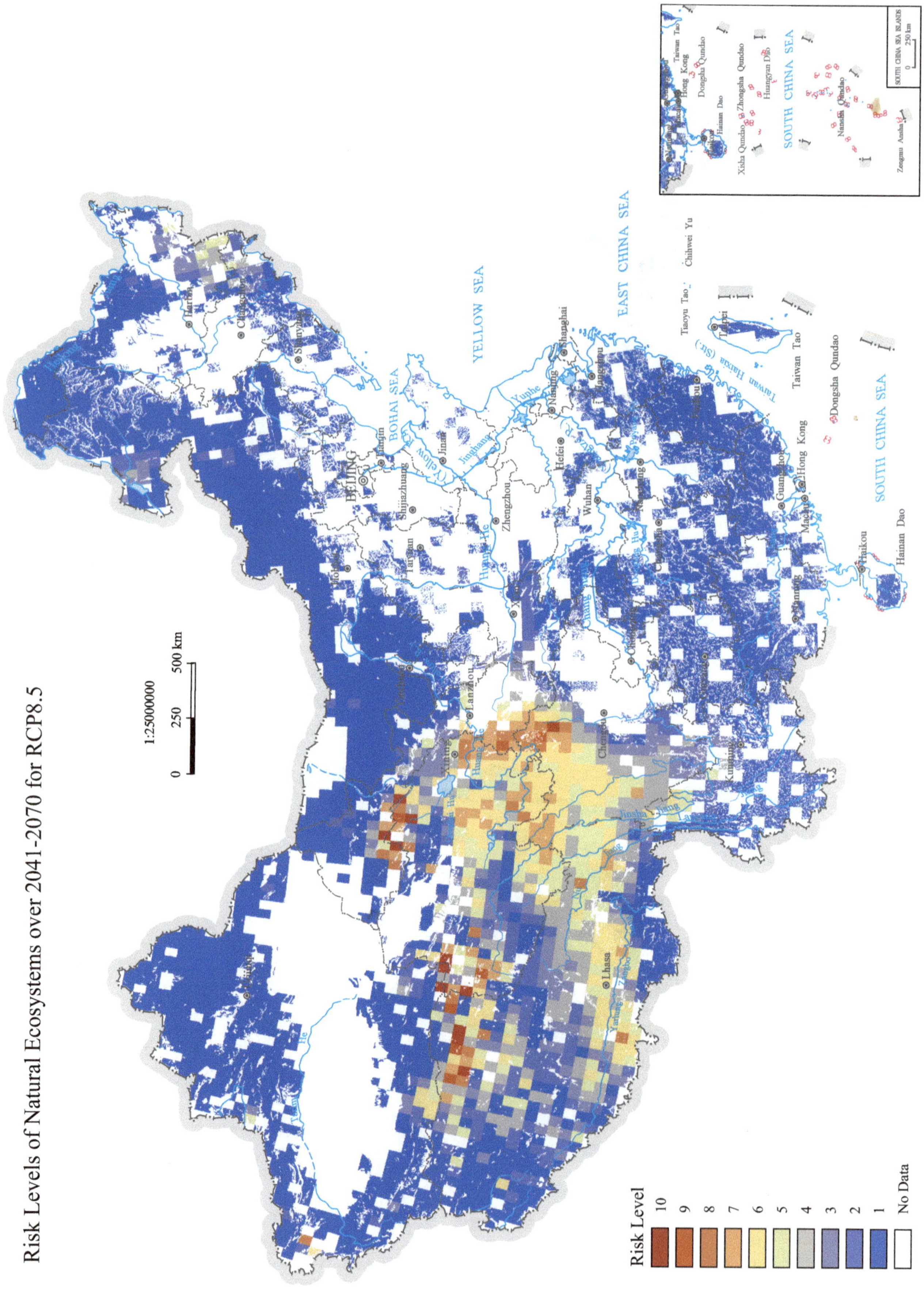
Risk Levels of Natural Ecosystems over 2041-2070 for RCP8.5
1:25000000
0 250 500 km
Risk Level
10
9
8
7
6
5
4
3
2
1
No Data
YELLOW SEA
EAST CHINA SEA
BOHAI SEA
SOUTH CHINA SEA
BEIJING
Shanghai
Lhasa
Chengdu
Lanzhou
Zhengzhou
Wuhan
Hefei
Nanjing
Shijiazhuang
Taiyuan
Harbin
Shenyang
Xining
Haikou
Hong Kong
Hainan Dao
Taiwan Tao
Dongsha Qundao
Chihwei Yu
Tiaoyu Tao
Taipei
Xisha Qundao
Zhongsha Qundao
Huangyan Dao
Nansha Qundao
Zengmu Ansha

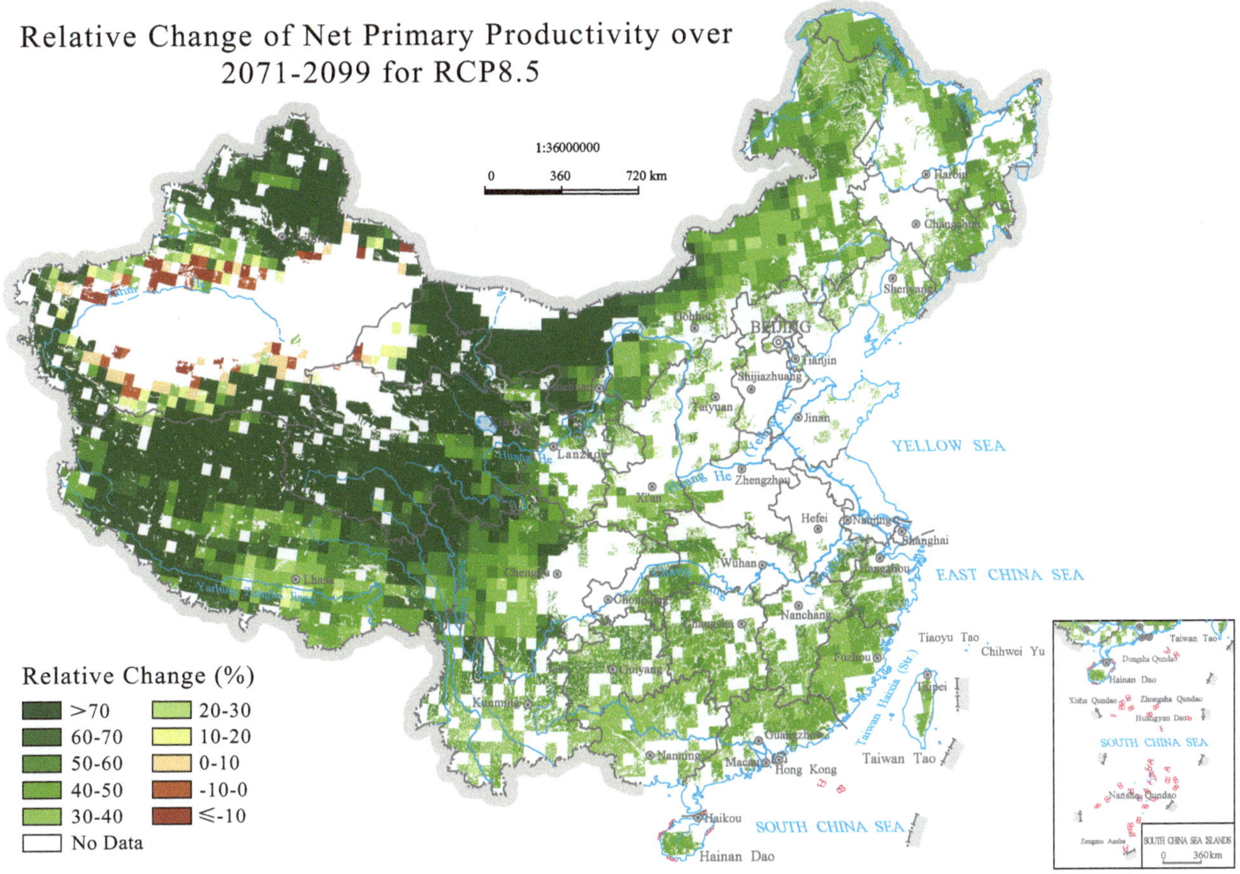
Relative Change of Net Primary Productivity over 2071-2099 for RCP8.5
1:36000000
0 360 720 km
Relative Change (%)
>70
60-70
50-60
40-50
30-40
20-30
10-20
0-10
-10-0
≤-10
No Data
BEIJING
Tianjin
Shijiazhuang
Taiyuan
Jinan
Zhengzhou
Hefei
Nanjing
Shanghai
Wuhan
Hangzhou
Nanchang
Fuzhou
Taipei
Guiyang
Kunming
Nanning
Guangzhou
Macao
Hong Kong
Haikou
Hainan Dao
Lhasa
Lanzhou
Xi'an
Harbin
Changchun
Shenyang
Hohhot
YELLOW SEA
EAST CHINA SEA
SOUTH CHINA SEA
Tiaoyu Tao
Chihwei Yu
Taiwan Tao
SOUTH CHINA SEA ISLANDS

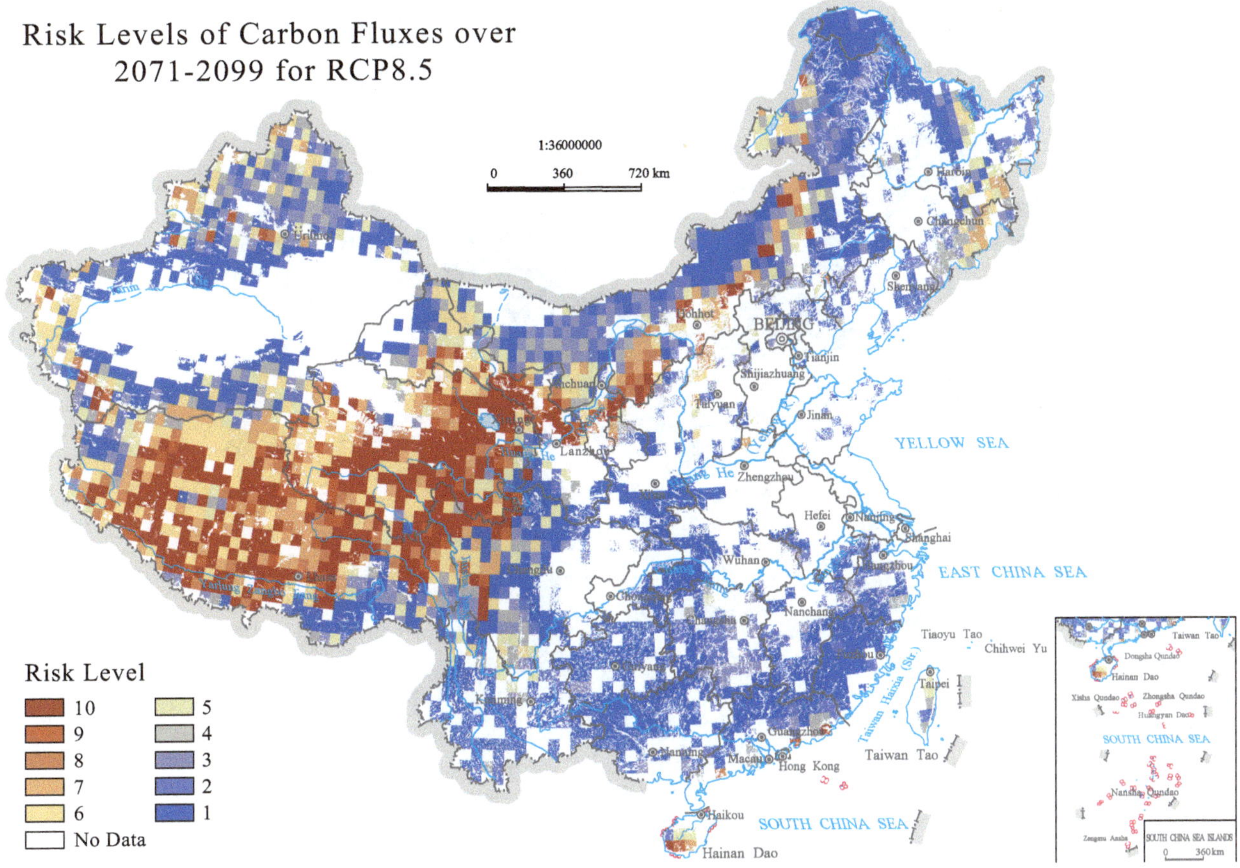
Risk Levels of Carbon Fluxes over 2071-2099 for RCP8.5
1:36000000
0 360 720 km
Risk Level
10
9
8
7
6
5
4
3
2
1
No Data
BEIJING
Tianjin
Shijiazhuang
Taiyuan
Jinan
Zhengzhou
Hefei
Nanjing
Shanghai
Wuhan
Nanchang
Taipei
Guangzhou
Macao
Hong Kong
Haikou
Hainan Dao
Lanzhou
Yinchuan
Harbin
Changchun
Shenyang
Hohhot
YELLOW SEA
EAST CHINA SEA
SOUTH CHINA SEA
Tiaoyu Tao
Chihwei Yu
Taiwan Tao
SOUTH CHINA SEA ISLANDS

Risk Levels of Carbon Stocks over 2071-2099 for RCP8.5

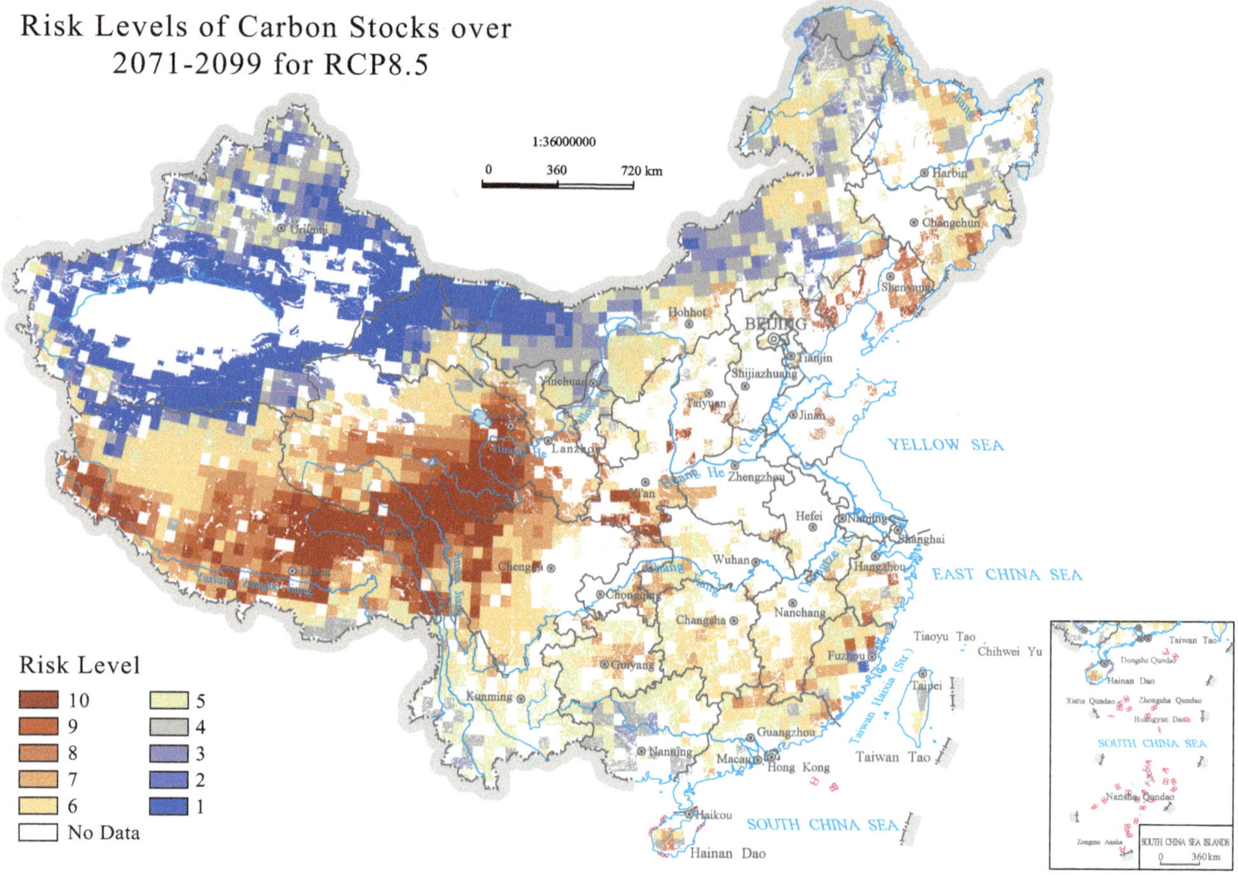

Risk Levels of Water Fluxes over 2071-2099 for RCP8.5

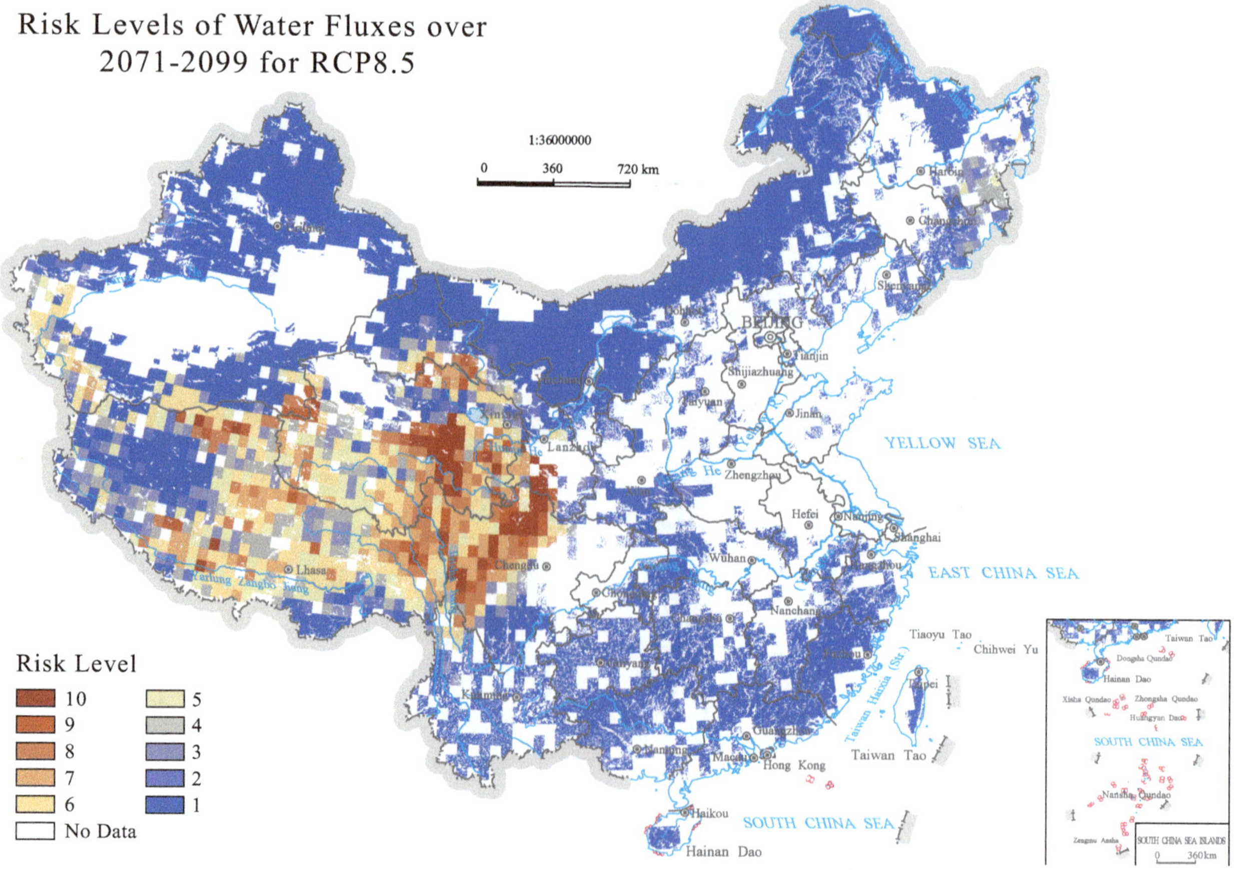

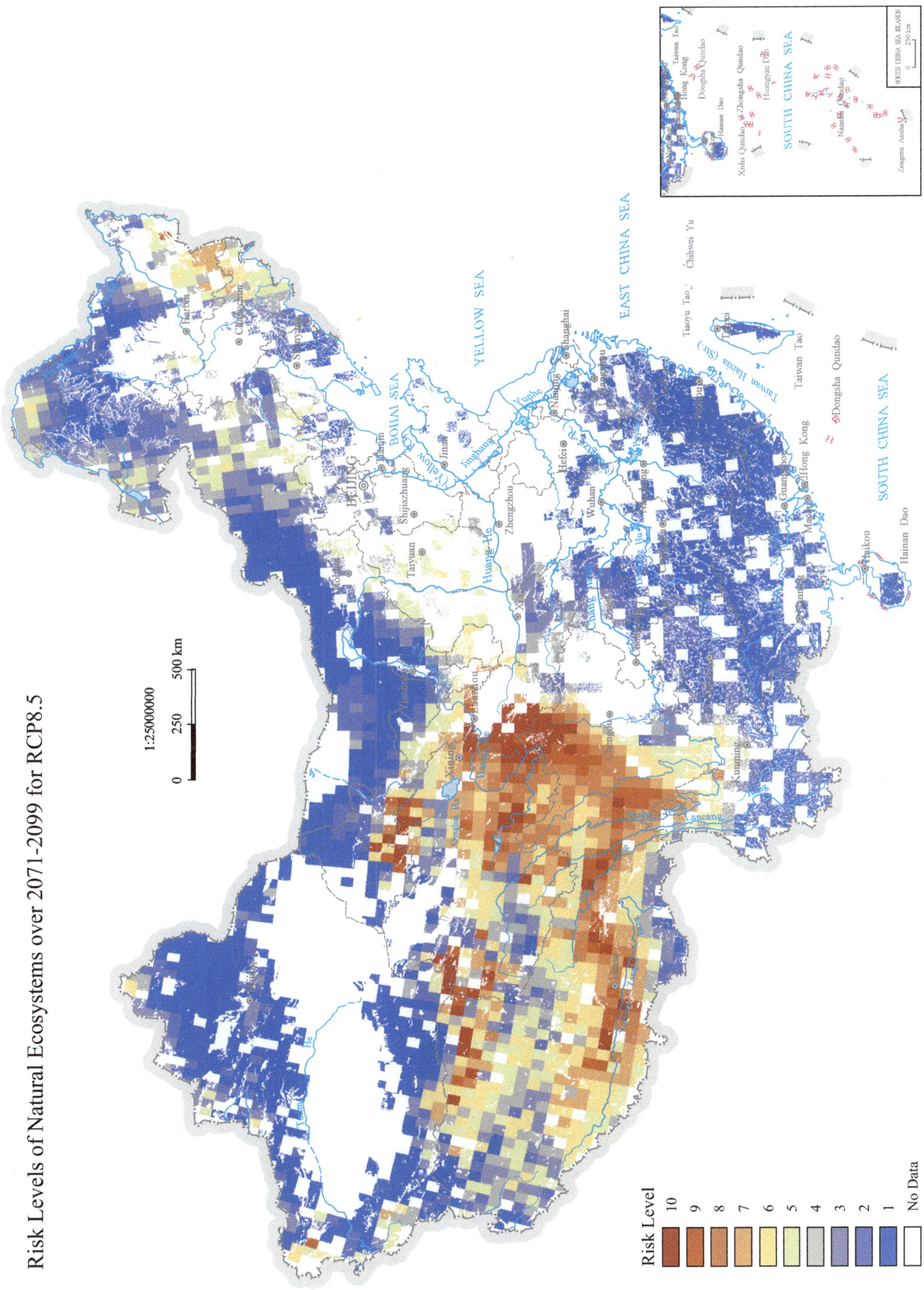
Risk Levels of Natural Ecosystems over 2071-2099 for RCP8.5
1:25000000
0 250 500 km
Risk Level
10
9
8
7
6
5
4
3
2
1
No Data
BEIJING
Tianjin
Shijiazhuang
Taiyuan
Jinan
Zhengzhou
Hefei
Wuhan
Nanjing
Shanghai
Changchun
Shenyang
Harbin
Xining
Chengdu
Kunming
Guangzhou
Hong Kong
Macao
Haikou
Taipei
YELLOW SEA
BOHAI SEA
EAST CHINA SEA
SOUTH CHINA SEA
Taiwan Tao
Dongsha Qundao
Hainan Dao
Diaoyu Dao
Chiwei Yu
Xisha Qundao
Zhongsha Qundao
Nansha Qundao
Huangyan Dao
Zengmu Ansha
SOUTH CHINA SEA ISLANDS
250 km
Huang He
Chang Jiang
Yarlung Zangbo
Lancang
Jinsha

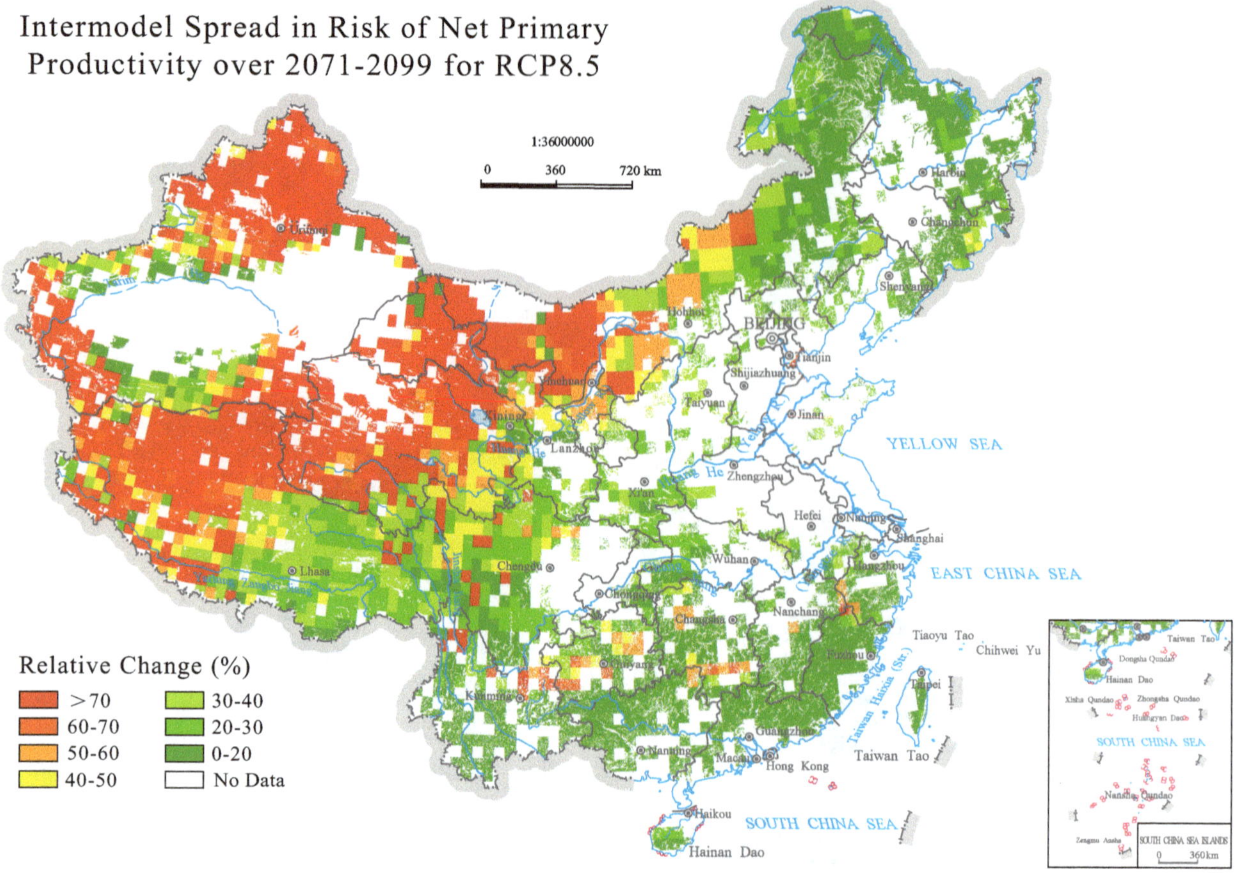
Intermodel Spread in Risk of Net Primary Productivity over 2071-2099 for RCP8.5
1:36000000
0 360 720 km
Relative Change (%)
>70
60-70
50-60
40-50
30-40
20-30
0-20
No Data
Harbin
Changchun
Shenyang
Hohhot
BEIJING
Tianjin
Shijiazhuang
Taiyuan
Jinan
Yinchuan
Xining
Lanzhou
Zhengzhou
Xi'an
Hefei
Nanjing
Shanghai
Wuhan
Hangzhou
Chengdu
Chongqing
Changsha
Nanchang
Fuzhou
Guiyang
Kunming
Nanning
Guangzhou
Macau
Hong Kong
Taipei
Haikou
Lhasa
YELLOW SEA
EAST CHINA SEA
SOUTH CHINA SEA
Taiwan Tao
Tiaoyu Tao
Chihwei Yu
Hainan Dao
SOUTH CHINA SEA ISLANDS

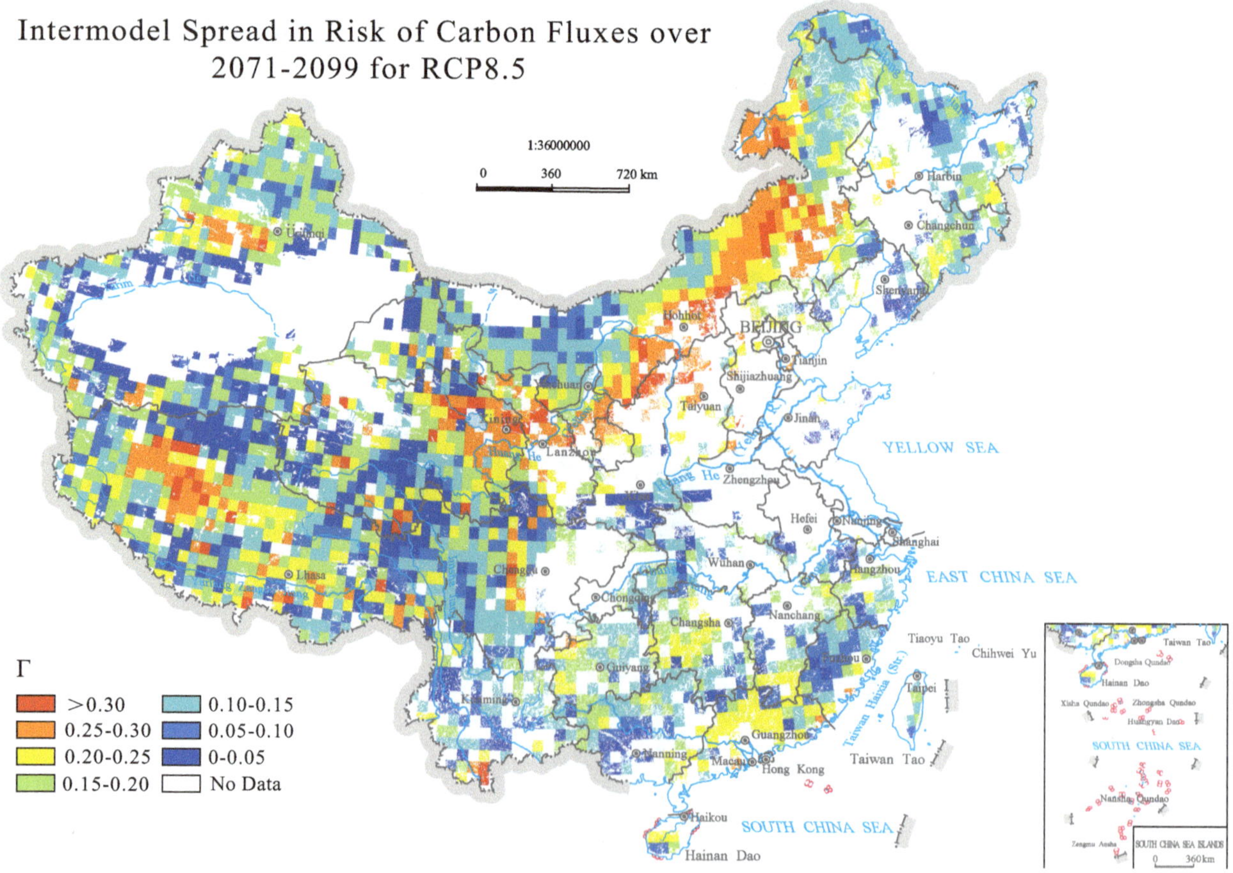
Intermodel Spread in Risk of Carbon Fluxes over 2071-2099 for RCP8.5
1:36000000
0 360 720 km
Γ
>0.30
0.25-0.30
0.20-0.25
0.15-0.20
0.10-0.15
0.05-0.10
0-0.05
No Data
Harbin
Changchun
Shenyang
Hohhot
BEIJING
Tianjin
Shijiazhuang
Taiyuan
Jinan
Lanzhou
Zhengzhou
Hefei
Nanjing
Shanghai
Wuhan
Hangzhou
Chongqing
Changsha
Nanchang
Fuzhou
Guiyang
Kunming
Nanning
Guangzhou
Macau
Hong Kong
Taipei
Haikou
Lhasa
YELLOW SEA
EAST CHINA SEA
SOUTH CHINA SEA
Taiwan Tao
Tiaoyu Tao
Chihwei Yu
Hainan Dao
SOUTH CHINA SEA ISLANDS

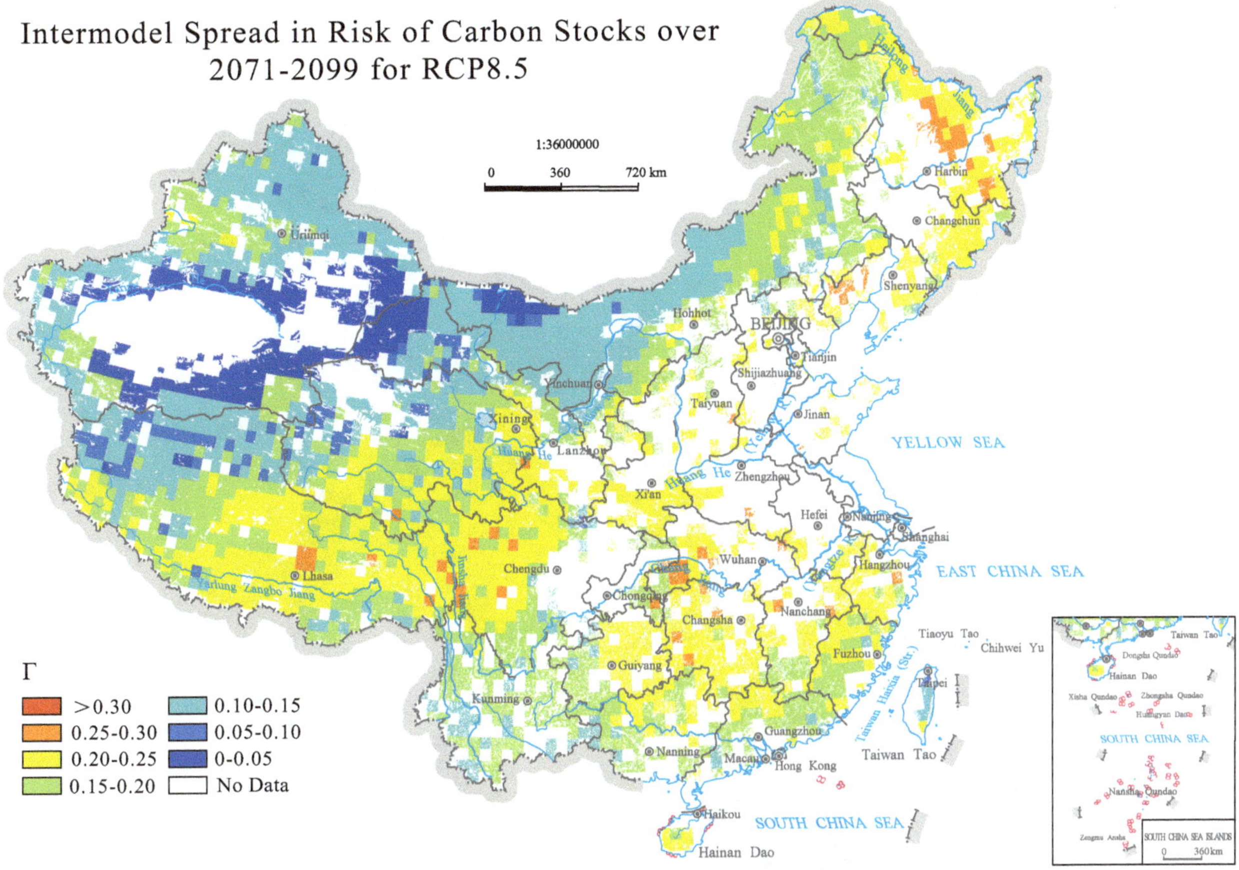
Intermodel Spread in Risk of Carbon Stocks over
2071-2099 for RCP8.5
1:36000000
0 360 720 km
Γ
>0.30
0.25-0.30
0.20-0.25
0.15-0.20
0.10-0.15
0.05-0.10
0-0.05
No Data
Urümqi
Harbin
Changchun
Shenyang
Hohhot
BEIJING
Tianjin
Shijiazhuang
Taiyuan
Yinchuan
Jinan
Xining
Lanzhou
Zhengzhou
Xi'an
Hefei
Nanjing
Shanghai
Wuhan
Hangzhou
Chengdu
Lhasa
Chongqing
Changsha
Nanchang
Fuzhou
Guiyang
Kunming
Taipei
Guangzhou
Nanning
Macao
Hong Kong
Haikou
Hainan Dao
Taiwan Tao
Tiaoyu Tao
Chihwei Yu
YELLOW SEA
EAST CHINA SEA
SOUTH CHINA SEA
Yarlung Zangbo Jiang
Huang He
SOUTH CHINA SEA ISLANDS
0 360km

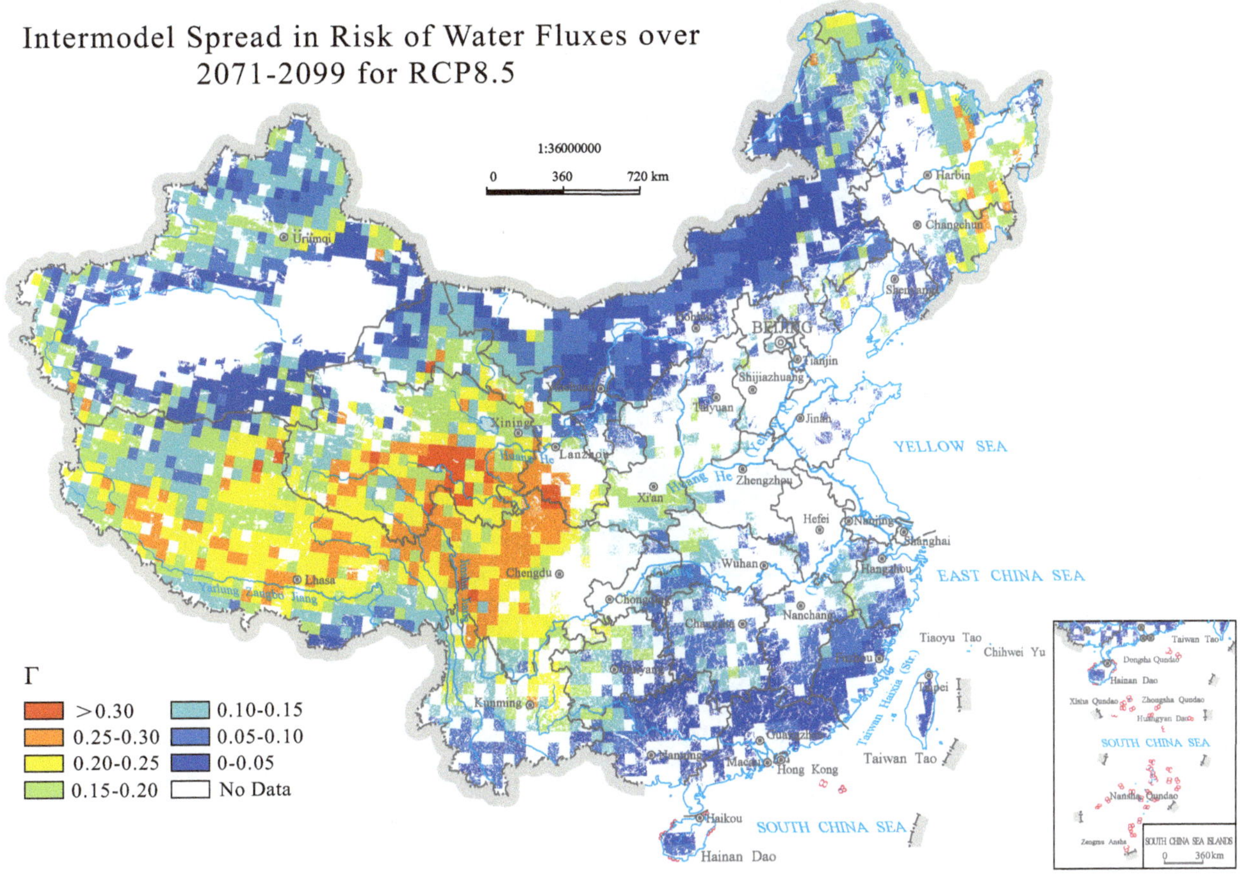
Intermodel Spread in Risk of Water Fluxes over
2071-2099 for RCP8.5
1:36000000
0 360 720 km
Γ
>0.30
0.25-0.30
0.20-0.25
0.15-0.20
0.10-0.15
0.05-0.10
0-0.05
No Data
Urümqi
Harbin
Changchun
BEIJING
Tianjin
Shijiazhuang
Taiyuan
Xining
Lanzhou
Zhengzhou
Xi'an
Hefei
Nanjing
Shanghai
Wuhan
Hangzhou
Chengdu
Lhasa
Nanchang
Kunming
Taipei
Macao
Hong Kong
Haikou
Hainan Dao
Taiwan Tao
Tiaoyu Tao
Chihwei Yu
YELLOW SEA
EAST CHINA SEA
SOUTH CHINA SEA
SOUTH CHINA SEA ISLANDS
0 360km

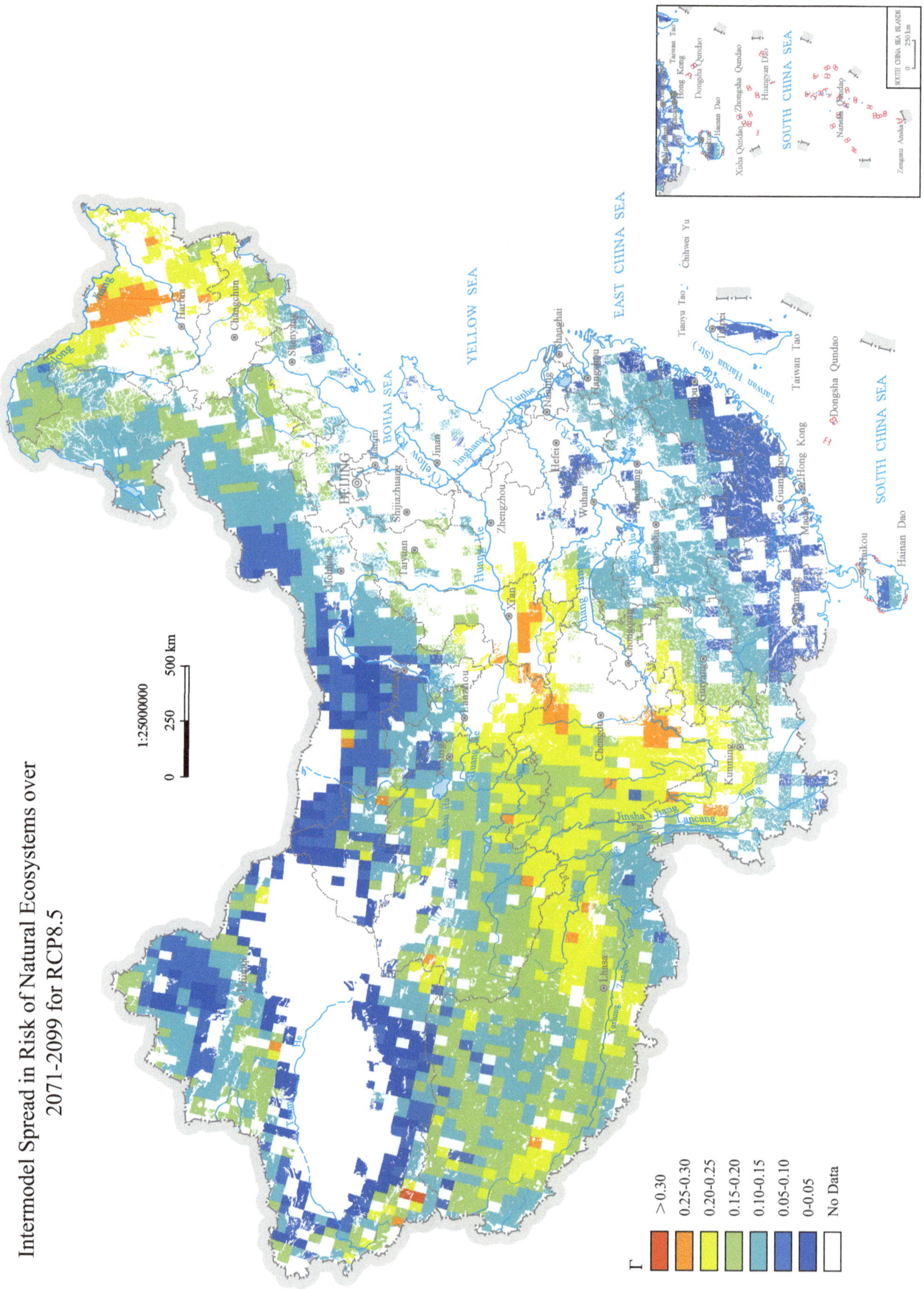
Intermodel Spread in Risk of Natural Ecosystems over
2071-2099 for RCP8.5
1:25000000
0 250 500 km
Γ
>0.30
0.25-0.30
0.20-0.25
0.15-0.20
0.10-0.15
0.05-0.10
0-0.05
No Data
BEIJING
BOHAI SEA
YELLOW SEA
EAST CHINA SEA
SOUTH CHINA SEA
Taiwan Tao
Hainan Dao
Dongsha Qundao
Hong Kong
Macao
Shanghai
Nanjing
Hefei
Wuhan
Zhengzhou
Xi'an
Chengdu
Chongqing
Kunming
Guiyang
Lhasa
Haikou
Taipei
Shijiazhuang
Taiyuan
Jinan
Tianjin
Harbin
Changchun
Shenyang
Guangzhou
Jinsha Jiang
Lancang
Diaoyu Dao
Chiwei Yu
Xisha Qundao
Zhongsha Qundao
Nansha Qundao
Huangyan Dao
SOUTH CHINA SEA ISLANDS

References

Donohue, R. J., Roderick, M. L., McVicar, T. R., & Farquhar, G. D. (2013). Impact of CO_2 fertilization on maximum foliage cover across the globe's warm, arid environments. *Geophysical Research Letters, 40,* 3031–3035.

Heyder, U., Schaphoff, S., Gerten, D., & Lucht, W. (2011). Risk of severe climate change impact on the terrestrial biosphere. *Environmental Research Letters, 6,* 034036. doi:10.1088/1748-9326/6/3/034036.

IPCC (Intergovernmental Panel on Climate Change). (2014) *Climate change 2014: impacts, adaptation, and vulnerability.* In *Contribution of working group ii to the fifth assessment report of the intergovernmental panel on climate change.* Cambridge, United Kingdom and New York, NY USA: Cambridge University Press.

Ni, J. (2011). Impacts of climate change on Chinese ecosystems: Key vulnerable regions and potential thresholds. *Regional Environmental Change, 11,* S49–S64.

Piao, S., Fang, J., Ciais, P., Peylin, P., Huang, Y., Sitch, S., et al. (2009). The carbon balance of terrestrial ecosystems in China. *Nature, 458,* 1109–1114.

Scheffer, M., & Carpenter, S. M. (2003). Catastrophic regime shifts in ecosystems: Linking theory to observation. *Trends in Ecology Evolution, 18,* 648–656.

Tao, F., & Zhang, Z. (2010). Dynamic responses of terrestrial ecosystems structure and function to climate in China. *Journal Geophysical Research, 115,* G03003. doi:10.1029/2009JG001062.

Taylor, K. E., Stouffer, R. J. & Meehl, G. A. (2012). An overview of CMIP5 and the experiment design. *Bulletin of the American Meteorological Society, 93,* 485–498. Under Geophysical Research, 115, G03003. doi:10.1029/2009JG001062

Wang, H. (2014). A multi-model assessment of climate change impacts on the distribution and productivity of ecosystems in China. *Regional Environmental Change, 14,* 133–144.

Yin, Y., Tang, Q., Wang, L., & Liu, X. (2016). Risk and contributing factors of ecosystem shifts over naturally vegetated land under climate change in China. *Scientific Reports, 6,* 20905.

Integrated Environmental Risks

6

Xingcai Liu, Yuanyuan Yin, Xinchuang Xu, Qiuhong Tang and Huijuan Cui

6.1 Introduction

Most environmental risks have been recognized as top global risks in terms of their likelihood and impact (WEF 2015, 2016). In recent years, environmental risks facing society are apparently systematic in nature and greatly impaired ecosystem and resulted in huge losses to socioeconomic domain (Briggs 2008). Thus, integrated assessment of environmental risks is necessary for risk management. Several integrated risk indices have been developed to describe global risk patterns or trends (Dilley 2005; Shi 2009; Joern et al. 2014). For example, the United Nations Development Programme (UNDP 2004) has developed a disaster risk index (DRI) to assess the global integrated risk by combining the exposure of human to the hazards such as flood, drought, and cyclone. The United Nations University (Joern et al. 2014) has promulgated a world risk index (WRI) for the assessment and estimation of disaster risk for a country. WRI takes into consideration both external and internal factors, i.e., threats by natural hazards, climate change, and societal conditions. WRI was calculated for 171 countries by combining the exposure of individual countries to natural hazards, and the socioeconomic and ecological conditions of these countries. Shi and Kasperson (2014) used a total risk index (TRI) and a multi-hazard risk index (MRhI) to estimate the world risk with respect to annual mortality and affected population.

Many assessments have reported that China is being or will be under relatively high environmental risks (Dilley 2005). In this chapter, an integrated assessment of environmental risks focuses on the risks estimation for human, society, and natural system under changing climate was presented. To achieve this, the climate change risks in multiple sectors were assessed using multiple climate impact models (Warszawski et al. 2014; Tang and Oki 2016). The risk assessments of heat wave, agricultural production, and ecosystem shift were then combined into one integrated assessment. It is expected to present a more comprehensive and advanced assessment of China's environmental risks based on the climate impact assessments across sectors under new climate and socioeconomic scenarios developed for the Intergovernmental Panel on Climate Change (IPCC).

6.2 Methodology

In this study, we adopted a variation of the integrated environmental risk index (IERI) similar to the TRI after Shi and Kasperson (2014). First, three disaster hazards, i.e., the heat wave hazard, agricultural hazard in terms of crop yield, and ecosystem shift hazard, were estimated under various climate scenarios in the future. The corresponding risks were then estimated and normalized to the consistent risk levels (from 1 to 10 as risks increase) and finally, the IERI was calculated as the weighted sum of the three disaster risks using the following equation:

$$IERI = \sum_{i=1}^{n=3} r_i \times w_i \quad (6.1)$$

where r_i is the risk level of a disaster, w_i is the weight of the disaster. In this study, equal weights were used for IERI calculation.

X. Liu · Y. Yin · Q. Tang (✉)
Key Laboratory of Water Cycle and Related Land Surface Processes, Institute of Geographic Sciences and Natural Resources Research, Chinese Academy of Sciences, Beijing, 100101, China
e-mail: tangqh@igsnrr.ac.cn

X. Xu
The School of Resources Environment Science and Technology, Hubei University of Science and Technology, Xianning, 437100, China

H. Cui
Key Laboratory of Land Surface Pattern and Simulation, Institute of Geographic Sciences and Natural Resources Research, Chinese Academy of Sciences, Beijing, 100101, China

Q. Tang and Q. Ge (eds.), *Atlas of Environmental Risks Facing China Under Climate Change*,
IHDP/Future Earth-Integrated Risk Governance Project Series, DOI 10.1007/978-981-10-4199-0_6

Climate data from five general circulation models (GCMs) under four representative concentration pathways (RCPs) were used to drive the impact models to estimate the risks in China in the twenty-first century. The uncertainties cross GCMs and impact models were also addressed in terms of standard deviations of risk levels for each single RCP scenario. The estimations of the risks were described in Chaps. 3–5 in this atlas. The integrated environmental risk zones (IERZ) were then delineated based on the IERI estimates.

6.3 Results

The integrated environmental risks were estimated for the four RCPs, i.e. RCP2.6, RCP4.5, RCP6.0, and RCP8.5, respectively. The IERI values were also generalized to ten levels denoted as 1–10 from low to high risk. In general, there were relatively low risks in China under RCP2.6, and relatively high risks under RCP8.5. Both the maps of IERI and the associated uncertainty will be shown in the Maps section.

Six level-I IERZs, i.e., the Northwest-Low-Risk region, Northeast-Relatively-Low-Risk region, Tibet-Moderate-Risk region, Jin-Shaan-Moderate-Risk region, South-China-High-Risk region, and Huang-Huai-Hai-High-Risk region, were delineated using the top-down approach for China under RCP8.5 and RCP6.0 scenarios. Level-II IERZs were delineated from bottom to top by combining local counties that have similar environmental risks. According to the characteristics of local environmental risks, total 42 and 54 level-II IERZs were further identified under RCP8.5 and RCP6.0 scenarios, respectively.

6.4 Maps

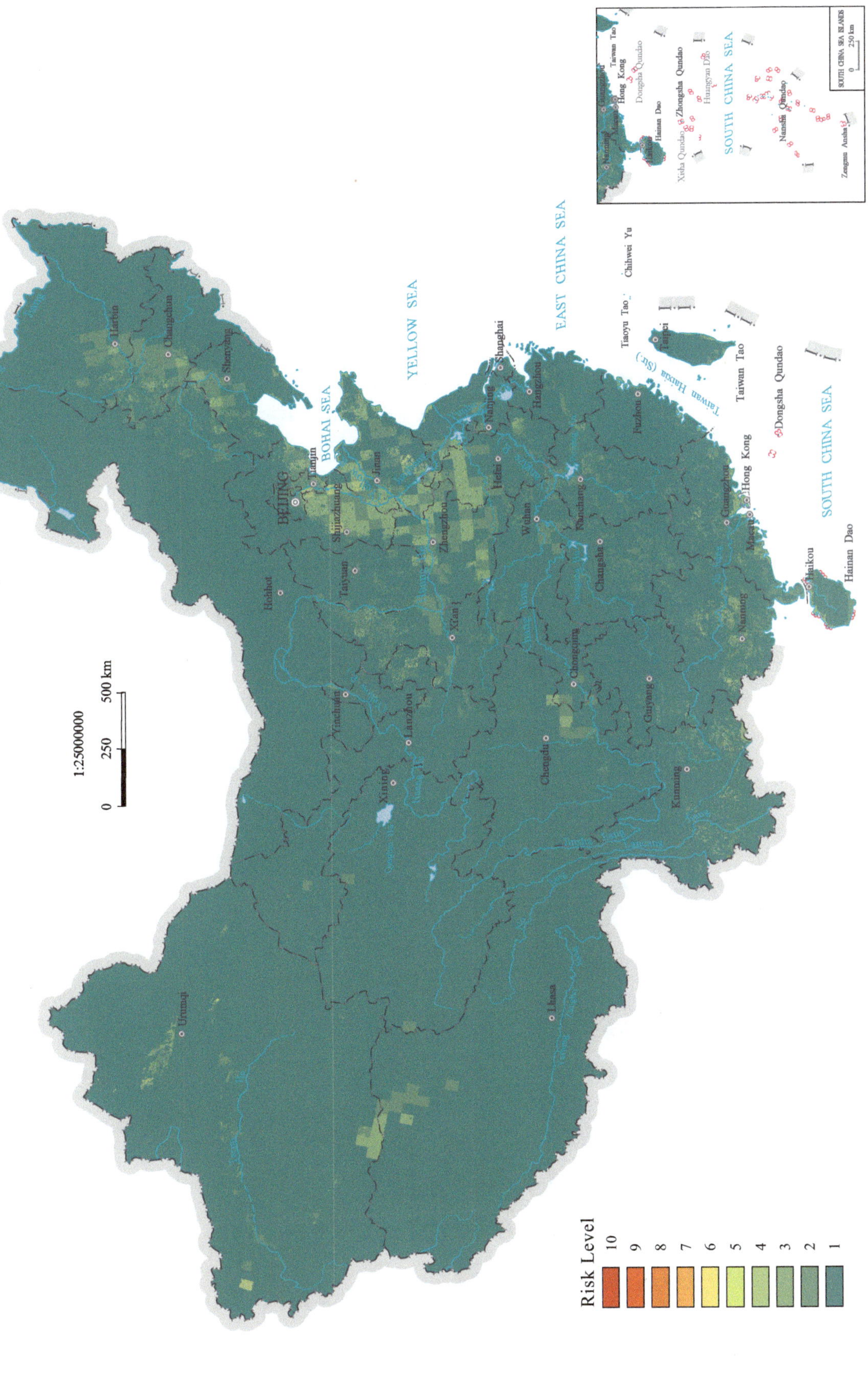

Integrated Environmental Risk over 2011-2040 for RCP2.6

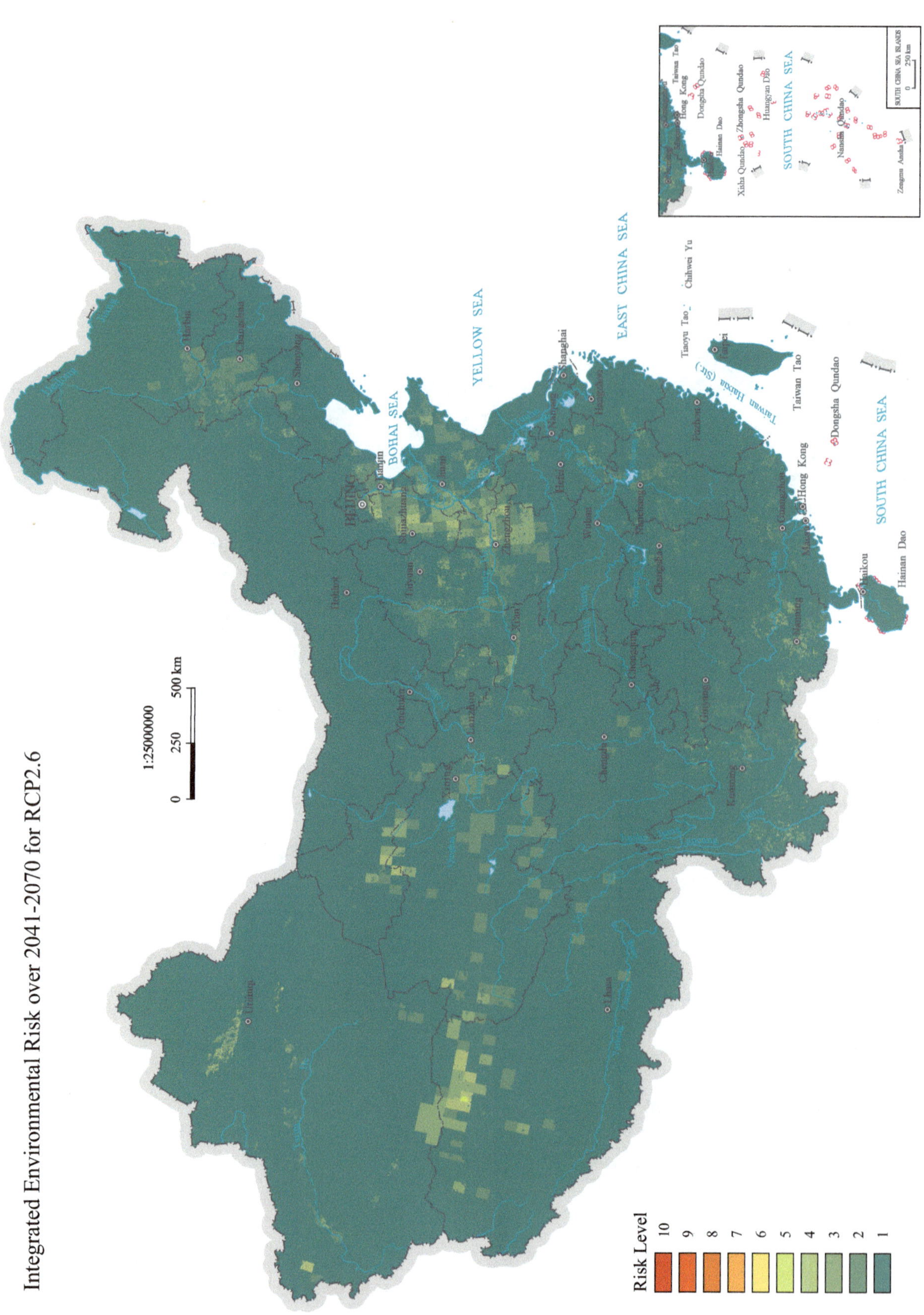
Integrated Environmental Risk over 2041-2070 for RCP2.6
1:25000000
0 250 500 km
Risk Level
10
9
8
7
6
5
4
3
2
1
BOHAI SEA
YELLOW SEA
EAST CHINA SEA
SOUTH CHINA SEA
BEIJING
Shanghai
Hong Kong
Taiwan Tao
Hainan Dao
Dongsha Qundao
SOUTH CHINA SEA ISLANDS

Integrated Environmental Risk over 2071-2099 for RCP2.6

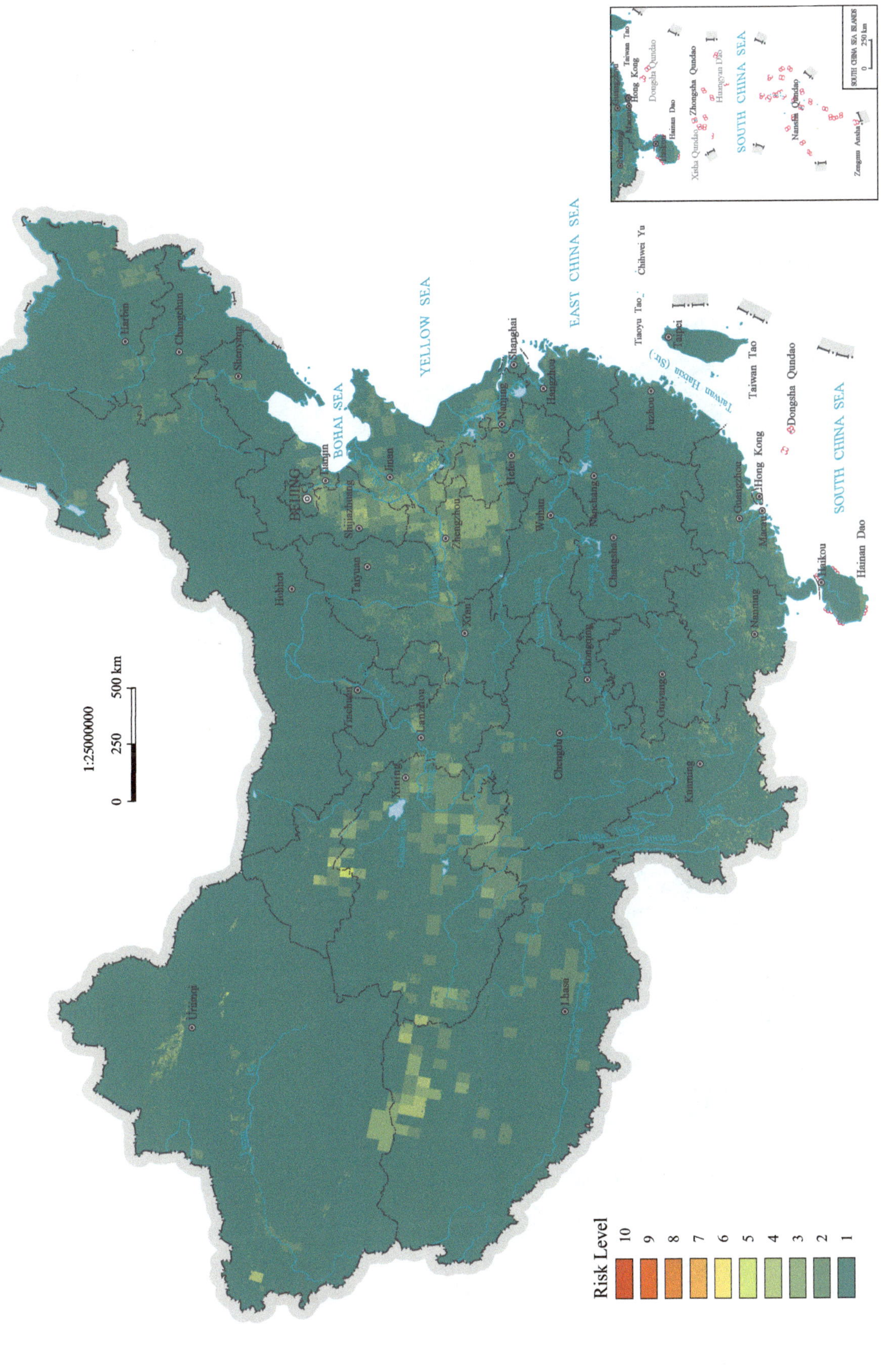

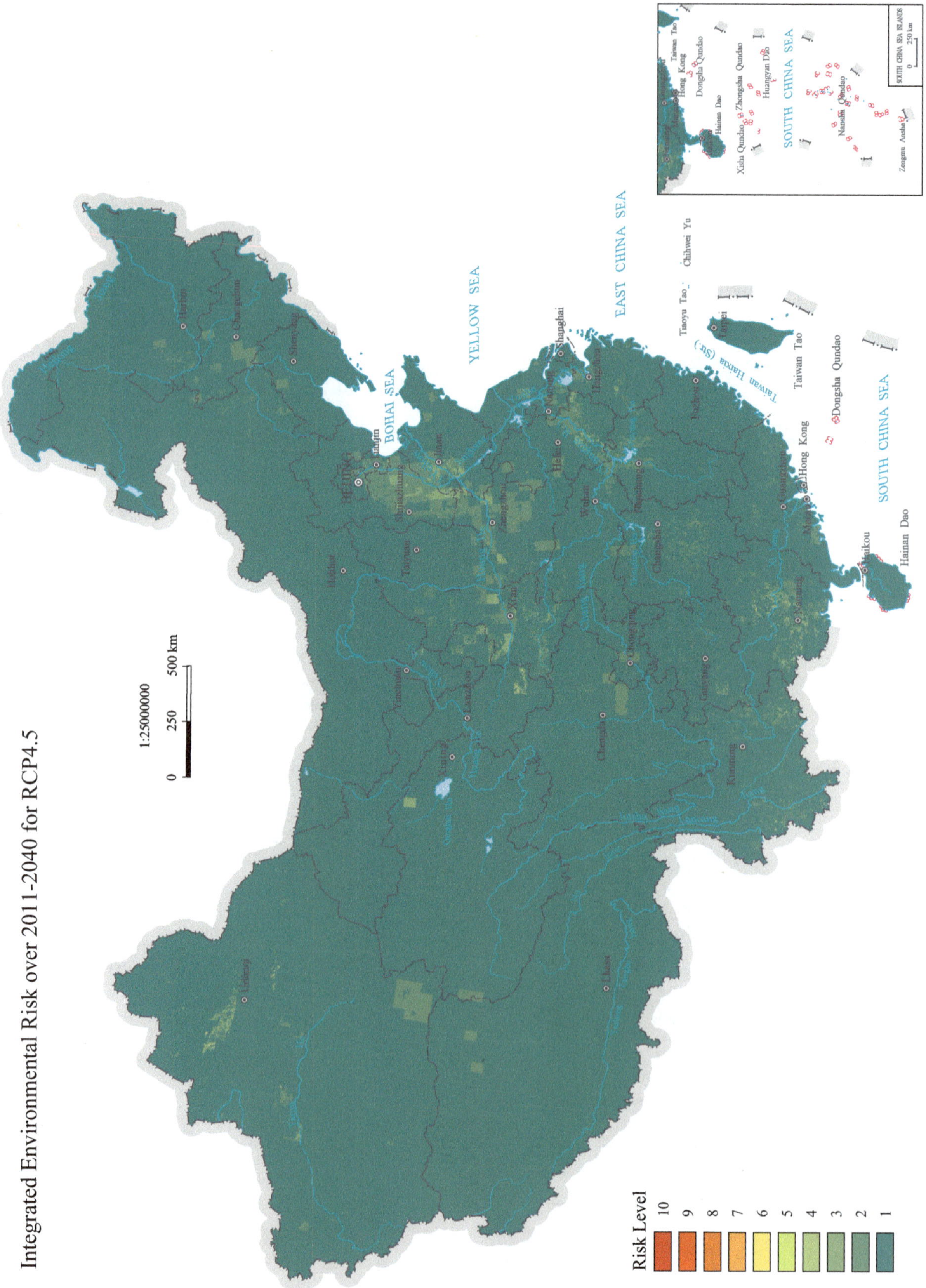

Integrated Environmental Risk over 2011-2040 for RCP4.5

Integrated Environmental Risk over 2041-2070 for RCP4.5

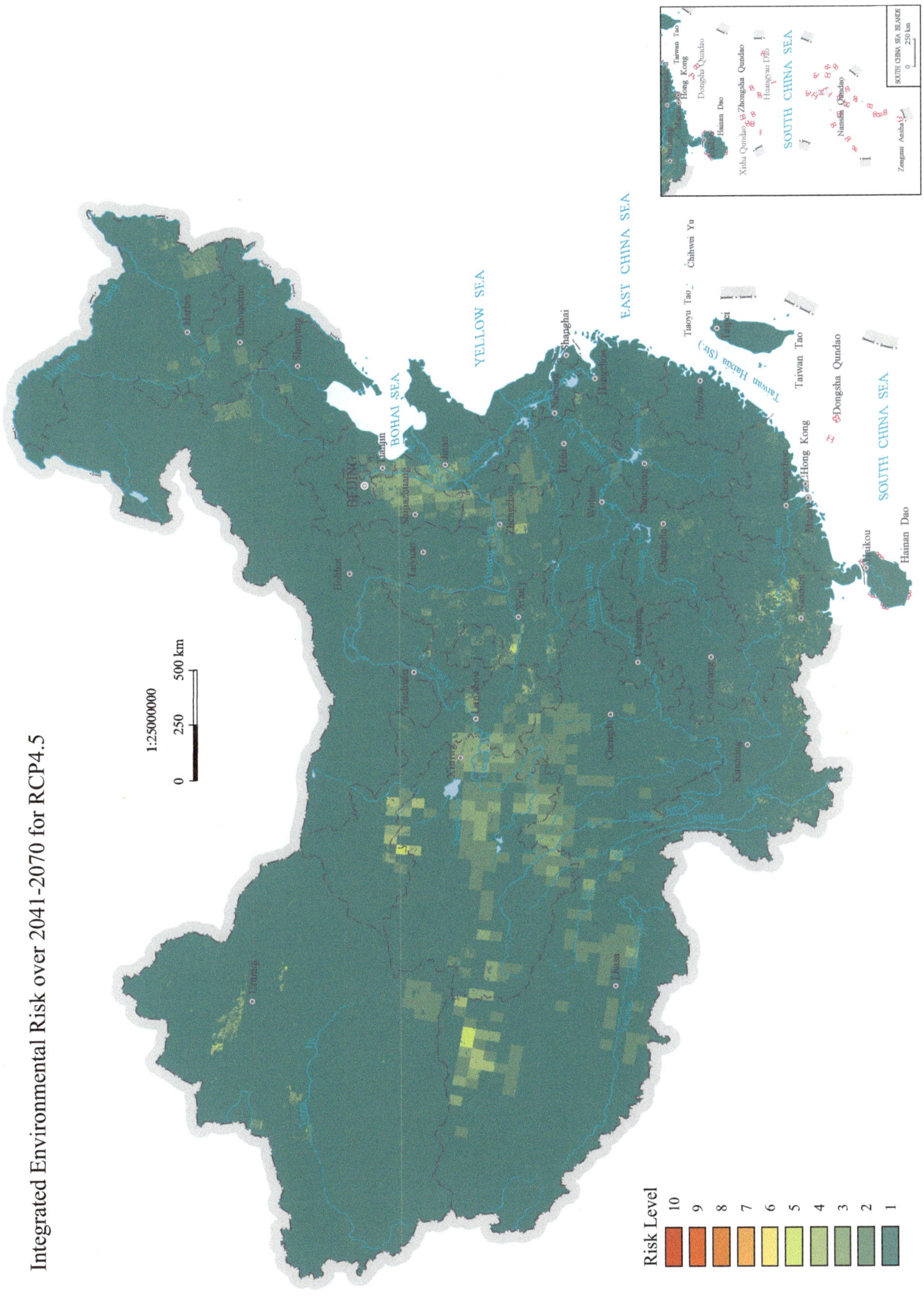

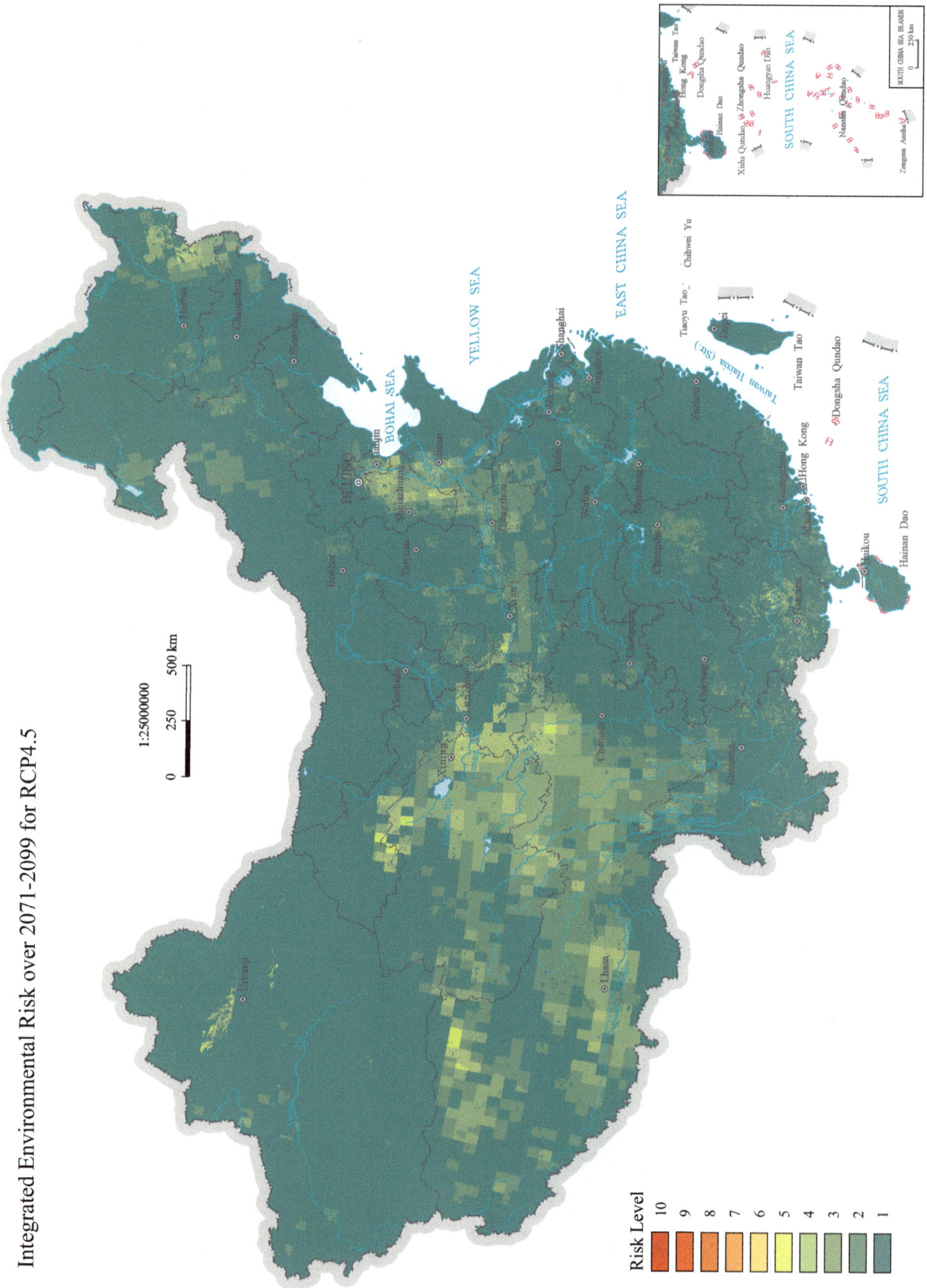
Integrated Environmental Risk over 2071-2099 for RCP4.5
1:25000000
0 250 500 km
Risk Level
10
9
8
7
6
5
4
3
2
1
BOHAI SEA
YELLOW SEA
EAST CHINA SEA
SOUTH CHINA SEA
Taiwan Haixia (Str.)
Taiwan Tao
Tiaoyu Tao
Chihwei Yu
Dongsha Qundao
Hong Kong
Hainan Dao
BEIJING
Shanghai
Lhasa

Integrated Environmental Risk over 2011-2040 for RCP6.0

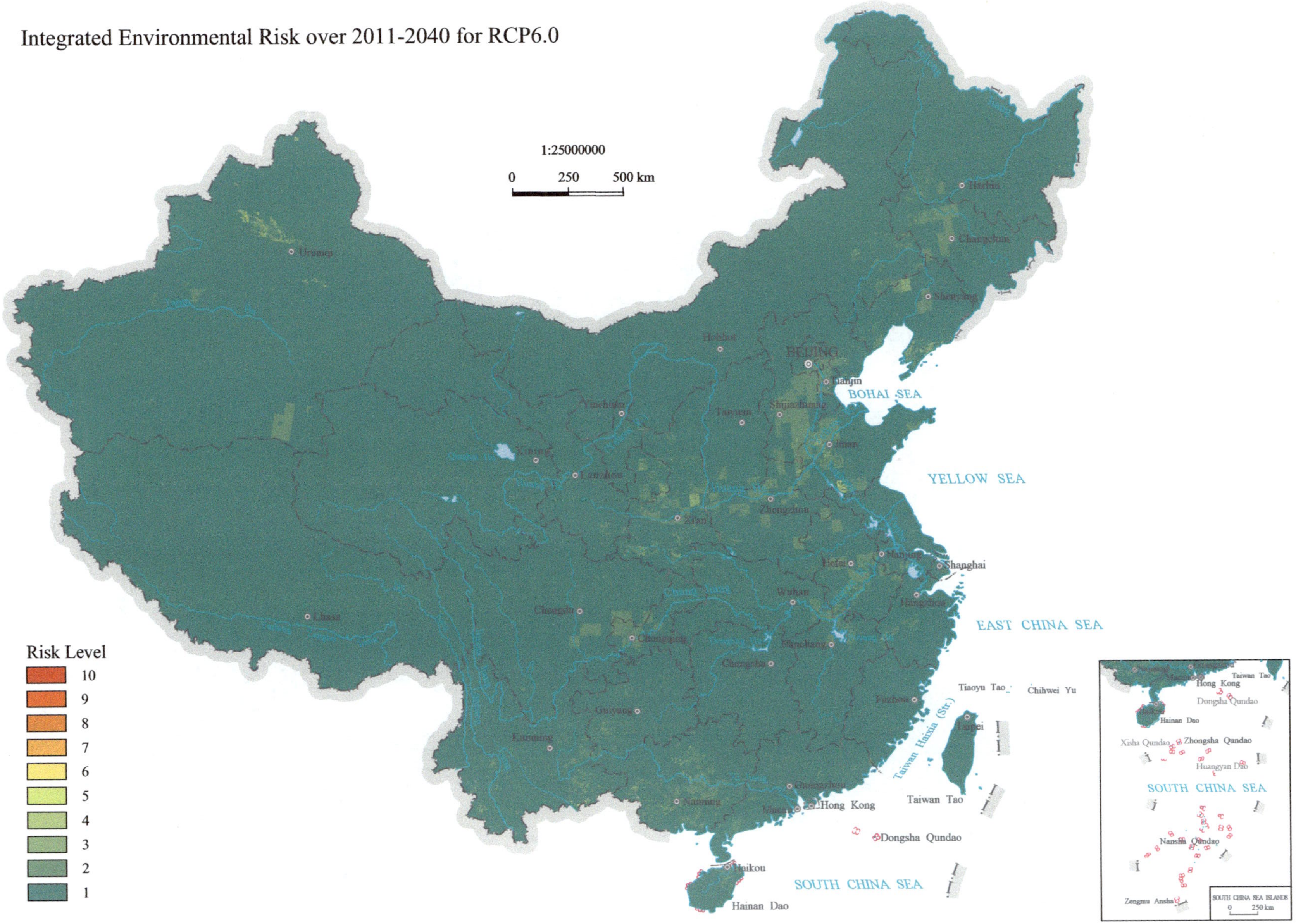

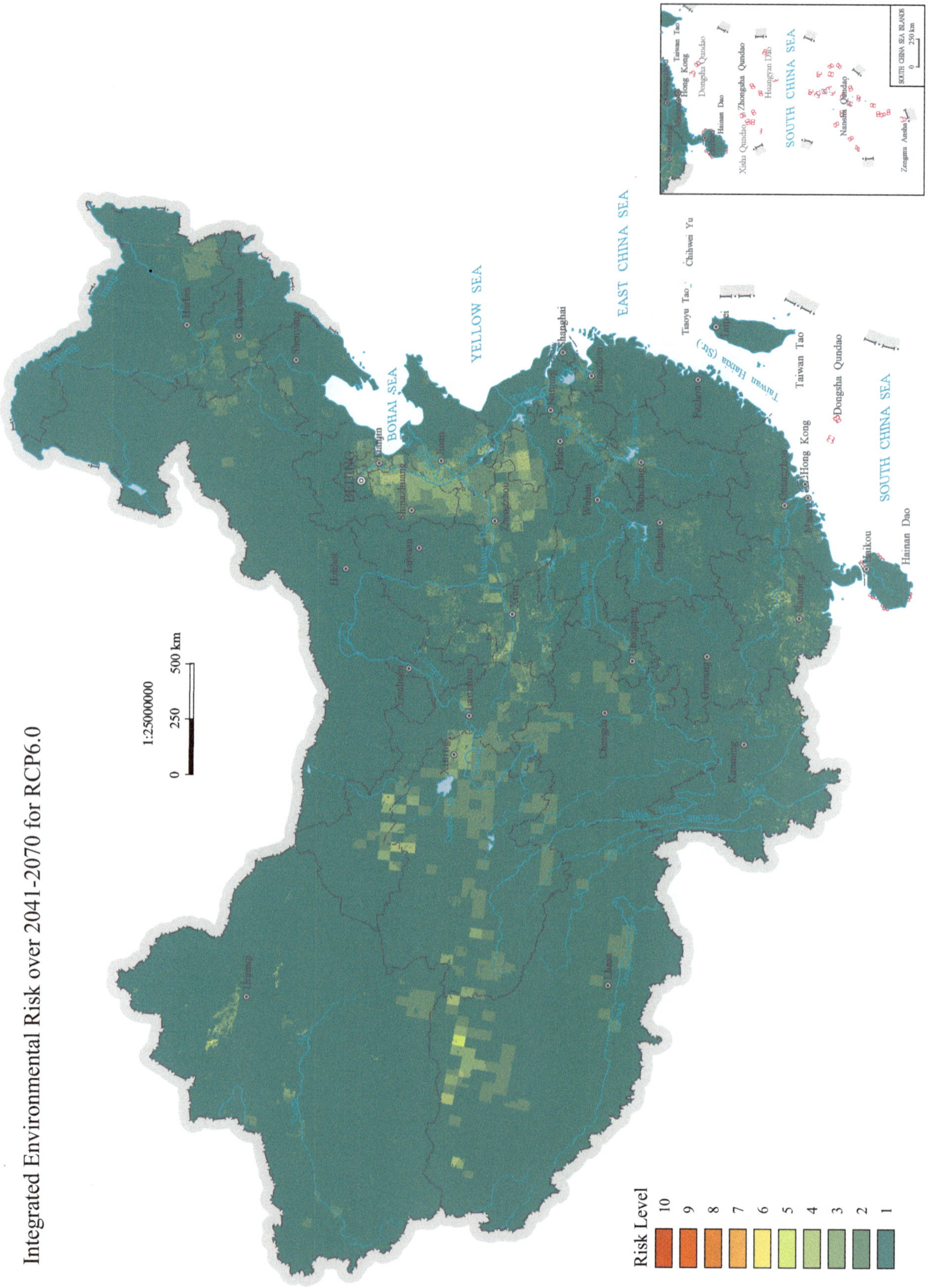
Integrated Environmental Risk over 2041-2070 for RCP6.0
1:25000000
0 250 500 km
Risk Level
10
9
8
7
6
5
4
3
2
1
BOHAI SEA
YELLOW SEA
EAST CHINA SEA
SOUTH CHINA SEA
BEIJING
Shanghai
Hong Kong
Taiwan Tao
Dongsha Qundao
Hainan Dao
Haikou
Taiwan Haixia (Str.)
Chihwei Yu
Taoyu Tao
Xisha Qundao
Zhongsha Qundao
Nansha Qundao
SOUTH CHINA SEA ISLANDS
0 250 km

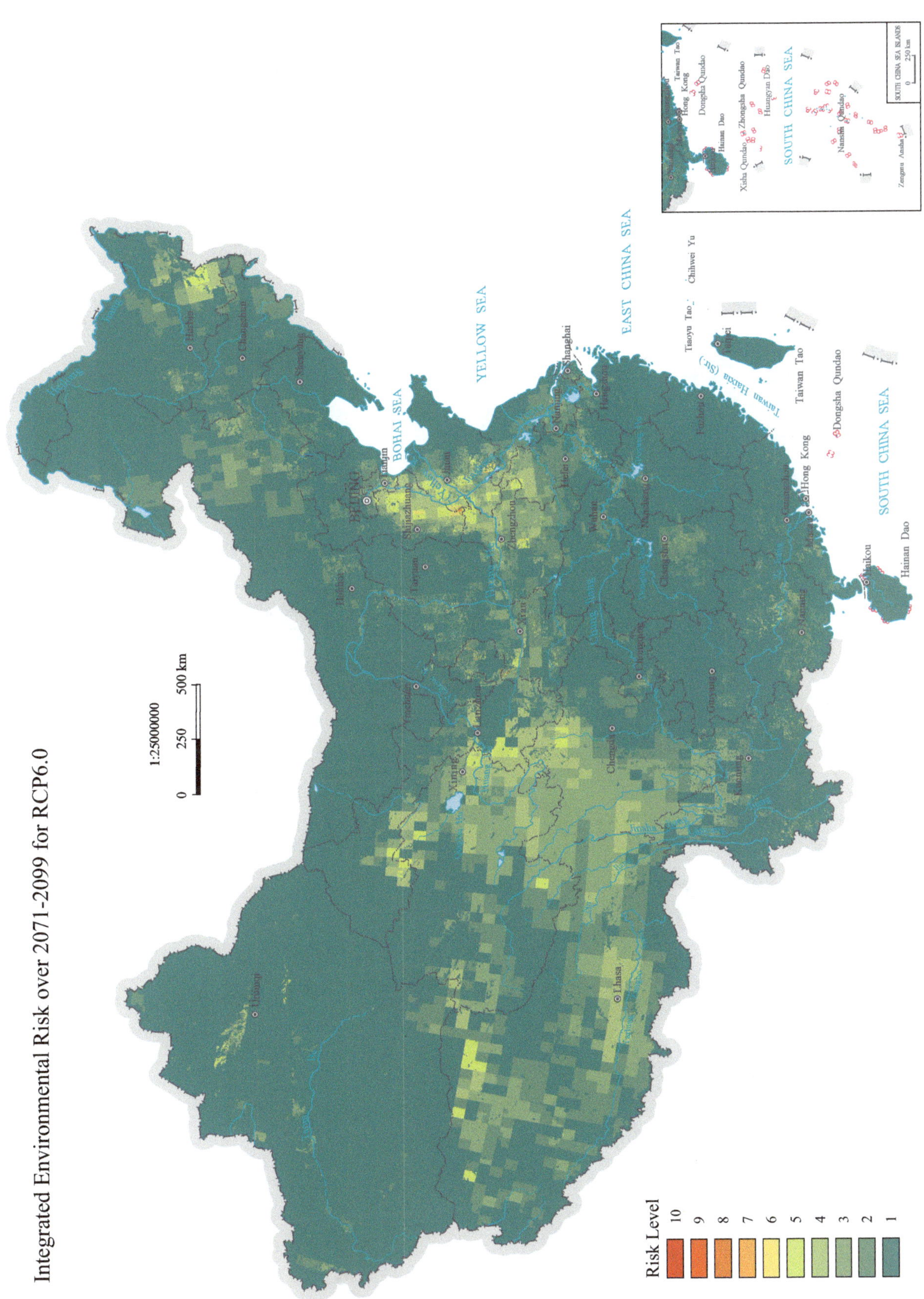
Integrated Environmental Risk over 2071-2099 for RCP6.0
1:25000000
0 250 500 km
Risk Level
10
9
8
7
6
5
4
3
2
1
BOHAI SEA
YELLOW SEA
EAST CHINA SEA
SOUTH CHINA SEA
BEIJING
Shanghai
Lhasa
Hong Kong
Taiwan Tao
Dongsha Qundao
Hainan Dao

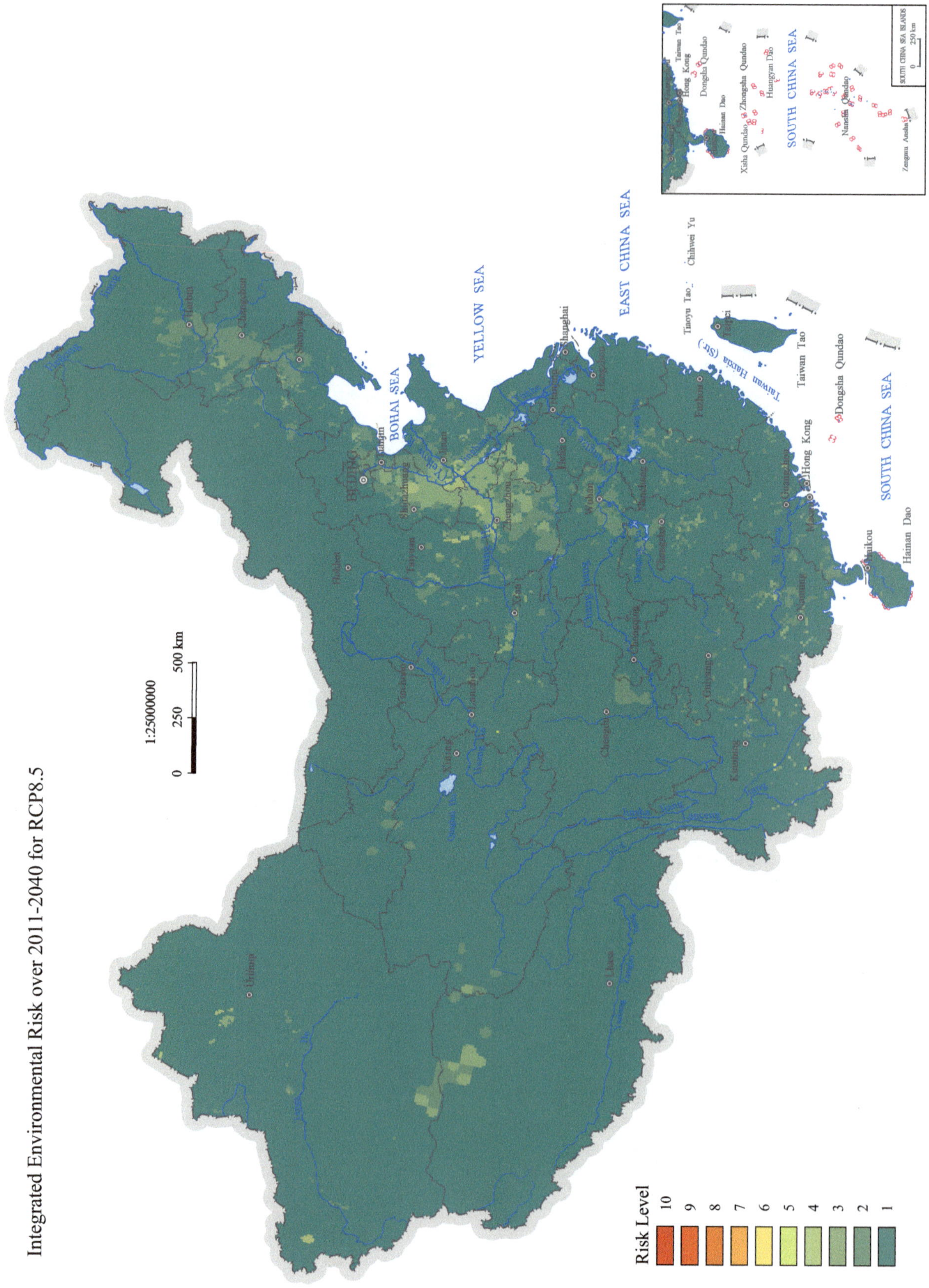

Integrated Environmental Risk over 2011-2040 for RCP8.5
1:25000000
0 250 500 km
Risk Level
10
9
8
7
6
5
4
3
2
1
BOHAI SEA
YELLOW SEA
EAST CHINA SEA
SOUTH CHINA SEA
Shanghai
Hong Kong
Taiwan Tao
Dongsha Qundao
Hainan Dao
Haikou

Integrated Environmental Risk over 2041-2070 for RCP8.5

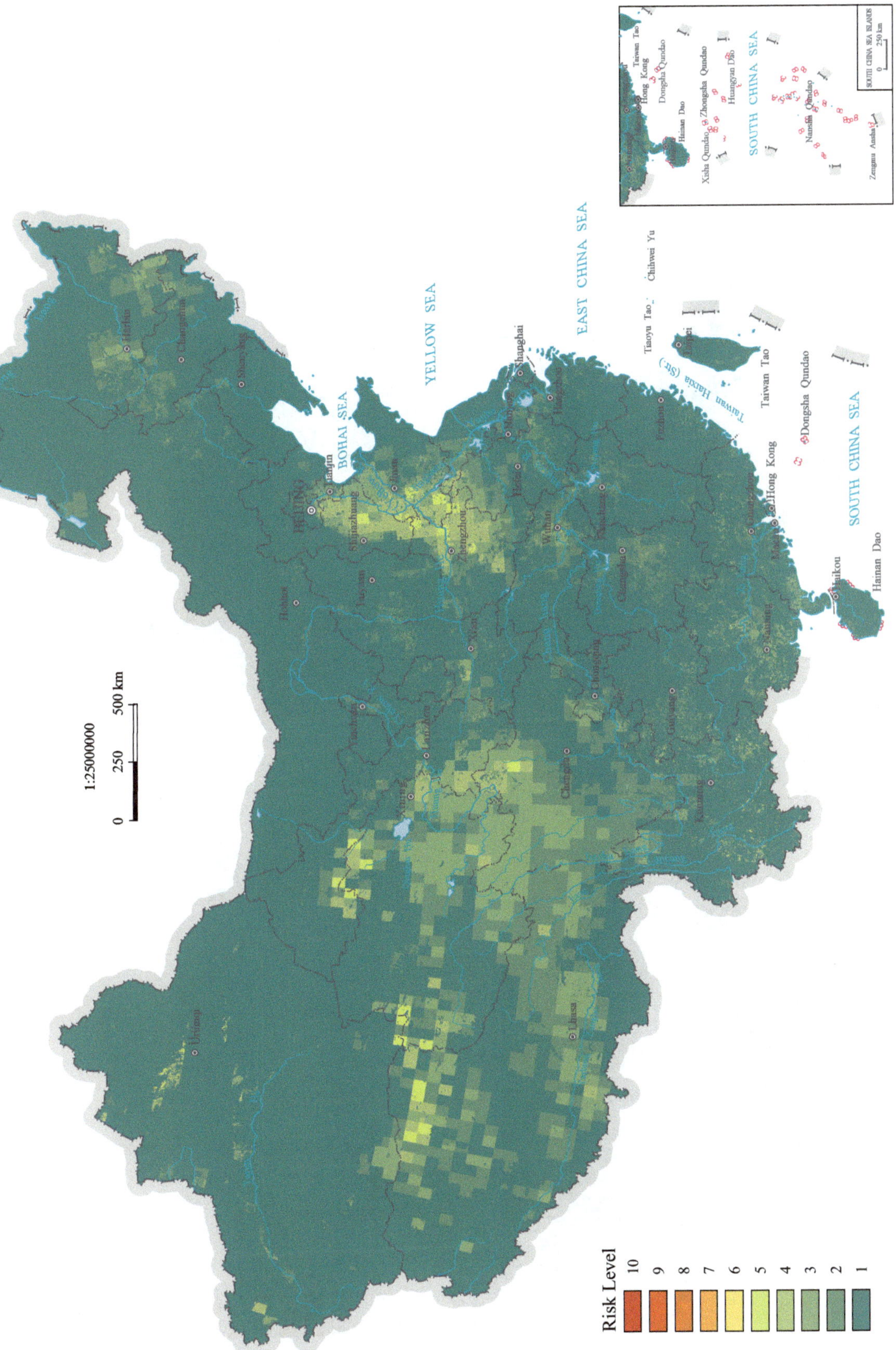

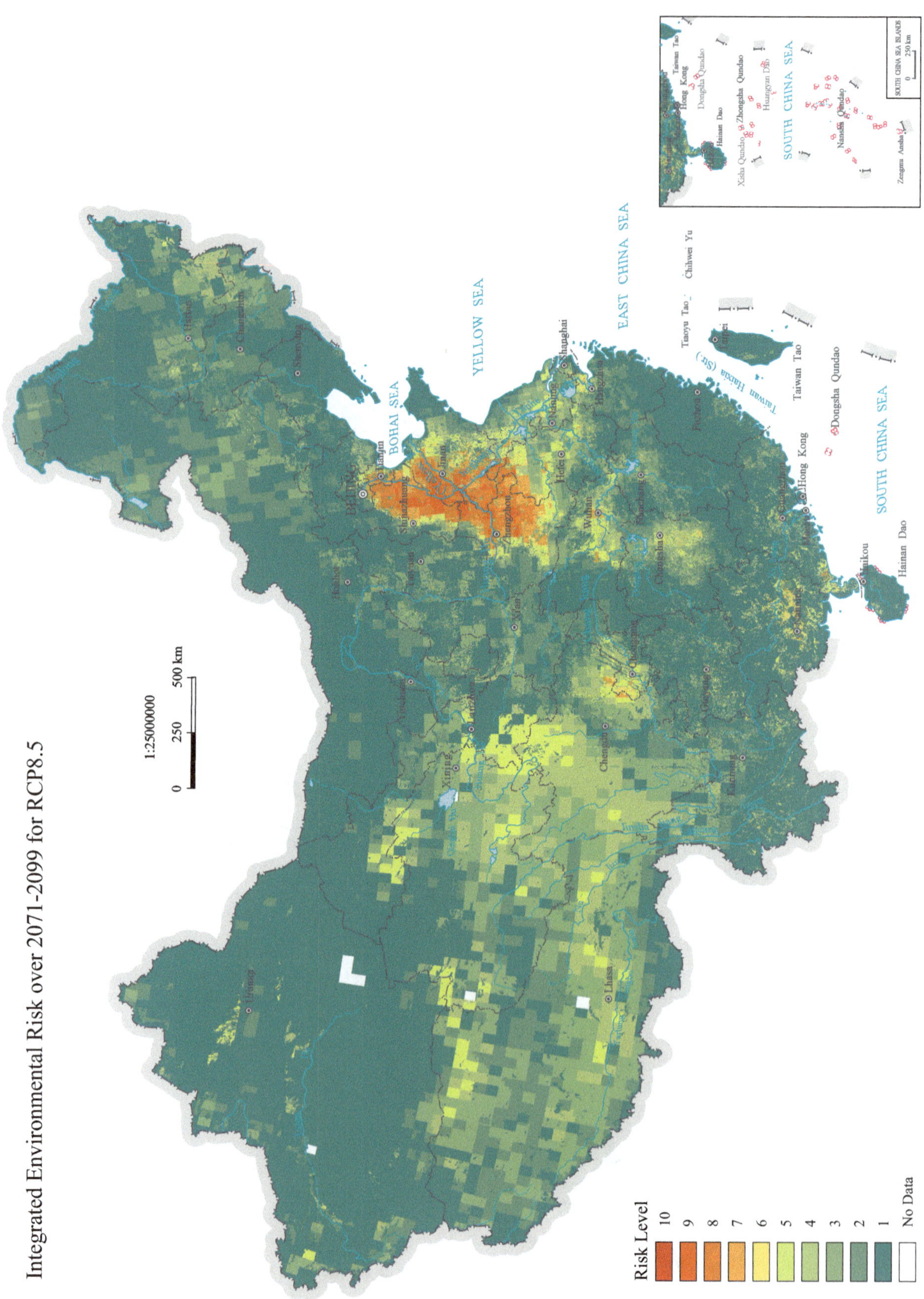
Integrated Environmental Risk over 2071-2099 for RCP8.5
1:25000000
0 250 500 km
Risk Level
10
9
8
7
6
5
4
3
2
1
No Data
BOHAI SEA
YELLOW SEA
EAST CHINA SEA
SOUTH CHINA SEA
Taiwan Haixia (Str.)
Taiwan Tao
Dongsha Qundao
Hainan Dao
Hong Kong
Shanghai
Wuhan
Lhasa
Urumqi
SOUTH CHINA SEA ISLANDS

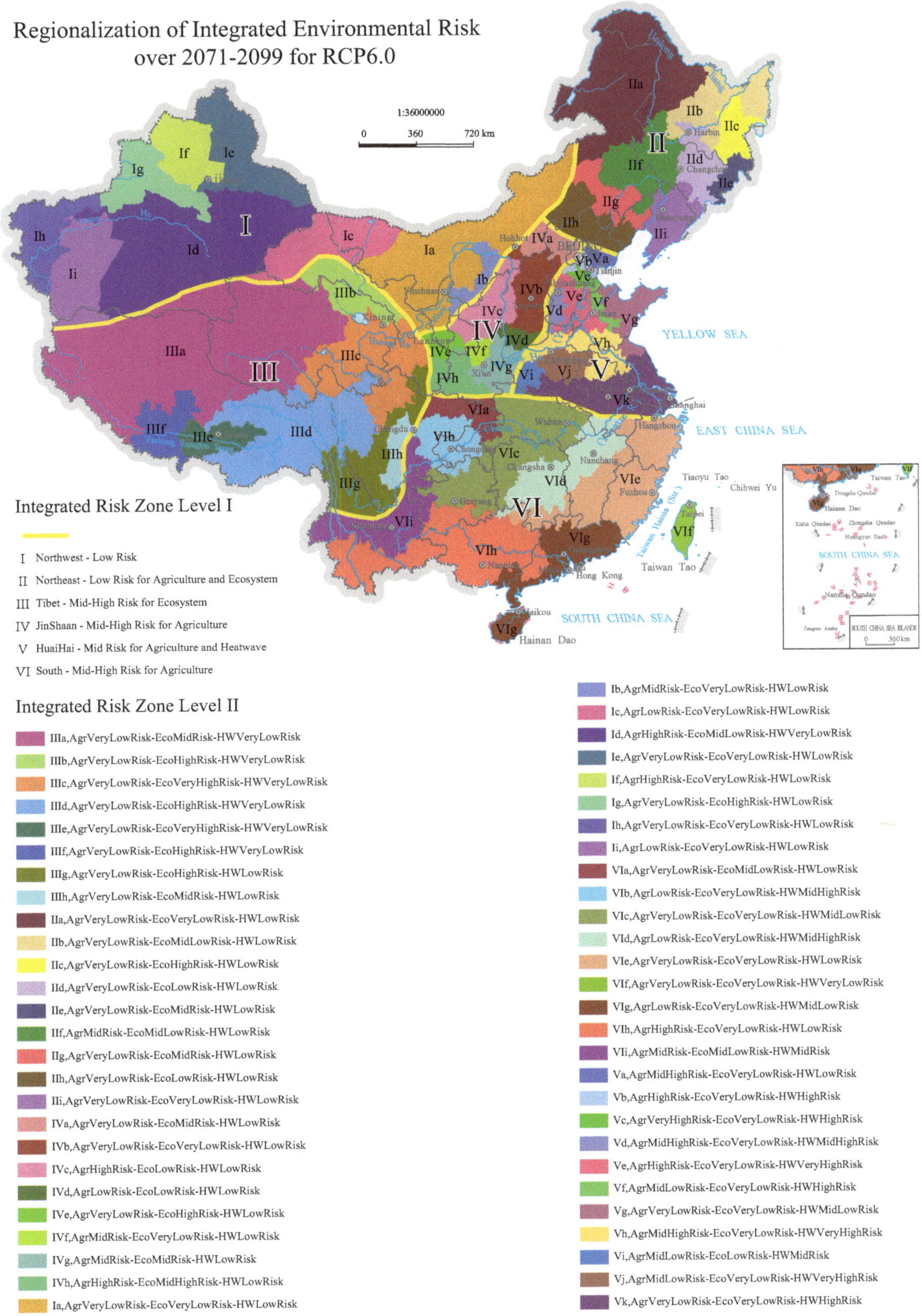
Regionalization of Integrated Environmental Risk over 2071-2099 for RCP6.0
1:36000000
0 360 720 km
I
II
III
IV
V
VI
YELLOW SEA
EAST CHINA SEA
SOUTH CHINA SEA
Tiaoyu Tao
Chihwei Yu
Taiwan Tao
Hong Kong
Hainan Dao
Integrated Risk Zone Level I
I Northwest - Low Risk
II Northeast - Low Risk for Agriculture and Ecosystem
III Tibet - Mid-High Risk for Ecosystem
IV JinShaan - Mid-High Risk for Agriculture
V HuaiHai - Mid Risk for Agriculture and Heatwave
VI South - Mid-High Risk for Agriculture
Integrated Risk Zone Level II
IIIa,AgrVeryLowRisk-EcoMidRisk-HWVeryLowRisk
IIIb,AgrVeryLowRisk-EcoHighRisk-HWVeryLowRisk
IIIc,AgrVeryLowRisk-EcoVeryHighRisk-HWVeryLowRisk
IIId,AgrVeryLowRisk-EcoHighRisk-HWVeryLowRisk
IIIe,AgrVeryLowRisk-EcoVeryHighRisk-HWVeryLowRisk
IIIf,AgrVeryLowRisk-EcoHighRisk-HWVeryLowRisk
IIIg,AgrVeryLowRisk-EcoHighRisk-HWLowRisk
IIIh,AgrVeryLowRisk-EcoMidRisk-HWLowRisk
IIa,AgrVeryLowRisk-EcoVeryLowRisk-HWLowRisk
IIb,AgrVeryLowRisk-EcoMidLowRisk-HWLowRisk
IIc,AgrVeryLowRisk-EcoHighRisk-HWLowRisk
IId,AgrVeryLowRisk-EcoLowRisk-HWLowRisk
IIe,AgrVeryLowRisk-EcoMidRisk-HWLowRisk
IIf,AgrMidRisk-EcoMidLowRisk-HWLowRisk
IIg,AgrVeryLowRisk-EcoMidRisk-HWLowRisk
IIh,AgrVeryLowRisk-EcoLowRisk-HWLowRisk
IIi,AgrVeryLowRisk-EcoVeryLowRisk-HWLowRisk
IVa,AgrVeryLowRisk-EcoMidRisk-HWLowRisk
IVb,AgrVeryLowRisk-EcoVeryLowRisk-HWLowRisk
IVc,AgrHighRisk-EcoLowRisk-HWLowRisk
IVd,AgrLowRisk-EcoLowRisk-HWLowRisk
IVe,AgrVeryLowRisk-EcoHighRisk-HWLowRisk
IVf,AgrMidRisk-EcoVeryLowRisk-HWLowRisk
IVg,AgrMidRisk-EcoMidRisk-HWLowRisk
IVh,AgrHighRisk-EcoMidHighRisk-HWLowRisk
Ia,AgrVeryLowRisk-EcoVeryLowRisk-HWLowRisk
Ib,AgrMidRisk-EcoVeryLowRisk-HWLowRisk
Ic,AgrLowRisk-EcoVeryLowRisk-HWLowRisk
Id,AgrHighRisk-EcoMidLowRisk-HWVeryLowRisk
Ie,AgrVeryLowRisk-EcoVeryLowRisk-HWLowRisk
If,AgrHighRisk-EcoVeryLowRisk-HWLowRisk
Ig,AgrVeryLowRisk-EcoHighRisk-HWLowRisk
Ih,AgrVeryLowRisk-EcoVeryLowRisk-HWLowRisk
Ii,AgrLowRisk-EcoVeryLowRisk-HWLowRisk
VIa,AgrVeryLowRisk-EcoMidLowRisk-HWLowRisk
VIb,AgrLowRisk-EcoVeryLowRisk-HWMidHighRisk
VIc,AgrVeryLowRisk-EcoVeryLowRisk-HWMidLowRisk
VId,AgrLowRisk-EcoVeryLowRisk-HWMidHighRisk
VIe,AgrVeryLowRisk-EcoVeryLowRisk-HWLowRisk
VIf,AgrVeryLowRisk-EcoVeryLowRisk-HWVeryLowRisk
VIg,AgrLowRisk-EcoVeryLowRisk-HWMidLowRisk
VIh,AgrHighRisk-EcoVeryLowRisk-HWLowRisk
VIi,AgrMidRisk-EcoMidLowRisk-HWMidRisk
Va,AgrMidHighRisk-EcoVeryLowRisk-HWLowRisk
Vb,AgrHighRisk-EcoVeryLowRisk-HWHighRisk
Vc,AgrVeryHighRisk-EcoVeryLowRisk-HWHighRisk
Vd,AgrMidHighRisk-EcoVeryLowRisk-HWMidHighRisk
Ve,AgrHighRisk-EcoVeryLowRisk-HWVeryHighRisk
Vf,AgrMidLowRisk-EcoVeryLowRisk-HWHighRisk
Vg,AgrVeryLowRisk-EcoVeryLowRisk-HWMidLowRisk
Vh,AgrMidHighRisk-EcoVeryLowRisk-HWVeryHighRisk
Vi,AgrMidLowRisk-EcoLowRisk-HWMidRisk
Vj,AgrMidLowRisk-EcoVeryLowRisk-HWVeryHighRisk
Vk,AgrVeryLowRisk-EcoVeryLowRisk-HWHighRisk

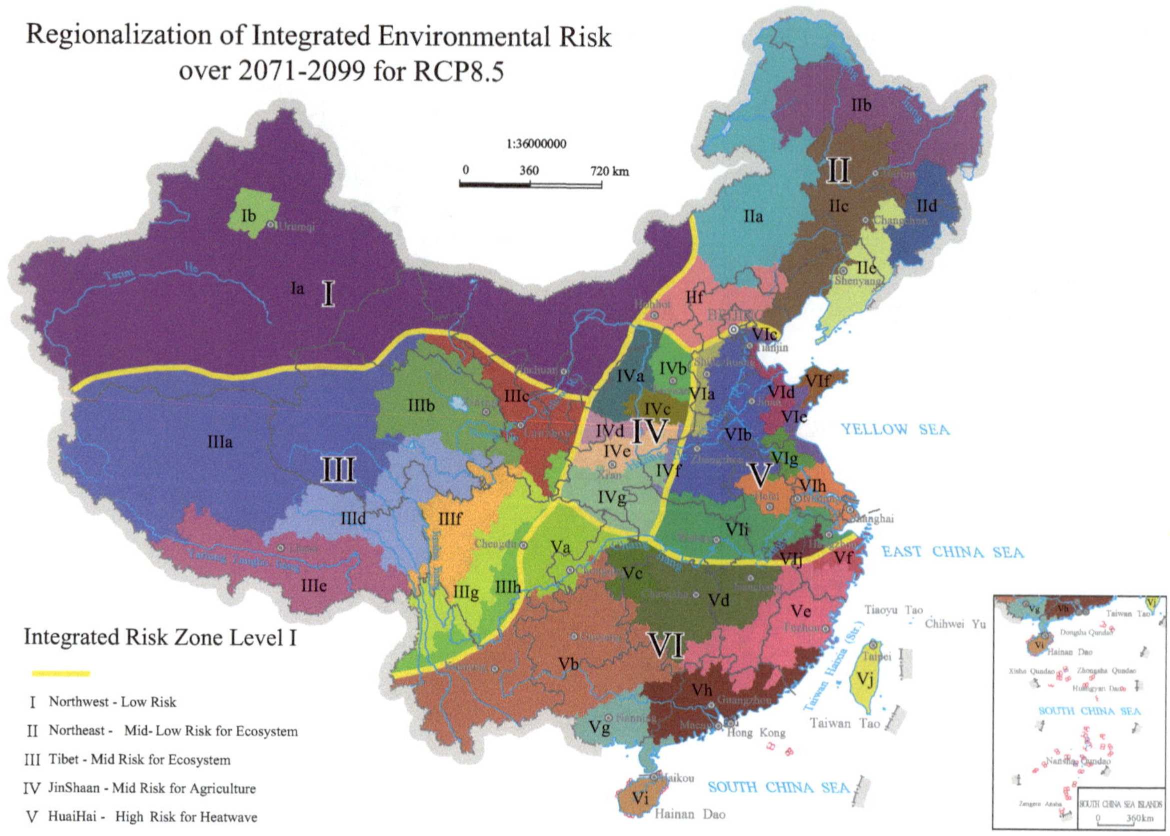

Integrated Risk Zone Level II

- IIIa,AgrVeryLowRisk-EcoHighRisk-HWVeryLowRisk
- IIIb,AgrLowRisk-EcoVeryHighRisk-HWVeryLowRisk
- IIIc,AgrLowRisk-EcoMidLowRisk-HWLowRisk
- IIId,AgrVeryLowRisk-EcoVeryHighRisk-HWVeryLowRisk
- IIIe,AgrLowRisk-EcoVeryHighRisk-HWVeryLowRisk
- IIIf,AgrLowRisk-EcoVeryHighRisk-HWVeryLowRisk
- IIIg,AgrLowRisk-EcoHighRisk-HWLowRisk
- IIIh,AgrMidLowRisk-EcoMidLowRisk-HWLowRisk
- IIa,AgrLowRisk-EcoMidRisk-HWLowRisk
- IIb,AgrLowRisk-EcoLowRisk-HWLowRisk
- IIc,AgrMidRisk-EcoLowRisk-HWLowRisk
- IId,AgrLowRisk-EcoHighRisk-HWLowRisk
- IIe,AgrLowRisk-EcoLowRisk-HWMidRisk
- IIf,AgrLowRisk-EcoVeryLowRisk-HWLowRisk
- IVa,AgrMidRisk-EcoLowRisk-HWLowRisk
- IVb,AgrLowRisk-EcoLowRisk-HWLowRisk
- IVc,AgrMidRisk-EcoLowRisk-HWLowRisk
- IVd,AgrLowRisk-EcoMidLowRisk-HWLowRisk
- IVe,AgrMidRisk-EcoVeryLowRisk-HWMidRisk
- IVf,AgrMidRisk-EcoLowRisk-HWHighRisk
- IVg,AgrLowRisk-EcoMidLowRisk-HWLowRisk
- Ia,AgrLowRisk-EcoLowRisk-HWLowRisk
- Ib,AgrMidRisk-EcoLowRisk-HWLowRisk
- VIa,AgrMidRisk-EcoLowRisk-HWHighRisk
- VIb,AgrVeryHighRisk-EcoVeryLowRisk-HWVeryHighRisk
- VIc,AgrHighRisk-EcoVeryLowRisk-HWHighRisk
- VId,AgrHighRisk-EcoLowRisk-HWHighRisk
- VIe,AgrMidRisk-EcoLowRisk-HWMidRisk
- VIf,AgrLowRisk-EcoLowRisk-HWLowRisk
- VIg,AgrHighRisk-EcoVeryLowRisk-HWHighRisk
- VIh,AgrMidRisk-EcoVeryLowRisk-HWVeryHighRisk
- VIi,AgrMidRisk-EcoLowRisk-HWHighRisk
- VIj,AgrVeryLowRisk-EcoLowRisk-HWMidRisk
- Va,AgrHighRisk-EcoVeryLowRisk-HWHighRisk
- Vb,AgrMidRisk-EcoLowRisk-HWLowRisk
- Vc,AgrVeryLowRisk-EcoLowRisk-HWLowRisk
- Vd,AgrMidRisk-EcoLowRisk-HWHighRisk
- Ve,AgrLowRisk-EcoMidLowRisk-HWMidLowRisk
- Vf,AgrMidRisk-EcoLowRisk-HWHighRisk
- Vg,AgrHighRisk-EcoLowRisk-HWMidRisk
- Vh,AgrMidRisk-EcoLowRisk-HWMidHighRisk
- Vi,AgrMidLowRisk-EcoLowRisk-HWLowRisk
- Vj,AgrLowRisk-EcoLowRisk-HWVeryLowRisk

Intermodel Spread in Integrated Environmental Risk over 2071-2099 for RCP2.6

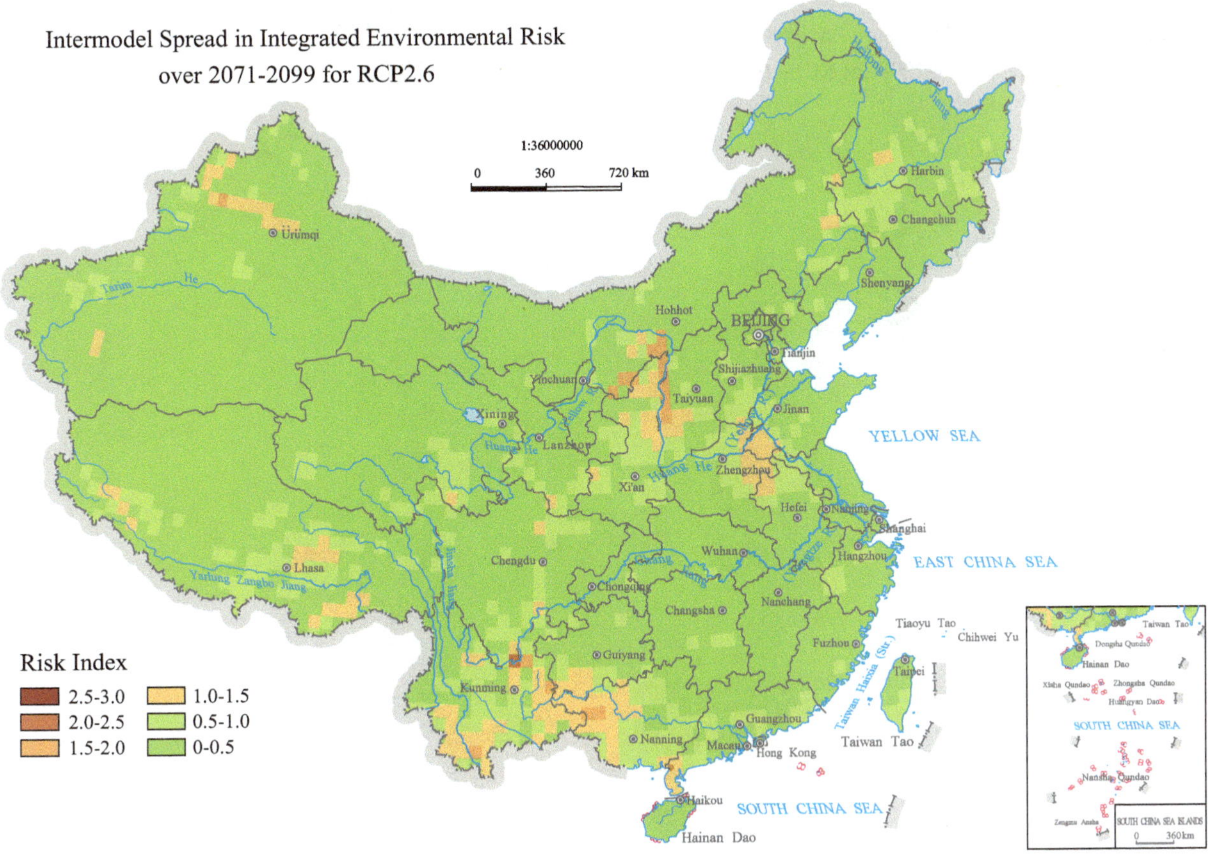

Intermodel Spread in Integrated Environmental Risk over 2071-2099 for RCP4.5

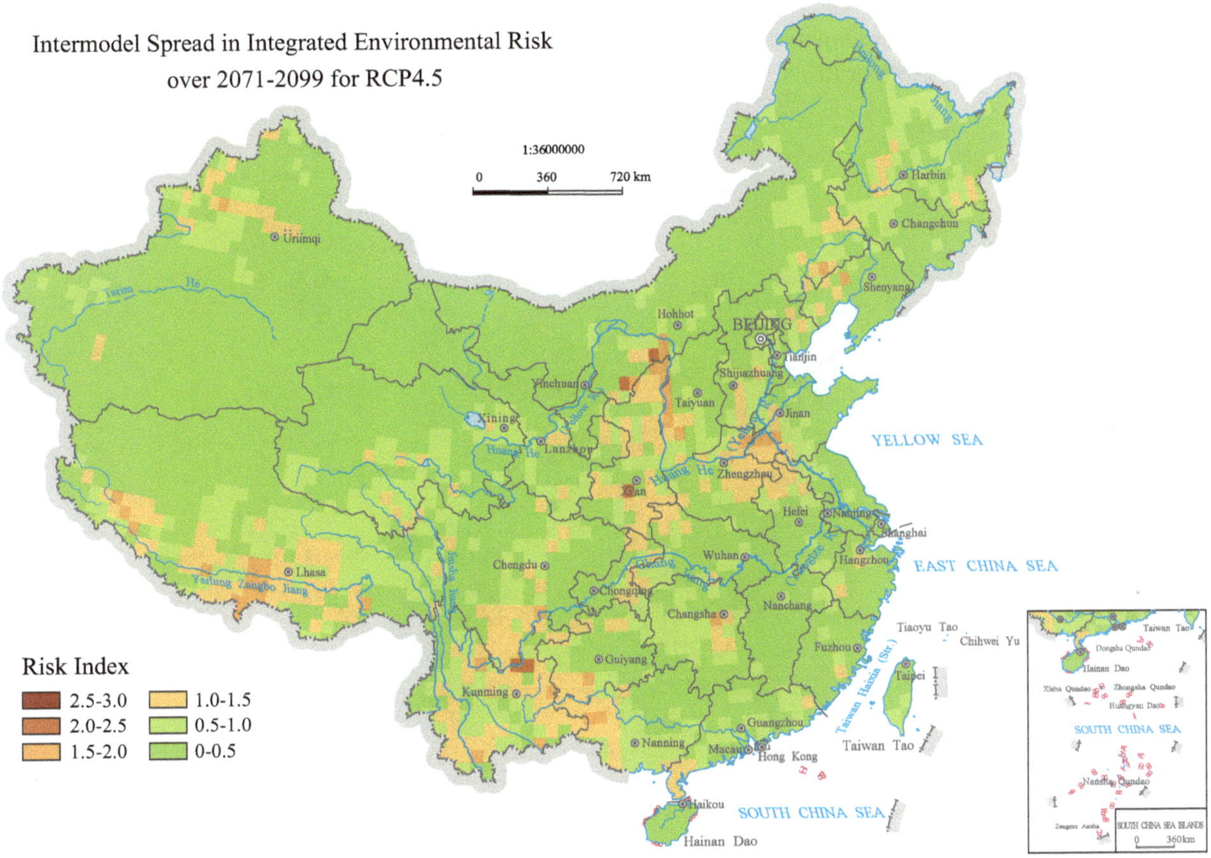

Intermodel Spread in Integrated Environmental Risk over 2071-2099 for RCP6.0

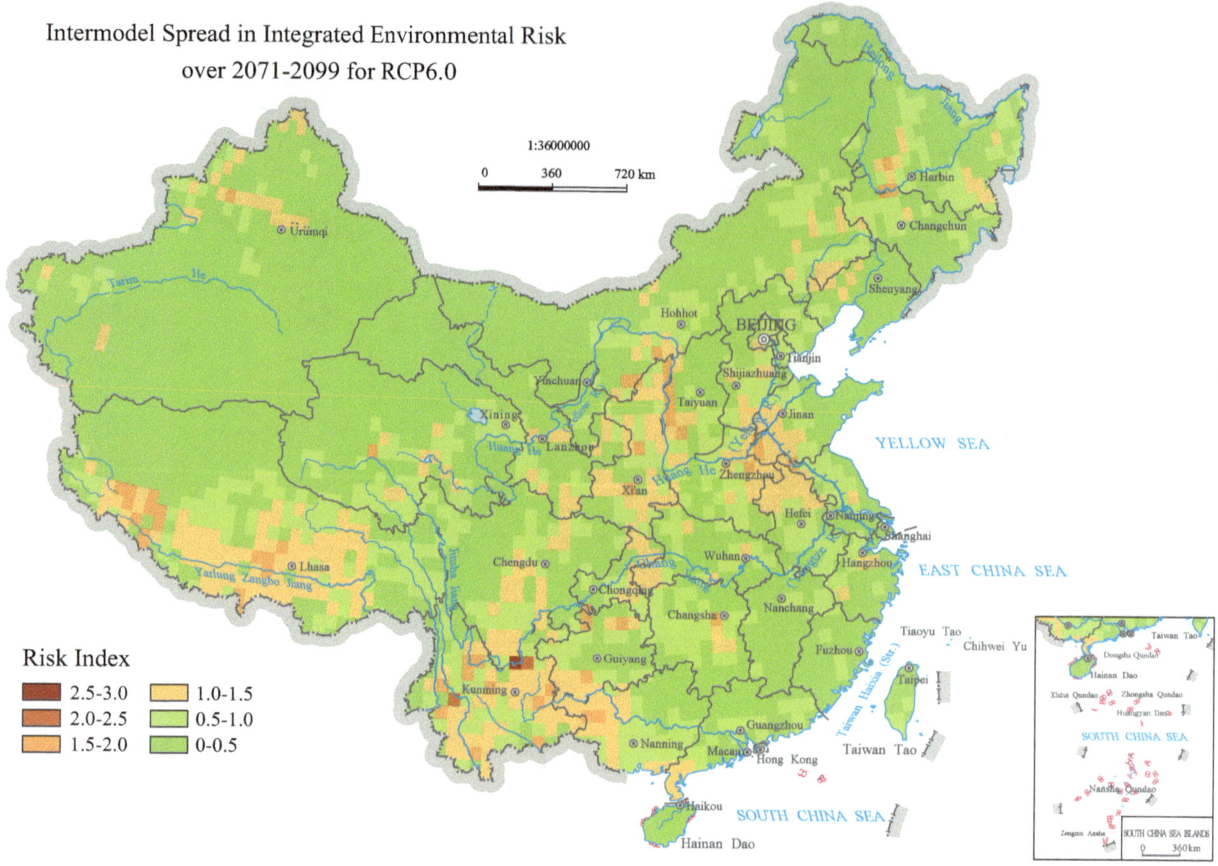

Intermodel Spread in Integrated Environmental Risk over 2071-2099 for RCP8.5

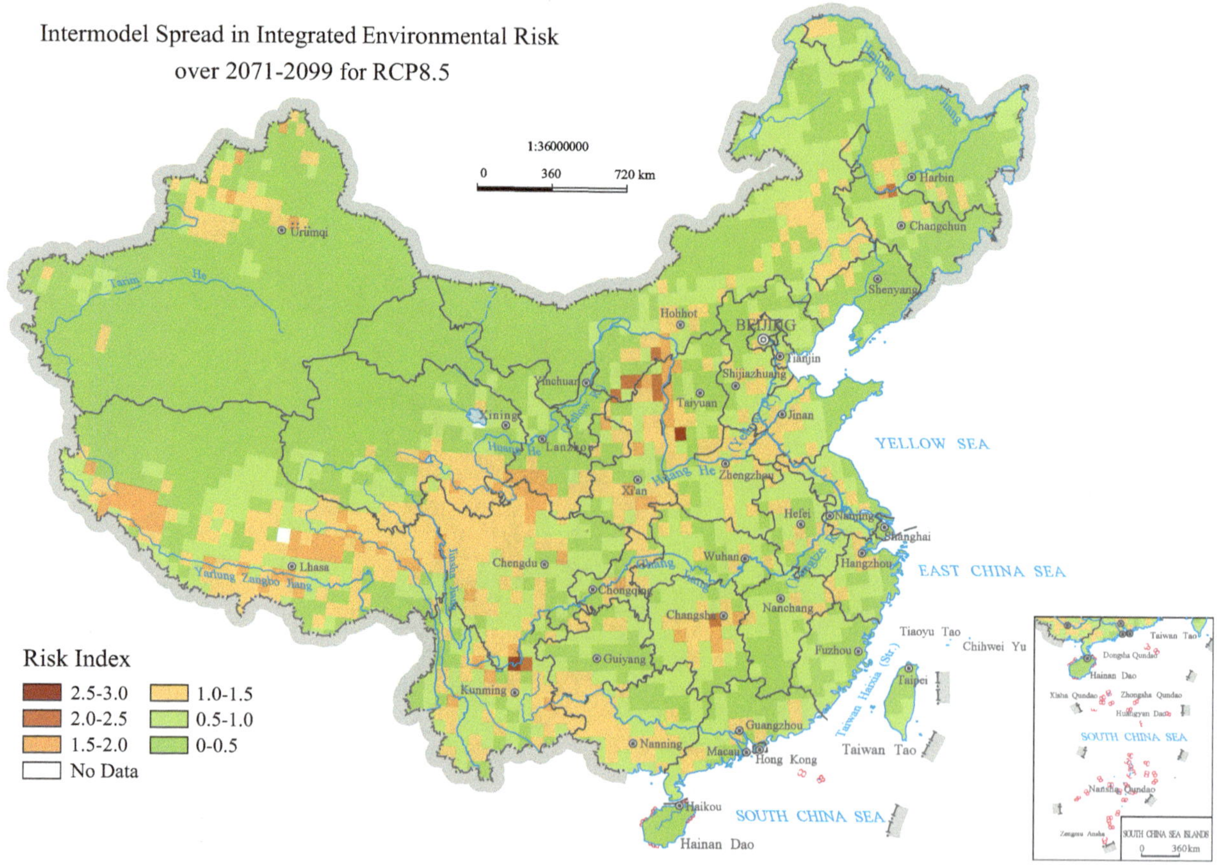

References

Briggs, D. J. (2008). A framework for integrated environmental health impact assessment of systemic risks. *Environmental Health, 7*(1), 1–17.

Dilley, M. (2005). *Natural disaster hotspots: A global risk analysis*. World Bank.

Joern, B., Matthias, G., Peter, M., Almuth, S., Thomas, S., Torsten, W., et al. (2014). *World risk report (2014)*. Bündnis Entwicklung Hilft and Institute for Environment and human Security: United Nations University (UNU-EHS), Berlin.

Shi, P. (2009). Theory and practice on disaster system research: The fifth discussion. *Journal of Natural disasters, 18*(5), 19. (in Chinese).

Shi, P., & Kasperson, R. (2014). *World atlas of natural disaster risk*. Heidelberg and Beijing: Springer and Normal University Press.

Tang, Q., & Oki, T. (Eds.). (2016). Terrestrial water cycle and climate change: Natural and human-induced impacts. American Geophysical Union (AGU) Geophysical Monograph Series (Vol. 221). Hoboken: Wiley.

Warszawski, L., Frieler, K., Huber, V., Piontek, F., Serdeczny, O., & Schewe, J. (2014). The Inter-Sectoral Impact Model Intercomparison Projection (ISI-MIP): Project framework. *Proceedings of the National Academy of Sciences of the United States of America, 111*, 3228–3232.

United Nations Development Programme (UNDP), Bureau for Crisis Prevention and Recovery. (2004). *Reducing disaster risk: A challenge for development*. New York: UNDP, Bureau for Crisis Prevention and Recovery.

World Economic Forum (WEF). (2015). Global risks 2015. Geneva: World Economic Forum.

World Economic Forum (WEF). (2016). The global risks report 2016. Geneva: World Economic Forum.

Appendix

Data sources and databases used in the Atlas of Environmental Risks Facing China Under Climate Change

Name	Data description	Period	Resolution/scale	Data sources	Usage
Base map	National boundaries, provincial boundaries, coastlines, rivers, and lakes	2016	1:25,000,000; 1:36,000,000	Star Map Press, Beijing, China No. of licensed maps: JS (2016)01-143	All of the maps
SRTM 90 m digital elevation data	SRTM 90 m is a global digital elevation data (DEM) produced by NASA originally, and is available for download from the National Map Seamless Data Distribution System, or the USGS ftp site. The SRTM 90 m DEM's have a resolution of 90 m at the equator	2008	3 arc second (approx. 90 m resolution)	The CGIAR Consortium for Spatial Information (CGIAR-CSI) http://srtm.csi.cgiar.org/	Topography
Climate change regionalization of China (1961–2010) data	The dataset was compiled and completed by Prof. Shi's group based on the tendency value and fluctuation value of temperature and precipitation from 1961 to 2010	2014	–	Beijing Normal University	Climate change regionalization
1:1,000,000 digital soil map of People's Republic of China	On the basis of the "1:1,000,000 Soil Map of the People's Republic of China" published by the National Soil Survey Office in 1995, the 1:1,000,000 digital soil map was compiled and completed by Prof. Shi's group through digitizing and edge matching. The soil attribute data are taken from the "Soil Series of China"	1995	1:1,000,000	Institute of Soil Science, Chinese Academy of Sciences	Soil types
1:1,000,000 digital vegetation map of People's Republic of China	On the basis of the "Vegetation Map of the People's Republic of China (1:1 000 000)" published in 1995, the 1:1,000,000 digital vegetation map was compiled and completed by Prof. Fang through digitizing	2007	1:1,000,000	Institute of Botany, Chinese Academy of Sciences	Vegetation types Ecological risks
	The map is published by the Geology Press by the end of	2007	1:60,000,000	Chinese Academy of Sciences	Vegetation regionalization

(continued)

Q. Tang and Q. Ge (eds.), *Atlas of Environmental Risks Facing China Under Climate Change*, IHDP/Future Earth-Integrated Risk Governance Project Series, DOI 10.1007/978-981-10-4199-0

The vegetation regionalization map of China	2007 (Edition Board of Vegetation Map of China, 2007).				
1 km grid land use and land cover dataset of China (2010)	The dataset is provided by Data Center for Resources and Environmental Sciences, Chinese Academy of Sciences (RESDC) (http://www.resdc.cn)	2010	1 km × 1 km	Global Land Cover Facility http://www.resdc.cn/data.aspx?DATAID=99	Land use and land covers Agricultural risks
Eco-geographical regionalization of China	The dataset is promulgated by the Ministry of Environmental Protection and the Research Center for Eco-Environmental Sciences, Chinese Academy of Sciences	2008	–	Ministry of Environmental Protection and the Research Center for Eco-Environmental Sciences, Chinese Academy of Sciences	Eco-geographical regionalization of China
Gridded population of the world, version 4 (GPWv4) (2010)	The Gridded Population of the World (GPW) models the distribution of the human population on a continuous raster surface. The dataset is constructed from national or subnational input administrative units. The purpose of GPW is to provide a spatially disaggregated population layer that is compatible with datasets from social, economic, and Earth science disciplines, and remote sensing data	2010	1 km × 1 km	Center for International Earth Science Information Network (CIESIN) Columbia University	Population density of China (2010)
1 km grid GDP dataset of China (2010)	The dataset was derived based on the comprehensive analysis and data integration with land use data which were extracted from satellite imageries with the resolution of 30 meters and the statistical GDP data of each county	2010	1 km × 1 km	Institute of Geographic Sciences and Natural Resources Research, Chinese Academy of Sciences	GDP density of China
Annual runoff depth and runoff coefficient of China	The dataset provides observation-based near-surface meteorological data and land surface hydrologic fluxes and states estimations from the Variable Infiltration Capacity (VIC) hydrological model	1952–2012	0.25° × 0.25°	Key Laboratory of Water Cycle and Related Land Surface Processes, Institute of Geographic Sciences and Natural Resources Research, Chinese Academy of Sciences http://hydro.igsnrr.ac.cn/	Annual runoff depth annual runoff coefficient
Global meteorological data	Daily precipitation, mean temperature, maximum temperature, minimum temperature. The data were bias-corrected by the ISI-MIP team and were derived from climate projections of five global climate models (GCMs), namely HadGEM2-ES, IPSL-CM5A-LR, MIROC-ESM-CHEM, GFDL-ESM2M, and NorESM1 under four RCPs	1971–2099	0.5° × 0.5°	German Federal Ministry of Education and Research (BMBF)—The ISI-MIP Fast Track project	Climate change projections in the twenty-first century Heat health risks

(continued)

Meteorological data of China	Maximum and minimum temperatures	1971–2010	Station	China Meteorological Administration	Mean daily minimum and maximum temperatures (1981–2010)
Global irrigated and rain-fed crop area data (MIRCA2000)	The MIRCA2000 data, provided by the Institutfür Physische Geographie, Goethe-Universität, consist of all major food crops, including wheat, rice, maize, and soybean. The datasets refer to the period of 1998–2002 and have been made available with a spatial resolution of 0.5° by ISI-MIP	2000	0.5° × 0.5°	German Federal Ministry of Education and Research (BMBF)—The ISI-MIP Fast Track project	Agricultural risks
Global simulated yield data	The simulated yield data were simulated by four global gridded crop model in ISI-MIP under four RCPs (see Chap. 4)	1971–2099	0.5° × 0.5°	German Federal Ministry of Education and Research (BMBF)—The ISI-MIP Fast Track project	Agricultural risks
Global ecosystem variables data	The variables include Net primary production (NPP), fire carbon, carbon contained in vegetation and soil, Transpiration, evaporation, runoff, and soil water content (SWC). The data were simulated by four global gridded vegetation model in ISI-MIP under four RCPs (see Chap. 5)	1971–2099	0.5° × 0.5°	German Federal Ministry of Education and Research (BMBF)—The ISI-MIP Fast Track project	Ecological risks

Further Readings

Center for International Earth Science Information Network—CIESIN—Columbia University. (2016). *Gridded population of the World, Version 4 (GPWv4): Population density*. Palisades. NY: NASA Socioeconomic Data and Applications Center (SEDAC). doi:10.7927/H4NP22DQ.

Edition Board of Vegetation Map of China. (2007). *Vegetation map of the People's Republic of China (1:1,000,000)*. Beijing, China: Chinese Academic of Science.

Hansen, M., DeFries, R., Townshend, J., & Sohlberg, R. (2000). Global land cover classification at 1 km resolution using a decision tree classifier. *International Journal of Remote Sensing, 21,* 1331–1365.

Huang, Y., Jiang, D,. & Fu, J. (2014). 1 km grid GDP data of China (2005, 2010). *Acta Geographic Sinica, 69*(Supplement), 140–143.

Portmann, F. T., Siebert, S., & Döll, P. (2010). MIRCA2000–Global monthly irrigated and rain-fed crop areas around the year 2000: A new high-resolution data set for agricultural and hydro-logical modeling. *Global Biogeochemical Cycles, 24,* 1–24.

Shi, P., Sun, S., Wang, M., Li, N., Wang, J., Jin, Y., et al. (2014). Climate change regionalization in China (1961–2010). *Science China: Earth Sciences, 57,* 2676–2689.

Warszawski, L., Frieler, K., Huber, V., Piontek, F., Serdeczny, O., & Schewe, J. (2014). The Inter-Sectoral Impact Model Intercomparison Projection (ISI-MIP): Project framework. *Proceedings of the National Academy of Sciences of the United States of America, 111,* 3228–3232.

Zhang, X., Tang, Q., Pan, M., & Tang, Y. (2014). A long-term land surface hydrologic fluxes and states dataset for China. *Journal of Hydrometeorology, 15*(5), 2067–2084.

Q. Tang and Q. Ge (eds.), *Atlas of Environmental Risks Facing China Under Climate Change*, IHDP/Future Earth-Integrated Risk Governance Project Series, DOI 10.1007/978-981-10-4199-0

FSC
www.fsc.org
MIX
Papier | Fördert
gute Waldnutzung
FSC® C083411

Kolibri 360 GmbH
Ferdinand-Jühlke-Straße 7
99095 Erfurt, Deutschland
produktsicherheit@kolibri360.de